IUTAM SYMPOSIUM ON SIMULATION AND IDENTIFICATION OF ORGANIZED STRUCTURES IN FLOWS

FLUID MECHANICS AND ITS APPLICATIONS
Volume 52

Series Editor: R. MOREAU
MADYLAM
Ecole Nationale Supérieure d'Hydraulique de Grenoble
Boîte Postale 95
38402 Saint Martin d'Hères Cedex, France

Aims and Scope of the Series

The purpose of this series is to focus on subjects in which fluid mechanics plays a fundamental role.

As well as the more traditional applications of aeronautics, hydraulics, heat and mass transfer etc., books will be published dealing with topics which are currently in a state of rapid development, such as turbulence, suspensions and multiphase fluids, super and hypersonic flows and numerical modelling techniques.

It is a widely held view that it is the interdisciplinary subjects that will receive intense scientific attention, bringing them to the forefront of technological advancement. Fluids have the ability to transport matter and its properties as well as transmit force, therefore fluid mechanics is a subject that is particulary open to cross fertilisation with other sciences and disciplines of engineering. The subject of fluid mechanics will be highly relevant in domains such as chemical, metallurgical, biological and ecological engineering. This series is particularly open to such new multidisciplinary domains.

The median level of presentation is the first year graduate student. Some texts are monographs defining the current state of a field; others are accessible to final year undergraduates; but essentially the emphasis is on readability and clarity.

For a list of related mechanics titles, see final pages.

IUTAM Symposium on

Simulation and Identification of Organized Structures in Flows

Proceedings of the IUTAM Symposium
held in Lyngby, Denmark,
25–29 May 1997

Edited by

J.N. SØRENSEN

Department of Energy Engineering,
Technical University of Denmark,
Lyngby, Denmark

E.J. HOPFINGER

LEGI-IMG,
Domaine Universitaire,
Grenoble, France

and

N. AUBRY

Mechanical Engineering Department,
New Jersey Institute of Technology,
Newark, U.S.A.

SPRINGER-SCIENCE+BUSINESS MEDIA, B.V.

A C.I.P. Catalogue record for this book is available from the Library of Congress.

ISBN 978-0-7923-5603-5 ISBN 978-94-011-4601-2 (eBook)
DOI 10.1007/978-94-011-4601-2

Printed on acid-free paper

Chairmen of the IUTAM Symposium:

J. N. Sørensen (DTU) E. J. Hopfinger (IMG)

Scientific Committee:

- R.J. Adrian (USA)
- N. Aubry (USA)
- J.P. Bonnet (France)
- G.G. Chernyi (Russia)
- E.J. Hopfinger (France)
- S. Kida (Japan)
- P. Orlandi (Italy)
- J.N. Sørensen (Denmark)

Organizing Committee:

- P. L. Christiansen (DTU)
- B. H. Jørgensen (DTU)
- M. P. Sørensen (DTU)
- J. N. Sørensen (DTU)

Sponsoring Organizations and Companies:

- Danish Centre for Nonlinear Dynamics (MIDIT)
- Danish Technical Research Council (STVF)
- Danish Natural Science Research Council (SNF)
- Thomas B. Thriges Foundation
- International Union for Theoretical and Applied Mechanics
- Department of Energy Engineering, DTU

Contents

PREFACE

The dynamics of transitional and turbulent flows is often dominated by organized structures with a life-time much longer than a characteristic time-scale of the surrounding small-scale turbulence. Organized structures may appear as secondary flows as a result of an instability but they persist in turbulent flows. They manifest themselves as eddies or localized vortices and play an important role in e.g. mixing and transport processes. Although the existence of organized structures has been revealed by many experiments and by numerical simulations they are somewhat elusive, as there is no consensus on how to define them and technically how to detect them.

In recent years several identification tools for analysing complex flows have been developed. These tools include various versions of the Proper Orthogonal Decomposition (POD) technique, wavelet transforms, pattern recognition, etc. At the same time, improvements in experimental techniques have made available data that further necessitate efficient detection methods. A prominent example is the Particle Image Velocimetry (PIV) technique from which complex spatio-temporal flow data can be obtained. An interesting feature of some of the identification techniques is that they form the basis for reduced models by which dynamical processes can be studied in details. From studies of dissipative dynamical systems it has been revealed that, in phase space, transitional and turbulent flows can be identified by their low-dimensional behaviour. Thus, employing data from experiments or numerical simulations to form modes residing on finite-dimensional attractors may dramatically reduce computing costs.

The present IUTAM symposium on Simulation and Identification of Organized Structures in Flows included topics such as:

- Decomposition techniques.
- Coherent structures and their relation to transport processes.
- Low-dimensional modelling.
- Structure definition and recognition.
- Experimental techniques (PIV, multiple hot-wire, etc.).
- Topological concepts.
- Computational approaches to the dynamics of organized structures.

At the IUTAM symposium, 58 papers were presented, 7 of which were

invited lectures. The invited lectures were given by J. Kim, J. Jimenez, F. Hussain, H. I. Andersson, G. R. Spedding, S. Tardu and M. N. Glauser. The invited lectures were chosen to address the topics:
- Boundary Layer Control.
- Dynamics of Coherent Structures.
- Structures in Rotating and Stratified Flows.
- Detection and Identification of Flow Structures.
- Low-Dimensional Modelling.

The present volume includes 44 papers that were reviewed by the editors and presented in the 10 sections listed below:
- Flow Control.
- Coherent Structures in Wall-Bounded Flows.
- Rotating Flows.
- Small-Scale Turbulence and 2-D Flows.
- Geostrophic and Stratified Flows.
- Topological Aspects.
- Experimental Techniques.
- Vortical Structures.
- POD, LSE and Other Techniques.
- Low-Dimensional Modelling.

The papers demonstrate that many recently developed identification techniques now are used routinely for analysing experimental data. Several papers concern the dynamics of vortical structures and related phenomenological models of the genesis and the dynamics of flow structures in near-wall turbulence. Although progress is demonstrated in this field, the papers reveal that there is still no consensus on definitions of structures based on clear physical meaning. It is shown that low-dimensional systems, obtained by e.g. the POD technique or truncation of the Euler equations, are capable of capturing essential phenomena related to the dynamics of organized flow structures. A future application of such methods lies in their ability to model and control flow structures employing only a few degrees of freedom. Another subject of great interest is the topological concept which now seems to have gained a renewed interest as a supplementary tool for analysing coherent structures.

On behalf of the editors
Jens N. Sørensen

I. Flow Control

TAMING NEAR-WALL STREAMWISE VORTICES: A MODUS OPERANDI FOR BOUNDARY LAYER CONTROL

JOHN KIM
Mechanical and Aerospace Engineering Department
University of California, Los Angeles, CA 90095-1597

Abstract

A brief overview of the recent progress achieved by the author's research group on turbulence control for drag reduction in turbulent boundary layers is given. Control schemes are developed based on the premise that the most effective way to control turbulent boundary layers is through proper manipulation of the near-wall streamwise vortices. Two different methods — neural networks and suboptimal control theory — are utilized to achieve this goal. The resulting feedback laws are applied to a low Reynolds-number turbulent channel flow. Numerical experiments indicate that both approaches yield substantial drag reduction. The application of a system theory approach to the feedback stabilization to delay the transition in a laminar channel flow is also discussed.

1. Introduction

The potential payoffs of managing and controlling turbulent flows that occur in various engineering applications are significant. For example, it is estimated that a 10% reduction in fuel cost of commercial aircraft would yield a 40% increase in the profit margin of an airline [1]. Understanding the transition process from laminar to turbulent flow is also a critical element in designing aerodynamically acceptable shapes.

The discovery of organized structures in turbulent flows during the past 30 years has brought much excitement from turbulence researchers in anticipation that this new information would lead to better understanding, and consequently, to better control of turbulence. Turbulence control of free shear flows has indeed benefited enormously from discovery of organized structures in free shear flows [2]. However, similar attempts in the control of wall flows — based on the observation that 1) the dynamical evolution of the organized structures in wall flows are responsible for the majority of turbulent energy production, and 2) the organized structures should be responsible for the high skin-friction drag in turbulent boundary layers — have been less successful. The role and dynamics of the large-scale structures and the instability mechanisms leading to the formation of such structures in the free shear flows are relatively well understood. In contrast, the underlying mechanism by which the organized structures are produced and sustained in the wall flows are less well-known. Furthermore, the organized structures in wall flows occur in a more random and chaotic manner and at much smaller scales, making them difficult to detect and manipulate.

Turbulence in wall-bounded shear flows is strongly affected by the near-wall region. Most production and dissipation of the kinetic energy take place in the buffer layer. The ubiquitous structural features in this region are low- and high-speed "streaks," which consist mostly of a spanwise modulation of the streamwise velocity, and streamwise vortices. Although we have learned a great deal about their kinematics, very little is known or agreed upon about the *dynamics* of streaks and streamwise vortices in spite of much effort and attention they have attracted during the past three decades (*e.g.*, see Ref. [3] for a review). There is no consensus, for example, about how such structures are generated and sustained nor about what determines their well-known scales. Hamilton *et al.* [4] and Waleffe [5] have proposed a dynamical model, in which streaks and streamwise vortices are maintained by a self-sustaining process. Jiménez and Pinelli [6] propose a similar regeneration mechanism based on their clever numerical experiments. Many other position papers by active investigators will appear on this particular issue in a forthcoming research monograph [7].

At present, our knowledge of near-wall structures is primarily limited to kinematic aspects: the streaks have a characteristic average spacing of about 100 wall units [1] ([8], [9], [10]), while the streamwise vortices are observed to have a characteristic average diameter of about 20-40 wall units [10]. The streaks are often linked to a sequence of events called the "bursting process," in which they move away from the wall, oscillate, and break up. The bursting process is believed to be the essence of the turbulence production mechanism [11]. The streaks have also been linked to the genesis of "horseshoe" vortices through a Kelvin-Helmholtz type roll-up (*e.g.*, Ref. [12]). Another interpretation of the streaks, one to which we subscribe,

[1]That is, $100\,\nu/u_\tau$, where ν is the kinematic velocity and $u_\tau = (\nu|dU/dy|_w)^{1/2}$ is the wall-shear velocity.

is that they are the wakes left behind the streamwise vortices as the latter are advected downstream at a speed close to the local mean flow [13].

Kim [13] has proposed the near-wall streamwise vortex as the most relevant turbulence structure from the perspective of drag reduction in turbulent boundary layers. This point of view is supported by the observation that streamwise vortices have been found to be responsible for both "ejection" and "sweep" events of the bursting process [3]. Recent studies have also shown that the high skin-friction regions in turbulent boundary layers are closely related to the near-wall streamwise vortices [14], [15]. These regions of high skin friction on the wall are created by the inrush of high-speed fluid toward the wall ("sweep"), which is induced by a strong streamwise vortex. Choi *et al.* [15] showed that a significant drag reduction is possible when the surface boundary condition is modified to suppress the near-wall streamwise vortices. They also relate the drag-reducing mechanism on a riblet surface to the restricted interaction between the riblet and streamwise vortices [16].

Our work on boundary layer control is therefore based on the premise that the most effective way to control turbulent boundary layers for drag reduction is through proper manipulation of the near-wall streamwise vortices. In our numerical experiments, we use surface blowing and suction as the control input, but other approaches could be used as well. For instance, in other applications, we have also used a wall-normal electromagnetic force as the control input, which resulted in similar effects. All numerical results discussed in this paper have been obtained by direct numerical simulations of turbulent and transitional channel flows using the numerical methods reported in Kim *et al.* [10].

Three examples of boundary layer control are given in this paper, each of which is an excerpt of a full paper. The reader is referred to these papers for detailed descriptions of the particular approach and results. In this paper, I shall use (x, y, z) for the streamwise, wall-normal, and spanwise coordinates, respectively, and (u, v, w) for the corresponding velocity components.

2. Application of a neural network to drag reduction

Choi *et al.* [15] used blowing and suction at the wall equal and opposite to the wall-normal component of velocity at $y^+ = 10$. They showed that this control effectively mitigated the streamwise vortices, giving approximately 25% drag reduction in a turbulent channel flow. Although the method employed in their work is impractical, since the information at $y^+ = 10$ is usually not available, it demonstrates that a control scheme that reduces the skin-friction by manipulating the near-wall streamwise vortices.

The objective of the work reported in Lee *et al.* [17] is to seek wall actuations, in the form of blowing and suction at the wall, which are dependent only on the wall-shear stress to achieve a substantial drag reduction. This requires knowledge of how the wall-shear stresses respond to wall actuations (*i.e.*, the correlation between the wall-shear stresses and the wall actuations). Because of the complexity of the solutions to the Navier-Stokes equations, however, it is not possible to find such a correlation in closed form or to approximate it in simple form. Instead, we used a neural network to approximate the correlation that then predicts the optimal wall actuation to achieve the minimum of the skin-friction drag. Lee *et al.* [17] describe how neural networks with a small number of shared weights were constructed and trained off-line, and then an on-line control scheme for drag reduction based on that neural network was implemented. A standard two-layer feed-forward network with hyperbolic tangent hidden units and a linear output unit was used. The functional form of the final neural network was:

$$v_{jk} = W_a \tanh\left(\sum_{i=-(N-1)/2}^{(N-1)/2} W_i \left.\frac{\partial w}{\partial y}\right|_{j,k+i} - W_b \right) - W_c \,,$$

$$1 \leq j \leq N_x \ \text{ and } \ 1 \leq k \leq N_z \,, \tag{1}$$

where W denotes weight, N is the total number of input weights, and the subscripts j and k denote the numerical grid point at the wall in, respectively, the streamwise and spanwise directions. N_x and N_z are the number of computational grid points in each direction. The summation is done over the spanwise direction. Seven neighboring points ($N = 7$), including the point of interest, in the spanwise direction (corresponding to approximately 90 wall units) were found to provide enough information to adequately train and control the near-wall structures responsible for the high skin friction. Note that the input to the neural network is $\partial w/\partial y$ at the wall, not $\partial u/\partial y$. Initially $\partial u/\partial y$ and $\partial w/\partial y$ at the wall at several instances of time were used as input data fields and the actuation at the wall was used for the output data of the network. Experimentally we found that only $\partial w/\partial y$ at the wall from the current time was necessary for sufficient network performance. It should be noted that a brute force application of a neural network, whose architecture and input parameters were designed without consideration for relevant flow physics, led to either no convergence or excessively long training time.

Applying this control scheme to direct numerical simulations of a turbulent channel flow at low Reynolds number resulted in about 20% drag reduction (Fig. 1). The computed flow fields were examined to determine the mechanism by which the drag reduction was achieved. The most salient feature of the controlled case was that the strength of the near-wall streamwise vortices was substantially reduced (Fig. 2). This result further substantiates the notion that a successful suppression of the near-wall streamwise vortices leads to a significant reduction in drag.

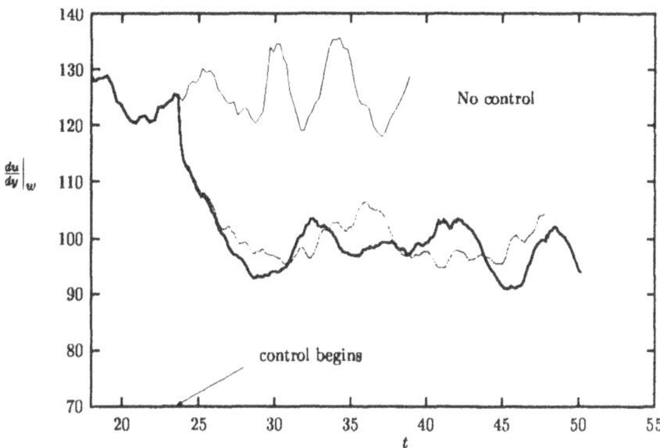

Figure 1. Mean wall-shear stress histories for various control schemes compared to the no-control case: ——— , no control; ━━━ , on-line control with neural network with 7x1 template; ········ , control with 7 fixed weights.

An examination of the weight distribution from the on-line neural network led to a very simple control scheme that worked equally well while being computationally more efficient. This simple control scheme indicates that the optimum blowing and suction at the wall should be in the form,

$$v_w \sim \overline{\frac{\partial}{\partial z}\frac{\partial w}{\partial y}\bigg|_w}, \tag{2}$$

where the overbar represents a local spatial average with high wavenumber components reduced. The converged weight distribution can be expressed analytically, thus making the implementation of this control scheme relatively easy. This control scheme produces a distribution of wall actuations that are very similar to those produced by the v-control of Choi *et al.* [15].

3. Application of suboptimal control theory to drag reduction

Most previous control work has been rather ad-hoc, in that it was mainly based on the investigator's intuition. More systematic approaches using an optimal control theory have appeared recently. Choi *et al.* [16] proposed a "suboptimal" control procedure, in which the iterations required for a global optimal control were avoided by seeking an optimal condition over

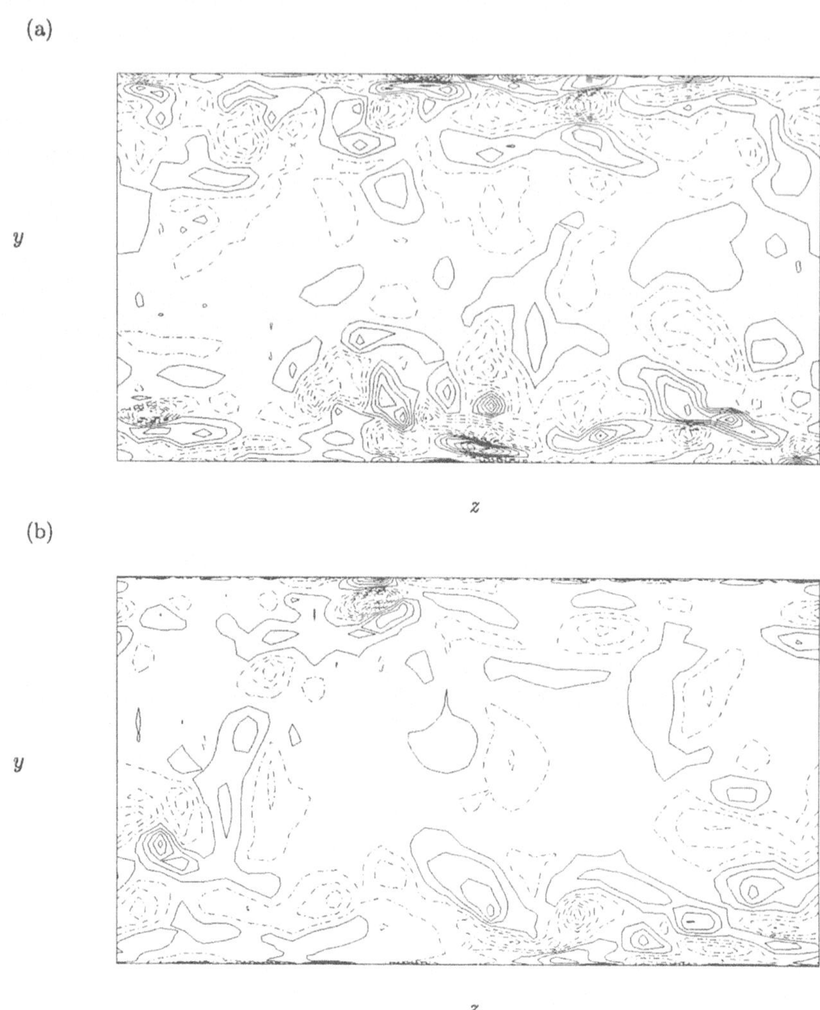

Figure 2. Contours of streamwise vorticity in a cross-flow plane: (a) no control; (b) control using 7 fixed weights. The contour level increment is the same for both figures. Negative contours are chain-dotted.

a short time period. The suboptimal control procedure was successfully applied to control of the Burgers equation. Bewley and Moin [18] were the first to apply the suboptimal control procedure to a turbulent flow and reported a drag reduction of about 17%. The procedure requires full velocity information throughout the flow in order to solve an adjoint problem, from which a feedback control input was then derived. In spite of this obvious drawback, the fact that a control theory applied to a turbulent flow resulted in a substantial drag reduction is encouraging, especially since their control procedure was derived rigorously from a control theory in which a predetermined cost functional was minimized.

In Lee *et al.* [19] we demonstrated that a wise choice of the cost functional coupled with a variation of the formulation can lead to a more practical control law. We showed how to choose a cost functional and how to minimize it to yield simple feedback control laws that only require quantities at the wall as input. The choice of the cost functional to be minimized is critical to the performance of the control. Two different cost functionals were chosen based on our observation of the successful control of Choi *et al.* [15]. As shown in Fig. 3, Choi *et al.*'s blowing and suction, which are equal and opposite to the wall-normal velocity component at $y^+ = 10$, effectively suppress a streamwise vortex by the counteracting up/down motion induced by the vortex. This blowing and suction creates locally high pressure in the near-wall region marked with '+', and low pressure in the region marked with '−', in Fig. 3. A crucial aspect of the present analysis is the observation that this blowing and suction *increases* the pressure gradient in the spanwise direction under the streamwise vortex near the wall. This suggests that we should seek blowing and suction that increases the pressure gradient in the spanwise direction near the wall for a short time period (*i.e.*, in the suboptimal sense) in order to achieve a drag reduction similar to that achieved by Choi *et al.* [15]. The cost functional $\mathcal{J}(\phi)$ to be minimized is then

$$\begin{aligned}\mathcal{J}(\phi) = & \frac{\ell}{2A\Delta t}\int_S\int_t^{t+\Delta t}\phi^2\,dt\,dS \\ - & \frac{1}{2A\Delta t}\int_S\int_t^{t+\Delta t}\left(\frac{\partial p}{\partial z}\right)_w^2 dt\,dS\,, \end{aligned} \tag{3}$$

where the integrations are over the wall, S, in space and over a short duration in time, Δt, which typically corresponds to the time step used in the numerical computation, and ℓ is the relative price of the control, since the first term on the right-hand side represents the cost of the actuation, ϕ (*i.e.*, the surface blowing and suction). Note that the minus sign in front of the second term, since we want to *maximize* the pressure gradient. It should be noted that the spanwise pressure gradient at the wall will be reduced eventually when the strength of the near-wall streamwise vortices are reduced through successful control. Here, blowing and suction that increase the spanwise pressure gradient for the next step are sought as a suboptimal control. The suboptimal control procedure described in Lee *et al.* [19] leads to

$$\phi = v_w \sim \overline{\frac{\partial^2 p_w}{\partial z^2}} \, , \tag{4}$$

where p_w denotes the wall pressure, and the overbar represents a local spatial average with high wavenumber components reduced.

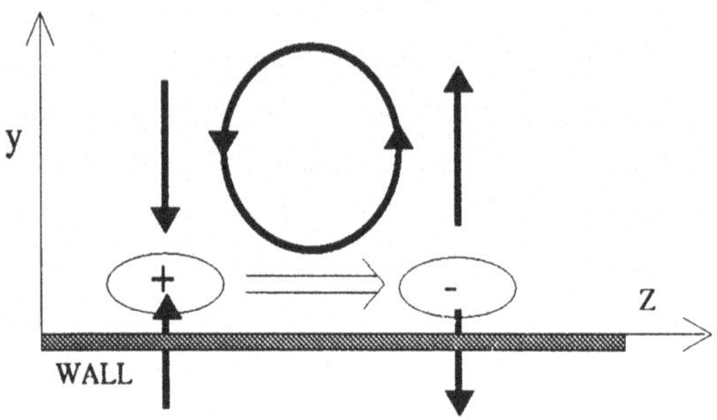

Figure 3. Schematic of a pressure field induced by a control based on $y^+ = 10$.

Another wall quantity that indicates similar changes of the near-wall dynamics that are due to the altered pressure field is the spanwise shear at the wall, $\partial w/\partial y$. Because of the added pressure gradient in the spanwise direction below the streamwise vortex, spanwise motion near the wall is also induced, thus increasing the spanwise shear stress at the wall (see Fig. 3). Thus another valid choice of the cost functional to be minimized is

$$\begin{aligned} \mathcal{J}(\phi) = \ & \frac{l}{2A\Delta t} \int_S \int_t^{t+\Delta t} \phi^2 \, dt \, dS \\ - \ & \frac{1}{2A\Delta t} \int_S \int_t^{t+\Delta t} \left(\frac{\partial w}{\partial y} \right)_w^2 dt \, dS \, . \end{aligned} \tag{5}$$

Following the procedure that led to Eqn. (4) yields the optimum actuation:

$$\phi = v_w \sim \overline{\left. \frac{\partial}{\partial z} \frac{\partial w}{\partial y} \right|_w} \, . \tag{6}$$

Equation (6) indicates that the optimum wall actuation should be proportional to a locally averaged spanwise derivative of the spanwise shear at the wall. Note that this result is very similar to that found in the application of a neural network in Sec. 3. The only difference in the two expressions is how the local average should be performed; see Lee *et al.* [19] for details.

Applying Eqns. (4) and (6) to a turbulent channel flow produced, respectively, about 16% and 22% drag reduction [19]. Turbulence characteristics of the controlled flow field are very similar to those obtained in Sec. 3.

4. Systems control-theory approach to transition control

In this section, we consider the problem of stabilizing laminar boundary-layer flows. The goal is to design a feedback control system that stabilizes the flow so that no unstable modes exist in the new system. This is a fundamental change in the way transition control is approached. Instead of allowing unstable modes to appear and then "canceling" them with out-of-phase modes, we develop here a control system that changes the system from being inherently unstable to being inherently stable. In essence, this approach changes the philosophy of the problem from thinking about how inputs can mitigate an inherently unstable system to thinking about how sensors and actuators can be added to form an entirely new *stable* system.

In Joshi *et al.* [20], a system theory framework is presented for the linear stabilization of a laminar channel flow. The linearized Navier-Stokes equations are converted into the state variable representation of the system using a Galerkin method. This representation relies on the fact that the motion of any finite-dimensional dynamic system can be expressed as a set of first-order ordinary differential equations in matrix form:

$$\frac{d\mathbf{x}}{dt} = A\mathbf{x} + B\mathbf{q} \tag{7}$$

$$\mathbf{z} = C\mathbf{x}\,. \tag{8}$$

The vector $\mathbf{x}$ is the state vector of the system, $\mathbf{q}$ is the input variable (*i.e.*, control input), and $\mathbf{z}$ is the output variable (*i.e.*, sensor output). The matrix A is the dynamic matrix of the system, and B and C are, respectively, the input and output matrices.

Joshi*et al.* [20] show that in addition to the well-studied system eigenvalues, the location of system zeros are important in linear stability control. The location of system zeros determines the effect of feedback control on both stable and unstable eigenvalues. In addition, system zeros can be used to determine sensor locations that lead to simple feedback control schemes. Joshi *et al.* showed that the channel flow can be stabilized using a simple, constant-gain feedback, integral-compensator controller. The goal of the control design was stability. By choosing proper sensor locations, they were able to achieve a stable, closed-loop system that was extremely robust to changing Reynolds numbers. It was also shown that a controller designed with linear theory also has a strong stabilizing effect on two-dimensional finite-amplitude disturbances. As a result, the secondary instabilities resulting from infinitesimal three-dimensional disturbances in the presence of a finite-amplitude two-dimensional disturbance cease to exist [20].

Joshi *et al.* [20] further illustrate that the system theory approach is a powerful tool both for designing control systems to stabilize flows, as well as understanding the physics of controlled transitional flows. Transfer-function models yield information on optimal sensor locations and sensor types. State variable models also show how each mode, both stable and unstable, is affected by feedback control. In addition, control theory concepts, such as observability and controllability, are used to explain possible pitfalls in flow control.

5. Conclusions

Successful applications of a neural network and a suboptimal control theory to a turbulent channel flow for drag reduction have been described. In the neural network approach, the network was able to identify a correlation between the wall-shear stress and the desired wall actuations. The optimum actuation, which is proportional to the spanwise shear stress at the wall, is not something one would normally consider for drag reduction. Apparently the properly trained neural network is capable of identifying the most relevant information from the given input and output. Although the present approach lacks the "ad-hoc" nature of earlier strategies, the design and application of the neural network nevertheless benefited from physical insight into the flow under consideration. In the suboptimal control-theory approach, the cost functional to be minimized was derived from an earlier control scheme that was designed to mitigate the interaction between streamwise vortices and the wall. The resulting feedback-control law was very similar to that found by the neural network. This is quite surprising, since the two approaches use totally different routes to get to a similar optimum actuation: the former uses a "black-box" approach to achieve the goal, whereas the latter tries to minimizes a quantity that is not directly related to the skin-friction drag. Both optimum actuations, which differ only slightly in the manner in which the input data are locally averaged, are related to the near-wall streamwise vortices. Thus our premise that proper manipulation of the near-wall streamwise vortices is the most effective way to control turbulent boundary layers is validated.

Although the control schemes presented in this paper are significant improvements over earlier approaches that require velocity information throughout the flow field, there are many technical issues that must be resolved before these control schemes can be implemented in real practice. Precise blowing and suction distribution over a surface, for example, is difficult to implement. Other approaches, such as surface movement by deformable walls, may prove to be more practical. Another issue worth mentioning is the time delay between sensing and actuation, which was not included in any of our numerical experiments. In a real situation, there will be a finite time between sensing and actuation, which must be accounted for.

We have explored new control strategies utilizing relatively new non-traditional technologies, and the results are promising. It is our belief that full exploitation of modern control theory should enhance our capability of taming turbulence for flow control.

Acknowledgments

This paper describes the work being carried out by the author's students and colleagues at UCLA. I thank them all. The financial support from Air Force Office of Scientific Research (Drs. James McMichael and Mark Glauser) and Office of Naval Research (Dr. Patrick Purtell) during the course of this work is gratefully acknowledged. The computer time has been provided by the NAS Program of NASA Ames Research Center, San Diego Supercomputer Center, and DoD High Performance Computer Center.

References

1. D. M. Bushnell, private communication, 1996.
2. C. M. Ho and P. Huerre, "Perturbed free shear layers," *Ann. Rev. Fluid Mech.* **16**, 365, 1984.
3. S. K. Robinson, "Coherent motions in the turbulent boundary layer," *Ann. Rev. Fluid Mech.* **23**, 601, 1991.
4. J. M. Hamilton, J. Kim and F. Waleffe, "Regeneration mechanisms of near-wall turbulence structures," by J. M. Hamilton, J. Kim and F. Waleffe, *J. Fluid Mech.* **287**, 317, 1995.
5. F. Waleffe, "On a self-sustaining process in shear flows," *Phys. Fluids* **9**, 883, 1997.
6. J. Jiménez and A. Pinelli, "Wall turbulence: How it works and how to damp it," AIAA PAper 97-2112, 4th AIAA Shear Flow Control Conference, June 29-July 2, 1997, Snowmass Village, Colorado.
7. Self-Sustaining Mechanisms of Wall Turbulence, Panton (ed.), Computational Mechanics Publication, 1997.
8. S. J. Kline, W. C. Reynolds, F. A. Schraub and P. W. Runstadler, "The structure of turbulent boundary layers," *J. Fluid Mech.* **30**, 741, 1967.
9. C. R. Smith and S. P. Metzler, "The characteristics of low-speed streaks in the near-wall region of a turbulent boundary layer," *J. Fluid Mech.* **129**, 27, 1983.
10. J. Kim, P. Moin and R. D. Moser, "Turbulence statistics in fully developed channel flow at low Reynolds number," *J. Fluid Mech.* **177**, 133, 1967.
11. H. T. Kim, S. J. Kline and W. C. Reynolds, "The production of turbulence near a smooth wall in a turbulent boundary layer," *J. Fluid Mech.* **50**, 133, 1971.
12. M. S. Acarlar and C. R. Smith, "A study of hairpin vortices in a laminar boundary layer. Part II: hairpin vortices generated by fluid injection," *J. Fluid Mech.* **175**,

43, 1987.

13. J. Kim, "Study of turbulence structure through numerical simulations: the perspective of drag reduction," in AGARD Report (R-786), AGARD FDP/VKI Special Course on "Skin Friction Drag Reduction," March 2-6, 1992, VKI, Brussels, Belgium.
14. A. G. Kravchenko, H. Choi and P. Moin, "On the relation of near-wall streamwise vortices to wall skin friction in turbulent boundary layers," *Phys. Fluids A* **5**, 3307, 1993.
15. H. Choi, P. Moin and J. Kim, "Active turbulence control for drag reduction in wall-bounded flows," *J. Fluid Mech.* **262**, 75, 1994.
16. H. Choi, P. Moin and J. Kim, "Direct numerical simulation of turbulent flow over riblets," *J. Fluid Mech.* **255**, 503, 1993.
17. C. Lee, J. Kim, D. Babcock, and R. Goodman, "Application of neural networks to turbulence control for drag reduction," *Phys. Fluids* **9**, No. 6, 1740, 1997.
18. T. R. Bewley and P. Moin, "Optimal control of turbulent channel flows," *Active Control of Vibration and Noise*, ASME DE-Vol. 75, 1994.
19. C. Lee, J. Kim and H. Choi, "Suboptimal control of turbulent channel flow for drag reduction," submitted for publication, 1997.
20. S. S. Joshi, J. L. Speyer and J. Kim, "A systems theory approach to the feedback stabilization of infinitesimal and finite-amplitude disturbances in plane Poiseuille flow," *J. Fluid Mech.* **332**, 157, 1997.

VISUALISATION OF COHERENT STRUCTURES IN MANIPULATED TURBULENT FLOW OVER A FENCE

ALEXANDER ORELLANO AND HANS WENGLE
Institut für Strömungsmechanik und Aerodynamik
Universität der Bundeswehr München
D-85577 Neubiberg, Germany

1. Introduction

The basic idea of the current project is that altering the dynamics of the large scale structures in the shear layer bounding a recirculation zone behind a flow obstacle should have a significant effect on macroscopic parameters such as recirculation length or pressure drag. In the case of a flow over a fence with laminar inflow the shear layer separating from the leading edge of the flow obstacle will undergo transition to turbulence and the main effect of a manipulation will result in delaying or accelerating the transition process. A successful manipulation in the case of fully turbulent inflow is more difficult to carry out, in particular if the perturbation cannot be introduced at the optimal location, i.e. as close as possible to the location of separation of the free shear layer. In fact, perturbation amplitudes of one order of magnitude higher than the free stream velocity had to be applied in an experimental study carried out by Schmidt (1997) to reduce the recirculation length by 50%. In that experiment the manipulation (time-periodic blowing/suction) has been applied through a cross-wind slot several obstacle heights in front of the fence. Similar experiments of manipulated flow over a fence have been carried out by Siller and Fernholz (1996). In the case of a transitional backward-facing step flow the amplitude of a similar blowing/suction perturbation at the edge of the step could be one order of magnitude lower than the free-stream velocity to achieve a 50 % reduction in the mean recirculation length, see Bärwolff et al. (1996).

In this paper, we will present an evaluation of large-eddy-simulations of a fully turbulent boundary layer flow over a surface-mounted fence at a Reynolds number of $Re_h = 3000$ (based on fence height and incoming free-stream velocity). For the non-manipulated case there are experimental data available from Larsen (1995). In our LES we obtained a mean reattachment length of $x_R/h = 12.5$ in comparison with the experimental value of $x_R/h = 12.0$. In addition, we present results for the same flow problem manipulated by periodic blowing and suction through a cross-wind slot in front of the obstacle (at $x/h = -3.0$) with an amplitude of $A = 0.5U_\infty$. For that perturbation amplitude the mean recirculation length has been reduced by 12 % compared to the non-manipulated reference case.

The two main objectives of this study are: (a) to evaluate the LES data for the two flow cases using proper orthogonal decomposition (or Karhunen-Loève decomposition) and (b) to compare time series of the two flow cases.

2. Numerical Simulation

The numerical scheme: The governing equations describing the flow quantities are derived from the integral conservation equations for mass and momentum. The resulting equations are solved numerically on a staggered and non-uniform cartesian grid, using second-order finite-differencing in time and space (explicit leap-frog for time discretization, central differencing for convection terms and time-lagged diffusion terms). The problem of pressure-velocity coupling is solved iteratively. The geometry of the flow obstacle is approximated by simply blocking out the corresponding grid cells within a cartesian grid. The sub-grid scale stresses, arising from the nonlinear convection terms have been evaluated by the Smagorinsky model (with $c_l = 0.1$) which relates the sub-grid scale stresses to the grid scale velocity field via an eddy viscosity model. In grid volumes next to the walls, we used for the mixing length the smaller value of $\kappa * x_n$ and $0.1 * (\Delta x * \Delta y * \Delta z)^{1/3}$, respectively ($x_n$ is the distance normal to the wall).

Computational domain and boundary conditions: Both flow cases have been simulated in a computational domain of size $(x/h, y/h, z/h) = (24.0, 6.0, 5.0)$ with $(NX, NY, NZ) = (236, 96, 74)$ grid points. The boundary conditions were periodic in lateral direction, a slip condition at the upper z-plane has been applied and at the outflow cross-section we applied zero-gradient boundary conditions. At the bottom wall and at the surfaces of the flow obstacle we have applied no-slip boundary conditions. The front face of the fence is located at $x/h = 0$. The inflow condition (at $x/h = -10.0$) was a stationary boundary layer profile (1/7-power law). Close to the inflow section (at $x/h = -9.0$) a series of vorticity generators is located on a span-wise line at the bottom wall to initiate a turbulent boundary layer with proper velocity fluctuations and length scales. The entry length of about 10 fence heights has been chosen to enable a sufficiently long downstream development for the turbulent boundary layer ($\delta/h = 0.5$

at $x/h = 0$) before significant upstream influence of the recirculation zone in front of the fence occurs. The manipulation is implemented via a time-periodic boundary condition for the vertical (w-) velocity component on a cross-wind line on the bottom wall at $x/h = -3.0$.

3. Evaluation of the Large-Eddy-Simulation

Instantaneous velocity field: Figure 1 shows time sequences of three-dimensional isosurfaces of the fluctuating vertical velocity component $w = +/-0.08$. On the left hand side (fig. 1a-1d) the non-manipulated case (case A) is presented and, on the right hand side (fig. 1e-1h), a time sequence over one forcing period of the manipulated case (case B) is shown. In front of and above the fence there are two neighbouring flow regions visible with opposite signs of the vertical velocity fluctuation indicating strong stream-wise vorticity.

Video sequences show that these stream-wise vortices are related to the streaks in the incoming turbulent boundary layer. In figure 1 (left) these streaks are visible as bright and dark streaky structures close to the bottom plate in front of the fence. Their span-wise separation is approximately 90 viscous units (of the incoming boundary layer). They are convected downstream while staying approximately at the same lateral position and they have a lifetime of approximately 5 to 10 non-dimensional time units ($T = h/U_\infty$). The longitudinal vortex structures immediately in front and above the fence are less strong in the case of 2D time-periodic forcing (case B). It seems that the production of stream-wise vorticity is weakened due to the partial destruction of the streaks at the location of perturbation ($x/h = -3.0$). Video sequences of the stream-wise velocity component in a vertical plane in stream-wise direction show that the streaks of the boundary layer do not enter the recirculation zone in front of the obstacle, they are convected over the fence. They play an important role for the three-dimensional distortion of the two-dimensional roll-up process. From figure 1 it can be concluded that the coherence of the flow structures increases in the case of forcing. This leads to a faster convergence of the series expansion presented in chapter 4.

Energy spectra: Although there is a quasi-2D roll-up process visible in case A no significant peak is apparent in the energy spectrum evaluated at fence height half a fence height downstream of separation (fig. 2). This is consistent with experimental findings of Siller and Fernholz (1996). The instability frequency scales with the momentum loss thickness at the separation point, but the momentum loss thickness varies in lateral direction and in time due to the fully turbulent incoming flow leading to an energy spectrum without dominant peaks as shown in figure 2. However, looking at time records of the vertical velocity component close to the edge of the fence a Strouhal number $Str = 1.2$ has been estimated. A Strouhal number of $Str = 0.6$ has been estimated further downstream close to the reattachment zone.

In case B the roll-up process in the shear layer has been phase locked and a strong peak at $Str = 0.6$ (forcing frequency) is visible in the energy spectrum (fig. 3). In addition, a second peak with $Str = 1.2$ (fundamental mode) is present in the energy spectrum (fig. 3). It is interesting that the primary roll-up process does not occur with the forcing frequency. The shear layer right after separation seems not to be receptive for the forcing frequency applied here. Nevertheless we can phase lock the primary roll-up process (fundamental mode) by forcing with its sub-harmonic mode for which the shear layer seems to be receptive further downstream. This indicates a dependence of the roll-up process right after the separation on the vortex merging and on the rollers further downstream.

Reattaching shear layer: Mainly as a conclusion from video sequences, figure 4 shows a schematic view of the different stages in the free shear layer bounding the separation region until it reattaches on the bottom plate. After the primary roll-up process of the shear layer (stage 1) a vortex pairing occurs (stage 2) resulting in a frequency which is half of the frequency of the roll-up process of stage 1. The convective velocity of the rollers in the first two stages is about half of the free-stream velocity. When the rollers approach the reattachment zone (stage 3) they grow in size and the wavelength of these vortices increases. Some parts of the lateral rollers slow down due to the reverse flow in the recirculation region and they merge with the following roller which leads to another kind of vortex pairing related to the recirculation/reattachment process.

4. Eigenmode-decomposition

The analysis of a turbulent flow by a Karhunen-Loève-decomposition (Lumley (1970)) is based on the idea that the flow field can be represented by a series expansion with spatial modes $\phi^n(x,y,z)$ and temporal modes $a^n(t)$: $u(x,y,z,t) = \sum_n a^n(t) \cdot \phi^n(x,y,z)$. The temporal modes $a^n(t)$ are the eigenvectors of the temporal correlation tensor $C(t,t')$ for the velocity components and the spatial modes can be calculated by projecting the velocity fields onto the temporal modes (Manhart, M. and Wengle, H., 1993). The method applied here is optimal in the sense that the series of eigenmodes converges more rapidly (in quadratic term) than any other representation. The method extracts first the structure that is the best correlated, in a statistical sense, with the background velocity field. The energetic dominant modes (i.e. the first few terms with the largest eigenvalues in the expansion) are then called *coherent structures.*

We used 2000 samples of the velocity field with a time-interval of $\Delta t = L_{ref}/U_{in} = 0.04$ and, in addition, 1000 samples with $\Delta t = 0.16$ to construct the correlation tensor for case B. For the non-manipulated reference case we used only 1200 samples with $\Delta t = 0.16$ to perform the eigenmode-decomposition which is sufficient for preliminary results only.

Spatial behaviour of the coherent structures: From the first four *fluctuating* velocity modes (n=2,3,4,5) containing most of the fluctuating energy the stream-wise velocity component only is shown in figure 5. The reference case (left row) is compared to the flow manipulated with a vertical oscillating jet on a cross-wind line (right row). The first four fluctuating modes of case A represent only 10 % of the fluctuating energy in contrast to case B where the first four modes contain 21 % of the fluctuating energy. This is due to the greater overall coherence of structures in the case of forcing (case B), see figure 1. Compared to case A a smaller number of modes is necessary to represent a certain amount of fluctuating energy. In fact the fluctuating energy of case B is mainly related to a two-dimensional roll-up process represented by mode 2 and 3.

The spatial modes 2 and 3 of case B exhibit a phase shift and together with their corresponding temporal modes (also being phase shifted, see fig. 7) they describe a two-dimensional coherent structure convected downstream. This dominant two-dimensional structure is exclusively present in the case of forcing and it is strongly related to the blowing/suction desturbation in front of the obstacle. The coherent structures with strong lateral vorticity contribute to a fast mixing of high-speed and low-speed fluid. This causes a faster thickening of the shear layer (compared to the non-manipulated flow case) and finally leads to a reduced recirculation length of the mean flow. The well defined lateral rollers are visible in the instantaneous flow states of the vertical velocity component, see figure 1.

Mode 2 and mode 3 of the non-manipulated case (fig. 5a-5d) seem to correspond to mode 4 and mode 5 of case B. It can be observed that streaky structures are apparent just above the fence with a span-wise distribution similar to the streaks of the incoming boundary layer. This indicates the important role of the streaks convected over the fence for the dynamics of the shear layer in terms of energy. It is interesting that in the non-manipulated case the 2D roll-up process does not appear in the low modes because of its relatively low energy content compared to the three-dimensional processes.

Temporal behaviour of the coherent structures: Fig. 6 and 7 show the temporal coefficients for the reference case A and the manipulated case B. It is evident that the perturbed flow includes coherent structures with well defined frequencies in contrast to the reference case that exhibit coherent structures (modes) with a broadband frequency (compare fig. 8 and 9).

The peak-frequency ($Str = 0.6$) of the mode-pair (2,3) of case B is equal to the forcing frequency indicating the strong dependence of this coherent structure on the oscillating jet in front of the obstacle. This frequency is the sub-harmonic of the frequency of the primary roll-up process ($Str = 1.2$) close to the edge of the fence. It is interesting that mode 3 exhibits a relatively weak peak at that (non-dimensional) frequency, see figure 9.

5. Conclusion

The fully turbulent flow over a fence contains large-scale structures that can be phase locked by manipulating the flow via a time-periodic oscillating (blowing/suction) jet on at a cross-wind line three obstacle heights in front of the obstacle. Corresponding structures can be detected applying a Karhunen-Loève-decomposition of the velocity field. The enhancement of the vortex roll-up leads to a faster growing of the separated and reattaching shear layer, and the mean recirculation length is reduced by 12 % if a perturbation amplitude of 50% of U_∞ is applied.

The stream-wise vortices of the incoming turbulent boundary layer are weakened in the case of forcing due to the destruction of the streaks by the blowing/suction procedure and they are convected over the flow obstacle. This stream-wise vortices are important for the 3D distortion of the 2D roll-up process. The 2D coherence of the structures of the shear layer is enhanced by forcing the flow periodically. Video sequences show that there is a vortex pairing related to the recirculation/reattachment process. This additional merging leads to a further growing of the structures near the instantaneous reattachment line and these enlarged structures are ejected into the new boundary layer developing after reattachment.

References

Bärwolff, G., Wengle, H., and Jeggle, H. (1996). Direct numerical simulation of transitional backward-facing step flow manipulated by oscillating blowing/suction. In Rodi, W. and Bergeles, G., editors, *Engineering Turbulence Modelling and Experiment 3*, pp. 219–228, Amsterdam. Elsevier Science Publishers.

Larsen, P. (1995). Database on tc-2c and tc-2d fence-on-wall and obstacle-on-wall test cases. Report AFM-ETMA 95-01, ISSN 0590-8809, TU Denmark.

Lumley, J. (1970). *Stochastic Tools in Turbulence.* Academic Press.

Manhart, M. and Wengle, H. (1993). A spatiotemporal decomposition of a fully inhomogeneous turbulent flow field. *Theoretical and computational fluid dynamics*, **5**, 223–242.

Schmidt, J. (1997). *Experimental and numerical investigation of separated flows.* PhD thesis, Technical University of Denmark.

Siller, H. and Fernholz, H. (1996). Turbulent separation regions in front and downstream of a fence. In S. Gavrilakis, L. M. and Monlewitz, P., editors, *Advances in Turbulence VI*, pp. 487–490, Dordrecht. Kluwer Academic Publishers.

6. Figures

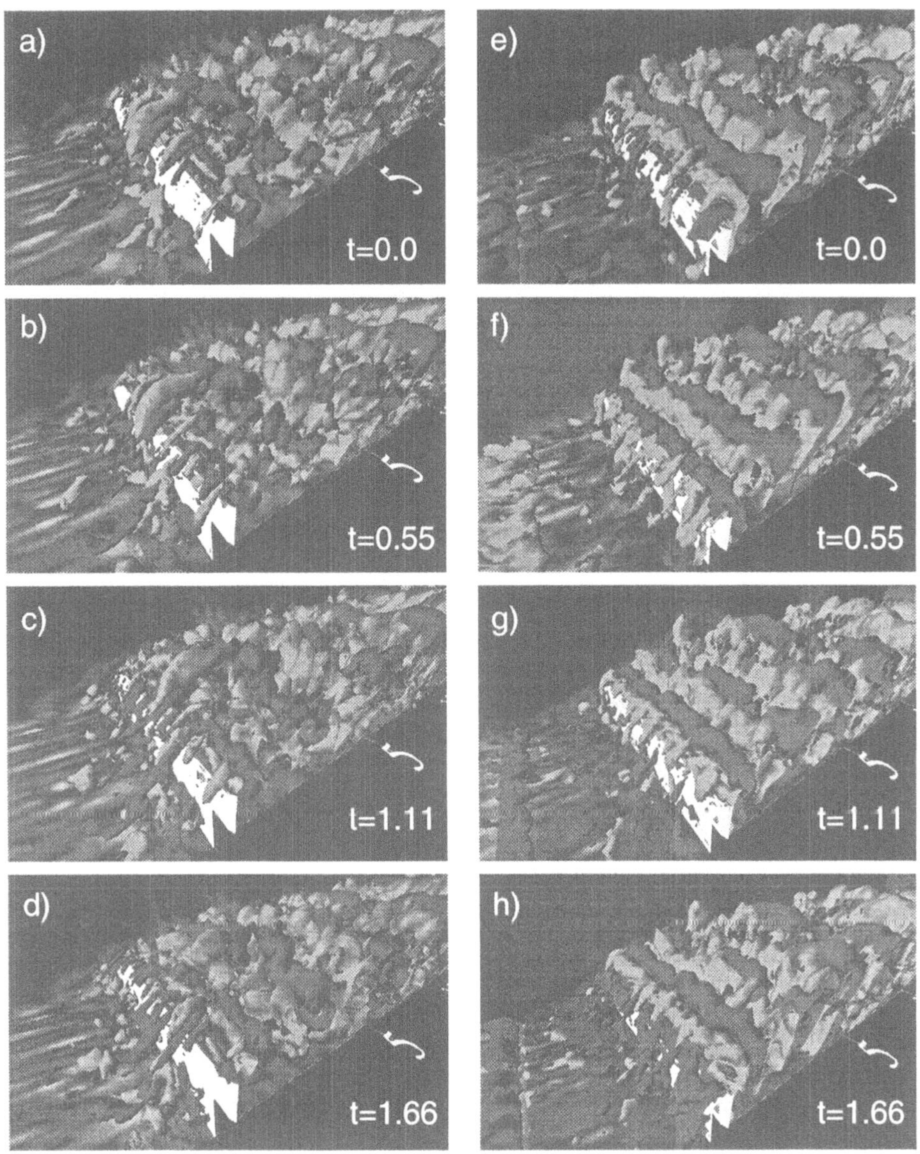

Figure 1 Isosurfaces $w = \pm 0.08$ (dark: negative, bright: positive)
a)-d): case A (non-manipulated), e)-h): case B (manipulated)

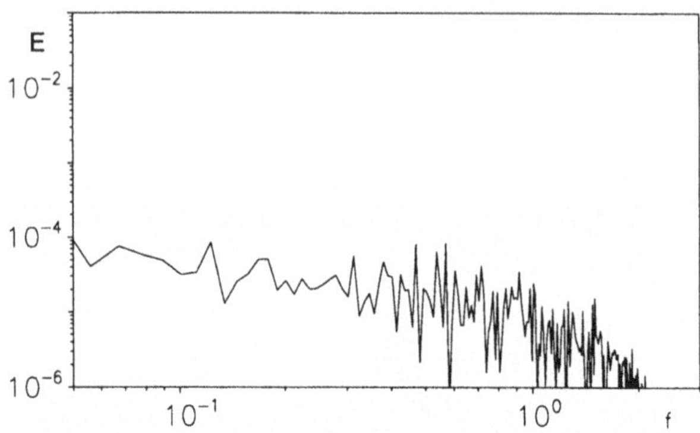

Figure 2 Energy spectrum of the velocity component u at $x/h = 0.43$ and $z/h = 1.0$ of case A (non-manipulated)

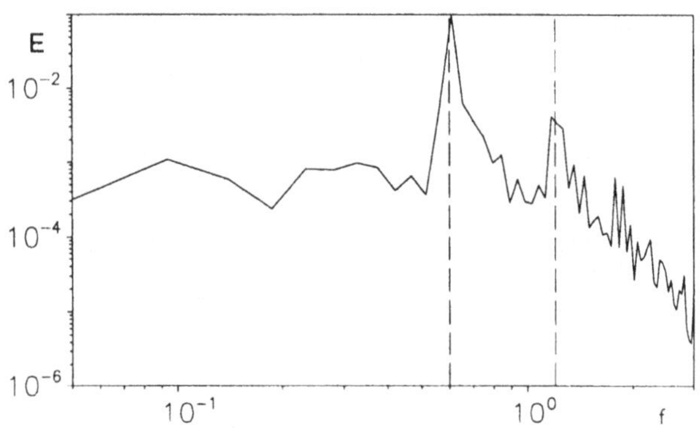

Figure 3 Energy spectrum of the velocity component u at $x/h = 0.43$ and $z/h = 1.0$ of case B (manipulated)

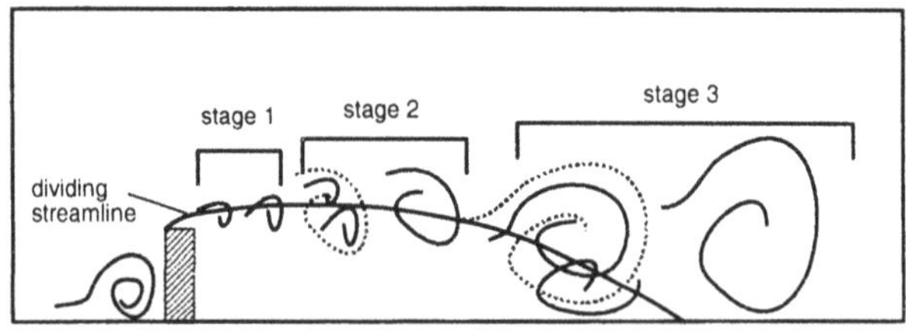

Figure 4 Sketch of the different stages in the reattaching shear layer

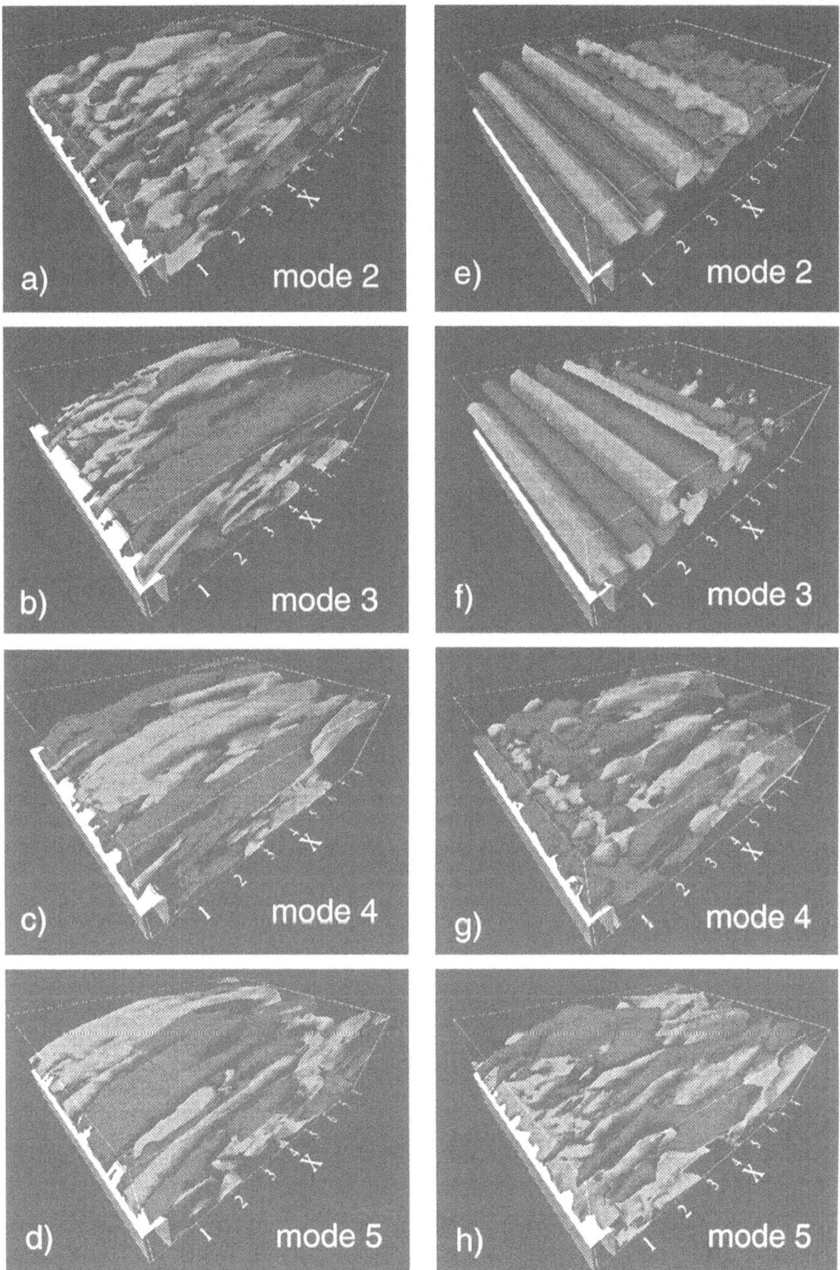

Figure 5 Spatial Karhunen-Loève modes: isosurfaces $u = \pm 0.03$
a)-d): case A (non-manipulated, e)-h): case B (manipulated)

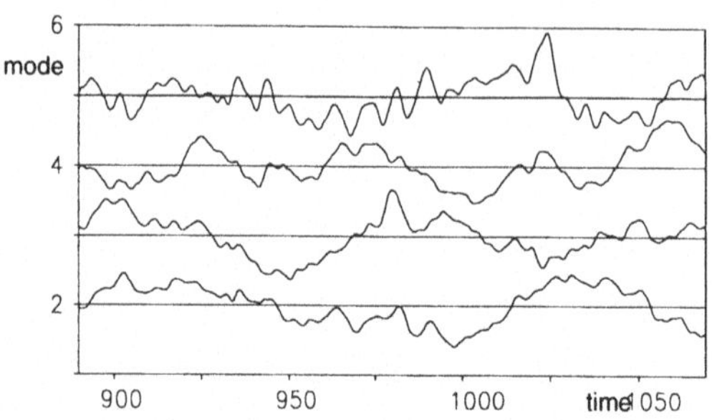

Figure 6 Temporal Karhunen-Loève modes of case A (non-manipulated)

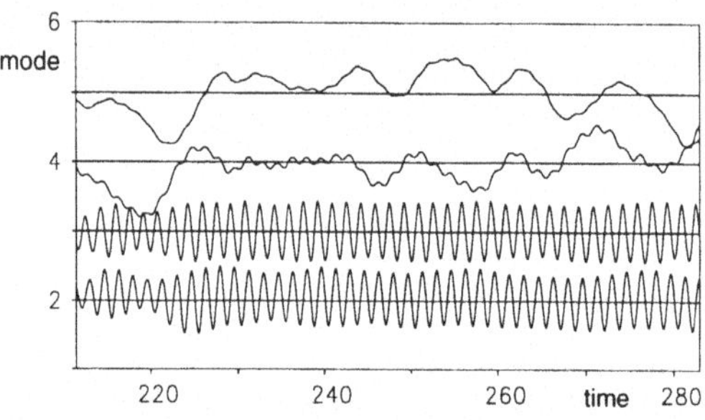

Figure 7 Temporal Karhunen-Loève modes of case B (manipulated)

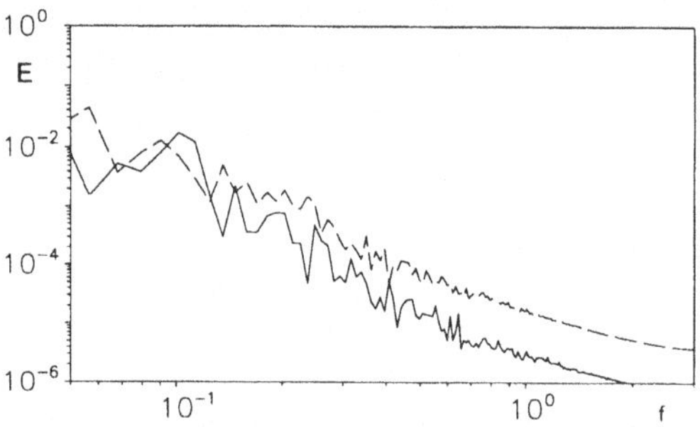

Figure 8 Energy spectrum of the temporal modes, case A (non-manipulated) mode 2 (full line), mode 3 (broken line)

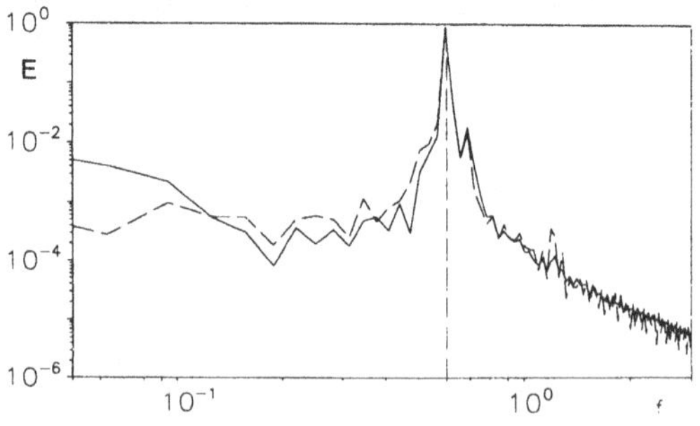

Figure 9 Energy spectrum of the temporal modes, case A (manipulated) mode 2 (full line), mode 3 (broken line)

EXPERIMENTAL STUDY OF TEMPORAL AND SPATIAL STRUCTURES IN FENCE-ON-WALL TESTCASE

P. S. LARSEN, J. J. SCHMIDT and U. ULLUM
Dept. of Energy Engineering,
Fluid Mechanics Section,
Technical University of Denmark
Building 404, DK-2800 Lyngby, Denmark

Abstract

The effects of periodic perturbations applied upstream of the fence by blowing/suction through a wall-slit are studied by LDA and PIV to establish velocity and turbulence fields, reattachment length and structures in the shear layer above the recirculation region. Perturbations, depending on frequency and amplitude, serve as an effective means of flow management, yielding a reduction of the downstream reattachment length as high as 56%. This reduction is caused by the promotion of coherent vortical structures resulting from multiple vortex-pairing processes and is most effective for sub-harmonic exitation. Results are presented for one Reynolds number, $Re_h = 1300$, both without perturbation and with 3 amplitudes of perturbation at two Strouhal numbers, $St_h = 0.15$ and 0.3. A correlation is proposed for the reduction in reattachment length.

1. Introduction

Large-scale structures in shear flows are believed to play a dominant role in entrainment and mixing processes. This is evident in the mixing layers bounding recirculating regions, such as those behind a step or behind a fence protruding from a wall. The formation and coherence of large structures may be enhanced by introducing external perturbations into a flow. Perturbations may be imposed by acoustic fields or by blowing/suction in jets or slits.

Typical flows in question include jets (Zaman and Hussain, 1980, Johnson and Nishi, 1990), plane shear layers (Miksad, 1972, Ho and Huang, 1982, Browand and Troutt, 1985, Weisbrot and Wygnanski, 1988, Wygnanski and Weisbrot, 1988, Husain and Hussain, 1995), and flows over steps (Bhattacharjee, et al., 1986, Hasan, 1992, Chun and Sung, 1996) and past obstacles, such as a wall-mounted fence of the present study (Orellano and Wengle, 1996, 1997, Siller and Fernholz, 1997).

Depending on the frequency, periodic perturbations may suppress or enhance vortex mergering (Husain and Hussain, 1995), sub-harmonic exitation being particularly effective in enhancing the mixing in shear layers by multiple merging (Ho and Huang, 1982). Active flow management by perturbations has attracted considerable interest, see e.g. Fernholz and Fiedler (1997), in fundamental turbulence and flow research as well as for practical applications in aeronautics, flow machinery, etc.

The present experimental study is concerned with the effect of upstream periodic perturbations on the extent and the structure of the downstream separated region for two-dimensional flow past a wall-mounted fence (at Re_h = 1300 and blockage ratio h/H = 0.133) as detailed by Schmidt (1997). The same problem has been studied by direct numerical simulation by Orellano and Wengle (1996, 1997) (at Re_h = 3000 and 10,000, and h/H = 0.2), and experimentally by Siller and Fernholz (1997) (at Re_h = 10,600, and h/H = 0.15).

2. Experiment

Test facility. For perturbation studies, the 40 x 10 mm fence was mounted at the wall 560 mm downstream of the inlet to a 600 mm wide by 300 mm high wind tunnel test section, Fig.1. The origin of the coordinate system is located at the trailing edge wall-corner of the fence. The bulk velocity U_0 = 0.5 m/s gave a Reynolds number based on fence height, h, of $Re_h = U_0 h/\nu$ = 1300. The flow at the inlet of the test section was weakly turbulent (3%).

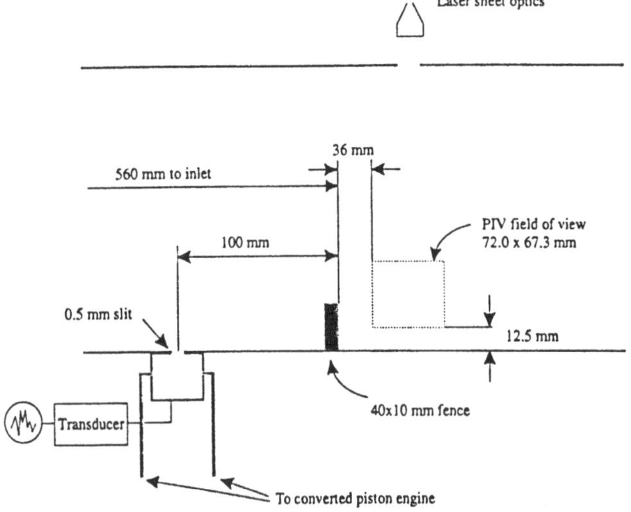

Figure 1. Test Section with 40 x 10 mm wall-mounted fence and 0.5 mm wide wall-slit at x/h = -2.5 for periodic pertubation flow from converted piston engine. LDA-data profiles recorded at x/h = 1, 2, 4, 6 and 8. Velocity vector maps recorded in PIV-field $0.90 < x/h < 2.70$ by $0.31 < y/h < 2.00$, where $(x,y) = (0,0)$ at trailing edge of the fence and wall.

Perturbations. The flow could be perturbed by oscillating blowing/suction through a 0.5 mm wide slit in the wall parallel to the fence, 100 mm (2.5 h) upstream of the trailing edge of the fence. The perturbation air was delivered by a modified piston engine driven by an electrical motor with adjustable speed. The frequency was monired by a hand-held tachometer. The amplitude was adjusted by a valve venting the manifold and was monitored by a pressure transducer yielding the standard deviation of the pressure oscillations as a measure of reproducibility and relative amplitude. In earlier experiments, perturbations were introduced by 5 wall jets, 4 mm dia, located 135 mm upstream of the fence. While attainable reductions in reattachment length were similar to those of slit experiments, jet experiments showed some spanwise distortions of velocity fields downstream of the fence.

LDA instrumentation. The two sides of the test section were made of glass which gave optical access for a two-component, three-beam back-scatter LDA system (Dantec 55X, 600 mm front lens, and a 58N20 Flow Velocity Analyzer). The optical system was mounted on a 3-axis traversing bench which could position the measuring volume to within 0.1 mm. The y-position of the measured profiles was corrected by analysing the data near the wall, as done to obtain skin friction and reattachment point from LDA data (Schmidt and Larsen, 1997). The effective measuring volume had a diameter of about 0.15 mm in the x,y plane and a length of 3.5 mm in the spanwise direction parallel to the fence. The flow was seeded with an atomized mixture of glycerol and water at the blower at the inlet of the wind tunnel.

The residence-time-averaged LDA data provided profiles of mean field U and V velocities, $\overline{u^2}$, $\overline{u\,v}$ and $\overline{v^2}$ as well as skewness and flatness factors S and F. Sample size was 2-3000 with a typical dead-time controlled sample rate of 20 Hz to ensure the data to be statistically uncorrelated. Time series (10,000 samples at 100-500 Hz) provided auto-correlations (by the time-slot method, Buchhave et al., 1979) and spectra (by resampling). The reattachment length was determined by first finding the x-positions of zero axial velocity during traverses parallel to the wall at 5 y-positions located 1-3 mm from the wall and then extrapolating these x-positions to the wall.

PIV instrumentation. The light source of the PIV system was a Continuum Surelite I-10 double-cavity Nd:YAG laser with an output of 200 mJ per pulse and a pulse duration of 4-6 ns. A Dantec 80X11 high-power light-guiding arm transmitted the pulsing laser beam to the test section where the 80X20 optics produced a light-sheet of 1.5 mm thickness with a divergence angle of 15° in the horizontal x,y mid-plane of the wind tunnel.

PIV images (covering the area $0.90 < x/h < 2.70$ by $0.31 < y/h < 2.00$, see Fig. 1) were captured by a 1k x 1k cross-correlation CCD camera (Kodak Megaplus ES1.0) and processed into vector maps by a Dantec FlowMap PIV 2000 Processor in real- time at 5 Hz. For the case of perturbations, conditional sampling was achieved by sending a delayed trigger signal from the piston engine to the FlowMap Processor (Ullum et al., 1997).

3. Results

Earlier data. The fence-on-wall testcase is well documented in an available database for Re_h = 1500, 2000, 2500 and 3000, recorded with a turbulence generating grid at

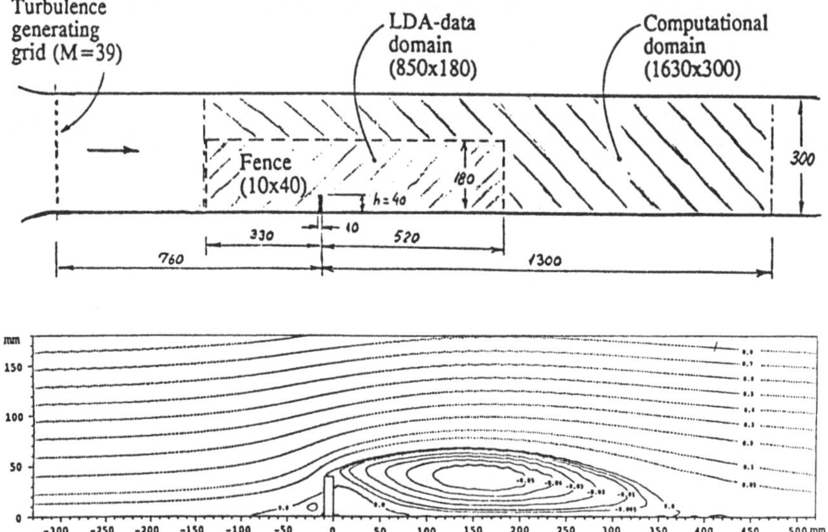

Figure 2. Earlier fence-on-wall testcase: LDA- and computational domains (top) and streamline plot from 2-D-smoothed LDA-data for Re_h = 1500 (bottom).

TABEL 1. Dimensions of recirculation regions

Characteristic lengths (mm)		L1	H1	L2	H2	L3	H3
Re_h =	1500	375	68	80	24.5	82	34
	2000	420	72	76	22.5	75	31
	2500	450	76	38	22	73	29
	3000	470	77	25	22	65	30

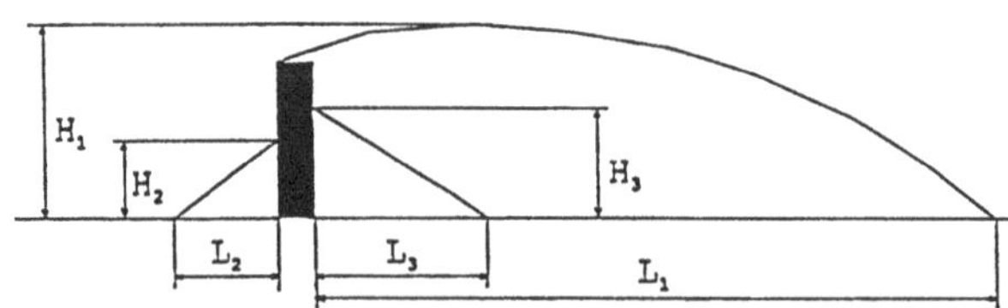

Figure 3. Definition of characteristic lengths of recirculation regions.

the inlet (Jørgensen and Marxen, 1993). As an example, streamlines from smoothed LDA data at $Re_h = 1500$ are shown in Fig.2, and measured dimensions of recirculation regions, defined in Fig.3, are shown in Table 1. For the present study at $Re_h = 1300$ without turbulence generating grid at the inlet, the primary reattachment length without perturbations was $L1 = X_{r,0} = 313$ mm.

Present experiments. Experimental results for $Re_h = 1300$ are presented for the 4 cases summarized in Table 2. They include the reference case of no pertubations and 3 cases of perturbations with small, medium and large amplitudes, the first two at $f = 1.87$ Hz and the last one at $f = 3.7$ Hz. We begin by discussing the choice of frequency and amplitude.

TABLE 2. Experimental results for 4 cases of study

Case case	Frequency f (Hz)	St_h -	P_E -	St_hP_E -	X_r mm	X_r/h -	$\Delta X_r/X_{r0}$ %
1	0	-	-	-	313	7.8	0
2	1.87	0.15	0.136	20.4	269	6.7	14
3	1.87	0.15	0.297	44.6	237	5.9	24
4	3.7	0.30	0.65	195	139	3.5	56

3.1. PERTURBATION FREQUENCY

For flow over a backward facing step, Bhattacharjee et al. (1986) found the most efficient Strouhal number based on freestream velocity and maximal height of the recirculation region to be in the range of St = 0.2-0.4, in agreement with the results of Browand and Troutt (1985). As seen from Fig.2 the maximal height of the recirculation region behind the fence is approximately twice the fence height which would give a most efficient Strouhal number of $St_h = hf/U_0 \approx 0.15$, the value used by Orellano and Wengle (1996) in their numerical study of the laminar separated flow over a fence at $Re_h = 250$.

Unperturbed flows. Presently the choice of perturbation frequency has been based on measured natural shedding frequencies of the unperturbed flow in the shear layer at different positions downstream of the fence. Time auto-correlations and one-dimensional power spectra of the axial velocity component are shown in Fig.4 (left). At $x/h = 1$, the first characteristic peak at time $t = 0.135$ is the average passage period of dominant structures (see Browand and Troutt, 1985). Hence the natural instability frequency of the flow at this point is $f = 7.4$ Hz. At $x/h = 2$, however, the first distinct peak in the correlation appears at $t = 0.27$ s which is twice the period observed at $x/h = 1$ and it corresponds to the frequency $f = 3.7$ Hz. This

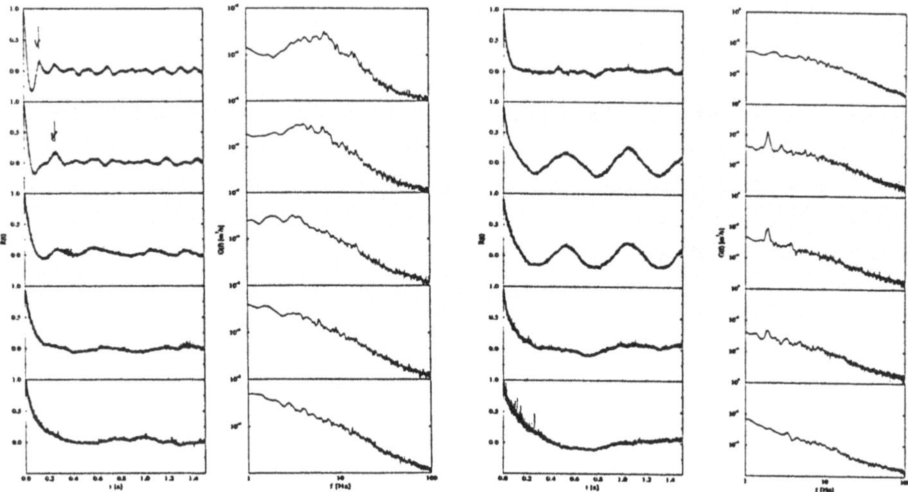

Figure 4. Time auto-correlations and corresponding spectra for axial velocity component at positions (from top down): $(x/h,y/h)$ = (1,1.5), (2,1.75), (4,2), (6,2) and (8,2). *Left:* Flow without perturbation. *Right:* Case of small amplitude (St_h = 0.15, P_E = 0.136, $\Delta X_r/X_r$ = 14%).

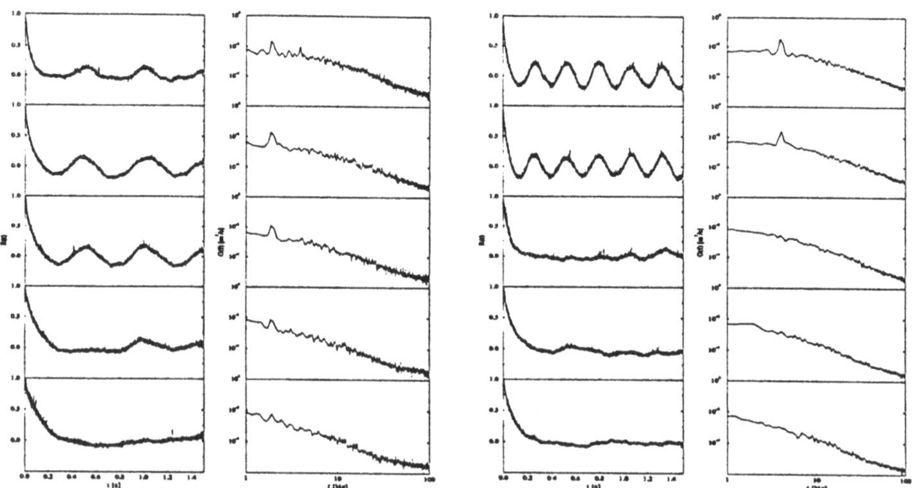

Figure 5. Time auto-correlations and corresponding spectra for axial velocity component at positions (from top down): $(x/h,y/h)$ = (1,1.5), (2,1.75), (4,2), (6,2) and (8,2). *Left:* Case of medium amplitude (St_h = 0.15, P_E = 0.136, $\Delta X_r/X_r$= 24%). *Right:* Case of large amplitude (St_h = 0.15, P_E = 0.136, $\Delta X_r/X_r$ = 65%).

indicates that pairs of vortices have merged between these two positions, a common feature of shear layers. The inverse cascade effect is apparent from the spectra. The above two frequencies correspond to a Strouhal number of $St_h = 0.6$ and 0.3, respectively.

It has been shown by Ho and Huang (1982) that forcing the flow at a subharmonic frequency will promote the merging of two or more vortices which is an important feature for enhancing the spreading rate of a shear layer. Therefore, perturbation Strouhal numbers chosen for further investigation are $St_h = 0.3$ and 0.15, the latter in accordance with the finding of Bhattacharjee et al. (1986) and also used by Orellano and Wengle (1996) in their numerical simulations.

Perturbed flows. Fig.4 (right) and Fig.5 (left and right) show correlations and spectra for three cases of flow past the fence at $Re_h = 1300$ with perturbations of small, medium and large amplitude, respectively, at $St_h = 0.15$, 0.15 and 0.3. At the lower frequency ($St_h = 0.15$), which would correspond to a double vortex pairing process, coherence of structures establishes most quickly at the higher amplitude but persist in both cases to at least $x/h = 4$. At the higher frequency ($St_h = 0.3$), which would correspond to a single vortex pairing process, coherence is established quickly due to the high amplitude and nevertheless die out before $x/h = 4$. Hence, the lower frequency appears to be the most effective.

Ensemble averages of conditionally sampled PIV vector maps (Ullum et al. 1997) taken at 4 different phases in the perturbation cycle for the case of $f = 1.87$ Hz show distinct vortical structures whose center move over distance $x/h = 1.23$ to 2.08 with velocity 0.25 m/s, which is one half of the freestream velocity. According to Fig.4 (right), from $x/h = 1$ to 2, the distinct frequency $f = 1.87$ Hz is established in this interval and it may be concluded that vortical structures are spaced $0.5/1.87 = 0.27$ m apart. Since they are confined to the shear layer which is at most 70-100 mm wide at this point (se later Fig.8) it is concluded that we deal with highly elongated structures as expected for multiple vortex pairing.

3.2. PERTURBATION AMPLITUDE

For a preliminary study of the influence of perturbation frequency and amplitude on the reattachment length, without actually measuring it, the axial velocity component was recorded at a monitoring point located close to the wall and upstream of the reattachment point for unperturbed flow. Fig.6 (left) for preliminary jet perturbations shows axial velocity at the monitoring point versus relative amplitude of perturbations for a number of frequencies. It indicates that $St_h = 0.15$ ($f = 1.87$ Hz) gives the most significant reduction in reattachment length for a given amplitude. The same conclusion can be drawn from Fig.6 (right) showing data from slit perturbations. Both figures also suggest that a maximal reduction in reattachment length may be expected for increasing amplitude.

As a quantitative measure of the amplitude of perturbations we introduce the dimensionless ratio of kinetic energy per unit spanwise length of perturbation flow (in terms of its rms-velocity at the wall) and bulk flow (over height of the fence),

$$P_E = (\rho \overline{v^2}_{rms}\ b)_{WALL\ SLIT}/(\rho U^2_0\ h)_{BULK} \ . \qquad (1)$$

Here b denotes the slit width which is replaced by the nozzle area divided by spanwise spacing of nozzles in case of wall jets. LDA data of the V-component of velocity measured 1 mm from the wall showed that the flow could be approximated as plug-flow with sinusoidal oscillations around $V = 0$, hence the averaged squared standard deviation of the slit or jet velocity V is used in P_E. Further, the product $St_h P_E$ is a measure of the power of perturbation flow relative to bulk flow, a parameter that proves useful in correlating effects of perturbations. Experimental values of parameters for the 3 cases of perturbations are listed in Table 2.

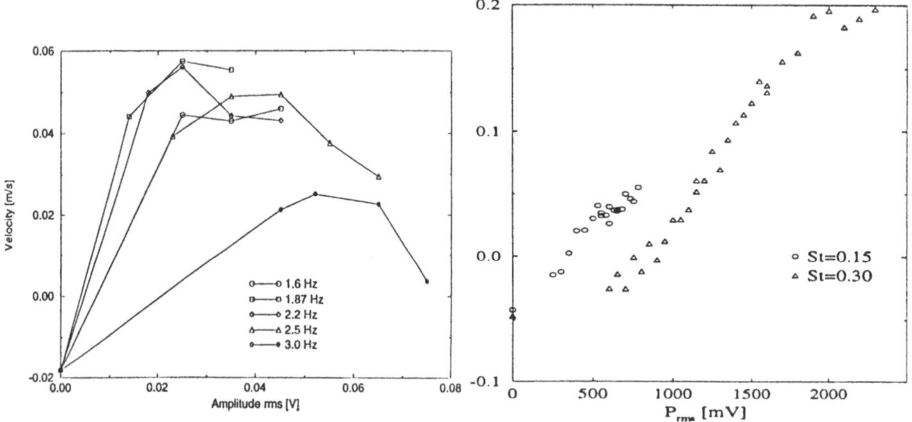

Figure 6. Axial velocity at monitoring point (m.p. at (x,y) mm) versus relative amplitude from pressure transducer at various frequencies. *Left:* Jet perturbations, m.p. at (300,6). *Right:* Slit perturbations, m.p. at (270,3).

3.3. REATTACHMENT LENGTH

The reattachment length found by LDA measurements for the 4 cases are listed in Table 2. The uncertainty on X_r/h is of the order of $\pm$ 0.1. The maximum reduction in separation length of 56% occurs for case 4 at $St_h = 0.30$ and $P_E = 0.65$. It is of the same magnitude as reported by Orellano and Wengle (1996) for numerical simulation of the laminar flow over a fence.

A plot of X_r/h versus $St_h P_E$ (Fig.7, left) suggests an exponential decrease from $X_{r,0}/h \approx 7.8$ towards an asymptotic value of $X_{r,a}/h \approx 3.1$ with decay constant $(St_h P_E)_0 \approx 0.08$, given by the empirical expression

$$(X_r - X_{r,a})/h = (X_{r,0} - X_{r,a})/h \cdot \exp[-St_h P_E/(St_h P_E)_0] \quad , \tag{2}$$

as shown in Fig.7 (right). The usefulness of the correlation parameter $St_h P_E$ remains to be verified through analysis of more data.

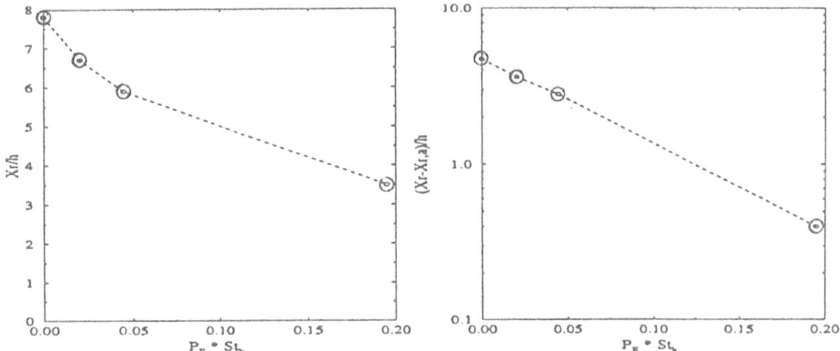

Figure 7. Reattachement length for 4 cases at 2 frequencies versus the parameter $St_h P_E$. *Left:* Raw data. *Right:* Semi-log plot of proposed empirical correlation, eq.(2).

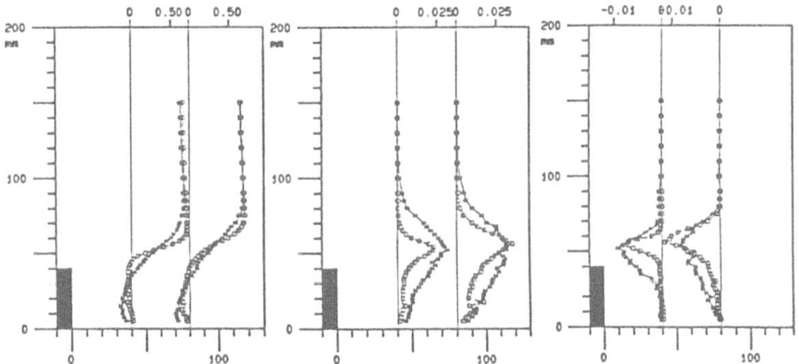

Figure 8. Profiles without perturbation (o) and with perturbation (*) at positions $X/h = 1$ and 2. Case of small amplitude ($St_h = 0.15$, $P_E = 0.136$, $\Delta X_r/X_r = 14\%$) *Left:* Axial velocity U. *Middle:* Turbulent kinetic energy k. *Right:* Turbulent shear stress $\overline{uv}$.

3.4. FLOW PROFILES

Profiles of mean fields of axial velocity U, turbulent kinetic energy k, computed from the two-component LDA measurements as $k = 0.75(\overline{u^2} + \overline{v^2})$, and the Reynolds stress $\overline{uv}$ are shown in Figs. 8 and 9 at various axial positions downstream of the fence for two cases of perturbation. In all cases the data are compared to those without perturbation.

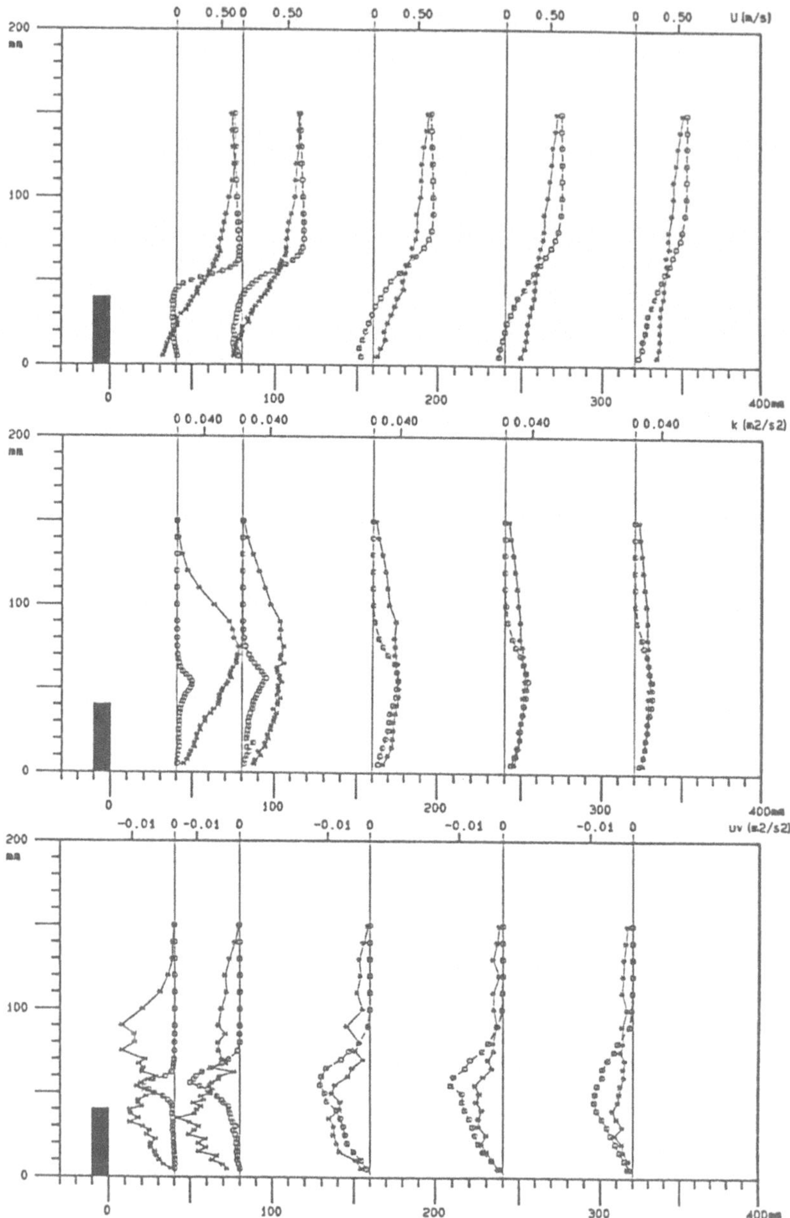

Figure 9. Profiles without perturbation (o) and with perturbation (*) at positions $X/h = 1, 2, 4, 6$ and 8. Case of large amplitude ($St_h = 0.15$, $P_E = 0.136$, $\Delta X_r/X_r = 65\%$). *Top:* Axial velocity U. *Middle:* Turbulent kinetic energy k. *Bottom:* Turbulent shear stress $\overline{uv}$.

The effect of perturbations on the spreading rate of the shear layer is clearly seen from the changes in axial velocity distributions and the associated increased levels of turbulent kinetic energy and shear stress. Sub-harmonic forcing is known to increase the spreading rate dramatically by stimulating the merging of several vortices (Ho and Huang, 1982). Presently the spreading rate is increased both towards the wall, where it causes a reduction of the separated region, and towards the freestream. These observations are somewhat in contrast to what was found for the backward facing step by Bhattacharjee et al. (1986) and what is commonly known from free shear flows, that the spreading is mainly increased towards the low-speed side of the flow when a perturbation is applied.

As expected, the effects of perturbations are most dramatic for large amplitude (Fig.9). However, the high levels of turbulent kinetic energi observed at $x/h = 1$ and 2 decay rapidly and reach essentially the level for flow without perturbation at $x/h = 4$ in the near-wall region while the decay is slow in the free stream. The initial high levels of turbulent shear stress follow those of turbulent kinetic energy, but downstream shear stress decreases more rapidly than for flow without perturbation due to the much faster recovery of the wall boundary layer.

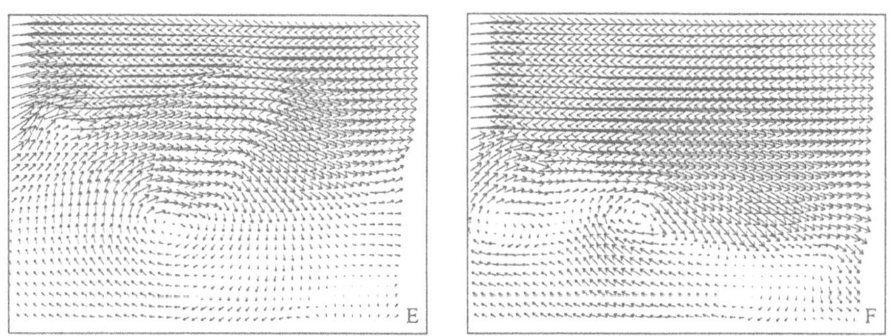

Figure 10. PIV velocity vector maps (jet experiment, small amplitude, $St_h = 0.15$). *Left:* Typical single vortical structure. *Right:* Vortex pairing event.

The preliminary data from PIV recordings in Fig.10 show a typical single vortical structure as well as a vortex pairing event. Vector maps sampled conditionally at selected phases of the periodic perturbation show, when averaged, a single vortical structure that moves downstream with half the bulk velocity (Ulrik et al., 1997). Although useful these planar vector maps do not give the three-dimensionality of structures. However, they may give an indication of the two-dimensionality of the flow. From the mean of a large number of vector maps the two-dimensional divergence, $Z = \partial U/\partial x + \partial V/\partial y$, was calculated. The results for cases without and with perturbations are shown in Fig.11 and they suggest the flow without perturbation is the more two-dimensional flow on the average, although there is a statistically significant positive value of divergence in the upper left corner of the PIV view (Fig.11, left). The two-dimensional divergence is more irregular and show larger values in the upper left corner for flow with perturbations (Fig.11, right).

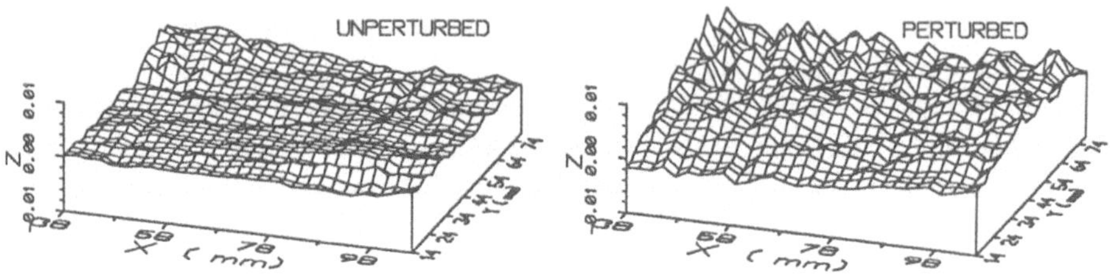

Figure 11. Distributions of two-dimensional divergence, $Z = \partial U/\partial x + \partial V/\partial y$ from average of 1000 unconditionally sampled PIV velocity vector maps (slit experiment, small amplitude, St_h = 0.15, P_E = 0.136, $\Delta X_r/X_r$ = 14%). *Left:* Without perturbation. *Right:* With perturbations.

4. Conclusions

The well documented fence-on-wall testcase has been studied experimentally by LDA and PIV to determine how periodic perturbations applied upstream of the fence affect the shear layer above the recirculation region and the attachment length. Experimental results for one Reynolds number are presented for 3 amplitudes and 2 frequencies of perturbation and compared to the case without perturbation.

Auto-correlations and spectra calculated from time series recorded at 5 axial positions in the shear layer show natural frequencies without perturbation and successive vortex pairing processes in response to sub-harmonic exitation. Velocity and turbulence profiles show the increased spreading of the shear layer. The resulting reduction in reattachment length is correlated empirically in terms of the parameter St_hP_E which represents the power of perturbation flow relative to bulk flow. This correlation suggests that reattachment length decreases exponentially towards a lower limit which is about 40% of the value without perturbation. It required a ten-factor increase in the parameter St_hP_E, from about 20 to 200, to change the reduction from 14% to 56%.

Acknowledgements – The study was partially supported by the CEC under contract AERO-CT92-0051 (ETMA project) and by the Danish Technical Research Council under grant STVF 5.26.16.31.

5. References

Bhattacharjee, S., Scheelke, B. and Troutt, T. R. (1986) Modification of Vortex Interactions in a Reattaching Separated Flow. AIAA Journal, 24, 623-629.

Browand, F. K., and Troutt, T. R. (1985) Turbulent mixing layer: geometry of large vorticies. J. Fluid Mech. 158, 489-509.

Buchhave, P., George, W. K., and Lumley, J. L. (1979) The measurement of turbulence with the laser-Doppler anemometer. Ann.Rev.Fluid Mech. 11, 443-503.

Chun, K. B., and Sung,,H.,J. (1996) Control of turbulent separated flow over a backward-facing step by local forcing. Exp. in Fluids, 21, 417-426.

Fernholz, H. H., and Fiedler, H. E. eds. (1997) EUROMECH Colloquium 361, Book of Abstracts, 17-19 March 1997, Hermann-Föttinger-Institut für Strömungsmechanik.

Hasan, M. A. Z. (1992) The flow over a backward-facing step under controlled perturbation: laminar separation. J. Fluid Mech. 238, 73-96.

Ho, C. M., and Huang, L. S. (1982) Subharmonics and vortex merging in mixing layers. J. Fluid Mech. 119, 443-473.

Husain, H. S., and Hussain, F. (1995) Experiments on subharmonic resonance in a shear layer. J.Fluid Mech. 304, 343-372.

Johnson, J. P., and Nishi, M. (1990) Vortex Generator Jets - Means for Flow Separation Control. AIAA J. 28, 989-994.

Jørgensen, O. A., and Marxen, U. (1993) Turbulent flow over ribs (in Danish), Report AFM EP 93-06, Master's thesis, Dept.of Fluid Mechanics, Techn.Univ.of Denmark.

Miksad, R. W. (1972) Experiments on the nonlinear stages of free-shear-layer transition. J. Fluid Mech. 56, 695-719.

Orellano, H., and Wengle, H. (1996) Numerical simulation and analysis of manipulated transitional flow over a fence, in Deville, Gavrilakis and Ryhming (eds.), *Computation of Three-dimensional Complex Flows*, NNFM Vol.53, Vieweg-Verlag, Braunschweig, Germany.

Orellano, A., and Wengle, H. (1997) Numerical simulations of manipulated flow over a fence and over a backward-facing step, in H. H. Fernholz and H. E. Fiedler (eds.), EUROMECH Colloquium 361, Book of Abstracts, 17-19 March 1997, Hermann-Föttinger-Institut für Strömungsmechanik.

Schmidt, J. J. (1997) Experimental and Numerical Investigation of Separated Flows. PhD dissertation, Technical University of Denmark, Feb. 1997.

Schmidt, J. J., and Larsen, P. S. (1997) Skin-friction and turbulent data for low Re-number bump-on-wall test case, in *Proceedings of the 2nd Int'l Sympos.on Turbulence, Heat and Mass Transfer*, K.Hanjalic (ed.) Delft Univ.Press, Addendum, 1-9.

Siller, H. A., and Fernholz, H. H. (1997) Control of the separated flow downstream of a two-dimensional fence, in H. H. Fernholz and H. E. Fiedler (eds.), *EUROMECH Colloquium 361, Book of Abstracts*, 17-19 March 1997, Hermann-Föttinger-Institut für Strömungsmechanik.

Ullum, U., Schmidt, J. J., Larsen, P. S., and McCluskey, D. R. (1997) Temporal evolution of the perturbed and unperturbed flow behind a fence: PIV analysis and comparison with LDA data. In Proceedings of 7th Int'l Conf. on Laser Anemometry Advances and Applications (to appear). Karlsruhe, Germany, 8-12 Sept.1997.

Weisbrot, I., and Wygnanski, I. (1988) On coherent structures in a highly excited mixing layer. J. Fluid Mech. 195, 137-159.

Wygnanski, I., and Weisbrot, I. (1988) On the pairing process in an excited plane turbulent mixing layer. J. Fluid Mech. 195, 161-173.

Zaman, K. B. M. Q., and Hussain, A. K. M. F. (1980) Vortex pairing in a circular jet under controlled exitation. Part 1 and 2. J. Fluid Mech. 101, 449-491 and 493-544.

II. Coherent Structures in Wall-Bounded Flows

DYNAMICS OF THE STRUCTURES OF NEAR WALL TURBULENCE[1]

JAVIER JIMENEZ AND ALFREDO PINELLI
School of Aeronautics, Universidad Politécnica
28040 Madrid, Spain[2]

Abstract. Numerical experiments on modified turbulent channels are used to differentiate between possible turbulence generation mechanisms in wall-bounded flows. It is shown that a regeneration cycle exists which is local to the near-wall region and does not depend on the outer flow. It involves the formation of velocity streaks from the advection of the mean profile by streamwise vortices, and the generation of the vortices from the instability of the streaks. Interrupting any of those processes leads to laminarisation of the wall. The production of secondary vorticity at the wall is not important in turbulence generation.

1. Introduction

The study of turbulence has been radically changed in the last decades by numerical experiments. Their superior diagnostic capability has provided new structural and statistical information on many flows that had resisted analysis for a long time. It is part of the purpose of this paper to argue that an even more useful feature of the numerical experiments is their ability to simulate the "wrong" physics, and thereby elucidate dynamical interactions among structures that cannot be easily clarified in any other way.

Turbulent flows are complex, and involve many interacting phenomena. It is often difficult to distinguish between the different effects, and almost impossible in physical situations to manipulate the flow so that a particular aspect is enhanced or isolated. Numerical experiments have fewer restric-

[1]Presented by invitation at the IUTAM Symposium on Simulation and Identification of Organized Structures in Flows, held at Lyngby, Denmark, on May 25-29, 1997.

[2]J.J. is also part of the Centre for Turbulence Research at Stanford University

tions, and permit the artificial manipulation of the boundary conditions, or even of the equations of motions, to test which phenomena are more important in a given case. In this paper we will be interested in wall-bounded flows, and we will give several examples of the use of modified numerical experiments in understanding the turbulence generation process in them.

It has long been understood that the near-wall region is crucial in the dynamics of attached shear flows, being the seat of the highest rate of turbulent energy production and of the maximum turbulent intensities. It is highly intermittent, with locally low Reynolds numbers, and it is controlled by the interaction of intense structures [17]. From the physical point of view, this region is one of the few genuinely different classes of turbulent flows, distinguishable from isotropic turbulence in the same way that solid surfaces are different from the solid state itself.

In technological applications, the region below $y^+ \approx 50$ acts as a boundary condition for the logarithmic law [11], and determines the magnitude of the wall stress[3].

The organisation of this paper is as follows. The near-wall structures, the interactions that have been proposed among them, and the evidence for the existence of a near-wall turbulence cycle are summarised in §2. Two proposed regeneration cycles are tested next by numerical experiments designed to block them. It it shown that blocking one of them increases the intensity of turbulence, while blocking the other leads to relaminarisation. In the final section the results are discussed and related to possible strategies for flow control.

2. Near-wall structures

The dominant structures of the near-wall region are the streamwise velocity streaks and the quasi-streamwise vortices. The former were first described in [12], and are long ($x^+ \approx 1000$) sinuous arrays of alternating streamwise jets, with an average spanwise separation of $z^+ \approx 100$ [20]. Where the jets point forward, the wall shear is higher than the average, while the opposite is true at the "low velocity" streaks where they point backwards. The quasi-streamwise vortices are slightly tilted away from the wall, and each one stays in the near-wall region for only $x^+ \approx 200$ [7]. Several vortices are associated with each streak, with a longitudinal spacing of the order of $x^+ \approx 400$ [10].

It has long been understood that the vortices cause the streaks by advecting the mean velocity gradient [2], and that this process results in a higher mean shear at the wall, being the immediate cause of the turbulent

[3]Wall units are defined as usual in terms of the wall stress τ, by normalising with the kinematic viscosity ν and with the friction velocity $u_\tau = (\tau/\rho)^{1/2}$

wall drag [15]. There is less agreement on the mechanism by which the vortices are produced, and three candidates will be discussed next.

Soon after the streaks were first observed, it was proposed that their instability was involved in turbulence production [12], especially after it was noted in [22] that the layers of wall-normal vorticity which separate the low- from the high-velocity streaks are subject to inflectional instabilities. Although the vortices created in this way are normal to the wall, they are tilted forward and intensified by the prevailing mean shear. This conceptual model was elaborated in [18, 9, 6], and is also behind the reduced dynamical system approximations of the wall region in [1]. We will refer to it as the *streak* cycle. It can be summarised as that the quasi-streamwise vortices act on the mean shear to create the streaks, which become unstable and break into wall-normal vortices, which are then tilted streamwise by the mean shear.

Another vortex regeneration mechanism, which does not involve the streaks, is based on the behaviour of a vortex when it approaches a no-slip wall, inducing a layer of vorticity of the opposite sign which rolls into new vortices [4, 14]. The proposed cycle begins when a quasi-streamwise vortex approaches the wall and creates a new vortex of opposite sign [21]. The new structure, which is already more or less parallel to the stream, leaves the wall under the induction of its parent, and is stretched and intensified by the mean shear. Note that this *wall* cycle is almost two-dimensional in the cross-flow plane, $z - y$, and that, except for using the mean shear as an energy source, depends only on the transverse no-slip condition at the wall, $w = 0$. As in the previous case, this cycle has been observed in real flows, mainly as the formation of secondary vortices near the tails of strong hairpins in otherwise laminar boundary layers [5, 19].

It is important to understand that the fact that both cycles occur experimentally does not prove that they are relevant in the random environment of fully developed turbulence, where they have to compete with other mechanisms. A better criterion would be to remove one by one the events to be tested from a fully turbulent flow. Those which result in the damping of turbulence would then be shown to be important in its maintenance.

Consider for example the possibility that no near-wall cycle is important and that vortices are formed directly by the intensification of perturbations coming from the outer flow. The experiment of removing the outer turbulent flow was performed in [10], where a simulation of a turbulent channel was shrunk in the spanwise and streamwise directions until it was reduced to its minimum possible dimensions. This "minimal" unit was small enough that wall structures could be simulated, but it left no space for the larger eddies of the core region. Near-wall turbulence, with very similar characteristics to those in regular channels, was maintained even when the core was laminar.

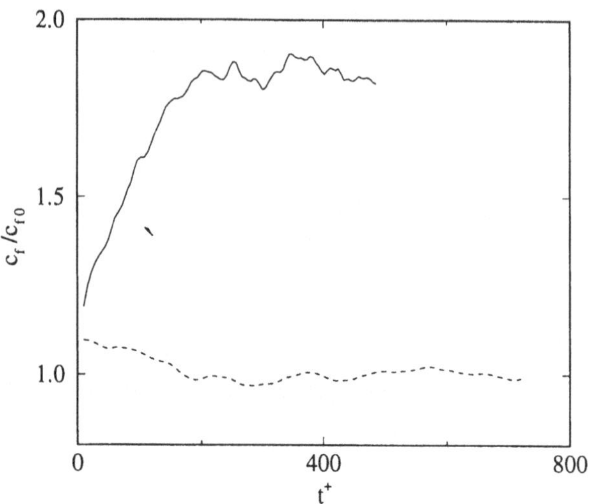

Figure 1. Time evolution of the skin friction for a natural turbulent channel (dashed line), and for one in which the transverse no-slip condition at the wall has been substituted by $\partial w/\partial y = 0$. Initial conditions are a fully developed channel with $Re_\tau = 120$.

That result showed that the amplification of outer perturbations is not crucial in the maintenance of wall turbulence, although it is clearly a plausible mechanism and probably happens to a certain extent in all wall flows.

3. Testing the cycles

3.1. THE WALL CYCLE

We have mentioned that the key event in the *wall* regeneration cycle is the formation of secondary streamwise vorticity at the wall, which depends on the transverse no-slip boundary condition, $w = 0$, while the amplification of turbulent fluctuations depends, on all plausible scenarios, on the mean velocity shear maintained by the streamwise no-slip condition, $u = 0$ [16]. In real flows both conditions are satisfied at the wall, but in numerically simulated ones they can be imposed independently.

A numerical experiment in which the *wall* cycle was tested in this way was presented in [8, 9], where the transverse no-slip condition was substituted by a zero-stress one, $\partial w/\partial y = 0$. If the *wall* cycle were the dominant turbulence regeneration mechanism, that substitution would inhibit the formation of new streamwise vortices, turbulence would be weakened, and the skin friction would drop with respect to that of a "natural" flow. Conversely, if the effect of secondary vorticity were not important, the main effect of the transverse no-slip condition would be to weaken the streamwise vortices through viscous friction at the wall. The streamwise vortex formation mechanism would not be affected by the removal of the friction , but the transverse dissipation would decrease. The vortices would then become stronger, increasing both the turbulent intensities and the skin friction.

The result of the experiment is shown in figure 1, which compares the time evolution of a natural turbulent channel with that of an anisotropic one with a transverse free-slip wall. In disagreement with the *wall* model for vortex generation, the skin friction is larger by a factor of about 1.8 in the anisotropic case. A similar experiment, using a different numerical code at a higher Reynolds number ($Re_\tau = u_\tau h/\nu = 180$, where h is the half-channel width) and a somewhat higher resolution, was carried out by Orlandi (private communication) with similar results.

These results strongly suggests that the *wall* cycle is not dominant in the regeneration of turbulence in the wall region, and that the main effect of the transverse no-slip condition at the wall is to limit the vortex intensity through viscous dissipation.

3.2. THE STREAK CYCLE

The *streak* cycle involves events in the interior of the flow and, in consequence, testing it implies modifying the equations of motion rather than just the boundary conditions. The ideal test would eliminate the streaks without directly perturbing the streamwise vortices. If the *streak* cycle were the key vortex regeneration mechanism, this would prevent the production of new vortices, existing ones would decay viscously, and turbulence would either be damped or decay altogether. On the other hand, if this were not the case, turbulence would either be enhanced or remain essentially unaffected, as in the two cases discussed above. Such a experiment is discussed [11] and summarised here.

3.2.1. *The definition of a streak*

Consider a numerical simulation of a channel in which the length of the computational box is short with respect to the typical longitudinal extent of the streaks. It was shown in [10] that turbulence could be sustained in boxes longer than about $L_x^+ \approx 400$, while natural streaks have streamwise coherence lengths which are several times larger. The present experiments are run in doubly periodic boxes of size $L_x^+ \times L_z^+ \approx 500 \times 300$, at initial Reynolds numbers $Re_\tau \approx 180 - 200$. Each wall therefore contains two or three velocity streaks which extend over the whole length of the box, each of which is associated on the average with one streamwise vortex of each sign. The box is wide enough for the large eddies of the core region and the one-point statistics agree well with those in larger computational boxes.

Since streaks span the length of the box and are roughly parallel to the mean flow, they can be approximately represented by that part of the

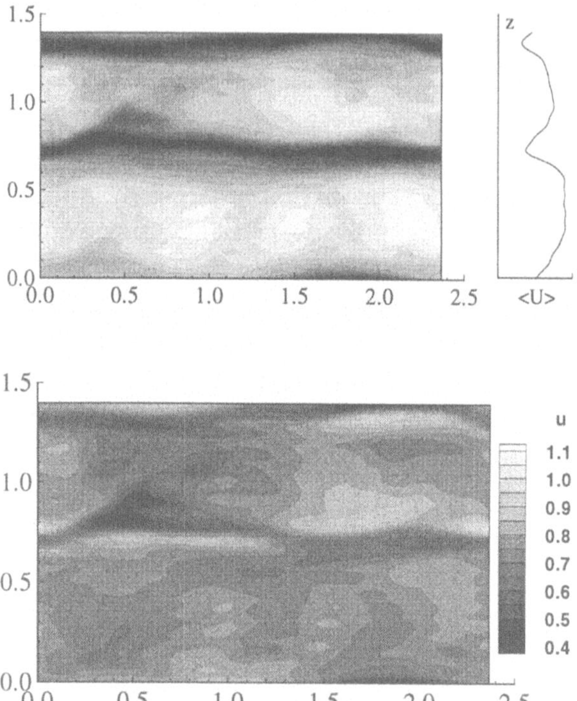

Figure 2. Definition of streaks. The top figure is the streamwise velocity in a plane $y^+ = 30$, with its streamwise mean represented on the right. The bottom figure is the same field after the streak component has been removed. Flow is from left to right. The gray scales are the same in both cases, and lighter shades correspond to higher values.

streamwise velocity u which depends on the transverse coordinates, y and z, but which is independent of x. Moreover, since we are only concerned with the spanwise modulation of u, we can define a "streak component" as the streamwise average of $\partial u/\partial z$ (figure 2),

$$\Omega_y(y, z) = \frac{\partial}{\partial z} \int_0^{L_x} u \, \mathrm{d}x = \int_0^{L_x} \omega_y \, \mathrm{d}x. \tag{1}$$

It was shown in [13] that the equations of motion can be written in terms of the normal vorticity ω_y and of the normal velocity v (plus the mean velocity profile for u), from where the other variables can be obtained using continuity. We can now decompose ω_y into "streak" and "incoherent" components,

$$\omega_y(x, y, z) = \Omega_y(y, z) + \tilde{\omega}_y(x, y, z). \tag{2}$$

The assumption behind the *wall* cycle is that the streak component (1) is responsible for the regeneration of the streamwise vortices, and that we should be able to manipulate wall turbulence by acting on it. Two observations are important. The first is that Ω_y does not contribute to the continuity equation, so that we can modify it without directly perturbing the transverse velocity components or the streamwise vortices. The second is that the incoherent component retains a large part of the energy of the longitudinal velocity fluctuations (figure 2.b).

3.2.2. *Damping the streaks*

The simplest experiment would be to damp the streaks with a filtering function below a given distance from the wall. The equations of motions would be integrated as usual but, after each step, Ω_y would be multiplied by

$$F(y) = \left[\frac{1+\alpha}{2} + \frac{1-\alpha}{2} \tanh 4(y^2/\delta^2 - 1)\right], \tag{3}$$

which leaves unchanged the upper wall, but damps the streaks in the lower one by a factor α.

It is actually more instructive to damp the source term of the streak component, rather than the streaks themselves. The evolution equation for Ω_y can be written as

$$\partial_t \Omega_y = \partial_z \langle F_1 v\, \omega_z - F_2 w\, \omega_y \rangle + \nu \nabla^2 \Omega_y, \tag{4}$$

where $\langle\rangle$ is the streamwise averaging operator in (1) and $F_j = F(y, \alpha_j)$, as in (3). The term multiplied by F_1 represents the advection of the spanwise vorticity by the wall-normal velocity, and is the one usually cited as being involved in the generation of the streaks from the mean velocity gradient. The term multiplied by F_2 represent the lateral advection of the normal vorticity by the spanwise velocity. In a natural flow $\alpha_1 = \alpha_2 = 1$, but in the numerical experiments either filter, or both, can be switched on. In this way the streaks are not damped directly, but eventually decay by the effect of viscosity.

The results are discussed in [11], but they broadly confirm the standard interpretation of the *streak* cycle. Damping F_2 has almost no effect on the flow, while damping F_1, and therefore the generation of the streaks by the vortices, relaminarises the flow (figure 3).

A detailed study of the evolution of the turbulent intensity profiles [11] shows that the streaks and the longitudinal vortices decay completely, and that the only remaining fluctuations are weak passive ones induced by the core turbulence. The same analysis shows that the rate of decay of the longitudinal vortices agrees well, after an initial transient, with that of unstretched cylindrical structures with radii of the order of those observed in natural near-wall simulations [13].

It is also possible to test, by varying the filter width δ, which is the region in which the instability of the streaks gives rise to the vortices. It

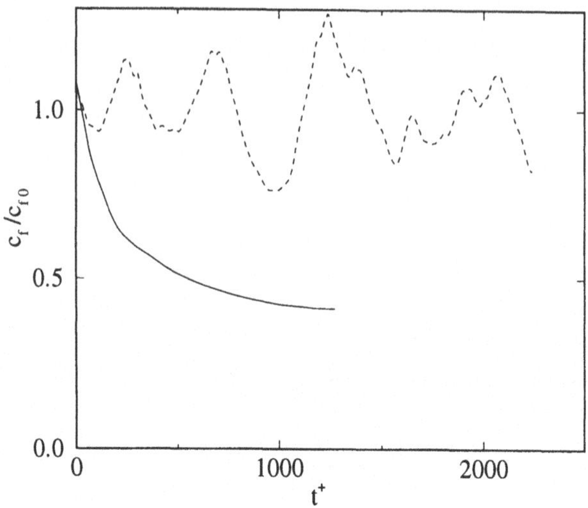

Figure 3. Time evolution of the skin friction in a natural turbulent channel (dashed line), and in one which has been filtered as in (4), with $\alpha_1 = 0.1$, $\alpha_2 = 1$ and $\delta^+ = 41$. Initial conditions are a fully developed channel with $Re_\tau = 190$.

turns out that filtering is not effective unless $\delta^+ \geq 30$, but that, as soon as the streaks are damped above that level, the flow laminarises. This agrees with the location of the maxima of ω_x' and v' in natural channels, which is about $y^+ = 20$ [13], and suggests that the vortex generation mechanism acts in this region. Note in particular that filters with $\delta^+ < 30$ modify the behaviour of the wall but have little or no effect on the flow, reinforcing the conclusion that the turbulence regeneration cycle resides away from the wall, rather than on it.

4. Conclusions

The main conclusion of this paper is the confirmation that wall turbulence is maintained by a cycle in which streamwise vortices extract energy from the mean flow to create alternating streaks of longitudinal velocity, and that the streaks in turn give rise to the vortices, presumably by inflectional instabilities. Two alternative possibilities, that the vortices are self-regenerating through the production of secondary vorticity at the wall, or that wall turbulence is predominantly fed by perturbations from outside, have been shown to be irrelevant in real turbulent flows, at least at the Reynolds numbers of the simulations.

It has also been shown that the generation cycle can be interrupted numerically at various places, leading to the decay of the fluctuations and to eventual laminarisation. Many attempts to control wall turbulence have been based in weakening the streamwise vortices, and thus presumably their effect in generating the streaks [3, 9]. An alternative strategy suggested by the present results would be to weaken the streaks themselves over a region of the order of $y^+ \approx 30$. Several perturbations have this effect, with the essential ingredient apparently being the decorrelation of the streaks from the generating effect of the existing streamwise vortices [11].

The final point of this paper is methodological, concerning the use of numerical experiments in exploring turbulence physics. We have shown that numerics can be used in ways which do not duplicate real experiments, and that it is precisely the ability of simulating "impossible" physics that makes numerical techniques a tool of choice in unravelling complex systems.

Part of this work was perform at the Centre for Turbulence Research, and supported by it. The rest was supported the Spanish CICYT under contract PB95-0159. A.P. was supported by a HCM postdoctoral fellowship from the European Commission, and by CICYT.

References

1. Aubry,N., Holmes,P., Lumley, J.L. & Sone,E. 1988 *J. Fluid Mech.*, **192**, 115-173.
2. Blackwelder, R.F. & Eckelmann, H. 1979 *J. Fluid Mech.* **94**, 577–594.
3. Choi, H., Moin, P. & Kim, J. 1994 *J. Fluid Mech.* **262**, 75-110.
4. Doligalski, T.L. & Walker, J.D.A. 1984 *J. Fluid. Mech.* **139**, 1–28.
5. Haidari, A.H. & Smith, C.R. 1994 *J. Fluid Mech.* **277**, 135-162.
6. Hamilton, J.M., Kim, J. & Waleffe, F. 1995 *J. Fluid Mech.* **287**, 317-348.
7. Jeong, J., Hussain,F., Schoppa,W. & Kim,J. 1997 *J. Fluid Mech.* **332**, 185-214.
8. Jiménez, J. 1992 *11th Australasian Fluid Mech. Conf.* (M.R.Davis & G.J.Walker eds.), Hobart, Australia. pp. 813-816.
9. Jiménez, J. 1994 *Phys. Fluids* **6**, 944-953.
10. Jiménez, J. & Moin, P. 1991 *J. Fluid Mech.* **225**, 221-240.
11. Jiménez, J. & Pinelli, A. 1997 *AIAA Paper* **97-2112**.
12. Kim, H.T., Kline, S.J. & Reynolds, W.C. 1971 *J. Fluid Mech.* **50**, 133-160.
13. Kim, J., Moin, P. & Moser, R. 1987 *J. Fluid Mech.* **177**, 133–166.
14. Orlandi, P. 1990 *Phys. Fluids* A **2**, 1429-1436.
15. Orlandi, P. & Jiménez, J. 1994 *Phys. Fluids* **6**, 634-641.
16. Perot, B. & Moin, P. 1995 *J. Fluid Mech.* **295**, 199-227.
17. Robinson, S.K. 1991 *Ann. Rev. Fluid Mech.* **23**, 601–639.
18. Sendstad, O. & Moin, P. 1992 *Report* **TF-57**. Dept. Mech. Eng., Stanford U.
19. Singer, B.A. & Joslin, R.D. 1994 *Phys. Fluids* **6**, 3724-3736.
20. Smith, C.R. & Metzler, S.P. 1983 *J. Fluid Mech.* **129**, 27-54.
21. Smith, C.R., Walker, J.D.A. , Haidari, A.H. & Sobrun, U. 1991 *Phil. Trans. R. Soc. Lond.* A **336**, 131-175.
22. Swearingen, J.D. & Blackwelder, R.F. 1987 *J. Fluid Mech.* **182**, 255-290.

THREE DIMENSIONAL CONFIGURATION OF A LARGE-SCALE COHERENT VORTEX IN A TURBULENT BOUNDARY LAYER

H. MAKITA
Dept. of Mechanical Eng., Toyohashi Univ. of Tech.
1-1 Tempaku-cho, Toyohashi 441, Japan

AND

K. SASSA
Dept. of Physics, Kochi Univ.
2-5-1 Akebono-cho, Kochi, 780, Japan

Abstract. Through one-point conditional hotwire measurements, a three-dimensional configuration was illustrated about a large-scale coherent vortex in a turbulent boundary layer. Coherent vorticity vector maps gave it a horseshoe sculpture. The horseshoe vortex induces four kinds of coherent structures; the strong outflow (ejection) between a pair of its legs, the inrush (sweep) outside the legs, the large-scale u-discontinuity upstream of the outflow and its head conforms to the turbulent bulge in the outer layer. Its coherent energy is mainly produced by the outflow. The legs play dominant roles in the energy and momentum transfer by the coherent motions and the head makes turbulent bulge with obscure structure. The coherent energy budget showed that coherent production, random production and random diffusion terms mainly contributed to the energy transfer about the large-scale coherent vortex.

1. Introduction

A turbulent boundary layer is a forest of coherent structures. Among them, small structures in the inner layer are known to contribute effectively to the production of turbulence energy. Their intensity and remarkable features make it relatively easy to detect them in the complicated turbulence field of the inner layer. The coherent motions such as the ejection and the sweep show strong correlation between u and v, and can be easily identified through the quadrant technique introduced by Wallace et al. (1972) and Willmarth and Lu (1972). The large-scale coherent structures in the outer layer, e.g., the turbulent bulge observed by Kovasznay et al. (1970) and the u-discontinuity or the back found by Brown and Thomas (1977) also play important roles as the interface between the turbulent boundary layer and the outside freestream. Nevertheless, we still have no consistent image to explain comprehensively the characteristic features of the coherent structures because they are quite weak structures and their detection is not so easy in the turbulent boundary layer. Then, their fluid dynamical characteristics and relation to the small structures have been scarcely clarified. Recently, DNS and PIV have given powerful tools in this field of the study. Using DNS, Robinson, Kline and Spalart (1990) dramatically illustrated instantaneous three-dimensional figures of pressure, shear stress and velocity vectors about the coherent structures. By PIV, Meinhart and Adrian (1995) found the existence of uniform momentum zones associated with the turbulent bulge, the quasi-streamwise vortex and the hairpin or cane-like vortex in the turbulent boundary layer. With these tools, we can draw so many detailed instantaneous figures for varieties of fluid dynamical quantities around the coherent structures but, at the same time, we are afraid to miss some fundamental features buried under too much information offered by the tools.

Against such a recent research trend that the detailed instantaneous figures are more desirable, we preferred to obtain a phase-averaged figure of the large-scale coherent structure because deducing a simple conceptual model is sometimes more favorable to explain basic physical features of the coherent structure than a group of the instantaneous realizations. Jeong et al. (1997) also focused on an ensemble-averaged structure and proposed a staggered array of quasi-streamwise vortices as a conceptual model of the coherent structure in the inner layer. The present experimental investigation aims to determine a three-dimensional configuration of the large-scale coherent structure in the outer boundary layer and to clarify its roles in the transfer mechanism of turbulence quantities. A coherent vortex was artificially developed in a turbulent boundary layer to make its detection easy by locking its occurrence in the turbulent boundary layer (Makita et al. 1989). The basic idea is taken from the bursting discovered by Kline et al. (1967). We have already reported that a tiny spanwise vortex initially given in a lower boundary layer grew up in the streamwise direction to be a large-scale coherent vortex having a head like the turbulent bulge (Makita and Sassa 1993).

2. Experimental procedure and data interpretation

The present experiment was made in a zero pressure gradient fully-developed turbulent boundary layer on a smooth flat glass plate, 420 mm wide and 5.7 m long, settled in a low turbulence wind tunnel. A small spanwise vortex was given at about $Y/\delta = 0.26$ in the lower boundary layer by injecting counter-rotating puffs from a pair of small mutually-facing slots on the flat plate at $X/\delta = 0$(Makita at al. 1989). Conditional measurements by an X or a triple wire probe were conducted referring to the electric control pulse of the puff injection. So, three-dimensional spatial data can be obtained by a sequence of one-point hotwire measurements. Measurement points were distributed on the $Y - Z$ plane at $X/\delta = 14$ where the large-scale vortex was assumed as fully developed in the decay stage. The decay stage was far longer than the preceding two stages, the growth stage and the self-preserving stage, as stated by the authors (1993), and the natural coherent structures in the actual boundary layers are mostly detected there in its whole life. The freestream velocity, U_∞, was 5 m/sec. The boundary layer thickness, δ, was about 44 mm at $X/\delta = 14$ and the Reynolds number, R_θ, was about 1830 there.

By the triple decomposition derived by Hussain and Reynolds (1970), we obtained the mean velocity, U, and the coherent components, u_c, v_c and w_c through the ordinary time averaging and 1024 times conditional averaging measured instantaneous velocity data. The coherent vorticity was numerically calculated from the coherent velocity data. We did not measure the random (incoherent) component of velocity and vorticity but obtained the random shear stress components, $\langle -u_r v_r \rangle$ and $\langle -u_r w_r \rangle$, by subtracting $\langle -u_c v_c \rangle$ and $\langle -u_r w_r \rangle$ from the conditionally averaged shear stress. The coherent energy equation was derived as follows.

$$
\begin{aligned}
\frac{\partial}{\partial t}\left(\frac{1}{2}u_{ic}^2\right) &+ U_j\frac{\partial}{\partial x_j}\left(\frac{1}{2}u_{ic}^2\right) + u_{jc}\frac{\partial}{\partial x_j}\left(\frac{1}{2}u_{ic}^2\right) \\
&= \langle u_{ir}u_{jr}\rangle\frac{\partial u_{ic}}{\partial x_j} - u_{ic}u_{jc}\frac{\partial U_i}{\partial x_j} - (\overline{u_{ic}u_{jc}} + \overline{u_{ir}u_{jr}})\frac{\partial u_{ic}}{\partial x_j} \\
&\quad - \frac{\partial}{\partial x_j}u_{ic}\langle u_{ir}u_{jr}\rangle + \frac{\partial}{\partial x_j}u_{ic}(\overline{u_{ic}u_{jc}} + \overline{u_{ir}u_{jr}}) \\
&\quad + \nu\frac{\partial}{\partial x_j}\left(u_{ic}\frac{\partial u_{ic}}{\partial x_j} + u_{ic}\frac{\partial u_{jc}}{\partial x_i}\right) - \frac{\partial}{\partial x_j}\frac{u_{jc}p_c}{\rho} \\
&\quad + \nu\frac{\partial u_{ic}}{\partial x_j}\left(\frac{\partial u_{ic}}{\partial x_j} + \frac{\partial u_{jc}}{\partial x_i}\right),
\end{aligned}
$$

where the capital letters show the time mean values and the subscripts c and r denote the coherent and the random components, respectively. The subscript indices, i and $j(=1, 2, 3)$, denote the Cartesian coordinate, X, Y and Z and the corresponding velocity components, u, v and w. From the first term of the left-hand side, each term shows the decay rate of coherent energy, the mean flow advection, the coherent advection, and of the right-hand side, the random production by the coherent motion, the coherent production by mean flow, the background production by coherent motion, the random diffusion, the background diffusion, the viscous diffusion, the pressure strain and the viscous dissipation, respectively. But for the pressure strain term, we obtained each term of the u-component of the coherent energy equation and drew three-dimensional maps of the quantities.

3. Results

The ensemble-averaged large-scale vortex has a horseshoe configuration symmetry to the $X - Y$ plane on the center axis, though it shows various asymmetric instantaneous configurations deformed by strong background turbulence, as pointed out by Robinson et al. (1989). Even for the averaged structure, Jeong et al. (1997) proposed asymmetric sculpture for the small scale coherent vortices in the inner layer. Nevertheless, we think, the averaged configuration of the large-scale coherent vortex must be symmetrical even for the natural coherent structure, because we can find no evidence that it should be asymmetrical in the outer layer. Furthermore, it makes no significant difference in fluid dynamical mechanisms whether the symmetric or the asymmetric models of the structures (Antonia and Bisset, 1991). So, we will show only the half side maps in the following figures.

Figure 1a shows the three-dimensional map of the coherent vorticity vector around the large-scale coherent vortex at $X/\delta = 14$. In this figure, the unit of the abscissa (50 msec/div) corresponds to about 4.5δ in X-direction. The vorticity vectors on the $T - Y$ planes at $Z/\delta = 0.2$ and 0.4 clearly show a quasi-streamwise vortex keeping its head upward. On the symmetry plane ($Z/\delta = 0$), the vorticity vectors turn to the negative Z direction at about $Y/\delta = 1.0$. These features of the vorticity field suggest that the grown-up large-scale coherent vortex has a horseshoe sculpture of which the head reaches about 1.2δ. The legs of the horseshoe are composed of a pair of upward stretched counter rotating vortex tubes with strong coherent vorticity in Fig. 1b. The vortex tubes are connected with each other at the head of the horseshoe and forms a turbulent bulge. Only weak spanwise vorticity exists there and the turbulent bulge acts as a weak passive structure in the outer layer.

Figure 2 plots the coherent velocity vectors in which the dashed line denotes the axis of the horseshoe vortex estimated from the vorticity map in Fig. 1. Strong 2nd quadrant event is observed between the legs of $Z/\delta \leq 0.2$

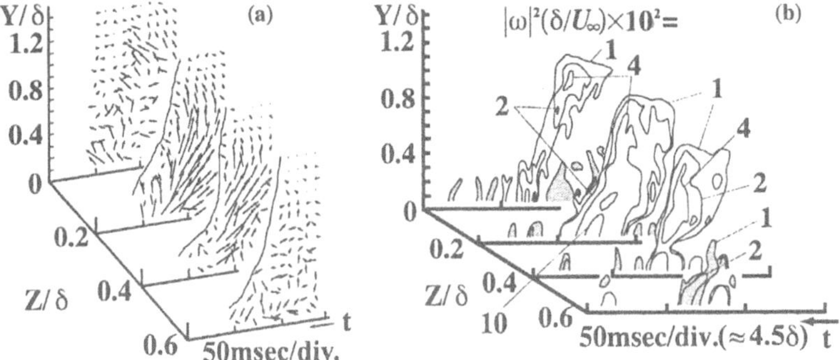

Figure 1. Three-dimensional coherent vorticity vector map (a) and contour maps of enstrophy $|\omega|^2 \cdot (\delta U_\infty) \times 10^{-2}$ around the large-scale vortex at $X/\delta = 14$. Hatched areas show the counter-rotating secondary vortex. (From Makita and Sassa, 1993).

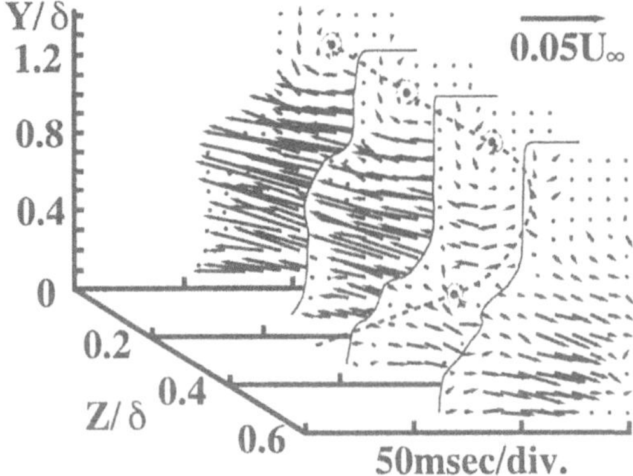

Figure 2. Coherent velocity vector map around the large-scale vortex. The broken curve denotes the axis of the large-scale vortex. (From Makita and Sassa, 1995)

and 4th quadrant event exists around $Z/\delta \sim 0.6$ outside them. We call these coherent motions the outflow and the inrush so as to distinguish from the ejection and the sweep in the inner layer. Yet there is no significant difference between them except their scale and intensity. At $Z/\delta = 0$, the velocity vectors are suddenly shortened on the upstream side of the outflow that corresponds to the δ-scale u-discontinuity observed by Brown and Thomas (1977). We can observe clockwise circulation in the turbulent bulge for $Y/\delta \geq 1.0$ entraining the freestream.

Figures 3a-d illustrate the contribution to momentum transfer on the cross sections of the head and the leg of the coherent vortex. In Fig. 3a, most of the coherent shear stress, $\langle -u_c v_c \rangle$, is accumulated around $Y/\delta = 0.6$ on the symmetry plane where the intense outflow exists. Though the distribution of the spanwise coherent vorticity, ω_{cZ}, at the head represents the cross section of the vortex tube, it in the lower layer for $Y/\delta \leq 0.6$ shows not the coherent vortex but only the high shear region induced by the coherent motion. The coherent shear stress is produced between the positive and negative vorticity regions. On the other hand in Fig. 3b, the distribution of the random shear stress, $\langle -u_r v_r \rangle$, representing momentum transfer by small scale turbulence almost corresponds to that of the coherent spanwise vorticity, which means that the random shear stress acts to relax the local shear induced by the coherent motion. Figure 3c compares the coherent velocity. The coherent vorticity and the coherent shear stress distributions on the horizontal plane at $Y/\delta = 0.4$. In this figure, the rotating motion shown by the vectors and the circular distribution of the vertical vorticity, ω_{cY} is clearly observed on the horizontal cross section of the leg. The alternately arranged islands of positive and negative coherent shear stresses, $\langle -u_c w_c \rangle$, show the four quadrant events by the coherent velocity, $(u_c,\ w_c)$. The negative vorticity distribution outside the positive one seems to show the counter rotating secondary vortex induced by the primary horseshoe vortex as demonstrated by Zhou et al. (1996). In Fig. 3d, the distribution of $\langle -u_r w_r \rangle$ resembles to that of ω_{cY}. The coherent motion is known to be directly diffused by the random fluctuation. Hussain and Zaman (1980) suggested that, contrary to the present results, the maximum point of the random shear stress is located on the saddle point of the coherent vorticity for periodic coherent vortices in a transitional jet. The difference may be because the present large-scale vortex exists alone and has no clear saddle point appearing between the adjacent coherent vortices.

In the coherent energy equation, the background production by coherent motion and the background diffusion are the terms involving ordinary mean values that cannot be canceled by the triple decomposition and their physical meaning is not so clear. Here, we can neglect them because they and the viscous terms are more than one order of magnitude smaller than the other terms. The advection terms are as large as the production and the diffusion terms, however, its total amount about the large-scale vortex is nearly zero. So, the present paper describes only about the production and the diffusion terms. The distribution of the coherent energy production at $Z/\delta = 0$ shown in Fig. 4a resemble that of the coherent shear stress in Fig. 3a. The u-component of coherent energy is mainly produced by the outflow and the inrush. The peak of the production locates in the region of the maximum coherent velocity of about $Y/\delta = 0.5$ at $Z/\delta = 0$.

Figures 4b and 4c show the maps of the vertical and the spanwise components of the random energy production by coherent motion. Their negative regions are distributed in the high shear regions around the head (Fig. 4b) and the leg (Fig. 4c), which tells that the coherent energy is deprived

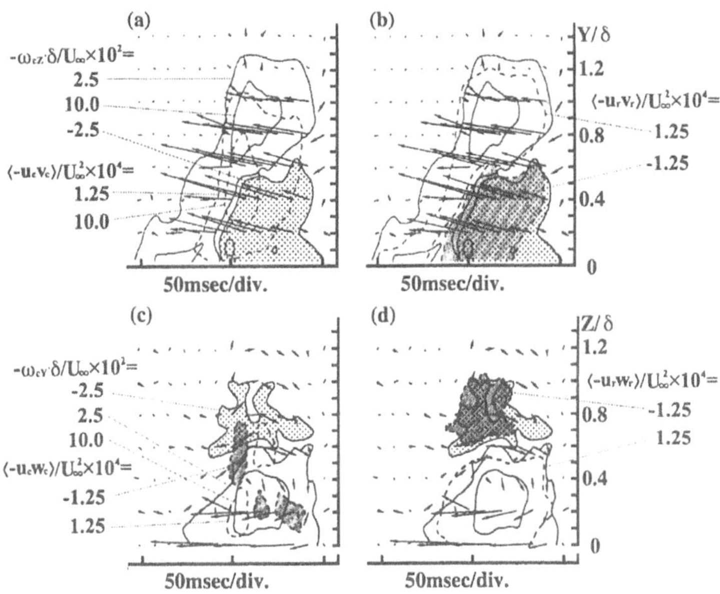

Figure 3. Cross sections of the large-scale vortex on $T-Y$ plane at $Z/\delta = 0$ (a, b) and $T-Z$ plane at $Y/\delta = 0.4$ (c, d). Arrows denote coherent velocity. Solid and broken curves denote contour lines of map ω_{cZ} and (a): $\langle -u_c v_c\rangle$; (b): $\langle -u_r v_r\rangle$; ω_{cY} and (c): $\langle -u_c w_c\rangle$; (d): $\langle -u_r w_r\rangle$. Hatched areas denote negative values.

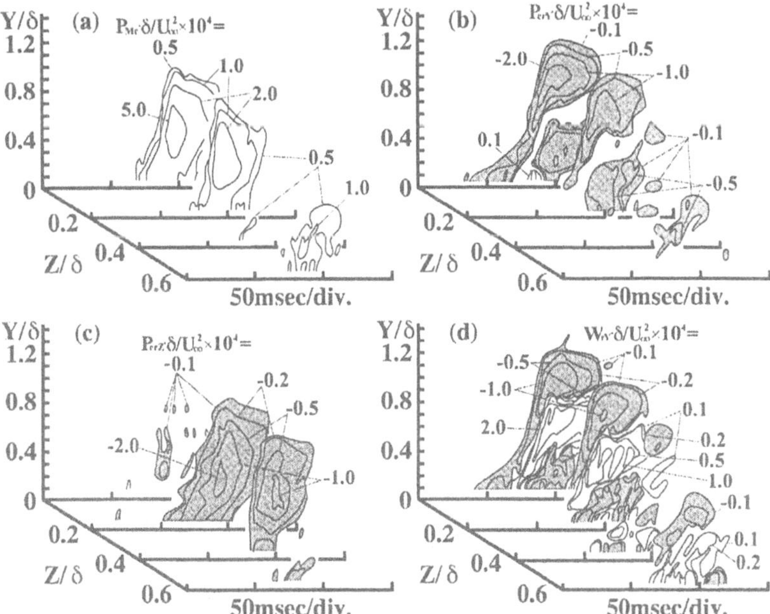

Figure 4. Contour maps of (a) coherent production by mean flow, (b) vertical component of random production by coherent motion, (c) spanwise component of random production by coherent motion and (d) diffusion by random motion. Hatched areas denote negative values.

to small scale turbulence through the collapse of the large-scale coherent vortex, i.e., the coherent vortex is presumed as the energy containing eddy in the hierarchy of the turbulence energy field. The map of Y-component of the diffusion by random shear stress is shown in Fig. 4d. The positive value is concentrated about the peak of the coherent velocity. This fact shows that the counter gradient diffusion occurs in the core region of the outflow, because the coherent energy diffused by the small scale turbulence is transferred from the neighboring high shear regions and accumulated there. In the most of the regions with not negligible orders of the induced velocity gradient, the small scale turbulence diffuses the coherent energy.

Figure 5 gives the spanwise distribution of the coherent energy budget about the large-scale coherent vortex. The integration of the terms was made on the $T-Y$ plane at each Z/δ. Apparently, the dominant terms are the coherent production by mean flow, the random production by coherent motion and the diffusion by random motion. The coherent production has large values for $Z/\delta \leq 0.2$ and around $Z/\delta = 0.6$, which shows the coherent energy of the large-scale vortex is produced inside and outside the leg where the outflow and the inrush exist as mentioned before. The random production has large negative values in the core of the leg at about $Z/\delta = 0.2$. The random diffusion gains coherent energy for $Z/\delta \leq 0.1$ and loses it for $0.2 \leq Z/\delta \leq 0.4$. The counter diffusion occurs in the core of the outflow but, taken overall, the large-scale coherent vortex loses its coherent energy by the random diffusion. The advection by mean flow slightly contributes to the energy budget. However, the contribution of the background production, the background diffusion, the viscous diffusion and the viscous dissipation are almost negligible, as also shown in the figure.

4. Discussion

In the present experiment, we initially gave a tiny disturbance of the spanwise vorticity into the bottom of a turbulent boundary layer to lock the occurrence of the coherent structure in time and space. The disturbance was automatically brought up in the shear layer to be a natural large-scale coherent vortex (Makita and Sassa 1993). When it was detected in the actual flow field, its instantaneous appearance was asymmetrical and randomly deformed affected by the background turbulence in the boundary layer. Based on the results of the present analysis, we propose a horseshoe sculpture in Fig. 6 as an ensemble averaged image of the large-scale coherent vortex having dimensions of about 1.2δ in height, 1.0δ in width and 5.0δ in streamwise length. We already have many evidences that the horseshoe or hairpin vortices are induced in a shear layer from some disturbances as reported by Acarlar and Smith (1987), Kim (1987), Haidari and Smith

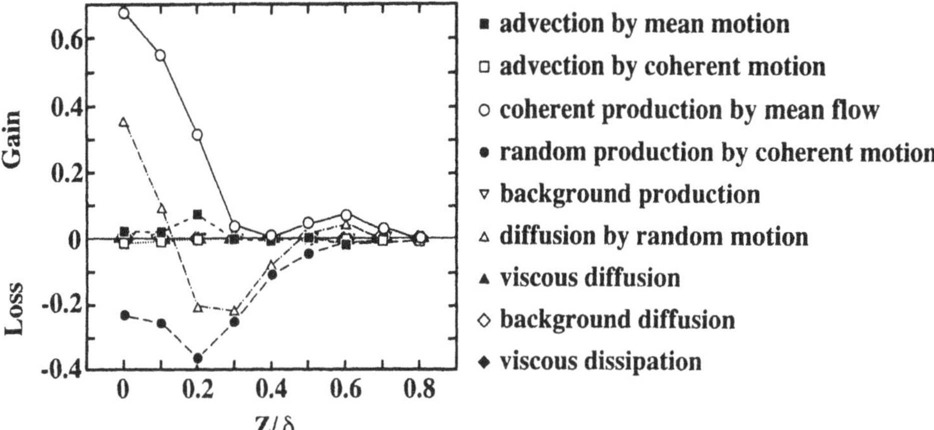

Figure 5. Spanwise distribution of coherent energy budget around the large-scale vortex.

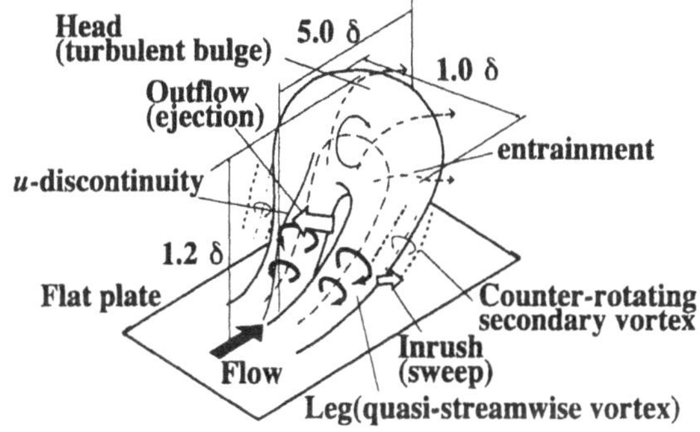

Figure 6. Conceptual model of the large-scale coherent vortex.

(1994), Zhou et al. (1996) and so on. Thus, it seems quite proper to adopt the horseshoe sculpture as the conceptual model of the large-scale coherent vortex in the outer layer. Ahead of the study, we imagined the roll-up of the spanwise vortex as the scenario of its growth, as observed Praturi and Brodkey (1978). Such a scenario is certainly available during the spanwise vortex rapidly develops upward in the short growth stage. We think, now, the dominant mechanism of the growth is the stretching of quasi-streamwise vortices at the legs by the mean shear. From this view point, our conceptual model resembles the double roller eddy proposed by Townsend (1970). The autogeneration of the secondary horseshoe vortex reported by Zhou (1996) is not clearly observed in the present study, however, the counter-rotating secondary vortices are observed to be induced outside the legs. We guess that the large-scale coherent vortex has the greater potential to produce the secondary vortex in the growth stage. Previously the authors (1993) showed that the decay stage occupied most of life time of about 20δ for the large-scale coherent vortex. Then, most of the natural large-scale coherent vortices in the outer layer were detected as only weak and passive structures, as stated by Kobashi and Ichijo (1986) through the measurement of correlations between wall pressure and velocities.

5. Conclusion

With the present model, we can give consistent explanation upon the occurrence of four well-known coherent structures in the outer turbulent boundary layer. Three of them are related to the legs; the outflow (ejection), the inrush (sweep) and the u-discontinuity (an upstream border of the outflow). The momentum transfer and the energy production are chiefly conducted by the outflow and the inrush. The coherent energy is produced mainly between the legs of the horseshoe vortex and diffused by the small-scale turbulence in the vortex tube forming the horseshoe. The large-scale coherent vortex behaves as the energy containing eddy in the energy cascade process. The turbulent bulge at the head of the horseshoe vortex conforms an obscure bridge between the two legs and entrains the momentum from the free stream.

References

Acarlar, M.S. and Smith, C.R., (1987) *J. Fluid Mech.*, **Vol.175**, pp.1.
Antonia, R.A. and Bisset, D.K., (1991) *Turbulence and Coherent Structures*. pp.141, Kluwer Academic.
Brown, G. and Thomas, A.S.W. (1977) *Phys. Fluids*, **Vol.20**, pp.s243.
Haidari, A.H. and Smith, C.R., (1994) *J. Fluid Mech.*, **Vol.277**, pp.135.
Hussain, A.K.M.F. and Reynolds, M.C. (1970) *J. Fluid Mech.*, **Vol.41**, pp.241.
Jeong, J. Hussain, F., Schoppa, W. and Kim, J., (1997) *J. Fluid Mech.*, **Vol.332**, pp.185.
Kim, J. (1987) *Turbulent Shear Flows 5*. pp.221, Springer Verlag.
Kline, S.J., Reynolds, W.C., Schraub,F.A. and Runstadler,P.W., (1967) *J. Fluid Mech.*, **Vol.30**, pp.741.
Kobashi, Y. and Ichijo, M., (1986) *Exp. Fluids*, **Vol.4**, pp.29.
Kovasznay, L.S.G., Kibens, V. and Blackwelder, R.F. (1970) *J. Fluid Mech.*, **Vol.41**, pp.283.
Lu, S.S. and Willmarth, W.W. (1973) *J. Fluid Mech.*, **Vol.60**, pp.481.
Makita, H. and Sassa, K. (1993) *Turbulent Shear Flows 8*. pp.23, Springer Verlag.
Makita, H. and Sassa, K., (1995) *Advances in turbulence V*. pp.346, Kluwer Academic.
Makita, H., Sassa, K., Abe, M. and Itabashi, A. (1989) *AIAA J.*, **Vol.27**, pp.155.
Meinhart, C.D. and Adrian, R.J., (1995) *Phys. Fluids*, **Vol.7 no.4**, pp.694.
Hussain, A.K.M.F. and Zaman, K.B.M.Q., (1980) *J. Fluid Mech.*, **Vol.101**, pp.493.
Praturi, A.K. and Brodkey, R.S., (1978) *J. Fluid Mech.*, **Vol.89**, pp.251.
Robinson, S.K., Kline, S.J. and Spalart, P.R. (1990) *Near-Wall Turbulence*. pp.218, Hemisphere.
Townsend, A.A., (1970) *J. Fluid Mech.*, **Vol.41**, pp.13.
Zhou, J., Adrian, R.J. and Balachandar, S., (1996) *Phys. Fluids*, **Vol.8 no.1**, pp.288.

FORMATION OF NEAR-WALL STREAMWISE VORTICES BY STREAK INSTABILITY

WADE SCHOPPA & FAZLE HUSSAIN
University of Houston
Department of Mechanical Engineering, Houston, TX 77204-4792

Abstract

Using direct numerical simulations of turbulent channel flow, we present new insight into the formation mechanism of near-wall longitudinal vortices. Instability of lifted, vortex-free low-speed streaks is shown to generate, upon nonlinear saturation, new streamwise vortices, which dominate near-wall turbulence production, drag, and heat transfer. The instability requires sufficiently strong streaks (y circulation per unit $x > 7.6$) and is inviscid in nature, despite the proximity of the no-slip wall. Streamwise vortex formation (collapse) is dominated by stretching, rather than rollup, of instability-generated ω_x sheets. In turn, direct stretching results from the positive $\partial u/\partial x$ (*i.e.* positive VISA) associated with streak waviness in the (x,z) plane, generated upon finite-amplitude evolution of the sinuous instability mode. Significantly, the 3D features of the (instantaneous) instability-generated vortices agree well with the coherent structures educed (*i.e.* ensemble averaged) from fully turbulent flow, suggesting the prevalence of this instability mechanism. Fundamental differences in the regeneration dynamics of minimal channel and Couette flows are revealed regarding (nonlinear) streak instability, vortex formation and evolution, and wall shear behavior.

1. Introduction

The boundary layers on transport vehicles and in industrial devices are invariably turbulent, with drastically increased drag and heat transfer at solid surfaces due to near-wall vortical *coherent structures* (CS). Viable control of near-wall turbulence, as yet largely unrealized in practice, has the potential to save billions of dollars per year in energy costs for engineering applications. Although massive efforts have been expended in developing drag reduction strategies, their engineering application has remained notably scarce, particularly for aircraft. The lack of success of boundary layer control to date without doubt reflects a currently limited understanding of CS initiation and evolution. In this paper, we develop a new mechanism of CS formation, well-supported by comparisons with near-wall turbulence.

The prominence of longitudinal vortices in near-wall turbulence is now well accepted (*e.g.* see Townsend 1956; Kline *et al.* 1967; Blackwelder & Eckelmann 1979; Robinson 1991), as is their critical role in elevating drag (Kravchenko *et al.* 1993) and heat transfer. The transport enhancing effect of near-wall vortices is easily understood. Due to their streamwise orientation, these vortices sweep near-wall fluid toward the wall on one flank and eject it away from the wall on the other. Drag and heat transfer are enhanced by the wallward motion, which steepens the wall gradients of streamwise velocity U and temperature respectively. Note that the gradient reduction on the outward motion side of vortices is relatively smaller, resulting in mean transport enhancement.

Our ensemble-averaged streamwise vortices, *i.e.* CS (Jeong & Hussain 1992; Jeong *et al.* 1997), display all previously classified near-wall features (Kline & Robinson 1990). Thus, the evolutionary dynamics of streamwise CS are the essence of near-wall turbulence. The central question addressed here is: how are streamwise vortices generated? Several widely disparate formation mechanisms have been proposed, many quite plausible and self-consistent, yet currently lacking convincing validation. Thus, a formidable challenge is to identify the naturally and frequently occurring dynamics.

Vortex formation must recur for turbulence to be sustained; *i.e.* existing vortices must ensure subsequent vortex regeneration. Of the numerous proposed regeneration mechanisms, most involve either: (i) the direct action (induction) of existing vortices nearby ("parent - offspring" scenarios), or (ii) local instability of a quasi-steady base flow, without directly requiring existing vortices. Note that recurring instability (ii) requires a feedback mechanism, by which previous vortices generate an unstable base flow and thus play only an indirect role.

A widely cited parent-offspring mechanism involves the generation of new vortices near existing hairpins, behind the (spanwise) arch and beside each of the (streamwise) legs (see Smith & Walker 1994 for a review). In contrast to hairpin generation, Brooke & Hanratty (1993) propose that an opposite-signed offspring vortex forms immediately underneath a parent vortex, whose downstream end has lifted from the wall. Vortex formation is also often attributed to 2D Kelvin-Helmholtz-type rollup of near-wall ω_x sheets (*e.g.* Jimenez & Orlandi 1993), with opposite sign of the streamwise vortex existing overhead, generated by the no-slip condition.

Of the numerous instability mechanisms developed to explain near-wall vortex formation, there is considerable disagreement as to the mechanisms of instability and feedback. For example, centrifugal (Brown & Thomas 1977) and Craik-Leibovich (Phillips & Wu 1994) instabilities, direct resonance of oblique modes (Jang *et al.* 1986), and shear-driven streak instabilities (Kim *et al.* 1971; Swearingen & Blackwelder 1987; Hamilton *et al.* 1995) have all been proposed. Unfortunately, close comparisons of the physical-space (nonlinear) evolutions of these instabilities with well-documented near-wall turbulence structures (Robinson 1991) are unavailable.

Here we demonstrate (via DNS) that instability of streaks – without any initial (parent) vortex – directly generates new streamwise vortices, internal shear layers, and arch vortices. The instability-generated streamwise vortices are found to correspond closely with the ensemble-averaged CS educed from near-wall turbulence (Jeong *et al.* 1997), suggesting the dominance of our proposed mechanism. Physical-space, vortex dynamics-based explanations for the observed vortex formation are also provided. In the following, we outline our computational approach (§2), and then demonstrate an underlying linear instability of lifted low-speed streaks (§3). The genesis of new vortices is illustrated in §4, including a brief description of the vortex dynamics involved. A detailed comparison of our results with recent studies of minimal Couette flow is presented in §5, followed by some concluding remarks and implications for boundary layer control (§6). Additional details may be found in Schoppa & Hussain (1997a).

2. Computational Approach

In the following, we address vortex regeneration using direct numerical simulations of the Navier-Stokes equations. Periodic boundary conditions are used in x and z, and the no-slip condition is applied on the two walls normal to y; see Kim *et al.* (1987) for the simulation algorithm details. The spatial discretization and Re are chosen so that all dynamically significant lengthscales are resolved (*i.e.* a finer computational grid does not markedly affect the solution); thus, no subgrid-scale turbulence model is necessary. Code validation and accuracy checks were performed by comparing the growth rates for simulated 2D and 3D Orr-Sommerfeld modes of the laminar (parabolic profile) flow with independent stability analysis (agreement within 1%).

To better isolate instability and the subsequent vortex formation, we use the minimum outer Reynolds number $Re=U_c h/\nu=2000$ (U_c is the centerline velocity of the $2h$ wide channel) and the minimum domain sizes in x and z for sustained channel flow turbulence – the so-called "minimal flow unit" of Jimenez & Moin (1991). For the simulations of isolated vortex regeneration, a constant volume flux is maintained, and 32x129x32 grid points are used in x, y, and z respectively.

3. Streak Instability

3.1 MINIMAL CHANNEL FLOW

Our own analysis of minimal channel regeneration suggests the presence of an underlying streak instability. During the quiescent phase of the regeneration cycle, when the wall shear stress is minimum, the buffer region contains only a lifted-up, long, low-speed streak, with no significant streamwise vortices or even ω_x. Shortly thereafter ($t^+\sim 40$ later), a new positive streamwise vortex is created (by instability, as shown here) in the buffer region from the vorticity sheet (predominantly $+\omega_y$) flanking the streak. Thus, the observed large temporal variations in integrated wall shear stress directly reflect the vortex regeneration: the drag is minimum during the quiescent phase, when near-wall vortices are very weak, and maximum once collapsed streamwise vortices (which bring high-speed fluid toward the wall to increase drag) are generated in the buffer layer. These observations suggest that "vortex-less" low-speed streaks are unstable and serve as an agent of vortex regeneration. In full-domain flows as well, extremely long ($\Delta x^+\sim 1000$) streaks are prevalent, and many regions along individual streaks are devoid of any streamwise vortices (Robinson 1991).

To isolate instability of vortex-less streaks, we consider a base flow of the form

$$U^+(y^+,z^+)=U_0^+(y^+)+(\Delta u_0^+/2)\cos(\beta^+ z^+)(y^+/30)\exp(-\sigma y^{+2}+0.5); \qquad (1)$$
$$V^+=W^+=0$$

as a first approximation, where U_0^+ is the turbulent mean velocity profile. The streak's normal circulation per unit length Δu_0^+, spanwise wavenumber β^+ and transverse decay σ are chosen to approximate a typical U distribution, shown in Fig. 1(a) for minimal channel flow. The corresponding (y,z) distribution of (1) with $\Delta u_0^+=11.2$, $\beta^+=0.06$ (*i.e.* streak spacing $\Delta z^+=100$), and $\sigma=0.00055$ (*i.e.* maximum streak ω_y at $y^+=30$) is shown in Fig. 1(b) and closely resembles the instantaneous realization in Fig. 1(a). Note that the base flow base flow (1) contains no ω_x and is a steady solution of the Euler equations.

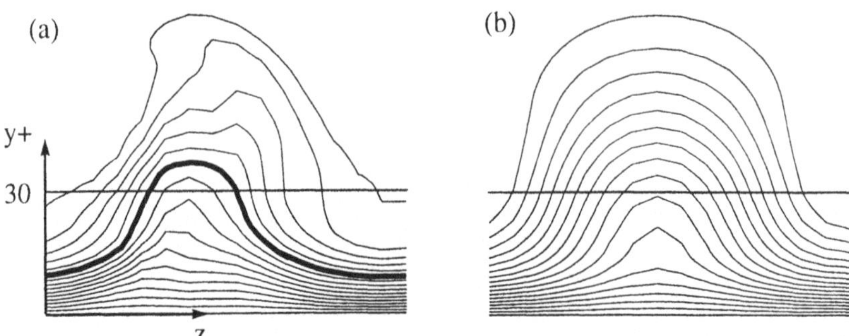

Figure 1. Lifted low-speed streak at the quiescent phase of minimal channel regeneration, illustrated by (a) a typical cross-stream distribution of U and (b) the analytical base flow (1) used for stability analysis. The bold contour shown in (a) is the $0.55U_c$ contour.

3.2 LINEAR STABILITY ANALYSIS

With the base flow (1) frozen and Re=2000, we find exponential growth of linear amplitude sinuous perturbations (*i.e.* streak bending in z commonly observed), indicating that lifted streaks (1) are indeed linearly unstable. The instability growth is dominated by $E_{10}(t)$ (z harmonics are also present, but less energetic), the volume-integrated energy in Fourier modes with a z-wavenumber of 0 (mean in z) and an x-wavenumber of α (x-fundamental mode). Interestingly, enhanced growth of E_{10} with increasing Re reflects an inviscid instability mechanism, found to be quite similar in nature to oblique instability modes of free shear layers. Consequently, viscous effects and the no-slip wall play no destabilizing role. This raises the question: how does viscosity, obviously crucial near the wall, enter the instability dynamics?

We find that the viscous damping of instability is quite strong for a streak spacing of Δz^+=100, the popularly accepted value. Since the peak E_{10} occurs at a linear amplitude (*i.e.* 3D perturbation amplitudes near machine accuracy; see Fig. 2b), the typical nonlinear (*i.e.* finite-amplitude) saturation is not occurring here. Instead, attenuation is due primarily to cross-diffusion (*i.e.* viscous annihilation, a kind of planar reconnection) of the opposite-signed ω_y flanking the low-speed streak. In fact, ω_y is reduced to 68% of its initial value by the E_{10} saturation time, indicating that the (exponential) streak decay rate due to cross-diffusion (Fig. 2d) is non-negligible (approximately half the instability growth rate; Fig. 2b).

We now consider the instability scaling at higher Re, keeping Δz^+=100 fixed. As shown in Figs. 2(b,d), both instability growth and streak diffusion (annihilation of normal circulation Δu^+) scale well in (inner) wall units. Although this is perhaps not surprising because of the absence of outer vortices in these flows, the possibility of autonomous inner-scaling dynamics clearly exists. These results demonstrate that streak instability grows similarly at higher Re (perhaps even at very high Re), provided that the streak velocity profile is self-similar in wall units. By considering the dimensional time evolution, one can see how this is possible. As Re is increased, the wall vorticity Ω_w (*i.e.* $U(y)$ slope) increases (according to the Blasius skin friction law), and the (dimensional) streak spacing decreases. Consequently, the streak annihilation by cross-diffusion is faster at higher Re (Fig. 2c), but the instability growth rate is also enhanced due to concomitant increased wall vorticity (Fig. 2a), their balance maintaining a nearly constant

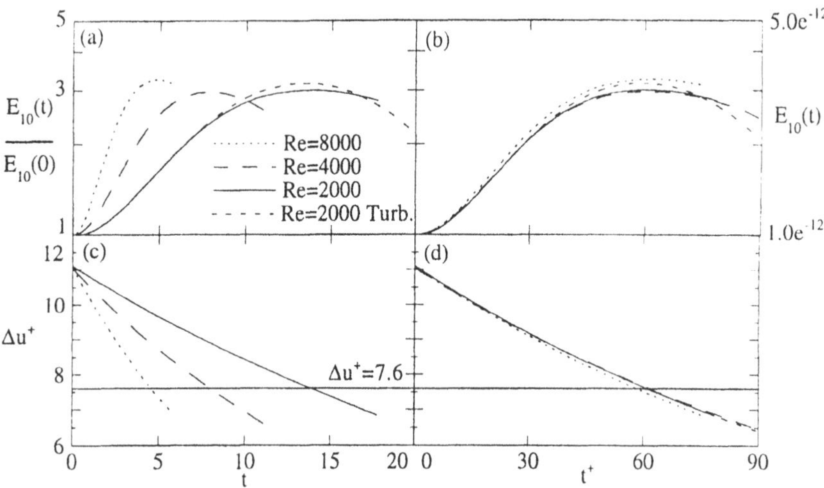

Figure 2. Temporal evolution of (a,b) E_{10} and (c,d) streak y circulation as a function of Re, in (a,c) dimensional and (b,d) wall time units, for a constant streak spacing $\Delta z^+=100$. The data collapse in (b,d) illustrate the inner scaling and balance of streak instability and viscous streak annihilation, showing Re invariance. Except for the case with a turbulent mean profile, the Reichardt relation is used for $U_o^+(y^+)$ in (1).

E_{10} amplification. Figures 2(b,d) also confirm the stabilizing role of viscous cross-diffusion across streaks; saturation occurs in each case at a critical normal circulation of Δu^+=7.6. Note that Δu^+, being a measure of the tilt angle of the vortex lines (in (y,z)) on either side of a streak, represents the extent of streak lifting (*i.e.* the crest amplitude of u contours in Fig. 1). Thus, sufficient lift-up of the low-speed streak into the buffer layer is required for instability to occur. In the following, we focus on the more computationally tractable Re=2000 case, noting that the streak instability is generic to higher Re.

4. Vortex Formation Mechanism

Having confirmed that (one-walled) streaks with sufficient y circulation (Δu^+) are indeed linearly unstable, we now consider the subsequent nonlinear evolution using DNS. We consider a "clean" flow initialization, containing only vortex-less streaks (1) and a sinusoidal w perturbation of amplitude w'/U_c =1% at y^+=30. Results clearly illustrate the genesis of streamwise CS, near-wall shear layers, and arch vortices, suggesting that streak instability is the dominant mechanism of vortex generation and turbulence production.

4.1 STREAMWISE VORTICES

Most significantly, as the mode grows to a nonlinear, new collapsed streamwise vortices are directly created (Fig. 3a-c). At early times, instability growth is characterized by increased circulation of flattened ω_x sheets, with the spanwise symmetry of the linear eigenmode approximately maintained. Subsequently, as nonlinear effects (described below) become prominent, $+\omega_x$ begins to concentrate on the $+z$ flank of the low-speed streak (Fig. 3b). By symmetry, the ω_x distribution at a half wavelength in x away is obtained by z reflection and sign inversion; thus, $-\omega_x$ is generated on the $-z$ flank here. As this ω_x amplification continues, collapsed (*i.e.* with compact cross-section) streamwise

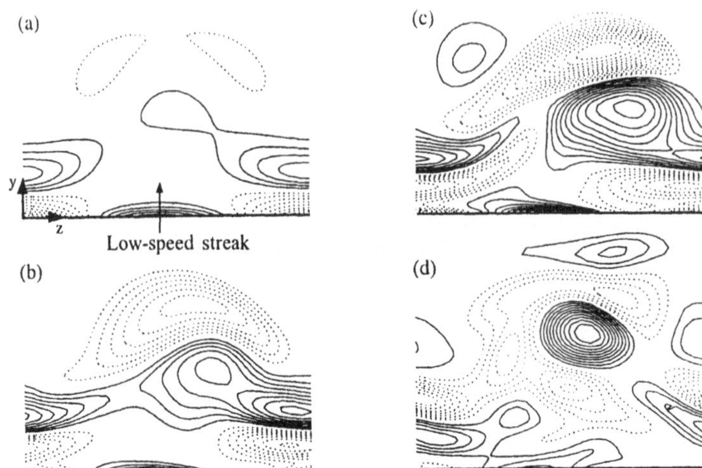

Figure 3. Streamwise vortex formation due to finite-amplitude streak instability, illustrated by cross-stream distributions of ω_x at (a) $t^+=17$, (b) $t^+=51$, (c) $t^+=103$, (d) $t^+=928$. Planes in (b) and (c) are tracked with the instability phase speed of approximately $0.6U_c$.

vortices quickly emerge (Fig. 3c). This genesis of new vortices from ω_x layers is strikingly similar to that frequently observed in minimal channel flow. Previous studies (*e.g.* Jimenez & Orlandi 1993) presumed that the layer simply rolls up due to its own (2D) advection. Our results (discussed below) indicate that the vortex formation is not in reality a (Kelvin-Helmholtz type) roll up process; the formation is inherently 3D, dominated by intense ω_x stretching. Even well past their initial formation, streamwise vortices and hence turbulence continue to be sustained (*e.g.* Fig. 3d), indicating the importance of streak instability to turbulence sustenance.

The 3D geometry of the instability-generated vortices (Figs. 4a,b) (say, the x-overlapping of tilted, opposite-signed streamwise vortices on either side of a low-speed streak) agrees well with the typical flow structure during the active phase of minimal channel regeneration. Most significantly, this vortex geometry (maintained upon evolution except for increasing overlap) is strikingly similar to that of 3D CS educed (from more than 100 vortex realizations) in full-domain turbulence (Fig. 5), which has been shown to capture all important near-wall events (Jeong *et al.* 1997). Irregularities (*e.g.* kinks) of the base flow streaks and finite-amplitude incoherent turbulence will surely occur, causing variations in vortices from one realization to another. If an underlying instability mechanism is present, it should be revealed by ensemble averaging over a large number of base flow/perturbation combinations, *i.e.* by CS eduction. The close correspondence of Figs. 4 and 5 indicates that this is in fact the case, serving as strong evidence that this vortex formation process is a dominant mechanism in fully developed near-wall turbulence.

Since the newly generated vortices are predominantly streamwise (Fig. 4a), the essential dynamics of vortex formation are those of ω_x, whose inviscid evolution is governed by

$$\frac{\partial \omega_x}{\partial t} = -u\frac{\partial \omega_x}{\partial x} \underbrace{- v\frac{\partial \omega_x}{\partial y} - w\frac{\partial \omega_x}{\partial z}}_{\text{Self-induction}} + \underbrace{\omega_x \frac{\partial u}{\partial x}}_{\text{Stretching}} + \underbrace{\frac{\partial v}{\partial x}\frac{\partial u}{\partial z} - \frac{\partial w}{\partial x}\frac{\partial u}{\partial y}}_{\text{Tilting}} . \quad (2)$$

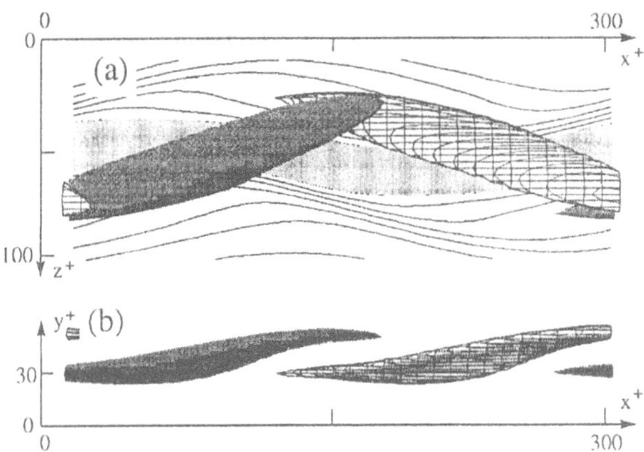

Figure 4. Streamwise vortices' (x,z) plane tilting, x-overlapping, and location relative to a low-speed streak in (a) top view, (b) side view. The 80% isosurfaces of $+\omega_x$ and $-\omega_x$ at $t^+=103$ are (dark) shaded and hatched respectively; contours of u at $y^+=20$ are overlaid in (a), with low levels of u light-shaded to demarcate the low-speed streak. Note the striking resemblance of this instantaneous realization with the ensemble-averaged CS (Fig. 5).

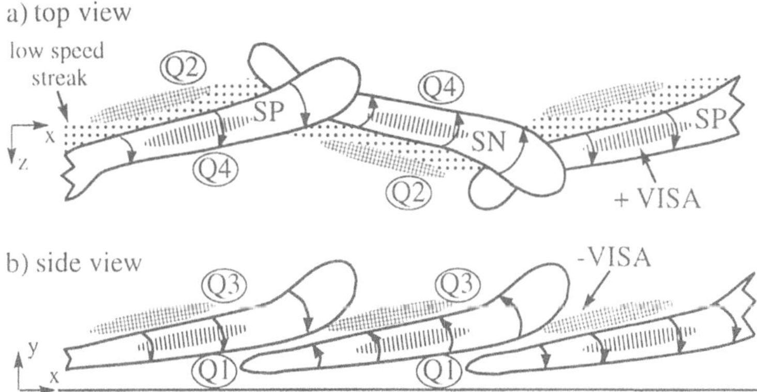

Figure 5. Near-wall educed CS and associated coherent events (adapted from Jeong *et al.* 1997); including ±VISA events ($\pm\partial u/\partial x$); quadrant *Re* stresses Q1, Q2 (ejection), Q3, and Q4 (sweep); and a kinked low-speed streak.

In Fig. 6, we observe that the circulation of the elongated near-wall ω_x layers (Fig. 6a) increases due to vortex line tilting, given by the latter production term $-(\partial w/\partial x)(\partial u/\partial y)$ (Fig. 6c), which dominates the former. Although typically largest in magnitude over all other, the $-(\partial w/\partial x)(\partial u/\partial y)$ term actually generates a flattened tail in the near-wall ω_x layer (C in Fig. 6c), *not* a vortex. Contrary to prior speculation, these layers do not roll up due to their self-advection – a purely 2D mechanism. In fact, the cross-stream transport (B in Fig. 6b) actually opposes the rollup process, due to the opposite-signed ω_x immediately overhead (SN in Fig. 6a). In reality, vortex formation is due to direct stretching of $+\omega_x$ on the $+z$ flank of the low-speed streak (also, $-\omega_x$ amplification on the $-z$ flank, at a half x wavelength away), evident from nearly circular regions of $+\omega_x\partial u/\partial x$ there (D in Fig. 6d). We find that this local ω_x stretching is sustained in time and is mainly responsible for the

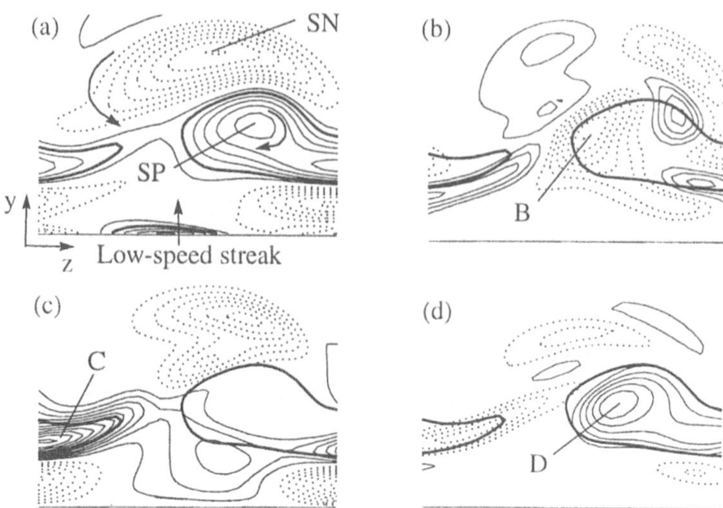

Figure 6. Distributions of (a) ω_x, and selected terms of the ω_x evolution equation: (b) self-induction (cross-stream), (c) the $-(\partial w/\partial x)\ (\partial u/\partial y)$ tilting term, and (d) direct stretching ($\omega_x \partial u/\partial x$); (a-d) are at an intermediate time during vortex formation (t^+=51). The bold line in each panel identifies the ω_x layer.

eventual vortex collapse, whose location coincides with the $+\omega_x \partial u/\partial x$ peak (*cf.* Figs. 3c, 6d).

In turn, the positive $\partial u/\partial x$ responsible for vortex collapse by stretching is a simple consequence of low-speed streak waviness, illustrated in Fig. 4(a). Recall that streak waviness is generated by (linear) sinuous streak instability. Once this waviness grows to a finite size, strong $+\partial u/\partial x$ develops downstream of the streak crests, causing direct stretching of positive (SP) and negative (SN) ω_x existing there. Since a large velocity difference exists across the streak flanks (with vorticity comparable to the mean velocity gradient at the wall), a sizable value of $+\partial u/\partial x$ is quickly generated by the rapidly growing (initially exponentially) streak wave. The initial ω_x sheets (Fig. 3a) then suddenly collapse (Fig. 3c) due to localized stretching (Fig. 6d), overcoming viscous diffusion which would otherwise cause their annihilation (on a similar timescale as the collapse). Note that these dynamics are also captured as (ensemble-averaged) +VISA events (*i.e.* $+\partial u/\partial x$) existing within the CS core (Fig. 5), indicating that this vortex generation process is indeed a dominant one.

4.2 INTERNAL SHEAR LAYERS & ARCH VORTICES

The significance of (nonlinear) streak instability is not limited to streamwise vortex formation; it also captures the genesis of new internal shear layers and spanwise arch vortices. Internal shear layers, indicated by wall-detached sheets of ω_z, form alongside (in z) the generated streamwise vortices (Figs. 7a,b). Subsequently, the downstream "end" of the internal shear layer rolls up into a new (locally spanwise) arch vortex (Figs. 7d,f). A surprising result is that the downstream tips of (newly generated) streamwise vortices tilt and propagate outward to form arches (Fig. 7c,e). Note that this process, *i.e.* streak instability → streamwise vortices → arch vortices, is contrary to the mechanism proposed by Robinson (1991), *i.e.* streak instability → arch vortices → streamwise

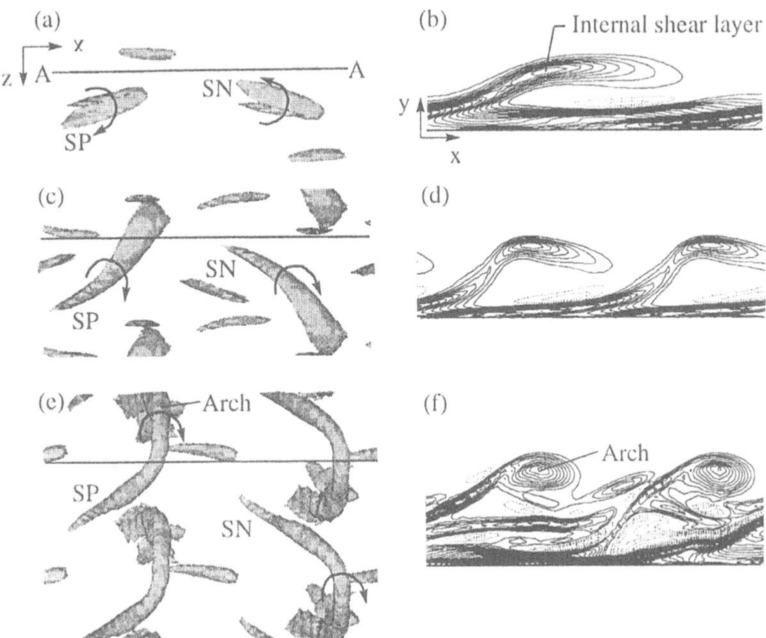

Figure 7. Genesis of internal shear layers and arch vortices due to nonlinear evolution of streak instability, illustrated by actual DNS data. (a,c,e): the evolutions of vortices SP and SN (top view) represented by λ_2 isosurfaces; (b,d,f): corresponding contours of ω_z in the section A-A (the straight line in a,c,e), indicating internal shear layer and arch vortex formation. The (periodic) z domain is expanded for clarity in (c) and (e).

vortices. Direct formation of arches through instability is unlikely, since the corresponding instability would involve varicose modes, found to be stable for relevant streak distributions (Schoppa & Hussain 1997a). In minimal channel flow, we find that arches without legs are commonly created, not by instability; but by viscous annihilation of a leg originally attached to an arch (like the vortices in Fig. 7e).

5. Comparison with Minimal Couette Flow

Although the regeneration identified in minimal *Re* Couette flow by Hamilton *et al.* 1995 and Waleffe 1995 (hereinafter HKW and W95 respectively) also involves repeated streak formation and streak instability, we find that the corresponding dynamics are fundamentally different from our mechanism. While similar near-wall dynamics are expected in unconstrained Couette and channel flows at sufficiently high *Re*, minimal *Re* Couette flow involves streaks which entirely fill the narrow gap between the walls. This narrow gap constraint, absent in our case, significantly alters the regeneration dynamics and (nonlinear) streak instability, as shown below.

5.1 REGENERATION CYCLE COMPARISON

To prevent further confusion, we emphasize: from a CS/vortex dynamics viewpoint, the regeneration cycles in minimal channel and minimal Couette flows are fundamentally different, although both flows sustain "turbulence". Thus, our results and analysis for the

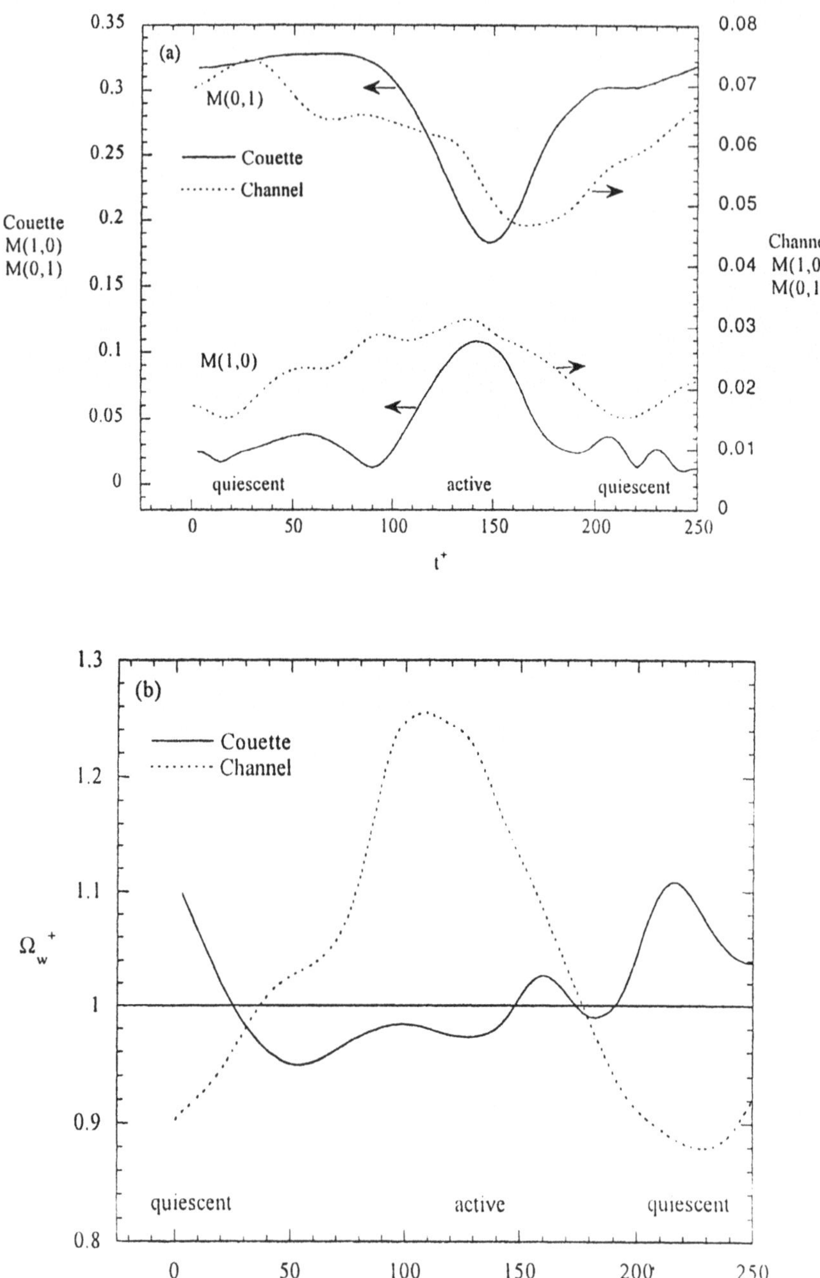

Figure 8. Evolutions of (a) modal amplitudes and (b) wall-integrated shear stress during one regeneration cycle of (turbulent) minimal channel and Couette flows. The quiescent and active phases of channel regeneration are indicated, and the streak formation and breakdown phases of Couette regeneration should be compared to the active→quiescent and quiescent→active processes respectively.

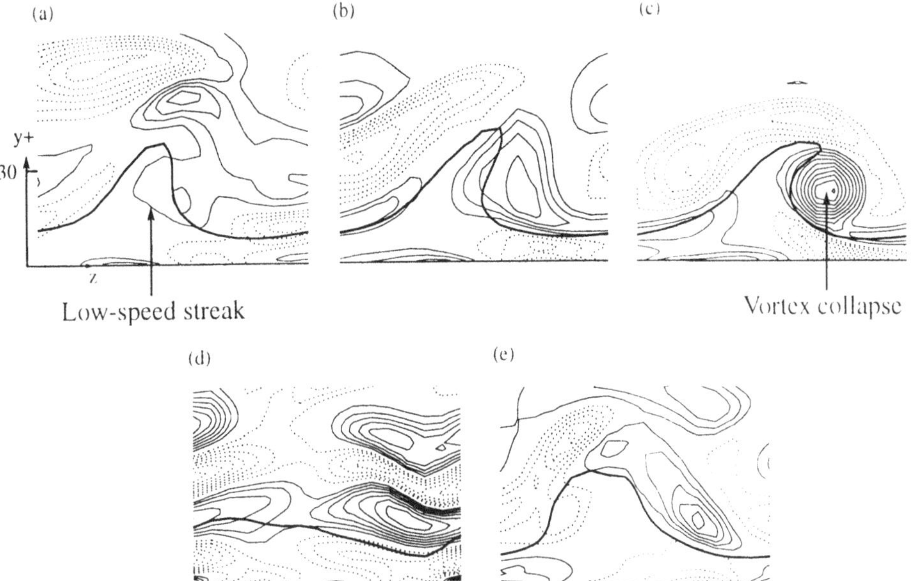

Figure 9. Evolution of instantaneous ω_x (dotted contours negative) from (a) quiescent to (c) active, back to (e) quiescent phases of minimal channel flow, illustrating streamwise vortex formation; (a-e) correspond to t^+=0, 14, 43, 156, 211 respectively. The (y,z) plane moves downstream with the near-wall propagation velocity of $0.55U_c$. The u=$0.55U_c$ contour is denoted by a bold line, and the contour increment is $\delta\omega_x^+$=0.05 in each frame.

former are naturally distinct from those of the latter in HKW and W95. To illustrate, we present below a side-by-side comparison of representative regeneration cycles in (turbulent) minimal channel and minimal Couette flows.

5.1.1 *Minimal channel flow*

Our own analysis of the Jimenez & Moin (1991) minimal channel database reveals a close correspondence of the quiescent→active regeneration process to our (nonlinear) streak instability evolution. For channel regeneration, the quiescent and active phases of regeneration are identified by minima and maxima of wall-integrated ω_z respectively (Fig. 8b). In terms of the volume-integrated modal amplitudes $M(0,1) \equiv (E_{01})^{1/2}$ (*i.e.* mean in x, fundamental in z) and $M(1,0) \equiv (E_{10})^{1/2}$ (*i.e.* fundamental in x, mean in z), during the quiescent→active process, M(0,1) decreases and M(1,0) increases (Fig. 8a).

From a structural viewpoint, at the quiescent phase, the buffer region typically contains only a lifted-up, long low-speed streak, with no significant streamwise vortices or even ω_x (Fig. 9a). Shortly thereafter (t^+~100 later), a new positive streamwise vortex is created in the buffer region (Figs. 9b,c) from the vorticity sheet (predominantly $+\omega_y$) flanking the streak, constituting the quiescent→active regeneration process. Note the striking correspondence of this process with the nonlinear evolution of vortex-less streak instability, both in a statistical (*cf.* Figs. 2b and 8a) and a structural (*cf.* Figs. 9a-c and 3a-c) sense. Past the active phase, the new streamwise vortex and its opposite-signed neighbors are stretched by the mean shear, causing them to stack on top of each other. At this point, the mutual induction of the stacked opposite-signed vortices (typically three) causes them to flatten into vorticity layers (Fig. 9d), much like head-tail formation in vortex dipoles. These thin, opposite-signed ω_x layers are then annihilated by viscous cross-diffusion, so that the quiescent configuration (Fig. 9e; *cf.* Fig. 9a) is regained, *i.e.* lifted streaks with weak ω_x and no strong vortices immediately near the wall.

In summary, the observed large temporal variations in integrated wall shear stress (Fig. 8b) directly reflect the vortex regeneration: the drag is minimum during the quiescent phase, when near-wall vortices are very weak (Figs. 9a,e), and maximum once collapsed streamwise vortices (whose induced v brings high-speed fluid toward the wall to increase drag) are generated in the buffer layer (Fig. 9c). The close correspondence of minimal channel regeneration with our mechanism (described above) confirms the operation of an underlying instability of essentially vortex-less streaks. In full-domain flows as well, extremely long ($\Delta x^+ \sim 1000$) streaks are prevalent, and many regions along individual streaks are devoid of any streamwise vortices (Robinson 1991).

5.1.2 *Minimal Couette flow*

To compare corresponding results for Couette regeneration, we have independently repeated a minimal Couette flow simulation similar to that described in HKW. The statistics of a typical regeneration cycle are depicted in Figs. 8(a,b), attained after long-time evolution of a finite-amplitude initial disturbance in a minimal domain (L_x=1.75π; L_z=1.2π) at Re=400. In HKW, the regeneration cycle is divided into two phases, based on the temporal trend of M(0,1) (Fig. 8a): (i) *streak formation* when M(0,1) increases and (ii) *streak breakdown* when M(0,1) decreases. Furthermore, growth in M(1,0) accompanies streak breakdown, attributed by HKW to an instability process. Thus, in terms of instability growth and M(1,0) amplification, the streak breakdown phase of Couette flow should be compared with the quiescent→active process of channel flow.

As shown in Fig. 8(a), similar trends occur in the modal amplitudes M(1,0) and M(0,1) over the course of a single regeneration cycle in both Couette and channel flow, despite more irregularity of the channel data. However, the corresponding wall shear stress evolutions (Fig. 8b) are surprisingly different and in fact indicate opposite trends. For instance, Ω_w reaches its maximum after streak formation and before streak breakdown for Couette regeneration, and Ω_w is near its minimum during streak breakdown (Figs. 8a,b). That is, amplification of M(1,0) does not increase the wall shear stress in Couette flow! This is quite unlike the situation in channel flow (Figs. 8a,b); in fact, Jimenez & Moin (1991) successfully use Ω_w to identify individual regeneration cycles.

This discrepancy in the wall shear trends of Couette and channel regeneration reflect a fundamentally different structural evolution in the two cases. In the following, we distinguish between streamwise "vortices" in instantaneous (y,z) planes (*e.g.* Fig. 3c, Fig. 9c) and streamwise "rolls" – large-scale, opposite-signed ω_x cells in x-averaged flows. Of course, different instantaneous vortical flows with distinct transport characteristics can have identical roll distributions.

For Couette flow, after streak formation and before streak breakdown, instantaneous ω_x contours (Figs. 10a,e) reflect relatively weak, cellular regions of opposite-signed ω_x, resembling the x-averaged rolls in most cross-sections. During Couette streak breakdown, one of the vortices is preferentially strengthened at a given x location (*e.g.* the negative roll in Fig. 10b), leading to strong, x-dependent (sinuous) spanwise motion. Subsequently, the x-dependent vortices (generated from pre-existing roll-like vortices) become sheared into z-aligned layers of ω_x (Fig. 10c), reflecting locally parallel $w(y)$ shear layers (*i.e.* ω_x is predominantly $\partial w/\partial y$); at this point, the streak breakdown is

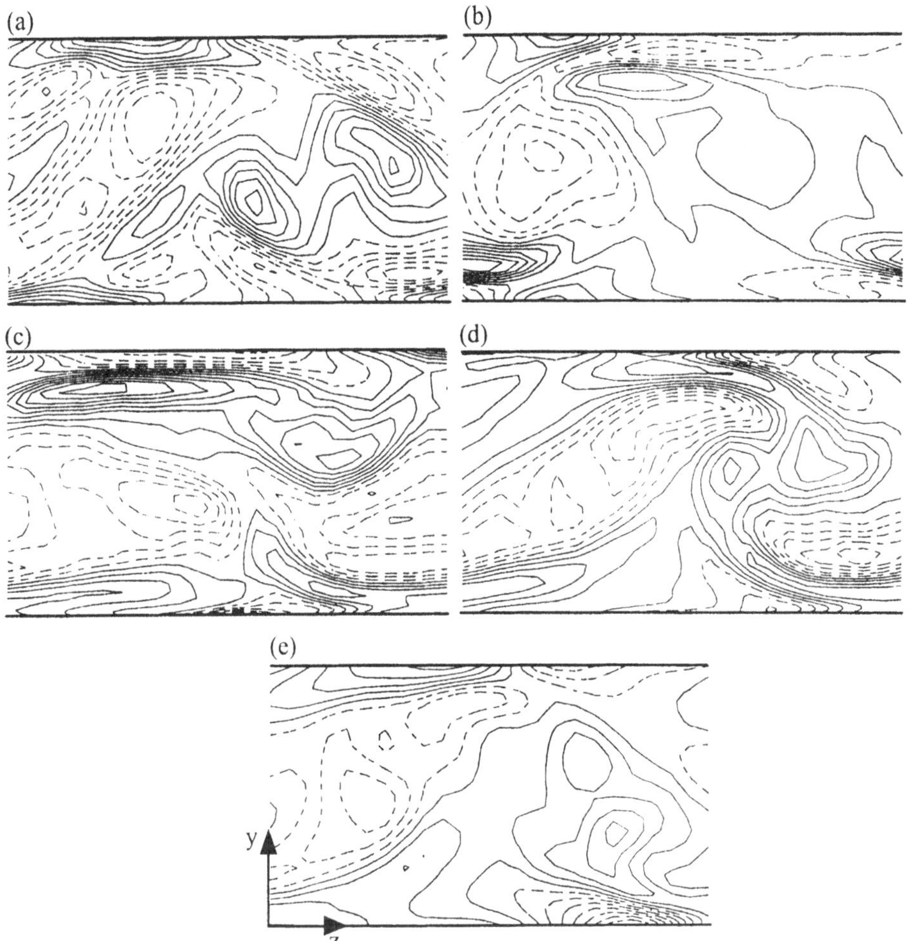

Figure 10. Evolution of instantaneous ω_x (dotted contours negative) during (a)→(c) streak breakdown and (c)→(e) streak formation processes of minimal Couette flow; (a-e) correspond to t^+=0, 90, 150, 180, 300 respectively. The (y,z) plane is fixed in each frame, and the contour increments $\delta\omega_x^+$ are (a) 0.03, (b) 0.05, (c d) 0.06, and (e) 0.02.

complete (see Fig. 8a). During the subsequent streak formation phase, like-signed ω_x layers coalesce and partially roll up (Fig. 10d), becoming reorganized once again into quasi-2D, counter-rotating vortices (Fig. 10e), whose induction reforms straight streaks.

This process and the corresponding structural evolution, masked in x-averaged flows, are obviously distinct from those outlined for channel regeneration in §5.1.1. In particular, the quiescent phase of channel regeneration (Figs. 9a,c) is characterized by vortex-less streaks near the wall; short remnants of previous individual vortices exist outside of the buffer region, but no side-by-side, counter-rotating rolls are present. Furthermore, M(1,0) amplification generates (at saturation) new compact-core streamwise vortices near the wall, rather than the violent "breakdown" and $w(y)$ shear layers (Fig. 10c) produced in Couette flow.

These structural differences shed insight into the anomalous wall shear behavior for Couette regeneration. We first note that the x-averaged *roll* circulation is maximum immediately after streak breakdown (*i.e.* near the time of peak M(1,0)), due to the

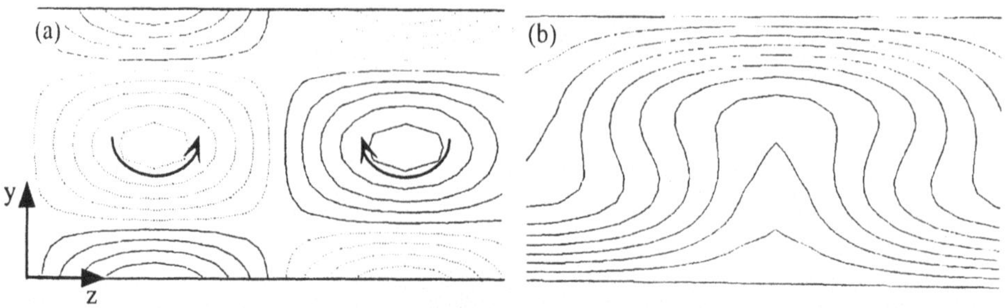

Figure 11. Base flow distributions of (a) $\Omega_x(y,z)$ and (b) $U(y,z)$ for minimal Couette flow stability analysis. Note that type I instability contains only streaks (b), while type II instability contains both the rolls (a) and the streaks (b) they generate.

nonlinear mechanisms described in detail in HKW and W95. In contrast, the wall shear stress does not begin to increase until the streak formation is complete (Figs. 8a,b). This counterintuitive result is not explainable as a simple time lag (advection time) between peak roll strength and peak wall shear. For rolls of peak strength to generate even *new* streaks, only approximately 15 outer time units are required; the lag between roll strength and wall shear is significantly longer (see Fig. 8b). Furthermore, no corresponding lag between peak M(1,0) and wall shear is observed during channel regeneration (Figs. 8a,b). Instead, explanation of the wall shear evolution requires consideration of instantaneous structures during regeneration (Fig. 10). As noted above, M(1,0) growth due to streak/roll instability is characterized by strong spanwise motions ($w(y)$ shear), rather than newly formed vortices with drag-producing v (as in channel regeneration). Thus, Couette instability acts to shear streaks apart (hence the description "streak breakdown") and thus does not act to increase the wall shear, in contrast to the streak strengthening by new vortex formation in channel flow. Hence, the strong x-averaged "rolls" observed after streak breakdown are actually dominated by instantaneous $\partial w/\partial y$ layers, not instantaneous vortices. The wall shear increases only after these local shear layers roll up and merge into roll-like vortices (Figs. 10a,e), well past the stage of streak instability and peak roll strength.

5.2 LINEAR STABILITY ANALYSIS COUETTE FLOW

To add quantitative support to this comparison, we consider the linear instability (analogous to the analysis in §3.2) of two Couette base flows: (type I) streaks without rolls (as in our mechanism), and (type II) streaks with preexisting counter-rotating rolls. While the heart of the channel/Couette regeneration distinction lies in the corresponding nonlinear evolutions, we review here the linear instability results for completeness. The streak $U(y,z)$ distribution (Fig. 11b) is obtained by evolution ($Re \equiv U_w h/\nu = 400$) of a linear Couette velocity profile due to its least stable cross-stream mode (Fig. 11a; see W95 for details); a maximum normal velocity V of $0.03U_w$ is used for the rolls of type II instability, shown in Fig. 11(a). For both cases, we consider the streak spacing of $L_z = 2\pi/1.67$ (approximately 125 wall units) analyzed by HKW and W95. Since the streamwise periodicity length $L_x = 2\pi/1.14$ in HKW is nearly stable for both the type I (see W95) and type II base flows considered here, we increase the disturbance wavelength to $L_x = 2\pi/1.05$ to ensure significant instability amplification. The instability of $L_x = 2\pi/1.14$

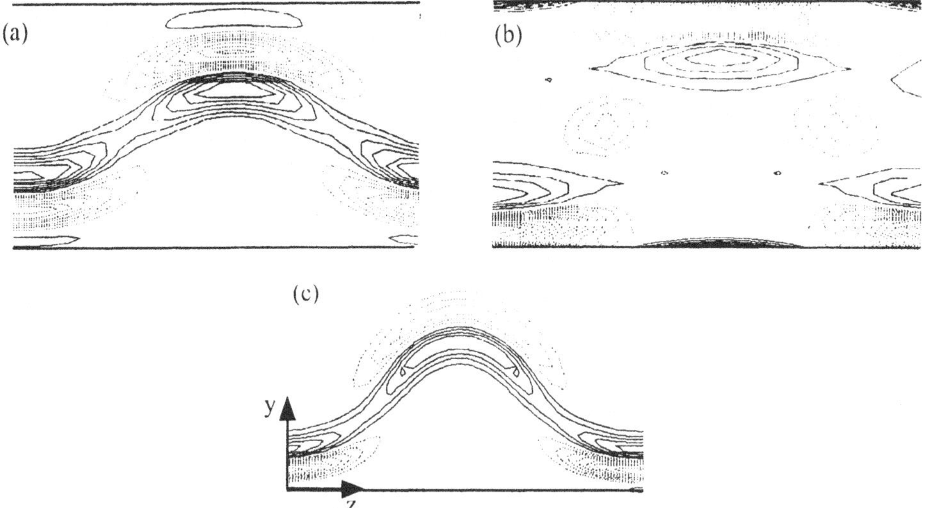

Figure 12. Perturbation $\omega_x(y,z)$ distribution for the sinuous eigenmode of (a) type I and (b) type II Couette instability and (c) channel streak instability, at a fixed x location.

wavelengths for the type II base flow considered in HKW reflects sensitivity in the short-wave cutoff to both the base flow details and the x wavelength (see also W95).

The type I sinuous instability mode (documented in detail in W95), shown in Fig. 12(a), is characterized by sheets of perturbation ω_x surrounding the streak crests and troughs. The structure of this linear eigenmode is similar to that for our one-walled streaks (Fig. 12c, from analysis in §3.2), reflecting the fact that the linear streak instability is inviscid in nature, and not strongly affected by the presence of the no-slip and no-penetration conditions at the wall(s) (one wall for channel streaks; two walls for Couette streaks) bounding the streak. Nevertheless, we show below that the presence of the second wall for Couette streaks drastically alters the subsequent nonlinear evolution of vortex-less streak instability.

The type II eigenmode (Fig. 12b) also shows sheets of ω_x surrounding the streak crests and troughs, but also two circular regions of opposite-signed perturbation ω_x centered within the roll core, not present when rolls are absent (type I). Note that the roll core perturbation preferentially strengthens the negative roll and weakens the positive roll at this x location; vice-versa for x a half wavelength away. This behavior is in fact observed during streak breakdown in turbulent Couette flow, as noted above (see Fig. 10 b), indicating operation of type II instability. Thus, type II instability extracts 3D energy not only from the streaks, but also from the rolls, distinguishing it from (type I) vortex-less streak instability.

5.2 NONLINEAR EVOLUTION: COUETTE FLOW

Extending the type I and II Couette instabilities into the nonlinear regime (via DNS), the corresponding ω_x evolutions in a typical (fixed) cross-stream plane are shown in Fig. 13 both during instability growth (*i.e.* M(1,0) amplification) and after its saturation. Note that the type I and II flows differ only in the inclusion of relatively weak rolls

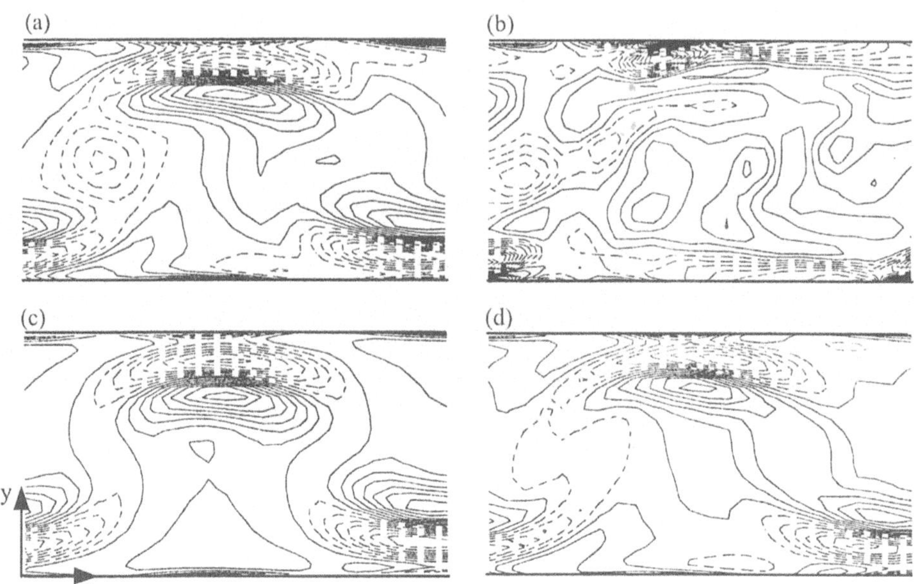

Figure 13. Evolution of instantaneous ω_x (at fixed x) for nonlinear evolution of Couette streak instabilities: (a,b) type II (with rolls) and (c,d) type I (without rolls), shown (a,c) during instability growth (t^+=30) and (b,d) after instability saturation; (b) t^+=120, (d) t^+=150.

(V_{max}=0.03U_w) in the initial conditions for type II; both instabilities are excited by a low-amplitude x-sinusoidal w disturbance (w'=0.01U_w).

In the presence of even weak preexisting rolls (type II), instability acts to preferentially strengthen one roll at a given x (*e.g.* the negative roll in Fig. 13a) and weaken the (opposite-signed) other. Note the close correspondence of this evolution with the flow structure during streak breakdown (Fig. 10b), indicating operation of type II instability in the turbulent Couette flow. Subsequently, a breakdown of the streak/roll flow occurs, characterized by $w(y)$ shear layers (Fig. 13b), also similar to turbulent Couette regeneration (Fig. 10c). Additionally, a sharp reduction in M(0,1), characteristic of Couette streak breakdown (Fig. 8a), occurs for type II, but not for type I, instability evolution (Fig. 14), further confirming the necessity of preexisting rolls (albeit weak) for Couette streak breakdown. As discussed earlier, channel regeneration does not involve such instability-driven streak breakdown, but rather instability-generated instantaneous streamwise vortices. Furthermore, our results show that preexisting vortices are unnecessary for channel regeneration.

Interestingly, despite the similarity of the streak instability eigenmode for one-walled (channel) and two-walled (Couette type I) streaks (Figs. 12a,c), the corresponding nonlinear evolutions are fundamentally different (*cf.* Figs. 13c,d with Figs. 3b,c). In particular, streamwise vortices are not formed by type I instability even past instability saturation (Fig. 14). Instead, the instantaneous type I flow is characterized by thin vorticity sheets (Fig. 13d), reflecting locally quasi-parallel $w(y)$ shear layers, not vortices. Although nonlinearity generates "rolls" in x-averaged flows, these rolls are much weaker than those observed during streak breakdown (*i.e.* in turbulent Couette flow), and they clearly do not reflect instantaneous vortices (Fig. 13d). Logically, one must conclude that

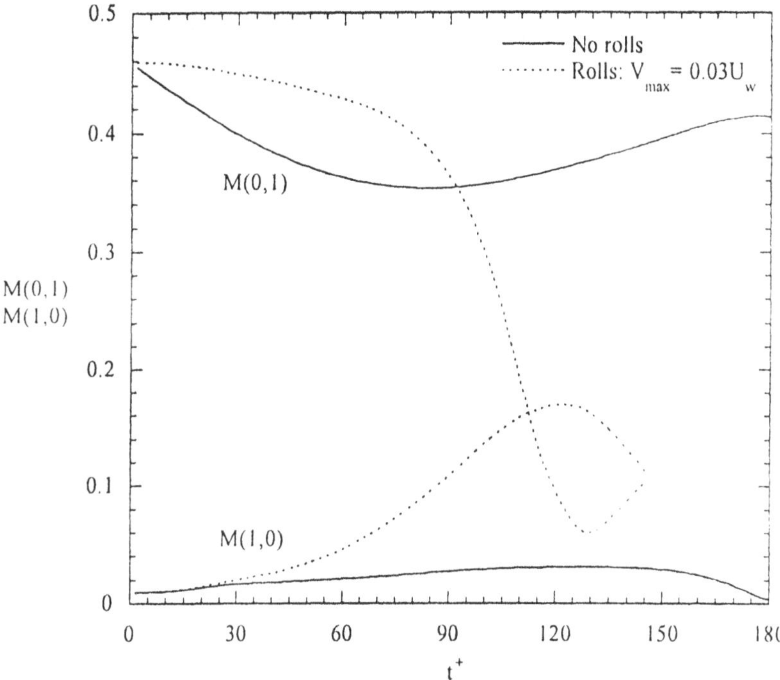

Figure 14. Evolution of modal amplitudes during nonlinear evolution of type I and type II Couette streak instabilities.

the presence of the second wall in the Couette case suppresses the vortex formation mechanism outlined above for one-walled streaks in channel flow.

6. Concluding remarks

To summarize, we have shown that (nonlinear) instability of ejected low-speed streaks, initially without any vortices whatsoever, directly generates new streamwise vortices, internal shear layers, and arch vortices. The resulting 3D vortex geometry is identical to that of the dominant CS, educed from fully developed near-wall turbulence, which in turn capture all important, extensively reported near-wall events. This serves as strong evidence that vortex-less streaks are the main breeding ground for new streamwise vortices, commonly accepted as dominant in turbulence production. In turn, the geometry of the newly generated vortices constitutes a built-in mechanism which sustains ejected streaks against their otherwise rapid self-annihilation due to cross-diffusion. Vortex-less streaks, the vehicle for instability-based vortex formation, are expected to arise inherently in full-domain turbulence due to the differential advection of vortices and the streaks they generate (see Schoppa & Hussain 1997a for details).

Since vortex formation and turbulence production are critically reliant on lifted low-speed streaks, large-scale (relative to the natural streak spacing) control of streaks is a potentially effective approach to drag reduction, noting the tiny scale of near-wall structures in most engineering situations. We have found that large-scale drag reduction is in fact effective via either counterrotating vortex generators or colliding spanwise wall jets (Schoppa & Hussain 1997b). Our control approach is particularly attractive from a practical standpoint, in that no sensors are required (necessary for adaptive control) and large-scale (hence more durable and feasible) actuation is effective.

This research was supported by AFOSR grant F49620-97-1-0131 and the NASA Graduate Fellowship grant NGT-51022 of W.S. Supercomputer time was provided by the NASA Ames Research Center.

7. References

Blackwelder, R.F. & Eckelmann, H. 1979 Streamwise vortices associated with the bursting phenomenon. *J. Fluid Mech.* **94**, 577.

Brooke, J.W. & Hanratty, T.J. 1993 Origin of turbulence-producing eddies in a channel flow. *Phys. Fluids A* **5**, 1011.

Brown, G.L. & Thomas, A.S.W. 1977 Large structure in a turbulent boundary layer. *Phys. Fluids* **20**, 5243.

Hamilton, J., Kim, J. & Waleffe, F. 1995 Regeneration mechanisms of near-wall turbulence structures. *J. Fluid Mech.* **287**, 317.

Jang, P.S., Benney, D.J. & Gran, R.L. 1986 On the origin of streamwise vortices in a turbulent boundary layer. *J. Fluid Mech.* **169**, 109.

Jeong, J. & Hussain, F. 1992 Coherent structures near the wall in a turbulent channel flow. In *Proc. of Fifth Asian Congress of Fluid Mech.*, Taejon, Korea (eds. K.S. Chang & H. Choi), p. 1262.

Jeong, J., Hussain, F., Schoppa, W. & Kim, J. 1997 Coherent structures near the wall in a turbulent channel flow. *J. Fluid Mech.* **332**, 185.

Jimenez, J. & Moin, P. 1991 The minimal flow unit in near-wall turbulence. *J. Fluid Mech.* **225**, 213.

Jimenez, J. & Orlandi, P. 1993 The rollup of a vortex layer near a wall *J. Fluid Mech.* **248**, 297.

Kim, H.T., Kline, S.J. & Reynolds, W.C. 1971 The production of turbulence near a smooth wall in a turbulent bounary layer. *J. Fluid Mech.* **50**, 133.

Kim, J., Moin, P. & Moser, R.D. 1987 Turbulence statistics in fully developed channel flow at low Reynolds number. *J. Fluid Mech.* **177**, 133.

Kline, S.J., Reynolds, W.C., Schraub, F.A. & Rundstadler, P.W. 1967 The structure of turbulent boundary layers. *J. Fluid Mech.* **30**, 741.

Kline, S.J. & Robinson, S.K. 1990 Quasi-coherent structures in the turbulent boundary layer: Part 1. Status report on a community-wide survey of the data. In *Near-Wall Turbulence* (eds. S.J. Kline & N.H. Afgan). Hemisphere.

Kravchenko, A.G., Choi, H. & Moin, P. 1993 On the relation of near-wall streamwise vortices to wall skin friction in turbulent boundary layers. *Phys. Fl. A* **5**, 3307.

Phillips, W.R.C. & Wu, Z. 1994 On the instability of wave-catalysed longitudinal vortices in strong shear. *J. Fluid Mech.* **272**, 235.

Robinson, S.K. 1991 Coherent motions in the turbulent boundary layer. *Ann. Rev. Fluid Mech.* **23**, 601.

Schoppa, W. & Hussain, F. 1997a Genesis and dynamics of coherent structures in near-wall turbulence. In *Self-sustaining Mechanisms of Wall Turbulence* (ed. R. Panton). Computational Mechanics Publications, p. 385.

Schoppa, W. & Hussain, F. 1997b Effective drag reduction by large-scale manipulation of streamwise vortices in near-wall turbulence. Presented at *4th AIAA Shear Flow Control Conference* (Snowmass CO), AIAA Paper No. *AIAA 97-1794*.

Swearingen, J.D. & Blackwelder, R.F. 1987 The growth and breakdown of streamwise vortices in the presence of a wall. *J. Fluid Mech.* **182**, 225.

Townsend, A.A. 1956 *The Structure of Turbulent Shear Flows*. Cambridge University Press.

Waleffe, F. 1995 Hydrodynamic stability and turbulence: beyond transients to a self-sustaining process. *Stud. Appl. Math* **95**, 319.

III. Rotating Flows

ORGANIZED STRUCTURES IN ROTATING CHANNEL FLOW

H.I. ANDERSSON
Division of Applied Mechanics
Department of Applied Mechanics, Thermodynamics and Fluid Dynamics
Norwegian University of Science and Technology
N-7034 Trondheim, Norway

1. Introduction

Coherent structures are known to play an essential role in wall-bounded turbulent shear flows, e.g. boundary layers and channel flows. Typical coherent structures with life time at least an order of magnitude longer than the small-scale turbulence are the low-speed streaks in the viscous sublayer and the internal shear layers in the buffer region. The pioneering flow visualizations of coherent structures by Kline et al. (1967) and Kim et al. (1971) triggered the exploration of such structures in wall bounded turbulence. Although this area of research originally relied on laboratory experiments, the advent of direct numerical simulations has further advanced the field.

The present contribution is concerned with organized structures in pressure-driven Poiseuille flow and shear-driven Couette flow subjected to system rotation. The Coriolis body force, which arises from the imposed system rotation, may have important implications on the flow field. Depending on the orientation and magnitude of the rotation vector $\vec{\Omega}$, i.e. the angular velocity of the coordinate system, different flow phenomena may occur. The relative importance of the Coriolis force and inertia is conventionally expressed in terms of a rotation number Ro = $2\Omega L/U$ in industrial fluid dynamics and as a Rossby number $U/2\Omega L$ in geophysical and astrophysical fluid dynamics. Here, L and U denote a characteristic length and velocity, respectively, in the rotating frame-of-reference. While Ro is a dimensionless measure of the *magnitude* of the imposed system rotation, different *orientations* of the rotation vector both with respect to any solid boundary and the vorticity vector may make the impact of the Coriolis force fundamentally different. This is among the topics addressed in the lecture notes compiled and edited by Hopfinger (1992).

Laminar Poiseuille flow and Couette flow in plane channels are among the prototype flows in classical fluid mechanics. When the channel is rotated about a spanwise axis, i.e. in orthogonal mode rotation, so that the rotation vector $\vec{\Omega}$ is either parallel or antiparallel to the walls and the vorticity vector, the flow is susceptible to a roll-cell instability first considered by Hart (1971) and Lezius & Johnston (1976). The development of an array of regularly spaced roll-cells aligned with the primary flow is the rotational analogue to the centrifugal instability due to streamline curvature. The existence of pairs of counter-rotating longitudinal vortices or roll-cells even in turbulent channel flows subjected to rotation was

TABLE 1. Characteristics of the DNSs.

Flow	Re	Re_τ	Ro_{max}	domain	points
Poiseuille	2900*	194	0.5	$4\pi h \times 2h \times 2\pi h$	$128 \times 128 \times 128$
Couette	1300	82.2*	1.0	$10\pi h \times 2h \times 4\pi h$	$256 \times 70 \times 256$

* varies somewhat with Ro

first observed experimentally by Johnston et al. (1972) in Poiseuille flow and numerically by Bech & Andersson (1996a,b) in Couette flow. These longitudinal roll-cells or Taylor-Görtler-like (TG) vortices, which result from an imbalance between the Coriolis force and the pressure gradient in the wall-normal direction, are in principle of infinite streamwise extent. In comparison with the *localized* low-speed streaks and internal shear layers, which are generally present in wall-bounded turbulence, the TG-vortices are *global* in nature.

The objective of this paper is to summarize the major observations made from a series of computer experiments on rotating channel flows, emphasizing the influence of the Coriolis force on the organized flow structures. First, however, the computational approach will be briefly described in the next section. Then, in the third section, roll-cell formation in turbulent Poiseuille and Couette flow will be considered. However, the localized flow structures, which are present even in non-rotating channel flow, are also affected by the Coriolis force and this is dealt with in Section 4. Finally, possible implications of the organized structures for conventional turbulence modelling are addressed in the closing section.

2. Direct Numerical Simulation

Unlike traditional prediction methods for turbulent flows, the direct numerical simulation (DNS) approach does not rely on semi-empirical or statistical closure models. Instead, the unsteady Navier-Stokes equations governing the turbulent motion of an incompressible fluid are represented on a three-dimensional computational mesh and integrated numerically without any further assumptions than those involved in the replacement of the partial differential equations by the non-linear set of discretized momentum equations. However, in order to resolve all the eddy scales present in a turbulent flow, the spatial resolution of the computational mesh must be of the order of the tiniest scales of the turbulence. The directly simulated flow fields can then be considered as numerical realizations of real turbulence, which exhibit both complex non-linear dynamics and vortex stretching. The numerically generated turbulence therefore contains detailed information on spatial coherent structures as well as on the time-dependency and statistics of the instantaneous velocity and pressure fields.

With the view to provide further insight into rotational-induced flow phenomena, fully developed Poiseuille flow and Couette flow between two infinite parallel planes have been considered by Kristoffersen & Andersson (1993) and Andersson & Kristoffersen (1995) and by Bech & Andersson (1996a, 1996b, 1997), respectively. Some essential characteristics of the simulations are given in Table 1, where the two alternative Reynolds numbers are defined as $Re = UL/\nu$ and $Re_\tau = U_\tau L/\nu$, U_τ being the wall-friction velocity. The length scale L is taken as the channel half-width h, while the velocity scale U is half the velocity difference between the walls for the Couette flow and the bulk mean velocity for the Poiseuille flow. In both cases the value of the inner variable $y^+ = yU_\tau/\nu$ at the centreline of the channel is equal to Re_τ.

3. Global Structures

The turbulent analogue to the Taylor-Görtler-like roll-cells in rotating laminar Poiseuille flow was first observed experimentally by Johnston et al. (1972) and later detected also in the large-eddy simulations of Kim (1983). It is noteworthy that the spanwise rotating Poiseuille flow exhibits at the same time a cyclonic (suction) side and an anticyclonic (pressure) side where the mean flow vorticity $-dU/dy$ is parallel or antiparallel, respectively, to the imposed background vorticity 2Ω. The spanwise array of longitudinal vortices originates from the anticyclonic side, which is destabilized by the system rotation. However, a truly steady roll-cell pattern was neither observed by Johnston et al. nor by Kim. Kristoffersen & Andersson (1993) performed DNSs for several rotation numbers Ro up to 0.5 and found that, except for Ro = 0.15, the roll-cells gradually changed their shape, size and strength, although the time scale of these changes was long relative to the turbulence time scale. Nevertheless, it was concluded that the number of vortex pairs seemed to increase with Ro while the vortices being gradually shifted towards the anticyclonic side of the channel.

At the particular rotation number 0.15, however, a fairly persistent roll-cell pattern was observed. The projection of the instantaneous velocity field u_i' into the cross-sectional plane in Figure 1a suggests the presence of some large-scale vortical structures. In order to facilitate the analysis of the simulated flow field, the triple decomposition

$$u_i'(x,y,z,t) = U_i(y) + \tilde{u}_i(y,z) + u_i(x,y,z,t) \qquad (1)$$

analogous to that applied by Moser & Moin (1987) is adopted. Here, U_i is the one-componential one-dimensional flow which is obtained by averaging over x, z and t, while $\tilde{u}_i$ represents the three-componential two-dimensional secondary flow field associated with the roll-cells and u_i denotes the 3D turbulent fluctuations. The decomposition (1) implicitly assumes that the Taylor-Görtler-like vortices are coherent over times and streamwise distances substantially larger than the corresponding scales of the underlying turbulence. Thus, since $\partial\tilde{u}/\partial x = 0$, the mass conservation equation simplifies to

$$\frac{\partial \tilde{v}}{\partial y} + \frac{\partial \tilde{w}}{\partial z} = 0 \qquad (2)$$

and the associated streamlines are depicted in Figure 1b. The quite distinct flow pattern suggests that the time scale of the vortices is much larger than the averaging time $5.6\ h/U_\tau$.

The vortex pattern observed at rotation numbers other than 0.15, however, did not survive an averaging time of this order.

To further explore the vortical flow induced by system rotation, the purely shear-driven plane Couette flow was considered by Bech & Andersson (1996a, 1996b, 1997). In contrast with the rotating Poiseuille flow, the mean velocity distribution U(y) in the Couette flow increases monotonically from one wall to the other, thereby exhibiting an inflectional point at the centre. Thus, both sides of the channel become either cyclonic or anticyclonic simply by changing the sense of rotation.

When the turbulent Couette flow was subjected to weak system rotation $Ro = \pm 0.01$, qualitatively different impact of the Coriolis force could be observed depending on the sense of rotation, see Bech & Andersson (1996a). Persistent roll-cells occurred only in the case of anticyclonic rotation, cf. Figure 2, but it turned out that the turbulence structure was nearly unaffected by this low rotation. The kinetic energy associated with the roll-cells amounted to 50 per cent of the turbulence energy, but only 1/10 of this stemmed from the cross-flow $\tilde{v}$ and $\tilde{w}$ depicted in Figure 2. More surprising was the observation that the kinetic energy of the turbulence was reduced in comparison with the non-rotating flow, irrespective of the sense of rotation.

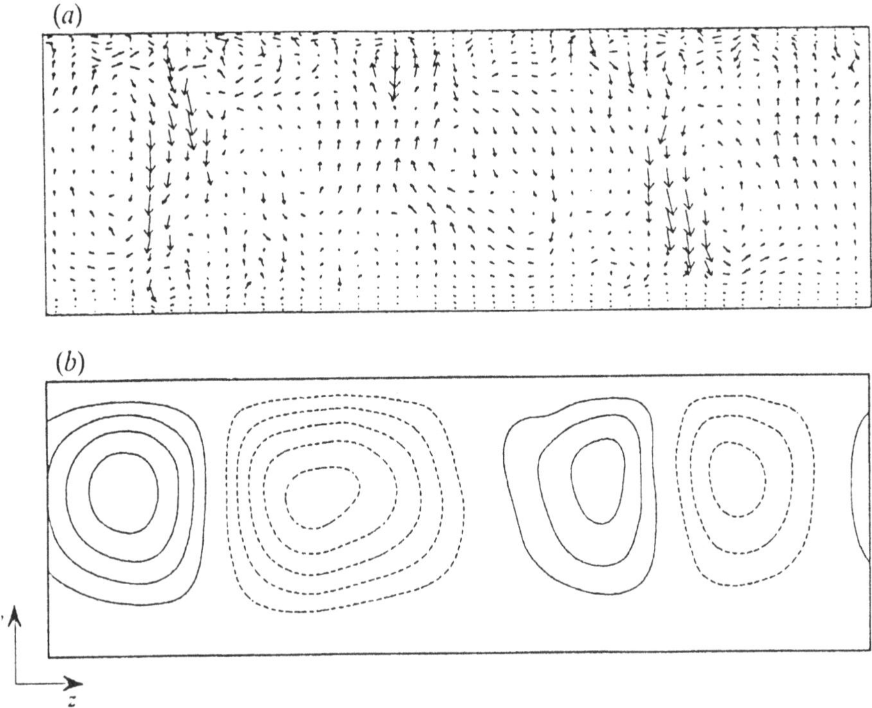

Figure 1. Plane Poiseuille flow at Ro = 0.15. The anticyclonic (pressure) side is at the top of the figures. From Kristoffersen & Andersson (1993). a) Instantaneous velocity vectors in a cross-sectional plane. Each vector represents an average over 18 computational cells; b) Secondary flow streamlines averaged over 5.6 h/U_τ. Solid and broken lines denote clockwise and counter-clockwise motion, respectively.

At the intermediate rotation numbers 0.10 and 0.20 in Figure 2, the Taylor-Görtler-like vortices were far more energetic than at Ro = 0.01 and contained four times the energy contained in the turbulence. It should be noticed that these roll-cell patterns are by far more regular and intense than those of the Poiseuille flow displayed in Figure 1. This could be expected since the Coriolis force acts destabilizing on both sides of the Couette channel, whereas the Poiseuille flow becomes unstable only near the anticyclonic pressure side.

For the high rotation number Ro = 0.50, a roll-cell breakdown set in and only reminiscences of the vortical structures were left. The energy content of the turbulence was now 3 times higher than for Ro = 0.10 and exceeded the turbulence level in the non-rotating case by nearly 50 %; cf. Bech & Andersson (1996b). The breakdown of the persistent roll-cell pattern at Ro = 0.50 was accompanied by a substantial amplification of the streamwise vorticity. This disappearance of the Taylor-Görtler-vortices is consistent with a recent Poiseuille flow simulation by Lamballais et al. (1996), in which such large-scale rolls did not seem to be present at their highest rotation rate.

Inspection of the two-point velocity correlations $R_{ii}(r)$, notably in the spanwise direction, revealed another manifestation of the persistent roll-cell pattern both in the Poiseuille flow (Kristoffersen & Andersson 1993) and the Couette flow (Bech & Andersson 1997). More surprising was the peculiar oscillations in $R_{11}(r_x)$, which was interpreted as a signature of

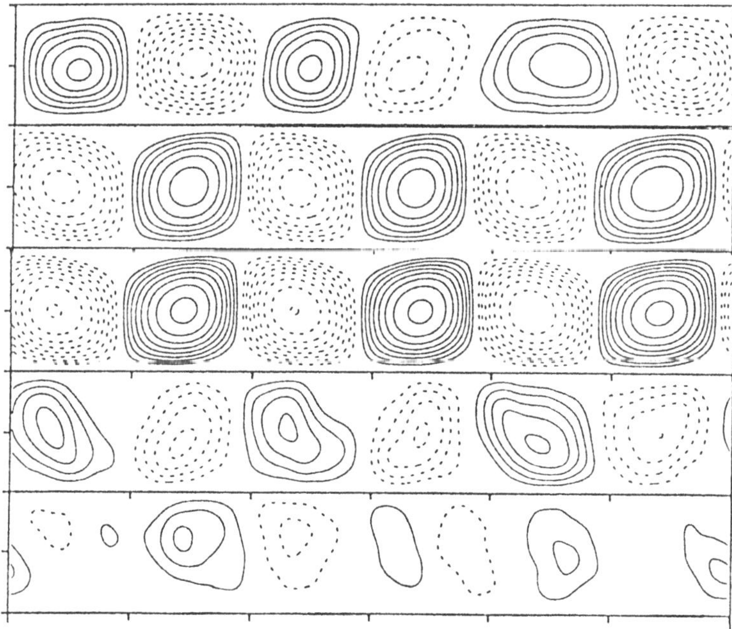

Figure 2. Secondary flow streamlines in plane Couette flow subjected to anticyclonic rotation. From Bech & Andersson (1996b). From top to bottom: Ro = 0.01; Ro = 0.10; Ro = 0.20; transient roll-cell breakdown; Ro = 0.50. The increment between the contour lines for Ro = 0.01 is 1/4 of that for the other rotation numbers.

periodic variations of the roll cells in the streamwise direction at Ro = 0.20. It can be conjectured that a disordering of this mode occurs when the rotation number is further increased until the ultimate breakdown results. However, it can not at all be ruled out that other non-linear effects (cf. Floryan, 1991) come into play.

Significant arbitrariness concerning the possible formation of Taylor-Görtler vortices in curved channel flow has been mentioned by Moser & Moin (1987) and more recently by Kobayashi & Maekawa (1995). It is likely to assume that the Taylor-Görtler-like roll-cells in a rotating channel are equally sensitive to upstream conditions and the spanwise width of the channel. In computer experiments these parameters are more easily controlled than in laboratory investigations.

4. Localized Structures

The attention has so far been focused on organized structures which arise from the presence of the Coriolis force in a rotating environment. However, localized structures which are present in non-rotating channel flows are not only directly affected by the Coriolis force but also indirectly by the rotational-induced Taylor-Görtler-like vortices or roll-cells.

The elongated streaky structures in the viscous sublayer can be analysed by various means. The spanwise two-point correlation coefficient of the streamwise velocity component u, for instance, exhibits a distinct (negative) minimum in the near-wall region. Twice the spanwise separation corresponding to this minimum is a measure of the mean spacing λ between the wall-layer streaks. On the anticyclonic (pressure) side of the rotating Poiseuille flow, Kristoffersen & Andersson (1993) observed that λ decreased substantially with increasing rotation. This is consistent with their flow visualizations at Ro = 0.15, which revealed a higher density of the streaks. Along the cyclonic (suction) side, on the other hand, elongated streaky flow structures were found only in some streamwise bands which coincided with the impingement zone of the roll-cells. Timelines generated by passive particles released from a spanwise line revealed calm and laminar-like regions in between the more vigorous bands associated with the roll-cells. This is in accordance with the illuminating hydrogen-bubble timelines displayed by Johnston et al. (1972).

Although the spatial correlations were obtained for the sum of the secondary flow $\tilde{u}_i$ and the turbulence u_i, the presence of the roll-cells was clearly visible both in Poiseuille flow (Kristoffersen & Andersson 1993) and Couette flow (Bech & Andersson 1997), notably in the spanwise correlation of $\tilde{v}+v$. Bech & Andersson (1996a) adopted the triple decomposition (1) and showed that also the spanwise correlation of the streamwise velocity was substantially affected by the roll-cells even at low rotation Ro = 0.01. In rapidly rotating Couette flow, however, Bech & Andersson (1997) found that the signature of any counter-rotating roll cells could no longer be discerned at Ro = 0.50.

The fractional contributions to the Reynolds shear stress $-\rho\overline{uv}$ can be obtained by means of quadrant-splitting analysis, i.e. according to the individual signs of the streamwise and wall-normal velocity components. Inspection of the Poiseuille flow at Ro = 0.15 and the Couette flow at Ro = 0.5 shows that the relative importance of fourth-quadrant or sweeping events, which are the dominating contribution in the near-wall region, is substantially reduced by the rotation, while second-quadrant ejections are practically unaffected. Recently, Bech & Andersson (1997) considered individual events with significant contributions to the time-averaged shear stress with a detection method similar to that of Guezennec et al. (1989). Whereas the number of detected Q2 events was nearly unaffected by rotation, the frequency of occurrence of Q4 events was reduced by some 30 per cent at Ro = 0.50. The ensemble averaged velocity field revealed that the fluctuating part of the Coriolis force, i.e. $-2\rho\vec{\Omega}\times\vec{u}$, has a greater impact in the wall-normal direction than in the mean flow direction since the magnitude of u exceeds that of v in the immediate vicinity of the walls. System rotation thus tends to enhance the instantaneous motion in the wall-normal direction, thereby increasing the inclination of the velocity vectors with respect to the wall, as shown in Figure 3. The conditionally averaged vorticity associated with the Q2 and Q4 events is anticyclonic and cyclonic, respectively. It is therefore not surprising that Q2 and Q4 events are affected differently by the imposed system rotation. The streamwise extent of the Q2 events became substantially greater than the Q4 events for rapid rotation and a pair of counter-rotating vortices with axes almost parallel to the x-axis could be observed downstream of the detection point for the ejections.

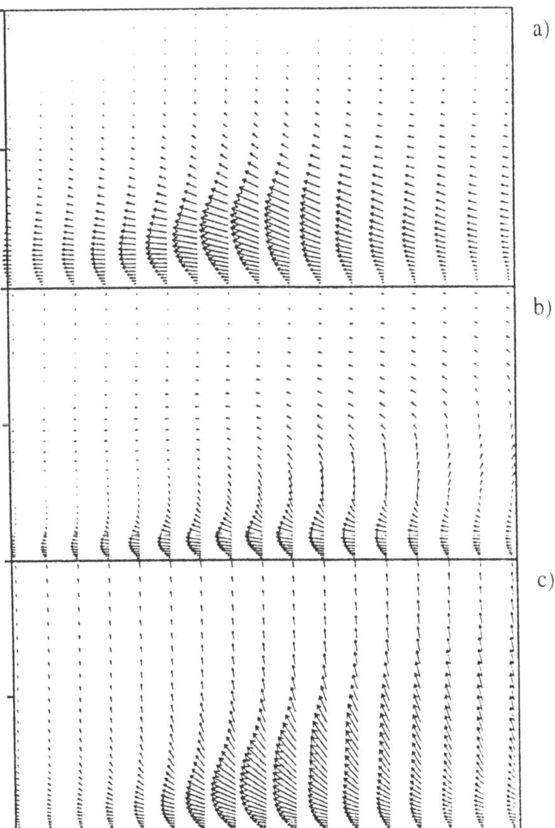

Figure 3. Plane Couette flow subjected to anticyclonic rotation. Ensemble-averaged velocity field for Q2-events (ejections) in the (x,y)-plane. From Bech & Andersson (1997). a) Ro = 0; b) Ro = 0.10; c) Ro = 0.50.

In the weakly rotating Couette flow (Ro = ± 0.01) Bech & Andersson (1996a) found that the turbulence energy was damped irrespective of the sense of rotation, i.e. not only when the rotation is cyclonic and the flow is stable with respect to roll-cell formation but also in the anticyclonic case in which counter-rotating roll-cells developed. By means of the triple-decomposition (1) the secondary flow and the turbulence could be investigated separately and it was found that the turbulence characteristics (e.g. the anisotropy invariant map for the Reynolds stresses) were practically unaffected at this low rotation rate. It was therefore interesting to examine the localized internal shear layers, which are mainly confined to the buffer region. These turbulence-producing flow structures were identified by the Variable Interval Space Averaging (VISA) technique, which detects an event when the local variance of u over a certain interval exceeds the overall variance. Comparisons with experiments and DNS data for Ro = 0 by Bech et al. (1995) showed that the frequency of occurrence of the VISA events was reduced due to weak system rotation, fully in accordance with the observed reduction of turbulent kinetic energy.

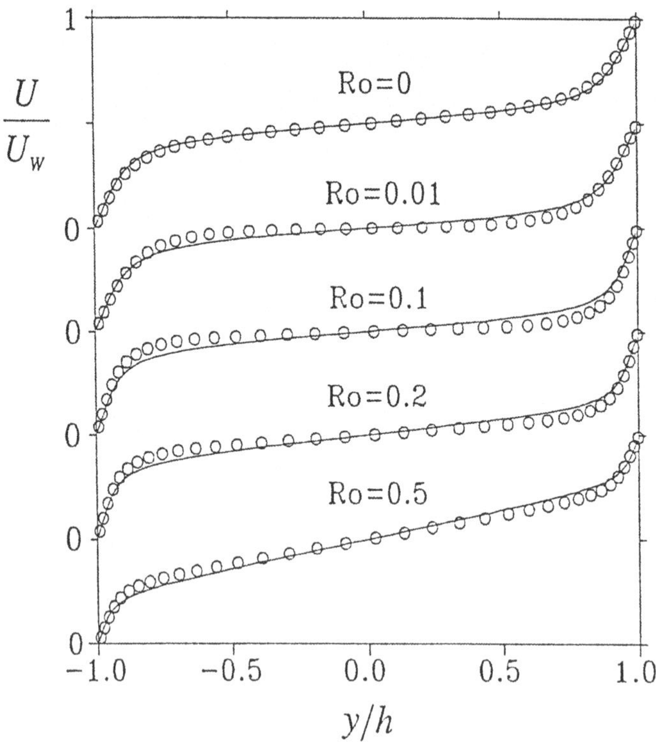

Figure 4. Plane Couette flow subjected to anticyclonic rotation. Model predictions of the mean velocity distribution (lines) and DNS data (symbols). From Pettersson & Andersson (1997).

5. Implications for Turbulence Modelling

Computerized flow analysis for engineering applications will in the foreseeable future be based on the Reynolds-averaged Navier-Stokes (RANS) equations and a suitable turbulence model and occasionally on so-called large-eddy simulations (LES) in which only the unresolved sub-grid scale turbulence is subjected to modelling. In spite of the fact that major contributions to the near-wall turbulence stem from localized structures (streaks and internal shear layers), one-point turbulence models based on the Reynolds stresses themselves have proved fairly successful in predictions of wall-bounded turbulence. However, in order to comply with the highly anisotropic conditions prevailing in the viscous sublayer, local characteristics like the Reynolds stress invariants are becoming essential elements in near-wall modelling. The effect of the localized structures on the averaged flow field thus appears to be reasonably well accounted for by the latest one-point closure models.

A different situation arises when global structures (e.g. counter-rotating Taylor-Görtler-like vortices) are embedded in the flow field. The triple decomposition (1) cannot be used within the framework of RANS. A first approach would therefore be to treat the secondary flow field $\tilde{u}_i(y,z)$ as a part of the turbulence, which then is subject to conventional modelling. Pettersson & Andersson (1997) followed this strategy and adopted a Reynold-stress model in which the near-wall effects are accounted for by elliptic relaxation. The one-dimensional mean velocity profile U(y) in Figure 4 responded adequately to the imposed system rotation. It is particularly noteworthy that the model predictions compare most favourably with the DNS data for Ro = 0 and 0.50, i.e. when the rotational-induced roll-cells are absent. Even the fairly sophisticated turbulence closure cannot properly accommodate the influence of the truly two-dimensional secondary flow.

A more refined and physically appealing approach is to treat the roll-cells as an integral part of a 2D three-componential mean flow $U_i(y) + \tilde{u}_i(y,z)$ governed by the RANS and let the turbulence model account only for the real turbulence u_i. This distinction between the organized global structures and the turbulence cannot be made unless the roll-cell pattern is persistent. The potential of this approach is now under investigation and tentative results for rotating Couette flow are promising. From a computational point of view, the rapidly rotating Couette flow (Ro = 0.50) is particularly attractive due to its one-dimensionality and one-componentality (Andersson & Bech 1997). In spite of the absence of roll-cells, the anomalous reversal of the conventional Reynolds stress and dissipation rate anisotropies makes this flow a physically challenging test for the assessment of turbulence closure models.

It is a pleasure to acknowledge the substantial contributions made by my former Ph.D. students K.H. Bech, R. Kristoffersen and B.A. Pettersson. The computer code used for the DNSs was made available by the Turbulence Unit at Queen Mary College in London and computing time was supported by The Research Council of Norway. Ms. I. Wiggen expertly prepared the camera-ready manuscript.

References

Andersson, H.I. & Bech, K.H. (1997) Statistics of rapidly rotating turbulent plane Couette flow, in *Proc. 11th Symposium on Turbulent Shear* Flows, Grenoble (in print).

Andersson, H.I. & Kristoffersen, R. (1995) Turbulence statistics of rotating channel flow, in F. Durst et al. (eds.), *Turbulent Shear Flows* 9, Springer Verlag, pp. 53-70.

Bech, K.H. & Andersson, H.I. (1996a) Secondary flow in weakly rotating turbulent plane Couette flow, *Journal of Fluid Mechanics* **317**, 195-214.

Bech, K.H. & Andersson, H.I. (1996b) Growth and decay of longitudinal roll cells in rotating turbulent plane Couette flow, in S. Gavrilakis et al. (eds.), *Advances in Turbulence VI,* Kluwer, pp. 91-94.

Bech, K.H. & Andersson, H.I. (1997) Turbulent plane Couette flow subject to strong system rotation, *Journal of Fluid Mechanics* (submitted).

Bech, K.H., Tillmark, N., Alfredsson, P.H. & Andersson, H.I. (1995) An investigation of turbulent plane Couette flow at low Reynolds numbers, *Journal of Fluid Mechanics* **286**, 291-325.

Floryan, J.M. (1991) On the Görtler instability of boundary layers, *Progress in Aerospace Sciences* **28**, 235-271.

Guezennec, Y.G., Piomelli, U. & Kim, J. (1989) On the shape and dynamics of wall structures in turbulent channel flow, *Physics of Fluids A* **1**, 764-766.

Hart, J.E. (1971) Instability and secondary motion in a rotating channel flow, *Journal of Fluid Mechanics* **45**, 341-351.

Hopfinger, E.J. (1992). *Rotating Fluids in Geophysical and Industrial Applications,* Springer Verlag.

Johnston, J.P., Halleen, R.M. & Lezius, D.K. (1972) Effects of spanwise rotation on the structure of two-dimensional fully developed turbulent channel flow, *Journal of Fluid Mechanics* **56**, 533-557.

Kim, H.T., Kline, S.J. & Reynolds, W.C. (1971) The production of turbulence near a smooth wall in a turbulent boundary layer, *Journal of Fluid Mechanics* **50**, 133-160.

Kim, J. (1983) The effect of rotation on turbulence structure, in *Proc. 4th Symposium on Turbulent Shear Flows,* Karlsruhe, pp. 6.14-6.19.

Kline, S.J., Reynolds, W.C., Schraub, F.A. & Runstadler, P.W. (1967) The structure of turbulent boundary layers, *Journal of Fluid Mechanics* **30**, 741-773.

Kobayashi, M. & Maekawa, H. (1995) Turbulent flow accompanied by Taylor-Goertler vortices in a two-dimensional curved channel, *Flow Measurement and Instrumentation* **6**, 93-100.

Kristoffersen, R. & Andersson, H.I. (1993) Direct simulations of low-Reynolds-number turbulent flow in a rotating channel, *Journal of Fluid Mechanics* **256**, 163-197.

Lamballais, E., Lesieur, M. & Metais, O. (1996) Effects of spanwise rotation on the vorticity stretching in transitional and turbulent channel flow, *International Journal of Heat and Fluid Flow* **17**, 324-332.

Lezius, D.K. & Johnston, J.P. (1976) Roll-cell instabilities in rotating laminar and turbulent channel flows, *Journal of Fluid Mechanics* **77**, 153-175.

Moser, R.D. & Moin, P. (1987) The effects of curvature in wall-bounded turbulent flows, *Journal of Fluid Mechanics* **175**, 479-510.

Pettersson, B.A. & Andersson, H.I. (1997) Near-wall Reynolds-stress modelling in noninertial frames of reference, *Fluid Dynamics Research* **19**, 251-276.

SIMULATION AND IDENTIFICATION OF ORGANIZED VORTICES IN ROTATING TURBULENT FLOWS

L. LOLLINI, C. CAMBON, M. MICHARD AND L. GRAFTIEAUX
Laboratoire de mécanique des fluides et d'acoustique
UMR 5509 CNRS - Ecole Centrale de Lyon - Université Lyon I
BP 163 - 69131 Ecully Cedex France

Abstract. Direct numerical simulations are performed to approach the experiment of Hopfinger, Browand & Gagne (Hopfinger *et al.*, 1982) in which turbulence is generated by an oscillating grid in a rotating tank. Effects of both rotation and confinement are analyzed, for various Rossby and Reynolds numbers. When the rotation is strong enough, we observe concentrated vortices having axes approximately parallel to the rotation axis. The Fourier-Fourier-Chebyshev pseudo-spectral code allows to take into account confinement between two parallel walls and the experimental oscillating grid is modeled by a forcing in an horizontal plane close to the lower wall. Both walls and the forcing plane are perpendicular to the vertical rotation axis.
A straightforward post processing of the numerical results for isolating coherent vortices would require the evaluation of vorticity which appear to be inappropriate because of its noisy and strongly inhomogeneous characters. Here, we propose an identification method based on the normalized angular momentum. Applied to the DNS velocity fields, it is a useful tool allowing the localization of both cyclonic and anticyclonic vortices with a very good accuracy.

1. General background

Rotation is known to modify the behavior of transitional or turbulent flows through the Coriolis force effect. Its stabilizing or destabilizing effects are known to exist but are not completely understood. Various physical or numerical experiments have been carried out, showing the influence of solid-body rotation on the stability of various flows (rotating turbulent flow in a plane channel, Kristoffersen & Andersson, 1993 ; turbulent Couette flow, Tillmark & Alfredsson, 1996 ; stability of barotropic vortices, Kloosterziel, 1990 ; stability of pure shear flow subjected to Coriolis force, Pedley, 1969). Recently, Leblanc & Cambon, 1997, using linear stability analysis, derived a general criterion allowing to predict the stability of plane flows subjected to Coriolis force. One can also refer to the review of Hopfinger & van Heijst, 1993 on the behavior of vortices in rotating frames.
The Coriolis force is always orthogonal to the velocity and thus bring no energy to the flow. So, its effects cannot be taken into account by the classical one-point closure models (see Cambon *et al.*, 1992 for possible improvements on single-point closure models). Reproducing the dynamical and structural effects of rotation requires the use of more complex models like two-points closure models or, of course, direct numerical simulations. The main

well known effect of the Coriolis force is the tendency to reach a quasi two-dimensional state (for strong rotation) with the emergence of vortices approximately parallel to the background rotation axis.
Regarding the homogeneous turbulence studies, several works showed the reduction of the dissipation rate due to the inhibition of the energy cascade by rotation. The role of the non-linear interactions, modified by rotation, in the transition from a 3D state to a 2D state have also been pointed out: these interactions can develop a moderate anisotropy consistent with the transition. Note that the modification of the dynamics by rotation at low Rossby number ultimately comes from the presence of inertial waves, having an anisotropic dispersion law, which are capable of changing the initial anisotropy of the turbulent flow and also can affect the non-linear dynamics. Contrary to a well admitted interpretation, the "Proudman theorem" (Proudman, 1916), only shows that the "slow manifold" (the stationary modes) is the two-dimensional manifold at small Rossby number, but it cannot explain the transition from 3D to 2D turbulence, which is a non-linear mechanism of transfer. From both experimental and DNS-LES approaches, it has been shown that two Rossby numbers are needed to describe the evolution of anisotropy (Ro^L based on the integral length scale and Ro^ω based on the Taylor micro-scale) and that the integral length scales anisotropy was a good statistical indicator to quantify rotation effects (Jacquin *et al.*, 1990, Cambon *et al.*, 1997). As an essential result, a clear transition 3D -2D was identified at $Ro^L \sim 1$ using anisotropic statistics, but without evidence on the emergence of strongly coherent vortices along the vertical direction.

In this work, we no longer study such homogeneous configuration but add rigid boundaries and a local forcing in physical space (in contrast with spectral forcing, see for example Yang, 1992 or Briggs *et al.*, 1996). The presence of both the parallel planes, orthogonal to the background rotation vector, and of the forcing, localized in physical space, may reinforce the structural effects of rotation (role of Ekman pumping, for instance).Consequently, the transition to a quasi two-dimensional state may be more abrupt, and not only driven by nonlinear dynamics.

2. The experiment of Hopfinger, Browand & Gagne (1982)

Done in the early 80's, this experiment includes the effects of rotation, forcing and confinement. The experimental facility consists of a cylindrical rotating tank in which a turbulence is generated by a grid oscillating in a horizontal plane (see figure 1). The rotation rate and the grid frequency are to be set up and lead to different regimes, with various ratio of inertial effect and rotation ones (related to different Rossby numbers). The turbulence involved is the so-called "diffusive turbulence", which have been extensively studied without rotation in the past (Bouvard & Dumas, 1967; Long, 1978; Thompson & Turner, 1975; Hopfinger & Toly, 1976; McDougall, 1979). When studying such a turbulence, we look at the spatial variation laws instead of the temporal decay laws defined by classical wind tunnel experiments with a fixed grid. For the diffusive turbulence, it is generally found that the turbulent kinetic energy and the characteristic length scale vary as follow:

$$k \propto x^{-n} \qquad L \propto x$$

where x is the distance to the source and $n \sim 2-3$. The experiments are undertaken with different frequencies and rotation rates which lead to different "grid Rossby numbers". Indeed, calling n the grid frequency and Ω the background rotation rate, HBG defined

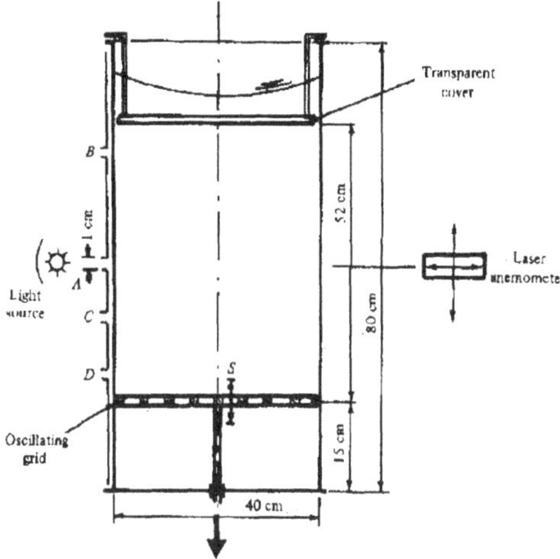

Figure 1. Cross section of the experimental apparatus of HBG.

TABLE 1. Different grid frequencies and rotation rates for the HBG's experiments

n (rad/s)	$6.63 \times 2\pi$, $3.30 \times 2\pi$
2Ω (rad/s)	0.2π, π, 2π, 20π

this number as $Ro_G = n/2\Omega$ which varies from 3.32 to 33.2 with the parameters given in table 1.

The main observations of the authors are the organization of the flow in the upper part of the tank, with emergence of organized vortices of which lifetimes are quite long (several system rotation characteristic times). The cyclonic ones seem to be more concentrated than the anticyclonic ones. When the Rossby number is decreased, anticyclonic vortices appear and occupy a ring of fluid near the outer boundary. The observations also show that the sizes of the vortices increase with the Rossby number (the faster the tank rotates, the smaller are the vortices). HBG found that the structural transition between the "near-grid region" and a rotation dominated region occurs at a distance from the grid where the local Rossby number is close to 0.2.

3. Numerical approach

Buffat and Pascal (Pascal, 1996, Pascal & Buffat, 1996) have developed a pseudo-spectral code solving the Navier-Stokes equations for a viscous, incompressible fluid between two parallel walls ("plane channel code"). The quantities $\boldsymbol{u}$, t, $\boldsymbol{r}$, P, ρ, ν, $\boldsymbol{f}$ and $\boldsymbol{\Omega}$ are, respectively, the velocity vector, time, the position vector, pressure, density, kinematic viscosity, optional body force and background rotation vector. Momentum conservation and incompressibility thus read

$$\left\{\begin{array}{l}\frac{\partial \boldsymbol{u}}{\partial t}+(\boldsymbol{\nabla}\times\boldsymbol{u}+2\boldsymbol{\Omega})\times\boldsymbol{u}=-\boldsymbol{\nabla}\Pi+\nu\boldsymbol{\nabla}^2\boldsymbol{u}\\ \boldsymbol{\nabla}.\boldsymbol{u}=0\\ \boldsymbol{u}=0 \text{ on the walls.}\end{array}\right. \quad (1)$$

Since we assume that the body force is conservative ($\boldsymbol{f} = -\boldsymbol{\nabla}\phi$), as the centrifugal force $\boldsymbol{\nabla}((\boldsymbol{\Omega}\times\boldsymbol{r})^2/2)$, the reduced pressure is given by

$$\Pi = P/\rho + \phi + \frac{1}{2}\boldsymbol{u}^2 - \Omega^2 r^2/2.$$

System (1) is solved using the pseudo-spectral method of Moser, Moin & Leonard which allow to satisfy simultaneously boundary conditions and incompressibility. The velocity field is expanded on a basis of divergence-free functions verifying the boundary conditions. The reader may refer, for further informations and details, to Pasquarelli *et al.*, 1987, and the book of Canuto *et al.*, 1988.

A few modifications have been done in order to simulate the HBG's experiment. The geometry used is shown on figure 2 ; the geometrical parameters L_1, L_2, L_3, S, and M are respectively the half of distance from wall to wall, the two box sizes in the homogeneous directions, the amplitude of the grid (actually twice the amplitude) and the grid mesh-size. The background rotation vector is so that $\boldsymbol{\Omega} = \Omega\boldsymbol{x}$ and n is 2π times the grid frequency. The forcing term we define hereafter acts in an horizontal plane localized at a distance $L_1/2$ ($x = x_g$) from the lower wall. The grid parameters appear in the forcing term as follow:

$$f_i(x_g, y, z, t) = \frac{1}{2}S\left[\delta_{i1}\cos\left(\frac{2\pi}{M}y\right)\cos\left(\frac{2\pi}{M}z\right)\sin(nt)+\beta\right] \quad (2)$$

where β is set at random at each point of the forcing plane with uniform density probability. In non-dimensional form, system (1) writes

$$\left\{\begin{array}{l}\frac{\partial \boldsymbol{u}}{\partial t}+(\boldsymbol{\nabla}\times\boldsymbol{u}+Ro^{-1}\frac{\boldsymbol{\Omega}}{\Omega})\times\boldsymbol{u}=-\boldsymbol{\nabla}\Pi+Re^{-1}\boldsymbol{\nabla}^2\boldsymbol{u}+\boldsymbol{f}\\ \boldsymbol{\nabla}.\boldsymbol{u}=0\\ \boldsymbol{u}=0 \text{ on the walls}\end{array}\right. \quad (3)$$

where $\boldsymbol{f}$ is now given by, with $S/2 = M$ (as in HBG's experiment),

$$f_i(x_g, y, z, t) = \delta_{i1}\cos(2\pi y)\cos(2\pi z)\sin(t) + \beta^*. \quad (4)$$

Its non-conservative deterministic part acts on the vertical momentum equation, on which is superposed the non-dimensionalized stochastic signal β^*. The two horizontal momentum equations are only forced by the white noise. The non dimensional numbers (namely the Rossby number Ro and the Reynolds number Re) are built on the background rotation rate, the grid mesh size, the viscosity and the grid frequency. Thus, they reads

$$Re = \frac{\nu}{nS^2} \quad \text{and} \quad Ro = \frac{n}{2\Omega} \;. \quad (5)$$

3.1. A FEW DEFINITIONS OF STATISTICAL QUANTITIES

Let us introduce statistical quantities needed for the post-processing and analysis of the D.N.S. results.

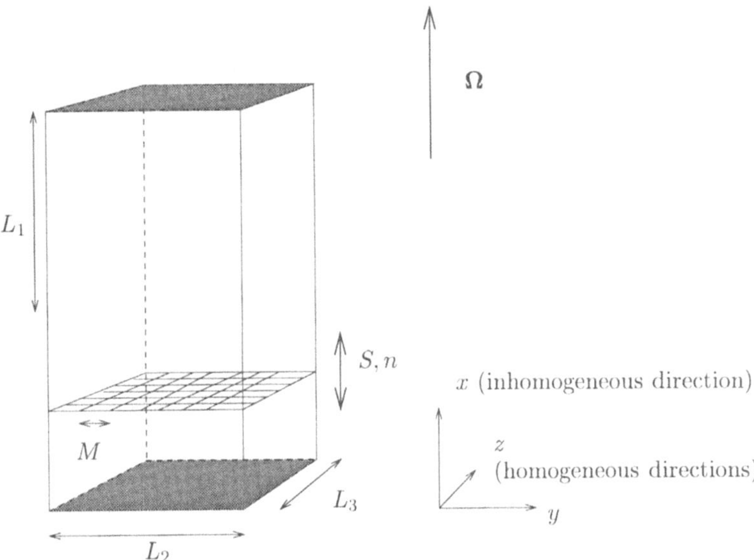

Figure 2. Sketch of the computational domain

Spatial averages computed in homogeneous planes $(\boldsymbol{x}_2, \boldsymbol{x}_3)$ are written $< \cdot >$. Using Reynolds decomposition, any variable U is splitted into a mean part $< U >$ and fluctuating part u,

$$U =< U > + u.$$

We will look at the root mean square quantities given by

$$u_{\text{rms}} = \left\langle \sqrt{(U- < U >)^2} \right\rangle .$$

Important informations are also brought by the integral length scales of turbulence, which characterize the energetic large structures of the flow. The nine integral length scales are given by

$$L_{ii,j}(\boldsymbol{x}) = \frac{1}{2} \int_{-\infty}^{\infty} \frac{\langle u_i(\boldsymbol{x}) u_i(\boldsymbol{x} + r\boldsymbol{n}_j) \rangle}{\langle u_i^2(\boldsymbol{x}) \rangle} dr \quad (i \text{ is not summed}). \tag{6}$$

with $\boldsymbol{n}_j$ the unit vector along the direction axis x_j. They are averaged on each homogeneous plane (parallel to the walls). Let us recall that only in homogeneous isotropic turbulence, a single scalar value is needed, in accordance with

$$L_{ii,j} = L_{\text{f}} \quad (i = j)$$

$$L_{ii,j} = L_{\text{g}} = \frac{1}{2} L_{\text{f}} \quad (i \neq j)$$

4. Identification and visualization of vortices

We introduce here a method first used by Michard *et al.*, 1996 for the post processing of P.I.V. experimental data. The Normalized Angular Momentum (N.A.M.) is an original tool, based on the geometrical properties of the flow, which modulus is maximum in the center of a vortex while it tends to zero far from it. To take into account the spatial consistency of the vortices and avoid the computation of purely local quantities like derivatives, we calculate, at each point P, the vector function $\boldsymbol{f}$ given by :

$$\boldsymbol{f}(\boldsymbol{x}_p) = \frac{1}{V}\int_{\boldsymbol{x}\in V} \frac{(\boldsymbol{x}-\boldsymbol{x}_p)\times \boldsymbol{u}(\boldsymbol{x})}{|\boldsymbol{x}-\boldsymbol{x}_p|\,|\boldsymbol{u}(\boldsymbol{x})|} d^3\boldsymbol{x} \tag{7}$$

where V is a volume surrounding P (see figure 3). The modulus of this vector varies between 0 and 1. We can weight it by the sign of the local vorticity in order to distinguish cyclones and anticyclones . Actually, P is computed with a restricted number of points M_i round the point P to reduce the computation time.
The ability of the N.A.M. function to identify a vortex core structure has been tested on

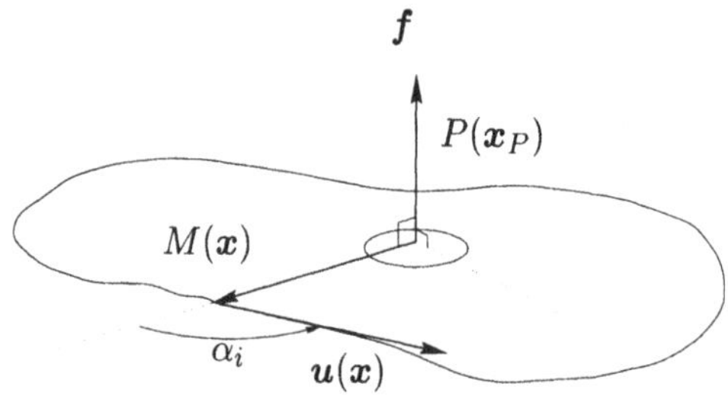

Figure 3. Normalized Angular Momentum : definition sketch.

several 2D and 3D laminar flows. We verified that it was identically zero in the case of the pure shear and that it was very efficient in the case of solid-body rotation or Taylor-Green cells. As a typical 3D test, the structure of the spherical Hill's vortex has been recognized by the criterion whereas vorticity distribution is not suitable. For further details about these tests, the reader may refer to Michard *et al.*, 1997. We expect the extrema of $|\boldsymbol{f}|$ to correlate with a vortex center or iso-surfaces of $|\boldsymbol{f}|$ to characterize a 3D vortical structures.

5. Simulation of the HBG's experiment

5.1. PARAMETERS AND PROCEDURE

Three simulations are presented. Referred to as Run I, II and III, they respectively correspond to a non-rotating case, a case with moderate rotation and a strongly rotating case (see table 2). They all started from the rest. In the rotating cases, that means that the grid starts in a uniformly rotating fluid with respect to the fixed frame of reference. After about ten grid periods, the total kinetic energy reaches a constant mean value but the computation must be continued for the building of various statistical quantities.

5.2. MAIN STATISTICAL RESULTS

The main observations concern the spatial evolution (evolution with respect to the distance to the grid) of the turbulence characteristics. Namely, we look at the spatial decay law of the turbulent kinetic energy and the increase of integral length scales. We observed that kinetic energy decays exponentially because of the blocking of the integral

TABLE 2. Parameters for the three simulations

Run	I	II	III
Grid Reynolds number	1110	10000	1100
Grid Rossby numbers	∞	50	3.32
Mesh M	1	1	1
Number of meshes	8×8	8×8	8×8
Box size	$13 \times 8 \times 8$	$13 \times 8 \times 8$	$13 \times 8 \times 8$
White noise intensity	0.25	0.25	0.25
Resolution	96^3	$96 \times 128 \times 128$	96^3

length scales due to presence of the upper wall. This phenomenon was first pointed out experimentally by Risso, 1994. Furthermore, we observe the same behavior with rotation. Anisotropic statistical quantities are consistent with the presence of organized vortices and this will be confirmed in the next section using the vortex identification method. For example, by comparison with the non-rotating case (Run I), we observe the increase of the vertical integral length scales (namely $L_{22,1}$ and $L_{33,1}$) which reflect an increasing spatial correlation in this direction (parallel to the background rotation axis). Note that the sudden preferential increase of these scales under rotation at $Ro^L \sim 1$ was the most typical sign of 2D-3D transition in homogeneous turbulence (Cambon *et al.*, 1997; Jacquin *et al.*, 1990). We also remarked that the horizontal root mean square velocity components increase in the rotating case, and especially in a region close to the upper wall. Furthermore, visualizations of the velocity field in planes orthogonal to the rotation axis seem to confirm the existence of these vortices.

Only in Run II a transition Rossby number close to the experimental one of Hopfinger, Browand & Gagne have been found ($Ro \approx 0.2 - 0.3$ between $X/M \approx 4$ and $X/M \approx 7$ on figure 4). We observed the emergence of an isolated cyclonic vortex at $X/M \approx 4$. This simulation has been undertaken at higher Reynolds and Rossby numbers and so lead to moderate local Rossby numbers. Figure 4 shows that the rotation was so strong in Run III so that the Rossby number is already less than $0.2 - 0.3$ close to the grid, where the structural transition occurs.

5.3. VISUALIZATION OF THE VORTICES

The normalized angular momentum is now used to identify the vortices appearing in the strongly rotating case. Note that, in our simulation, we know a priori the preferential orientation of the vortices axes (parallel to the background rotation) and, thus, the first component of $\boldsymbol{f}$ is relevant. This is shown on figure 5, where are plotted the three-dimensional iso-surfaces of the first component of $\boldsymbol{f}$ and vorticity's first component contours in a plane parallel and close to the upper wall. The vortices centers are accurately localized by the three dimensional iso-surfaces.

Let us now look at the modulus of the angular momentum (thus we now take into account the three components). Note that this scalar is able to identify and follow a vortex whatever his axis direction. The figure 6 shows a cyclonic and an anticyclonic vortex. Two-dimensional iso-surfaces of the vertical component of vorticity are also plotted on three planes. We can see that, even though the correlation is quite good between the "NAM"

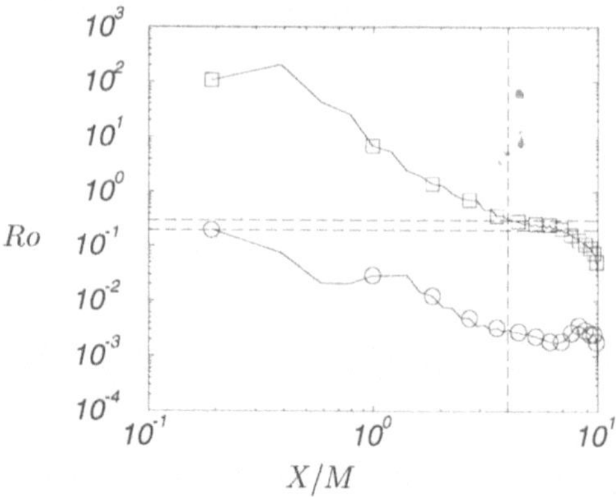

Figure 4. Spatial evolution of the Rossby number for Run *II* (□) and Run *III* (○). The grid is located at $X/M = 0$ while the upper wall is at $X/M = 10$.

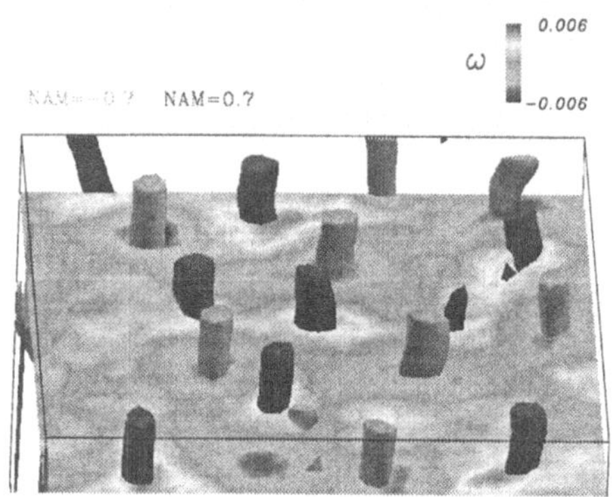

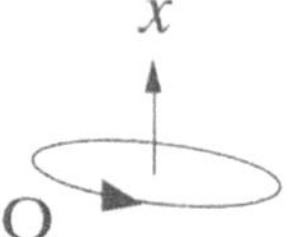

Figure 5. Run *III* ($Ro = 3.32$).

iso-surface and the regions of extrema of vorticity, the latter is "patchy" and seems to be a less accurate vortex identificator. The helicoidal motion we see on theses figures is very interesting. Indeed, this is due to the well known Ekman pumping effect (and perhaps could reflect the helical modes, that are the eigenmodes of inertial waves Cambon *et al.*, 1997). The visualization has been done in the upper half of the tank. In the cyclonic vortex, which is characterized by a low pressure region, the fluid is attracted and must dive because of both the presence of the wall and the divergence-free condition (top Ekman layer). Evidently, the fluid rises in the anticyclonic vortex. Note that the opposite phenomenon occurs close to the lower wall (bottom Ekman layer). These visualizations confirm the vertical extend of the vortices over the upper half of the tank.

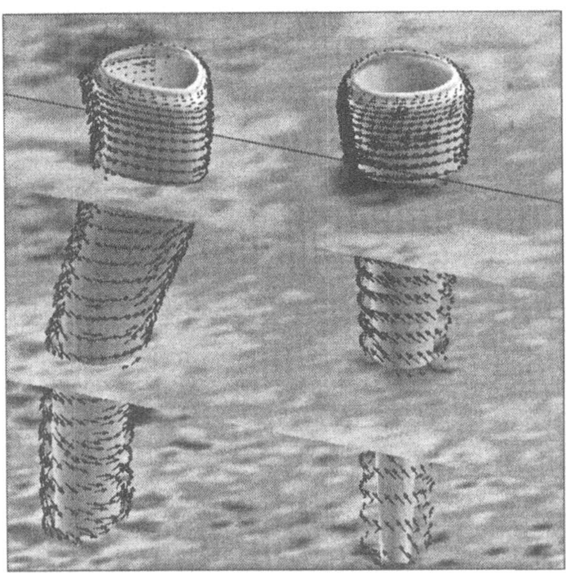

Figure 6. View of two vortices. The three-dimensional iso-surfaces of the N.A.M. modulus (0.45) cross three parallel planes on which 2D contours of the vertical vorticity component contours are plotted. The vortex on the left is cyclonic. Note the Ekman pumping effects.

6. Conclusion and future plans

The three simulations have brought some interesting results. The non-rotating case showed the particular behavior of the integral length scales and thus the exponential decay of kinetic energy instead of the classical power decay law. This phenomenon is due to the presence of the walls. The rotating cases, close to the experimental configuration of HBG, exhibits vortices with preferential orientation parallel to the background rotation axis. In the strongly rotating case, these vortices, which extend over the upper part of the tank, are cyclonic or anticyclonic with quite the same intensity and number for both kinds. Vertical motions are shown to be very well correlated with the presence of cyclonic or anticyclonic vortices, pointing out the well known "Ekman pumping" effect. The simulation at $Ro = 50$, performed to try to reproduce a situation where a thick enough "near grid region" and a "rotation dominated region" coexist, gave us a value of the transition Rossby number close to the experimental one of Hopfinger, Browand & Gagne. It also led to a single isolated cyclonic vortex. This confirmed the expected variation law of the number of vortices with the ratio Re/Ro (that is the Ekman number).
Direct numerical simulations of free decay of turbulence with and without rotation have been undertaken, between the two parallel walls, and will be compared to homogeneous results (periodic in the three directions) in order to evaluate the effects of confinement on the flow and the structural transition.
Even though it is not a physical quantity, the Normalized Angular Momentum is shown to be a very useful and accurate tool for localizing vortices, whatever their orientations. Indeed, in such simulations, where strongly inhomogeneous turbulence is studied, vorticity

depends on the smoothness of the numerical data and classical criteria like Weiss criterion (only checked for 2D flows) is not appropriate.

N. Grosjean is gratefully acknowledged for data-processing.

References

Bouvard, M. & Dumas, H. 1967. Application de la méthode du fil chaud à la mesure de la turbulence dans l'eau. *Houille Blanche*, 22:723–733.

Briggs, D.A., Ferziger, J.H., Koseff, J.R., & Monismith, S.G. 1996. Entrainment in a shear-free turbulent mixing layer. *J. Fluid Mech.*, 310:215–242.

Cambon, C., Jacquin, L., & Lubrano, J.L. 1992. Towards a new reynolds stress model for rotating turbulent flows. *Phys. Fluids*, 4:812–824.

Cambon, C., Mansour, N. N., & Godeferd, F. S. 1997. Energy transfer in rotating turbulence. *J. Fluid Mech.*, 337:303–332.

Canuto, C., Hussaini, M.Y., Quarteroni, A., & Zang, T.A. 1988. *Spectral Methods in Fluid Dynamics.* Springer-Verlag.

Hopfinger, E.J. & Toly, J.-A. 1976. Spatially decaying turbulence and its relation to mixing across density interfaces. *J. Fluid Mech.*, 78:155–175.

Hopfinger, E. J., Browand, F. K., & Gagne, Y. 1982. Turbulence and waves in a rotating tank. *J. Fluid Mech.*, 125:505–534.

Hopfinger, E. J. & van Heijst, G. J. F. 1993. Vortices in rotating fluids. *Ann. Rev. Fluid Mech.*, 25:241.

Jacquin, L., Leuchter, O., & Mathieu, C. Cambon J. 1990. Homogeneous turbulence in the presence of rotation. *J. Fluid Mech.*, 220:1–52.

Kloosterziel, R. C. 1990. *Barotropic vortices in a rotating fluid.* PhD thesis, University of Utrecht, The Netherlands.

Kristoffersen, R. & Andersson, H.I. 1993. Direct simulations of low-reynolds-number turbulent flow in a rotating channel. *J. Fluid Mech.*, 256:163–197.

Leblanc, S. & Cambon, C. 1997. On the three-dimensional instabilities of plane flows subjected to coriolis force. *To appear in Phys. Fluids.*

Long, R.R. 1978. Theory of turbulence in a inhomogeneous fluid induced by an oscillating grid. *Phys. Fluids*, 21(10):1887–1888.

McDougall, T.J. 1979. Measurements of turbulence in a zero-mean-shear mixed layer. *J. Fluid Mech.*, 94:409–431.

Michard, M., Graftieaux, L., Lollini, L., & Grosjean, N. 1997. Identification of vortical structures by a non-local criterion: Applications to p.i.v. measurements ans d.n.s results of turbulent rotating flows. *To appear in proceedings of the Eleventh Symposium on Turbulent Shear Flows, Grenoble.*

Michard, M., Simoens, S., Grosjean, N., & Safsaf, D. 1996. Caractérisation du mouvement de précession d'un tourbillon à l'aide de la vélocimétrie par images de particules. 5ème *Congrès Francophone de Vélocimétrie Laser, Rouen.*

Pascal, H. 1996. *Etude numérique d'une turbulence compressée et/ou cisaillée entre deux plans parallèles par simulation des grandes échelles (LES) : évaluation de modèles statistiques de turbulence.* Ecole Centrale de Lyon. Rapport de thèse.

Pascal, H. & Buffat, M. 1996. L.e.s. of turbulent flows conpressed and/or sheared between two walls on parallel computer using a "divergence-free spectral galerkin method. *Computational Fluid Dynamics*, Wiley & Sons Ltd:884–891.

Pasquarelli, F., Quarteroni, A., & Sacchi-Landriani, G. 1987. Spectral approximations of the stokes problem by divergence-free functions. *Journal of Scientific Computing*, 2:195–226.

Pedley, T.J. 1969. On the stability of viscous flow in a rapidly rotating pipe. *J. Fluid Mech.*, 128:97–115.

Proudman, J. 1916. On the motion of solids in a liquid possessing vorticity. *Proc. R. Soc. London*, A 92:408.

Risso, F. 1994. *Déformation et rupture d'une bulle dans une turbulence diffusive.* Thèse, I.N.P. Toulouse, Institut de mécanique des fluides de Toulouse.

Thompson, S.M. & Turner, J.S. 1975. Mixing across an interface due to turbulence generated by an oscillating grid. *J. Fluid Mech.*, 67:349–368.

Tillmark, N. & Alfredsson, P.H. 1996. Experiments on rotating plane couette flow. *Advances in Turbulence VI, Kluwer.*

Yang, G. 1992. *DNS of boundary forced turbulent flow in a non-rotating and a rotating system.* Cornell University. Ph. D. Thesis dissertation.

ON THE SPACE-TIME STRUCTURE OF AXISYMMETRIC ROTATING FLOWS

E.A. CHRISTENSEN [1], N. AUBRY [2] AND J.N. SØRENSEN [3]
[1] *Courant Institute of Mathematical Sciences, NYU*
251 Mercer Street, New York, NY 10012, USA

[2] *Department of Mechanical Engineering,*
Department of Mathematical Sciences and
Center for Applied Mathematics and Statistics, NJIT
University Heights, Newark, NJ 07102, USA

AND

[3] *Fluid Mechanics, Department of Energy Engineering, DTU*
Building 404, 2800 Lyngby, Denmark

Abstract. A fluid flow enclosed in a cylindrical container where fluid motion is created by rotation of the lid is studied. Through the no-slip boundary condition at the rotating lid, the fluid flow is centrifuged toward the cylinder wall, forming a strong vortex core along the axis of the container. Direct numerical simulations and spatio-temporal analysis of the axisymmetric flow are presented here at various Reynolds numbers in the early transition scenario. Slightly above the instability onset, the central vortex core forms a breakdown bubble which, at higher Reynolds numbers, undergoes a vertical beating motion. This is accompanied by the continued formation of axisymmetric spikes on the edge of the breakdown bubble which travel along the central vortex core toward the rotating end-wall. The data analysis is performed by using the proper orthogonal decomposition.

1. Introduction

Recently, there has been considerable interest in the dynamical behavior of very complex mechanical systems. In certain cases, one has realized that infinite dimensional dynamical systems can be reduced to systems of finite (sometimes low) dimension. There has been various approaches to such reductions, such as the theory of center manifolds (Carr (1981), Guckenheimer and Holmes (1983)), inertial manifolds (Temam, 1988) and techniques based on the Proper Orthogonal Decomposition (POD) (Aubry *et al.* (1988), Aubry *et al.* (1991), Aubry (1991), Berkooz *et al.* (1993)).

The issue is particularly challenging in fluid mechanics where it is known that, at high Reynolds numbers, flows systematically reach a very complex state, commonly referred to as turbulence. Even in that state, there is hope that the basic properties of the flow can be described by a low-dimensional dynamical system which either retains a few modes as it was successfully shown for the wall region of a turbulent boundary layer (Aubry *et al.* (1988), Sanghi and Aubry (1993)), or consists of modes which can all be deduced one from the other by mapping as it is predicted in the case of fully developed turbulence (Aubry *et al.* (1992A), Carbone and Aubry (1996)). If turbulence itself is far from being understood, transition to turbulence has remained a long standing issue as well. In this paper, we concentrate on understanding how complexity develops through the transition process for a particular flow.

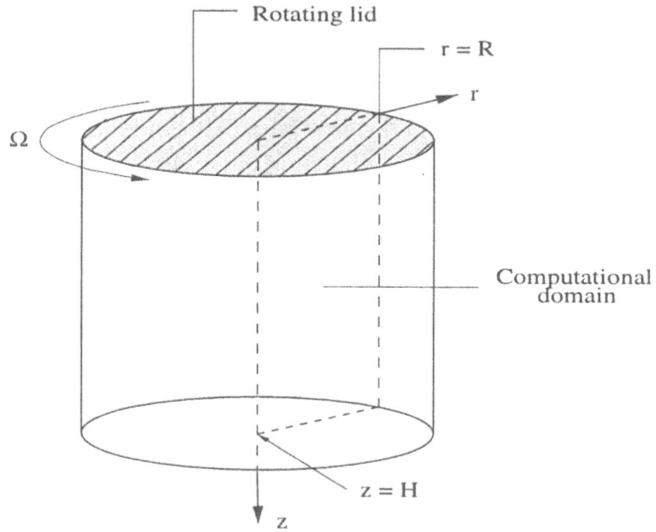

Figure 1. Cylinder with rotating lid with aspect ratio $H/R = 2$

The flow of interest here is the flow enclosed in cylindrical container of height H and radius R whose lid rotates at angular velocity Ω. The states of the system are uniquely defined by the Reynolds number, $Re = \Omega R^2/\nu$, the aspect ratio $\lambda = H/R$, ν being the kinematic viscosity of the fluid, and the initial condition (see figure 1).

Escudier (1984) showed experimentally that, in the axisymmetric, steady regime, the flow consists of a center vortex exhibiting up to three distinct breakdown bubbles while the numerical study of Sørensen (1992) showed that breaking of axisymmetry occurs in the Reynolds number range $Re = 3000 - 3500$ for an aspect ratio $H/R = 2$. Soong *et al.* (1993) experimentally observed asymmetric, unsteady behavior for $Re > 3000$ at the aspect ratio $H/R = 2.5$. In this paper, the aspect ratio is fixed to $H/R = 2$.

2. Numerical simulations

Since the flow is axisymmetric up to a certain Reynolds number (whose value depends on the aspect ratio), say $Re_S(H/R)$, we have first numerically integrated the axisymmetric equations of motion which are valid for $Re < Re_S(H/R)$. For this, we have considered the vorticity/stream function formulation of the Navier-Stokes equations in cylindrical coordinates (r, θ, z). In the following, (u_r, u_θ, u_z) denote the radial, azimuthal and axial velocity components, respectively, $\Gamma = rv$ refers to the circulation function, ψ is the stream function, and ω denotes the azimuthal vorticity function. The incompressible flow problem is closed by the no-slip boundary conditions (Sørensen and Ta Phuoc Loc, 1989). Direct numerical simulations of the flow at various Reynolds numbers revealed a complex axisymmetric transition scenario involving several types of bifurcations and hysteresis with up to three simultaneous branches (Sørensen and Christensen, 1995). Brown and Lopez (1990) also carried out detailed numerical simulations of the problem, in the bubble breakdown regime, as well as in the unsteady domain. Gelfgat *et al.* (1996) found that, for aspect ratios $H/R > 3$, the primary instability leading to unsteady flow behavior occurred at critical Reynolds numbers higher than those observed experimentally by Escudier (1984).

In order to investigate the three-dimensional (3-D) flows occurring at high Reynolds numbers, a three-dimensional Navier-Stokes solver is needed. For this, we implemented a discretized version of the Cauchy-Riemann vorticity-velocity ($\omega - V$) formulation of the equations which, together with a least mean squares formulation, made the system well-defined (Sørensen *et al.*, 1996). First results were obtained by running the 3-D code at the Reynolds numbers $Re = 2800$ and $Re = 3400$ with the initial condition consisting of the axisymmetric solution at $Re = 2400$ to which a small 3-D perturbation has been added. While the solution is axisymmetric at $Re = 2800$, the code clearly converged to a non-axisymmetric solution at $Re = 3400$. For $H/R < 3$ the early transition thus goes from a two-dimensional (2-D) steady flow through a 2-D unsteady flow and to a 3-D unsteady flow. In this case, the critical Reynolds number is smaller than the Reynolds number at which the flow becomes 3-D ($Re_{Cr} < Re_S$). It is interesting to note that experimental observations have shown that this is not the case at higher aspect ratios such that $H/R > 3$. In this domain of the bifurcation diagram, the 2-D steady flow becomes 3-D and unsteady simultaneously, that is the critical Reynolds number is equal to the Reynolds number at which the flow becomes 3-D ($Re_{Cr} = Re_S$).

3. Decomposition in Space and Time

We now investigate the space-time structure of a function, say $\zeta(r, z, t)$. Note that the values of ζ are considered *simultaneously* at all points (r, z) in space. A deterministic version of the Proper Orthogonal Decomposition (POD) can be described as follows (see Aubry *et al.* (1991), Aubry (1991)). We consider the operator $\mathcal{U}$ such that

$$(\mathcal{U}\varphi)(t) = \int \zeta(r, z, t)\varphi(r, z)drdz, \tag{1}$$

the adjoint operator being defined by

$$(\mathcal{U}^*\phi)(r, z) = \int \zeta(r, z, t)\phi(t)dt. \tag{2}$$

The spectral decomposition of $\mathcal{U}$, equivalent to the spectral decomposition of the matrix operator $\begin{pmatrix} 0 & \mathcal{U}^* \\ \mathcal{U} & 0 \end{pmatrix}$ leads to an orthonormal basis of spatial functions φ_M and an orthonormal basis of temporal functions ϕ_M such that

$$\begin{pmatrix} 0 & \mathcal{U}^* \\ \mathcal{U} & 0 \end{pmatrix} \begin{pmatrix} \varphi_M \\ \phi_M \end{pmatrix} = A_M \begin{pmatrix} \varphi_M \\ \phi_M \end{pmatrix}, \tag{3}$$

where $\langle \varphi_M, \varphi_P \rangle_{rz} = \langle \phi_M, \phi_P \rangle_t = \delta_{M,P}$ where the first brackets denote the integral with respect to r, z and the second brackets refer to the integral with respect to t. It is easy to see that the φ_M's can be considered as POD modes and the ϕ_M's as the normalized time dependent coefficients, or vice versa, the ϕ_M's can be viewed as POD modes and the φ_M's as the normalized space dependent coefficients. For this reason, the φ_M's are referred to as *topos* and the ϕ_M's are called *chronos* (Aubry *et al.*, 1991). The non-zero eigenvalues of the matrix operator can be ordered in a decreasing order, so that $A_1 \geq A_2 \geq \cdots \geq A_{M^*} > 0$. The spectral decomposition of $\mathcal{U}$ is equivalent to the biorthogonal decomposition of $\zeta(r, z, t)$ in terms of the eigenvalues, the topos and the chronos:

$$\zeta(r, z, t) = \sum_{M=1}^{M^*} A_M \varphi_M(r, z)\phi_M(t). \tag{4}$$

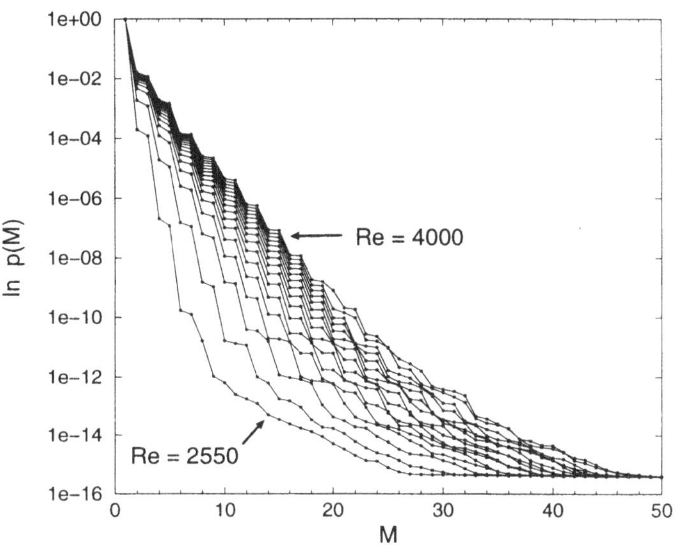

Figure 2. Eigenvalue spectra for Reynolds number value $Re = 2550, 2600, 2700, \ldots, 4000$.

While the eigenvalues of $\mathcal{U}$ are A_M, it is common to plot the energy spectrum A_M^2, which we can normalize by the total energy. The normalized eigenvalues can then be introduced:

$$p(M) = A_M^2 / \sum_{m=1}^{M^*} A_m^2. \tag{5}$$

The POD corresponds to the spectral decomposition of the operator UU^*, or U^*U. The spectral decomposition of the operator U allows us to extract the POD eigenfunctions and their time (or space) dependent coefficients at once.

The circulation-vorticity vector $(\Gamma(r,z,t), \omega(r,z,t))$ obtained from our numerical simulations of the rotating flow was decomposed and the energy spectrum is displayed in figure 2 at various Reynolds number values. We see that the eigenvalue spectrum is organized in pairs of quasi-degenerated eigenvalues at all Reynolds numbers considered here. We know that eigenvalue degeneracy is equivalent to the presence of a space-time symmetry in the flow (Aubry *et al.*, 1992B). It is clear that the number of degenerated eigenvalues increases with Reynolds number. For instance, we observe three pairs at $Re = 2550$, five pairs at $Re = 2600$ and nine pairs at $Re = 4000$.

As Reynolds number increases, more eigenvalues join the symmetric part of the spectrum and the spectrum broadens.

4. Unsteady Flow Structures

Like the circulation/vorticity vector, the extended vector $\Upsilon = (\Gamma, \omega, u_r, u_z, \psi)$ can be also decomposed according to (4), so that we can write

$$\Upsilon(r, z, t) = \sum_n A_n \Phi_n(r, z) \phi_n(t) \tag{6}$$

where the ϕ_n's are the same as in the decomposition of the circulation/vorticity vector. The spatial eigenfunctions of the extended vector can then be deduced from the projection,

$$\Phi_n(r, z) = \int \Upsilon(r, z, t) \phi_n(t) dt / A_n. \tag{7}$$

Visualization of the flow structures can then be performed by tracking fluid particles and integrating the system of equations

$$\frac{dr_p}{dt} = u_r, \quad \frac{dz_p}{dt} = u_z,$$

in time, where the velocity field is reconstructed with various truncations in Expansion (6). Figure 3 represents the flow visualization by particle tracking as reconstructed by one mode only, or the summation of modes 1 and 2, modes 2 and 3 and modes 1, 2, 3. The visualization obtained by superposing modes 1, 2, 3 coincides with the one obtained by direct numerical simulation. While the addition of modes 1 and 2, or modes 1 and 3, reproduces the existence of spikes formation and the right beating frequency of the bubble, the traveling character of the flow (combination of a right and a left traveling wave) is reproduced only if one retains the full quasi-generated pair, modes 2 and 3, in the model. This is consistent with the existence of a symmetry (or quasi-symmetry) in the flow, which represents the relative (traveling) motion of the spikes with respect to the vortex core. The traveling is, however, more complex than a simple translation motion in space and time, and may consist of a modulated traveling wave (Aubry and Lima, 1995), rather that a simple traveling wave.

Acknowledgments

One of the authors (EAC) acknowledges financial support from the Carlsberg Foundation (Grant 960204/20-1260), the Danish Natural Science Research Council (Grant 9400288) and a number of private Danish foundations. The computer simulations were performed on a Cray C90 vector processor Pittsburgh Supercomputing Center (Grant CTS950004P).

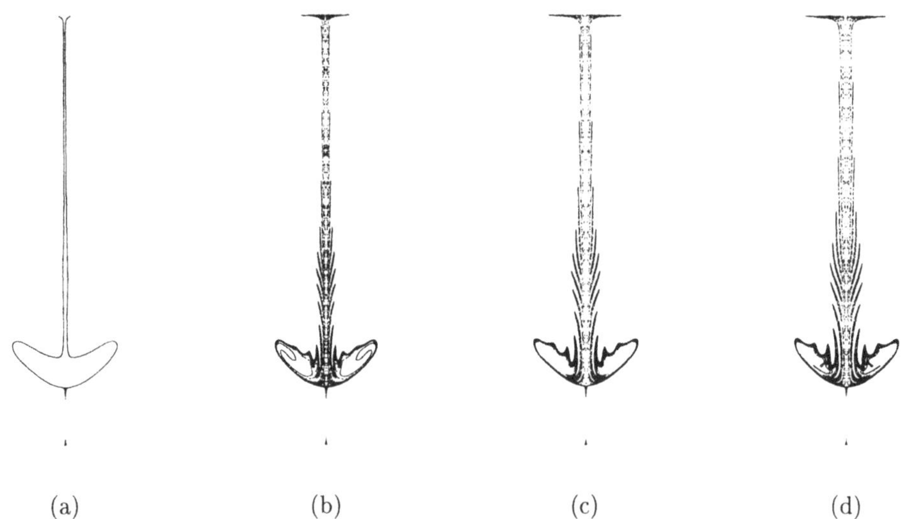

Figure 3. Particle tracking in the flow at $Re = 2550$ from various truncations including (a) mode 1, (b) modes 1 and 2, (c) modes 1 and 3, and (d) modes 1, 2 and 3 in eqn (6).

References

Aubry, N. (1991). On the hidden beauty of the proper orthogonal decomposition, *Theoretical and Computational Fluid Dynamics* **Vol. 2**, pp. 339.

Aubry, N., Guyonnet, R. and Lima, R. (1991). Spatio-temporal analysis of complex signals: theory and applications, *Journal of Statistical Physics* **Vol. 64**, pp. 683.

Aubry, N., Guyonnet, R. and Lima, R. (1992A). Turbulence spectra, *Journal of Statistical Physics* **Vol. 67**, pp. 203.

Aubry, N., Guyonnet, R. and Lima, R. (1992B). Spatio-temporal symmetries and bifurcations via biorthogonal decompositions, *Journal of Nonlinear Science* **Vol. 2**, pp. 183.

Aubry, N., Holmes, P., Lumley, J.L. and Stone, E. (1988). The dynamics of coherent structures in the wall region of a turbulent boundary layer, *Journal of Fluid Mechanics* **Vol. 192**, pp. 115.

Aubry, N. and Lima, R. (1995). The dynamics of spatio-temporal modulations, *Chaos* **Vol. 5**, pp. 578.

Berkooz, G., Holmes, P. and Lumley, J.L. (1993). The proper orthogonal decomposition in the analysis of turbulent flows, *Annual Review of Fluid Mechanics* **Vol. 25**, pp. 539.

Brown, G.L. and Lopez, J.M. (1990). Axisymmetric vortex breakdown. Part 2, *Journal of Fluid Mechanics* **Vol. 221**, pp. 553.

Carbone, F. and Aubry, N. (1996). Hierarchical order in wall bounded turbulence, *Physics of Fluids* **Vol. 8**, pp. 1061.

Carr, J. (1981). Applications of Center Manifold Theory. *Springer-Verlag, New York.*

Christensen, E.A., Brøns, M. and Sørensen, J.N. (1997). POD-based methods with applications to non-turbulent rotating flow in a closed cylinder, *Submitted to SIAM Journal of Scientific Computing.*

Escudier, M.P. (1984). Observations of the flow produced in a cylindrical container by a rotating endwall. *Experiments in Fluids* **Vol. 2**, pp. 189.

Gelfgat, A.Y., Bar-Yoseph, P.Z., and Solan, A. (1996). Stability of confined swirling flow with and without vortex breakdown. *Journal of Fluid Mechanics* **Vol. 311**, pp. 1.

Guckenheimer, J. and Holmes, P. (1983). Nonlinear Oscillations, Dynamical Systems, and Bifurcation of Vector Fields, AMS, **Vol. 42**, *Springer-Verlag, New York.*

Sanghi, S. and Aubry, N. (1993). Low dimensional models for the structure and dynamics in near wall turbulence, *Journal of Fluid Mechanics* **Vol. 247**, pp. 455.

Soong, C.Y., Young, D.L. and Zeng, R.B. (1993). Observations of periodic flow in a closed cylindrical container with rotating bottom. In Yaglom, A. and Tatarski, V. (Eds.), *10th National Conference on Mechanical Engineering*, Hsinchu, Taiwan, CSMEM.

Sørensen, J.N. (1992) Transition and instabilities in swirling flow. In *Spatial-temporal Structure and chaos in Heat and Mass Transfer Processes.*

Sørensen, J.N. and Christensen, E.A. (1995). Direct numerical simulation of rotating fluid flow in a closed cylinder. *Physics of Fluids* **Vol. 7**(4), pp. 764.

Sørensen, J.N., Hansen, M.O.L. and Christensen, E.A. (1996). Numerical investigation of symmetry breakdown in a cylindrical lid driven cavity. In *ECCOMAS96*, pp. 439, John Wiley & Sons, Ltd.

Sørensen, J.N. and Ta Phuoc Loc (1989). High-order axisymmetric Navier-Stokes code description and evaluation of boundary conditions, *International Journal for Numerical Methods in Fluids* **Vol. 9**, pp. 1517.

Temam, R. (1988). Infinite-Dimensional Dynamical Systems in Mechanics and Physics, AMS, **Vol. 68**, *Springer-Verlag, New York.*

OBSERVATIONS ON THE EARLY TRANSITION PROCESS IN A CLOSED CYLINDRICAL CONTAINER WITH ROTATING BOTTOM

A. SPOHN
Laboratoire d'Etudes Aerodynamiques-ENSMA UMR 6609
Site du Futuroscope, 86960 Futuroscope Cedex, France

1. Introduction

In recent years the flow generated by a rotating end wall inside a closed cylindrical container has become a new prototype flow to study the fundamental transition mechanisms in a confined dynamical flow system. Similar to the Taylor-Couette flow a large number of different steady and unsteady flow patterns can be observed. Experiments of Escudier (1984) and numerical simulations of Lugt & Abboud (1987), Lopez (1990) among others, illustrated the existence of multicellular flows, including flows with steady central recirculation bubbles very similar to vortex breakdown. More recently Lopez & Perry (1992) and Sørensen & Christensen (1995) examined in more detail how the bubbles become unsteady. According to both numerical studies periodic axial oscillations of the central recirculation bubbles mark the onset of the unsteady flow regime. However, the nature of this transition is not yet clear. Tsitverblit (1993) and Christensen *et al.* (1993) found unstable stationary solutions which coexist with stable unsteady solutions, indicating the existence of a supercritical Hopf bifurcation. In contrast to this Lopez & Perry (1992) found no such unstable steady solutions and, therefore, excluded the transition mechanism via a supercritical Hopf bifurcation. Experimental observations on the transition to unsteady flow remain scarce and are mainly based on the description of the bubble behavior with varying flow parameters. In this paper we present some new experimental observations which illustrate the transition process for different flow configurations with and without bubbles. By means of flow visualizations obtained with the electrolytic precipitation technique we identified three different transition scenario, corresponding to flow parameters with either no, one or two breakdown bubbles. Although in the case of one recirculation bubble the onset of the transition seems to be well predicted by previous experimental and numerical studies, we found that the observed unsteadiness has only transient character. After long observation times the flow becomes steady and non-axisymmetric. All of these observations suggest that the transition process inside the container flow is more complicated than previously suspected. Further investigations are needed to understand the influence of asymmetric perturbations on the flow behavior. In particular our experiments demonstrate the difficulty to evaluate the time scales necessary to achieve the steady flow state.

2. Experimental arrangement

A schematic of the experimental setup is presented in figure 1. The container with the cylindrical test section of 91.3 mm inner diameter was placed inside a rectangular water box of dimensions 460x460x600 mm. The flow depends on two parameters, the container aspect ratio H/R, where H and R are the height and the inner radius of the cylinder, and the Reynolds number $Re = \Omega R^2/\nu$, where Ω is the angular velocity of the rotating bottom and ν the kinematic viscosity of the fluid. The aspect ratio could be varied by changing the height of the cover above the rotating disc. As working fluid we used in most experiments tape water. The temperature of the working fluid was controlled by recirculating permanently the water inside the outer box through a constant temperature bath. In this way temperature fluctuations of the working fluid were reduced to values less than ± 0.2 °C. The rotating bottom was driven by a electric motor drive with the rotation speed Ω controlled continuously by a microprocessor. During the experiments the measured variations of Ω were less than 0.2 %. The Reynolds number Re could be varied between $1000 \leq Re \leq 40.000$ and the aspect ratio H/R between $0.5 \leq H/R \leq 4$. However, in this paper we focus on results obtained for Re in the interval [1000, 6000] and H/R within [1.75, 2.5]. The relation $H/R \cdot Gr/Re^2 = g\alpha\Delta T H^3/\nu^2 \cdot 1/Re^2$, which compares the importance of the thermal convection produced by temperature perturbations to the forced convection driven by the rotating disc, remained smaller than 0.14

To study the flow we used flow visualizations obtained with the electrolytic precipitation technique, described by Honji *et al.* (1980). In contrast to dye techniques this method allowed to release the tracer in a nearly perfectly axisymmetric way on the container wall. The singular points or lines of the flow field which determine the structure of the flow field were captured directly by the released streak surface. In order to cover all fixed parts of the container walls we tagged the flow at two different locations (see figure 1). A flat solder wire of 0.08 mm thickness was fixed on the cylin-

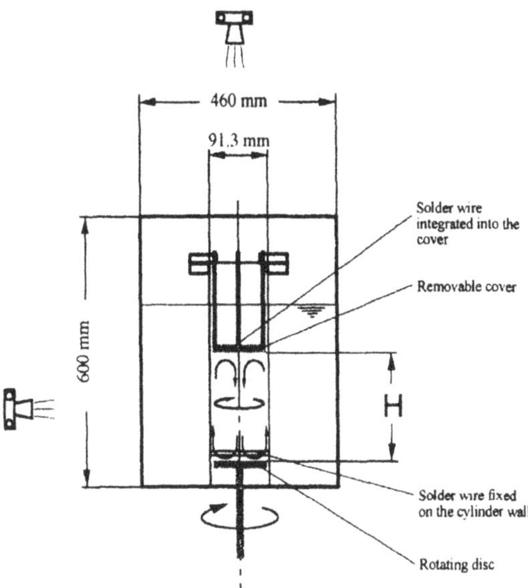

Figure 1. Schematic of the experimental setup. The positions of the two solder wires used for electrolytic precipitation is indicated. One solder wire is smoothly integrated into the rigid cover the other one is fixed at about 4 mm above the rotating bottom.

der wall at about 4 mm above the rotating bottom. This position allowed to visualize the flow structure near the cylinder walls. The second solder wire was smoothly integrated into the rigid cover. At this end we pressed the solder wire into a 1 mm large circular slit of 30 mm diameter. Results obtained with this arrangement showed the structure of the central vortex flow and in particular the flow around the recirculation bubbles. When we applied a constant or pulsed potential between 1.5 V and 15 V the wires released a white powder. The light scattered by this powder during its motion inside a sheet of light was used to visualize the flow. This sheet of light with about 1 mm thickness was produced by expanding the light beam of a 5 Watt Argon laser with a cylindrical lens. By turning the light sheet the flow could be studied in a vertical diametral plane or in different horizontal cross-sections at various heights z/R above the rotating bottom. A video camera was used to record the visualizations.

3. Results

In order to test the validity of our experimental installation we compared in a first series of experiments the observed flow states to previous experiments of Escudier (1984). Escudier used an installation with 200 mm inner diameter in combination with a glycerine water mixture of about 65 % glycerine as working fluid. Figure 2 shows that although we used water as working

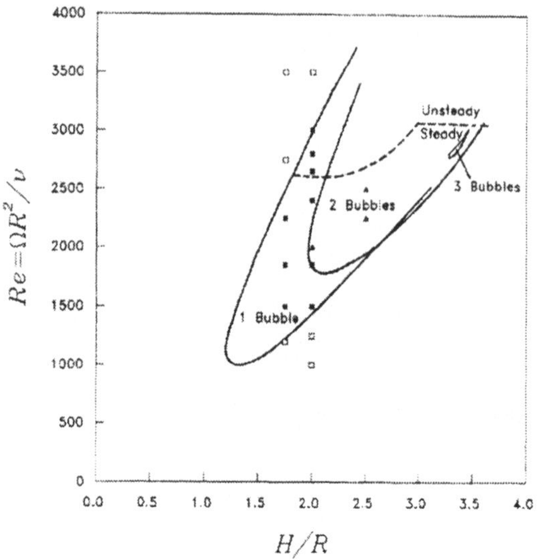

Figure 2. Comparison of flow states for some selected parameter combinations (H/R, Re) with the results obtained by Escudier (1984). Open symbols indicate that no recirculation bubble was found, closed symbols indicate the existence of bubbles.

fluid and an about 50% smaller test section our observations are in good agreement with his experiments. However, if we analyze the flow structure inside the recirculation zones and the boundary between steady and unsteady flow states some significant differences appear. Figure 3 shows the recirculation bubble for $H/R = 1.75$ and $Re = 1850$ in a vertical diametral and in a horizontal cross-section. Although the upper part of the bubble appears highly axisymmetric we recognize on the rear side of the bubble around the container axis non-axisymmetric folds of the streak surface.

The folds are not due to any detectable oscillations of the breakdown bubble since the same structure remains visible as long as diffusion and convection leave enough marker particles in the bubble region. If the Reynolds number is increased the breakdown bubbles disappear in agreement with the results of Escudier shown in figure 2 and the central core flow begins to oscillate as illustrated in figure 4(a-c) for $H/R = 1.75$ and $Re = 3500$. The indicated times were made dimensionless with the angular velocity Ω of the rotating bottom.

On the video frames 4(a-c) we recognize axial oscillations of the streak surface which are superimposed to an axial mean motion of the tracer towards the rotating bottom. Similar wave motions have been predicted by numerical simulations of Sørensen & Christensen (1995). However, surprisingly in our experiments these oscillations are damped out at later observations times. The corresponding final steady asymmetric flow state is shown in figure 4(d). If the Reynolds number is further increased the radial

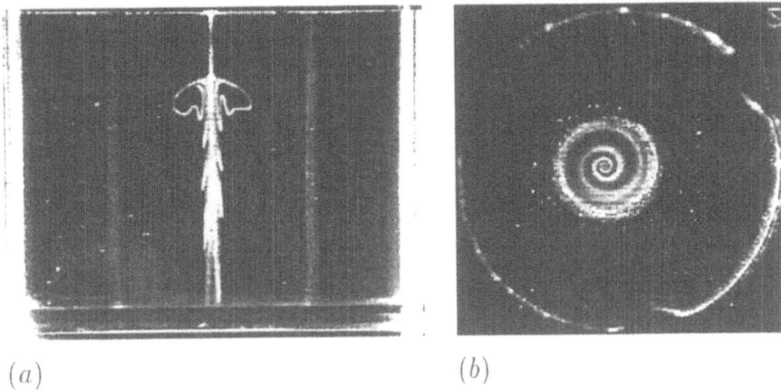

Figure 3. Recirculation bubble for $H/R = 1.75$ and $Re = 1850$, (*a*) in a vertical diametral plane (the tracer was released at the cover); (*b*) in horizontal plane (the tracer was released on the cylinder wall).

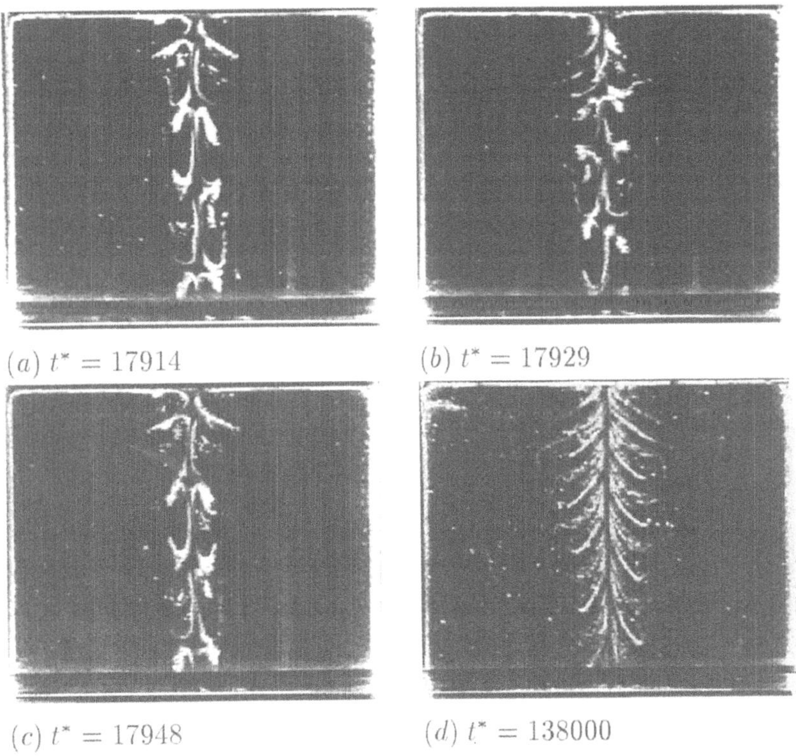

Figure 4. Axial oscillations of the core flow and final steady state for $H/R = 1.75$ and $Re = 3500$. The tracer was released on the cylinder wall .

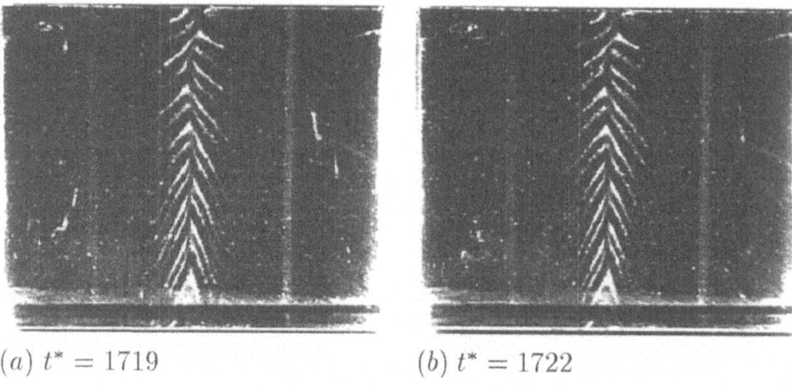

Figure 5. Flow structure for $H/R = 1.75$ and $Re = 6000$. The tracer was released on the container wall.

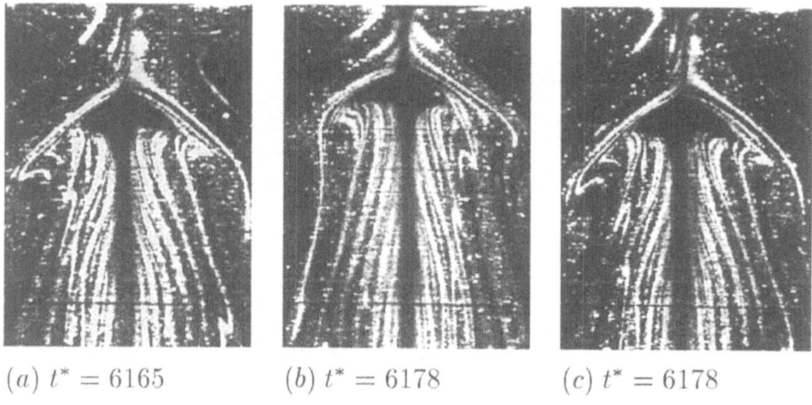

Figure 6. Axial oscillation of the breakdown bubble for $H/R = 2$ and $Re = 2800$. The tracer was released on the cylinder wall.

dimensions of the central core also increases as shown in figure 5. The comparison of figure 5(a) and 5(b) shows the remarkable steady structure of the core flow. Asymmetric flow separations appear near the upper corner between the cylinder wall and the rigid cover. While the location of these flow separations on the cylinder wall remains fixed in time, their neighborhood oscillates as indicated by the moving separated streak surface.

For parameter combinations $(H/R, Re)$ with initially one steady breakdown bubble axial oscillations of the bubble mark the onset to unsteady flow. Figure 6 shows one oscillation period in the rigid cover case for $H/R = 2$ and $Re = 2800$. The rear side of the bubble region is asymmetric.

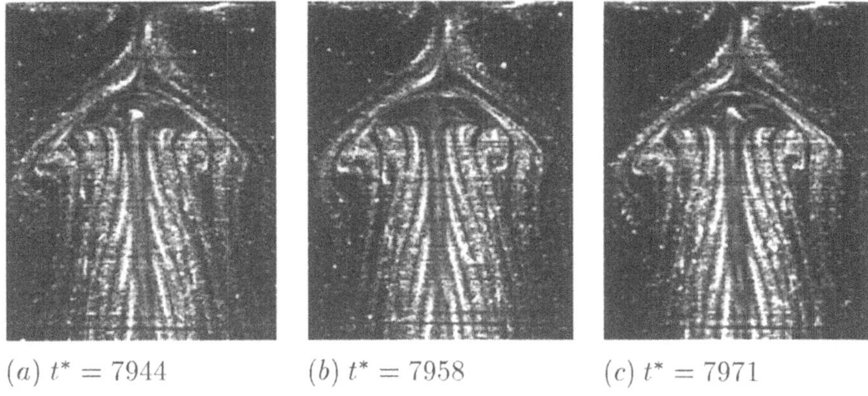

(a) $t^* = 7944$ (b) $t^* = 7958$ (c) $t^* = 7971$

Figure 7. Steady flow state after long observation times for $H/R = 2.00$ and $Re = 2800$. The tracer was released from the cylinder wall.

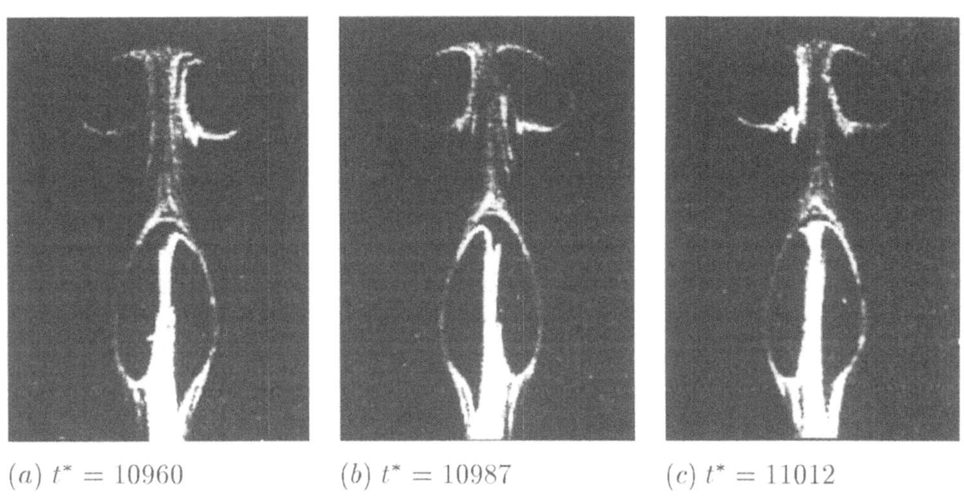

(a) $t^* = 10960$ (b) $t^* = 10987$ (c) $t^* = 11012$

Figure 8. Asymmetric oscillation of the lower breakdown bubble for $H/R = 2.5$ and $Re = 2500$. The tracer was released at the rigid cover.

The streak surfaces around the bubble give rise to the formation of hook like structures very similar to the filling and emptying process of the bubbles proposed by Lopez & Perry (1992). However, figure 6 also illustrates that the volume of the interior of the bubble remains nearly constant. In- and outflow of the central core region is restricted to the flow domain between the bubble rear side and the rotating bottom. After long observation times the oscillations are again damped and finally lead to the stationary asymmetric flow state visible in figure 7.

If two breakdown bubbles coexist the rear side of the lower bubble begins to oscillate asymmetrically at the onset of unsteadiness. Figure 8 shows such oscillations for $H/R = 2.5$ and $Re = 2500$. The upper bubble remained steady. Similar behavior was also observed by Escudier. He found, however, that such oscillations took place only for $H/R \geq 3.1$. It should be emphasized that this kind of oscillation was not considered by Escudier to distinguish steady and unsteady flow states.

4. Concluding remarks

The experiments presented show some interesting new aspects of the container flow which deserve further discussion. First all visualizations present asymmetric aspects of the bubble although the flow domain and the visualization technique are highly axisymmetric. We conclude that the container flow must be sensible to asymmetric perturbations. Therefore, it appears very difficult to realize physically the idealized model of perfectly axisymmetric flows. While recent numerical studies of Gelfgat *et al.*(1996) indicate that the container flow remains stable with respect to axisymmetric perturbations, only recently Sørensen *et al.*(1996) began to study the influence of asymmetric perturbations. Although such simulations are expensive the present results further emphasize the need for such investigations. Second the transition between steady and unsteady flow states turned out to be more complicated than indicated by previous experimental and numerical results. The major difference between our observation and the previous results are the much longer observation times. Escudier(1984) reported on very long observations times while numerical simulations of Lopez (1990) and others remain restricted to non-dimensional times of O(1000). The present results indicate that times at least of O(10000) are necessary to evaluate the steadiness. Unfortunately the exactly necessary time depends on the flow parameters making it thus very difficult to determine the boundary between steady and unsteady flow states. Experiments at higher Reynolds numbers with greater aspect ratios are currently under way in order to get more details on the asymmetric transition process.

References

CHRISTENSEN, E.A., SØRENSEN, J.N., BRØNS, M. and CHRISTIANSEN, P.L. (1993) Low-dimensional representations of early transition in rotating fluid flow, *Theo. Comput. Fluid Dyn.*, **Vol. 5**, pp. 259–267

ESCUDIER, M.P. (1984) Observations of the flow produced in a cylindrical container by a rotating endwall, *Exp.in Fluids*, **Vol. 2**, pp. 189–196

GELFGAT, A. Yu., BAR-YOSEPH, P.Z. and SOLAN, A. (1996) Stability of confined swirling flow with and without vortex breakdown, *J. Fluid Mech*, **Vol. 311**, pp. 1-36

HONJI, H., TANEDA, S. and TATSUNO, M. (1980) Some practical details of the electrolytic precipitation method of flow visualisation. Reports of the Research Insitute of Applied Mechanics, XXVIII, Kyushu University, Fukuoka, Japan.

LOPEZ, J.M. (1990) Axisymmetric vortex breakdown Part 1. Confined swirling flow, *J. Fluid Mech*, **Vol. 221**, pp. 533-552

LOPEZ, J.M. and PERRY, A.D. (1992) Axisymmetric vortex breakdown with and without temperature effects in a container with a rotating lid, *J. Fluid Mech*, **Vol. 179**, pp. 179–200

LUGT, H.J. and ABBOUD, M. (1987) Axisymmetric vortex breakdown Part 3. Onset of periodic flow and chaotic advection, *J. Fluid Mech*, **Vol. 234**, pp. 449–471

SØRENSEN, J.N., HANSEN, M.O.L. and CHRISTENSEN, E.A.(1996) Numerical investigation of symmetry breakdown in a cylindrical lid driven cavity. ECCOMAS96.

SØRENSEN, J.N. and CHRISTENSEN, E.A. (1995) Direct numerical simulation of rotating fluid flow in a closed cylinder, *Phys. Fluids*, **Vol. 7, no. 4**, pp. 764–778

TSITVERBLIT, N. (1993) Vortex breakdown in a cylindrical container in the light of continuation of a steady solution, *Fluid Dyn. Res.*, **Vol. 11**, pp. 19–35

SHEAR FLOW INSTABILITY IN A ROTATING FLUID LAYER

J.A. VAN DE KONIJNENBERG, A.H. NIELSEN, R. DE NIJS,
J. JUUL RASMUSSEN AND B. STENUM
Optics and Fluid Dynamics Department
Risø National Laboratory, P.O. Box 49,
4000 Roskilde, Denmark

1. Introduction

This contribution is concerned with experimental and numerical investigations of the instability of a forced, circular shear layer in a rotating fluid. The experiments were performed with a shallow layer of water in a parabolic tank, in which it is possible to model a geophysical beta-effect. The shear layer was produced by a secondary rotation of the inner (polar) part of the parabolic tank. Above a critical inner rotation speed, the shear layer becomes unstable, and is transformed into a chain of equally-signed vortices. We consider both cases with and without β-effect. We present measurements of the mode number as a function of the shear strength and dye visualizations of the vortex chains. The number of vortices is found to decrease with increasing strength of the shear. Our experimental results are compared with results from a numerical solution of the quasi-geostrophic equation (the Charney-Obukhov equation) in a geometry corresponding to the experimental situation and with a term modelling the experimental forcing.

The experiments are characterized by five physical quantities: the radius a of the differentially rotating disk, the depth H of the fluid layer, the kinematic viscosity ν of the fluid, the background rotation rate Ω and the differential rotation rate $\Delta\Omega$. In the experiments with a uniform depth, only H and $\Delta\Omega$ can change, leaving two independent parameters. The Rossby number $\mathrm{Ro} = \Delta\Omega/\Omega$ is a dimensionless measure of the inner rotation rate. The Ekman number $\mathrm{E} = \nu/\Omega H^2$ is a measure of the decay time of flow in a rotating system due to Ekman pumping. We may also characterize the flow by the Reynolds number $\mathrm{Re} = VL/\nu$ (Niino and Misawa, 1984) with V being half the velocity jump over the shear layer and L the thickness of the Stewartson layer. Here, we have $V = \frac{1}{2}a\Delta\Omega$ and $L = \mathrm{E}^{1/4}H$. According to the theory of Niino and Misawa, the Reynolds number is the single parameter determining whether or not the initial shear layer is stable. The Reynolds number may be expressed in terms of Ro and E according to Re $= a\mathrm{Ro}/2H\mathrm{E}^{3/4}$.

In a geophysical context, the β-effect corresponds to the northward gradient of the background vorticity $f = 2\Omega_e \sin\alpha$, with Ω_e the angular velocity of the earth, and α the geographical latitude. Thus, $\beta = \partial f/\partial y$, with y the northward direction. In the split-disk problem considered here, we introduce a β-effect as the gradient in the background vorticity in the direction towards the centre. A β-like effect in the experiments is obtained by a radial dependence on the fluid depth, achieved by a small change in the background angular velocity Ω. If viscosity is neglected in the present set-up, the potential vorticity $\omega - \beta r$ is a conserved quantity. It can be shown (Nezlin and Snezhkin, 1993) from conservation of potential vorticity that to lowest order in the Rossby number, the dynamics of fluid on a geophysical β-plane is similar to the dynamics of fluid in a shallow layer of water, with the corresponding value for β given by $2\Omega H^{-1}\partial H/\partial r$. The β-effect appears to stabilize the shear flow (Manin, 1989), (Manin and Chernous'ko, 1990). In the case we consider here, however, the width of the initial shear layer is so small that the β-effect influences the stability criterion only marginally. On the other hand, the β-effect is found to have a profound influence on the nonlinear evolution of the shear layer.

2. Experimental set-up

The experiments described in this paper were performed in a shallow layer of water in a rotating parabolic tank. The set-up has been described in detail by Laursen *et al.*. The shape of the parabolic tank corresponds to the free surface of a liquid rotating at 7.53 rad/s; this value will be referred to as the nominal angular velocity Ω_0. Having a central depth of 23 cm, the parabolic surface would fit in a cylinder with radius 28 cm and height 23 cm. A circular shear layer could be created by a differential rotation of the central part of the tank. The rotation of this central section may be cyclonic (in the same direction as the background rotation) or anticyclonic (opposite to it). The polar section has a radius a of 10 cm; the width of the slit between the inner disk and the main part of the tank is 0.3 mm. In the middle of the inner disk there is a hole with radius 1.5 cm. This hole is connected with a vertical tube in the non-rotating system, so it was possible to measure the central depth of the fluid layer while the tank was rotating. A circulation system with controllable flow rate makes it possible to sustain a flow from the periphery of the paraboloid to the centre, or vice versa. This facility was used to add fluid dyed with sodium fluorescein to the rotating layer.

For the present study, the paraboloid has been covered with a transparent lid in order to eliminate possible friction with the air in the laboratory. The paraboloidal section is surveyed by a video camera mounted in the rotating frame. The rotating fluid layer is illuminated by an ultraviolet light tube, mounted below the rigid lid. An ultraviolet filter was placed in front of the lens of the video camera, so that only fluorescent light is recorded.

Experiments were performed with a fluid depths of 1 and 2 cm, corresponding to values for the Ekman number of 1.3×10^{-3} and 3.3×10^{-4}, and Ekman decay times given by 3.6 and 7.3 s, respectively. The angular velocity of the inner disk was chosen in the interval from -2.2 to 2.2 rad/s, so the Reynolds number had values between -180 and 180. Note that in this study, we are referring the vertical depth, not the depth perpendicular to the sloping bottom of the parabolic vessel.

The experiments with β-effect were performed at a background rotation of 7.33 rad/s ($\beta < 0$) or at 7.73 rad/s ($\beta > 0$), with the depth at $r = a$ equal to 1 cm. In the geophysical situation, fluid moving away from either the North or the South pole acquires cyclonic vorticity. Regardless of the rotation sense of the vessel, this corresponds to a layer of water that is shallow in the centre, and deep at the periphery. Thus, the case $\beta > 0$ corresponds to the situation at either of the poles, whereas the case $\beta < 0$ has no geophysical analog.

3. Experimental results

Immediately after the onset of the inner rotation, the shear layer rolls up into a large number of vortices, generally more than ten. Depending on the fluid depth and the inner rotation rate, these vortices may merge into larger ones, until a pattern is formed that remains steady. In Fig. 1, a selection of a number of different symmetry modes found in experiments with $H = 2$ cm is shown. The pictures in this figure were made at least 10 minutes after the onset of the inner rotation; by this time any transient phenomena had disappeared. After the settling time, we slowly pumped in a small amount of dyed fluid. This dye is advected outwards through the Ekman layer, and is partially sucked up in the shear zone. Fig. 1 demonstrates a number of qualitative features. The first is the symmetry of the vortex chain. In particular, the modes with $n \geq 4$ turn out to be perfectly symmetric and steady, an indication of a distribution mechanism between the vortices. In the experiments with $n = 3$ and $n = 2$, the flow is no longer stationary. In the case $n = 3$, there were slight oscillations around a symmetric configuration; the vortices never seemed to obtain quite the same size, and the

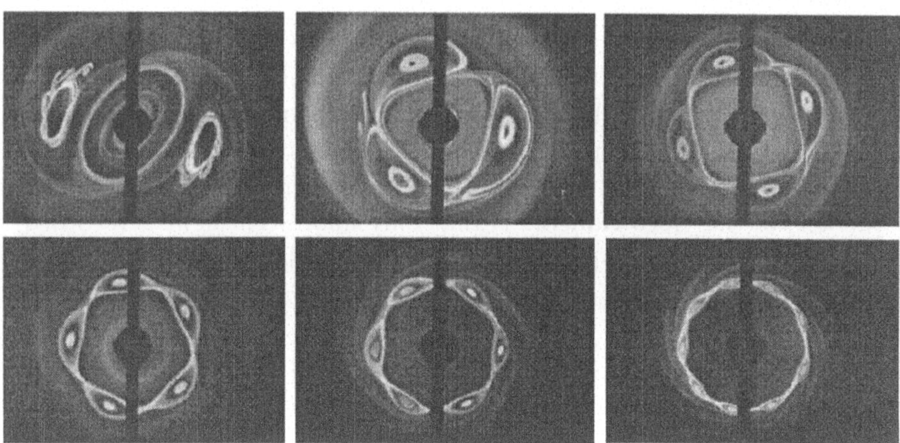

Figure 1. Stationary vortex distributions in the experiments for $\beta = 0$. From mode number 2 to mode number 9, the Reynolds number decreases from 180 to 20

triangular ring of dye forming the inner boundary of the vortices showed a periodic deformation. In the case $n = 2$, there was a periodic formation and decay of small inner vortices, showing up in Fig. 1a as undulations in the rings of dye around the vortex cores. These oscillations also appear for higher mode numbers if the depth is decreased. In some cases, a steady asymmetric flow appears: just above the critical value for $n = 2$, the two anticyclonic vortices have a slightly different shape. This asymmetry disappears if the speed of the inner disk is increased, and returns if the inner disk is slowed down again.

Another observation in Fig. 1 is that dye is concentrated along the periphery of the vortices and in the eye-like rings in the vortex centra, which seems not to be explainable with linear theory of Ekman layers. From that, one would indeed expect upward Ekman pumping at the edge of the inner disk, so that the dye entering the tank through the central hole will be trapped by the vortex chain. However, it is unclear why this upward flow occurs in the middle of the anticyclonic vortices. Oppositely, cyclonic vortices appear to expel the dye, and show up as dark patches against a dyed background.

In Fig. 2 we have plotted the number of observed modes in the final state versus the Reynolds number. For very small Reynolds numbers, the initial shear layer is stable. Every experiment was carried out at least three times; the experiments that yielded many vortices even four or five times. The data show that the number of vortices is largest just beyond the critical Reynolds number (Re = 16), and decreases to two as the Reynolds num-

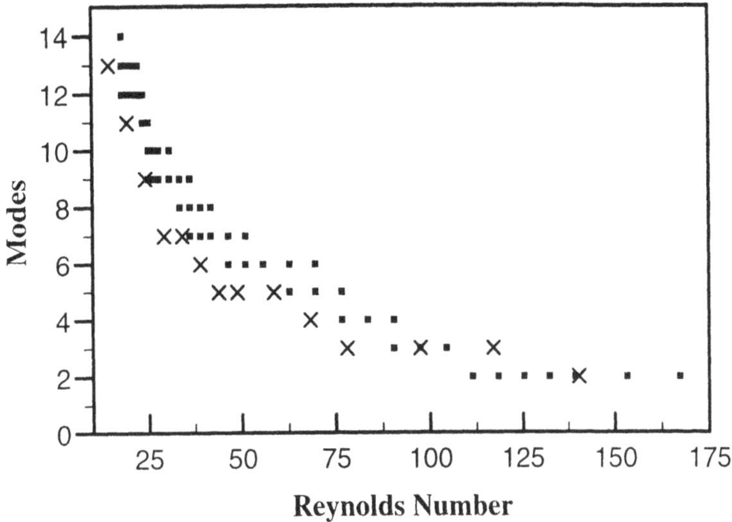

Figure 2. Number of modes vs. Reynolds number for $H = 1$ cm and $\beta = 0$. Filled squares correspond to experimental results, crosses to numerical simulations.

ber is increased. Furthermore, the figure shows a certain overlap between the plateaus corresponding to the different mode numbers. Apparently, an experiment with a certain combination of parameters may have different stable solutions, an observation also made by (Rabaud and Couder, 1983) and (Niino and Misawa, 1984). In fact, there is more hysteresis in the vortex chain than is apparent from Fig. 2. If the inner rotation is increased or decreased slowly starting from a certain mode number, the flow may remain longer in that mode than the horizontal plateaus in Fig. 2 suggest. The horizontal parts do not indicate the stability boundaries of the concerning modes, but rather the most likely outcome of the experiment as the inner disk is set into rotation suddenly.

The introduction of the β-effect affects the mode number as well as the stability of the vortex chain. However, it has no measurable effect on the critical Reynolds number. In a certain range of values for the Reynolds number, the vortex chain becomes unstable; vortices always keep merging and being formed. A visualization of such an unstable chain is shown in Fig. 3; the asymmetry connected with the instability is obvious. This instability occurs only for anticyclonic vortices in the 7.33 rad/s experiment ($\beta < 0$). and for cyclonic vortices in the 7.73 rad/s ($\beta > 0$) experiment, that is, only for vortices travelling in the "westward" direction, the direction of the Rossby wave propagation. For the inner rotation sense for which the mentioned instability does not occur, the number of vortices is higher than

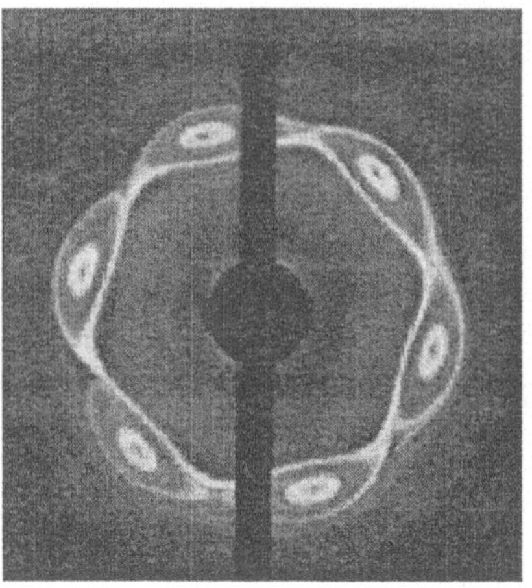
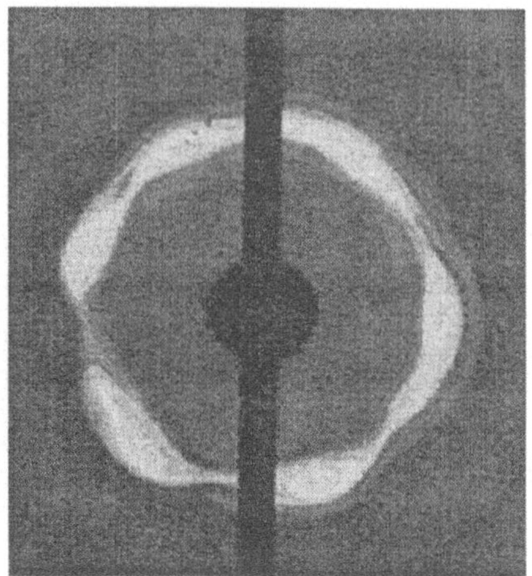

Figure 3. Steady and unsteady vortex distribution for a cyclonic inner rotation for $\Omega = 7.73$ rad/s ($\beta > 0$) and $\Omega = 7.33$ rad/s ($\beta < 0$), respectively.

in the case without β-effect. As on a rotating planet, the β-effect appears to confine the vortices to a narrow zonal region and lead to smaller length scales.

4. Numerical investigations

The equation that governs the dynamics in the shallow fluid layer in the experiment as well as models the atmosphere on a rotating planet is in the quasi-geostrophic approximation given by the Charney-Obukhov equation (Nezlin and Snezhkin, 1993). For infinite Rossby radius and including a forcing term this equation is given by

$$\frac{\partial \omega}{\partial t} + \vec{v} \cdot \nabla \omega - \beta v_r \quad = \quad \nu \nabla^2 \omega + \frac{\sqrt{\nu \Omega_0}}{H} (\omega^* - \omega) \tag{1}$$

where $\omega = (\nabla \times \vec{v}) \cdot \hat{z}$ is the scalar vorticity; the term proportional to $\omega^*(r) - \omega(r, \theta, t)$ represents both the forcing and decay of vorticity by the Ekman pumping mechanism. We model the forcing in terms of the azimuthal velocity $v_\theta^*(r)$ by

$$v_\theta^* = \frac{1}{2} r(\Omega_2 - \Omega_1) \tanh(\frac{r - a^*}{e}) + \frac{1}{2} r(\Omega_2 + \Omega_1) \tag{2}$$

where Ω_1 and Ω_2 are the angular velocity of the inner and outer disk, respectively, a^* is the radial position of the shear layer and e its width.

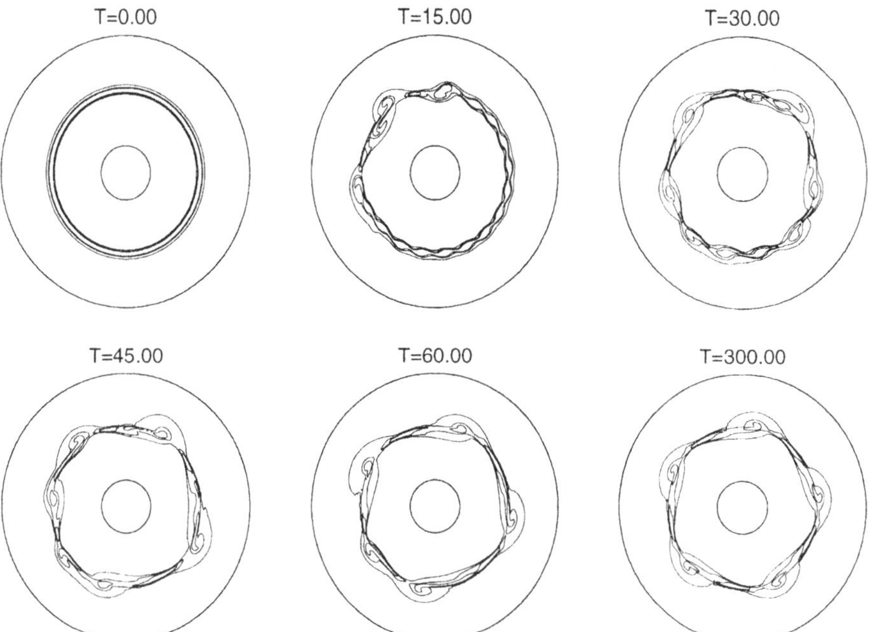

Figure 4. Vorticity distribution showing the creation of a mode 5 with an anticyclonic inner rotation. Continuous contour lines to positive values, dashed contour lines to negative values. Parameters: $\beta = 0$, Re $= 29.3$.

Equations (1) is solved numerically in an annular geometry with inner radius r^- and outer radius r^+, by means of a fully dealiased spectral code using N Fourier modes in the azimuthal direction and M Chebyshev polynomials in the radial direction. We used free-slip boundary conditions, since this gives the most realistic approximation to the experimental situation, in which there is no inner wall, and the outer wall is far away from the shear layer. Since Chebyshev polynomials are defined on the closed interval $[-1; 1]$ we map all length scales to this interval. For further details on the numerical method, see (Coutsias *et al.*, 1994; Nielsen, 1993; Coutsias and Lynov, 1991).

Having chosen the parameters for the forcing term in (2), and thereby determined the Reynolds number, we initiated the computation with the basic flow, which is the solution of (1) with the left-hand side equal to zero. In order to trigger the instabilities, we added a small noise perturbation localized at the shear zone. The width of the initial condition is calculated numerically in order to determine the Reynolds number.

5. Numerical results

In all the simulations presented below we have chosen $a^* = 1.5$ CU, $\Omega_0 = 7.53\ \mathrm{s}^{-1}$, $H = 0.15$ CU, $\epsilon = 0.03$ CU and $\nu = 2.25 \times 10^{-4}$ CU^{2}s^{-1}, where

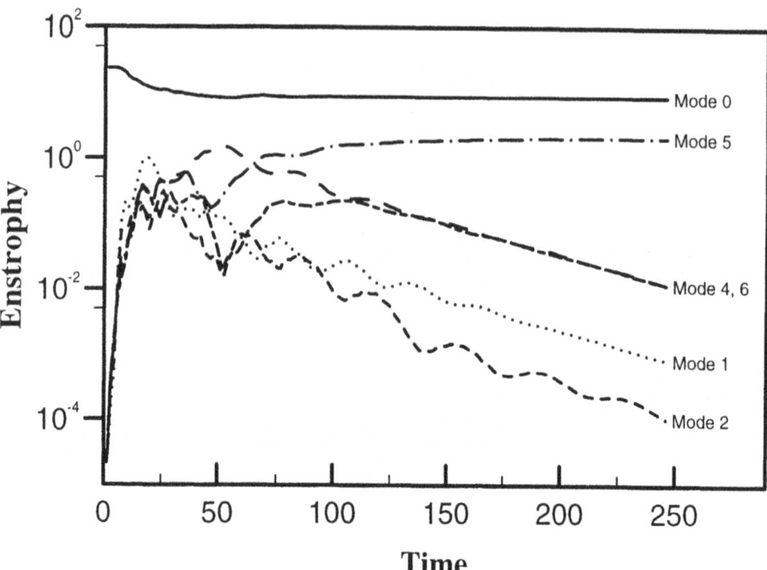

Figure 5. Temporal evolution of the enstrophy spectrum for the lowest 7 azimuthal modes of the system for the simulation in Fig. 4.

CU (Computer Unit) is a length scale used in the computation. The radial position of the shear zone in the experiment was $a = 10$ cm, so all lengths are mapped such that 1 cm = 0.15 CU.

In Fig. 4 we show the formation of a mode 5 evolving from the axisymmetric initial condition without β-effect. Since the Reynolds number is low the evolution is quite smooth. In order to clarify the dynamics for this evolution we have in Fig. 5 shown the enstrophy spectrum. This spectrum was obtained by integrating over the radial coordinate: $W(\theta, t) = \int_{r^-}^{r^+} \omega^2(r, \theta, t)\, r dr = \sum_{n=0}^{N} W_n(t) e^{in\theta}$. We note that the total enstrophy of the system is defined by $W(t) \equiv \int_D \omega^2(t) dA = \sum_{n=0}^{N} W_n(t)$. At $t = 0$ we see that nearly all the enstrophy is in the zero mode while all other azimuthal modes contain only their initial noise contributions. We can divide the evolution in three time intervals. First, we have a short time interval where all the unstable modes of the system grow exponentially. This is simply due to the fact that their amplitudes are so small that the nonlinear term in (1) is negligible. Then, when the amplitudes are large enough, approximately at $t = 15$, the nonlinear term becomes dominant and all modes start to interact. Finally, after $t = 100$ we observe that mode 5 stabilizes itself at a well defined enstrophy level, driving to zero all modes that are not an integer multiple of five. Thus, the forcing term in (2), is pumping enstrophy into the zero mode, the nonlinear term redistributes it to mode 5 and its higher harmonics, and from these modes enstrophy is removed by viscosity and Ekman damping.

In order to make a comparison with the results of the experiment we

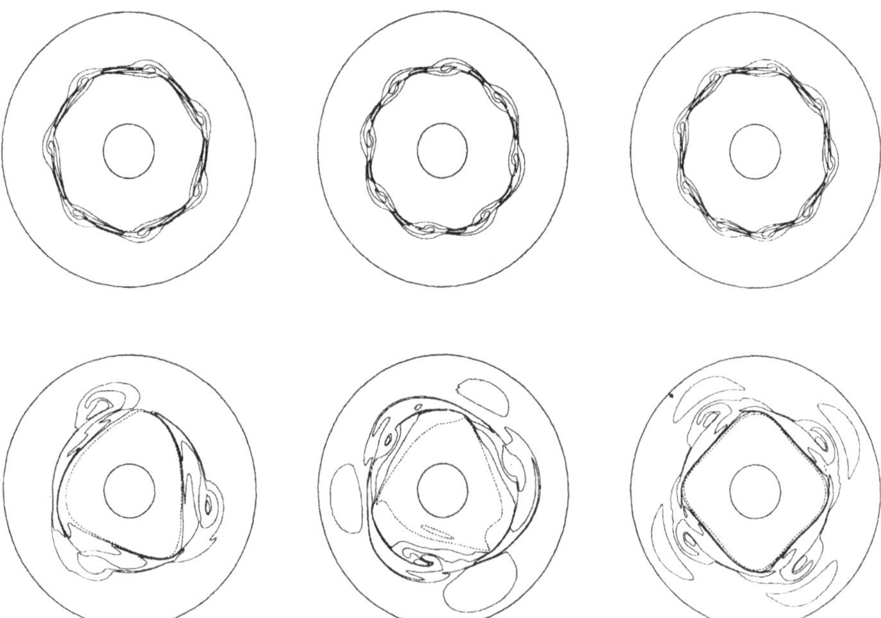

Figure 6. Vorticity distribution showing end state of 6 individual computer simulations with an anti-cyclonic inner rotation. The horizontal rows correspond to different Reynolds number (Re = 29.3 and 97.6); the vertical columns cooespond to different values of the beta parameter ($\beta = 0$, 3.012 and -3.012).

have plotted the observed number of modes versus the Reynolds number together with experimentally obtained values in Fig. 2. Each numerical observation originates from a stable configuration of individual computer simulations. We note that in general there is a good agreement between the two results, but the computer simulations tend to produce a lower mode number than seen in the experiment for the same Reynolds number. In view of the simple model for the Ekman pumping mechanism and the fact that the parabolic curvature is not taken into account in the numerical simulations, these results are quite satisfactory.

In order to examine the effect of the beta term we show in Fig. 6 the end states of 6 numerical runs with different Reynolds numbers and β-values. Including a beta term stabilizes the flow, and we observe a larger number of vortices for the same Reynolds number. Another observation is that the polygonal shape of the region inside the shear layer is more pronounced for $\beta < 0$, and less for $\beta > 0$. This difference is due to the increase in vorticity if a fluid element moves outwards (towards a vertex of the polygon) in the case $\beta < 0$, and a similar decrease if $\beta > 0$. This is explained by the conservation of potential vorticity, expressed by $\omega - \beta r =$ constant. Similarly, the undulating motion of a fluid parcel induces a vorticity pattern outside the vortex chain. For $\beta > 0$ one observes a patch of decreased vorticity at the azimuthal location between the vortices, whereas for $\beta < 0$ a patch of decreased vorticity appears at the location of the vortices. The vorticity modulation induced by the β-effect may partially explain the stability properties of the vortex chain. For $\beta < 0$, the slightly concave shape of the inner region tends to separate the vortices, and thus leads to a greater stability, and therefore also to a higher mode number. Oppositely, for $\beta > 0$ the more circular inner region may enhance interaction between the vortices, and contribute to the more unstable nature of the vortex chain in that case.

Another observation in Fig. 6 (also made in the experiments) is that for high Reynolds number the vortices appear to be unequal. Generally, we find that for 5 or more vortices present at the shear we obtain what can be called "pure" modes, like that of Fig. 4, where we only observe two azimuthal modes, note that the zero mode is always present. As seen from Fig. 6 (Re = 97.3, $\beta = 0$) mode 3 is not pure, since one of the vortices is perturbed. Examining its enstrophy spectrum reveals that this mode shows a periodic behaviour.

6. Acknowledgement

This work is supported by the Danish Natural Research Council (SNF) grant nr. 9600852 and by INTAS-RFBR 95-0988.

References

J.M. Chomaz. M. Rabaud, C. Basdevant and Y. Couder, Experimental and numerical investigation of a forced circular shear layer, J. Fluid Mech. **187**, 115–140 (1988).

E.A. Coutsias, K. Bergeron, J.P. Lynov and A.H. Nielsen, Self Organization in 2D Circular Shear Layers, *AIAA* 94-2407 (1994).

E.A. Coutsias and J.P. Lynov, Fundamental interactions of vortical structures with boundary layers in two-dimensional flows Physica D **51**, 482 (1991).

T.S. Laursen, M.O. Nielsen, M.L. Pedersen, J.J. Rasmussen, B. Stenum, M.V. Nezlin and E.N. Shezhkin, Set-up for Studies on Flows in Rotating Free Surface Shallow Water, submitted to Experiments in Fluids (1997).

D. Yu. Manin, Stability of two-dimensional shear flows in the presence of the beta effect and external friction, Izv. AN SSSR. Atmos. Ocean. Phys. **25** (8) (1989).

D. Yu. Manin and Yu. L. Chernous'ko, Experimental investigation of the stability of a quasi-two-dimensional jet flow produced in a rotating fluid by the method of sources and sinks, Izv. AN SSSR. Atmos. Ocean. Phys. **26** (5) (1990).

M.V. Nezlin and E.N. Snezhkin, Rossby Vortices, Spiral Structures, Solitons, Springer Verlag (1993).

A.H. Nielsen, Electrostatic turbulence in strongly magnetized plasmas, Ph.D. thesis, Risø National Laboratory, R-659 (1993).

H. Niino and N. Misawa, An experimental and theoretical study of barotropic instability, J. Atmos. Sci. **41**, 1992–2011 (1984).

M. Rabaud and Y. Couder, A shear-flow instability in a circular geometry, J. Fluid Mech. **136**, 291–319 (1983).

IV. Small Scale Turbulence and 2-D Flows

IDENTIFICATION OF COHERENT FINE SCALE STRUCTURE IN TURBULENCE

M. TANAHASHI, T. MIYAUCHI AND J. IKEDA

Department of Mechano-Aerospace Engineering,
Tokyo Institute of Technology
2-12-1 Ohokayama, Meguro-ku, Tokyo 152, Japan

1. Introduction

Theoretical description of intermittent character in small scale motion have been one of the most important subjects in turbulence research. Theorists have made efforts to establish a theory of fine scale structure by assuming various type of structure as a fine scale structure (Townsend, 1951; Tennekes, 1968; Lundgren, 1982; Pullin and Saffman, 1993; Saffman and Pullin, 1994). Most of them are based on an assumption that many tube-like or sheet-like vortices are embedded in turbulence randomly. Each vortex is considered to be an analytical solution of Navier-Stokes equations. Recent studies by direct numerical simulations of turbulence have found the evidence supporting theoretical expectations for coherent fine scale structures (Kerr, 1985; She *et al.*, 1990; Vincent and Meneguzzi, 1991; Ruetsch and Maxey, 1991; Ruetsch and Maxey, 1992; Jimenez *et al.*, 1993; Vincent and Meneguzzi, 1994). In homogeneous turbulence, high vorticity regions are supposed to be a candidate of fine scale structure (She *et al.*, 1990; Vincent and Meneguzzi, 1991; Jimenez *et al.*, 1993). However, definitions of fine scale structure based on vorticity magnitude are not clear and educed fine scale structures depend on the threshold. The objectives of this study are to specify fine scale structures in homogeneous isotropic turbulence and to investigate a scaling law and Reynolds number dependence of fine scale structures. DNS database of decaying homogeneous isotropic turbulence are analyzed by using a new identification method of fine scale structures.

TABLE 1. DNS database of homogeneous isotropic turbulence. Re_λ: Reynolds number based on Taylor microscale, Re_l: Reynolds number based on integral length scale, N: number of grid points, $S_{u'}$ and $F_{u'}$: skewness and flatness of longitudinal velocity derivative.

ID	Re_λ	Re_l	N	$S_{u'}$	$F_{u'}$
1	37.2	159	96^3	−0.484	4.10
2	37.1	163	216^3	−0.487	4.14
3	66.1	580	216^3	−0.495	4.52
4	87.9	700	216^3	−0.483	4.97

2. Identification of Coherent Fine Scale Structure in Turbulence

2.1. DNS OF DECAYING HOMOGENEOUS ISOTROPIC TURBULENCE

Direct numerical simulations of decaying homogeneous isotropic turbulence are conducted by using a spectral method. Aliasing errors due to nonlinear interactions are fully removed by a 3/2 rule. Numerical conditions are listed in Table 1. For three lower Reynolds number cases, calculations are conducted with the same conditions that have reported by Comte-Bellot and Corrsin (1971) for grid turbulence. Statistical properties obtained by these DNSs show a good agreement with the experimental results. Flatness and skewness factors of the velocity derivatives for the highest Reynolds number case coincide with previous experimental and numerical results (Van Atta and Antonia, 1980; Jimenez *et al.*, 1993).

2.2. IDENTIFICATION SCHEME

There are a lot of methods for identification of vortical structures in flows. The simplest method is usage of high vorticity or enstrophy regions. In previous studies related to fine scale structures in homogeneous turbulence, high vorticity regions have been used to identify the intermittent fine scale structure of turbulence (Kerr, 1985; She *et al.*, 1990; Vincent and Meneguzzi, 1991; Ruetsch and Maxey, 1991; Ruetsch and Maxey, 1992; Jimenez *et al.*, 1993; Vincent and Meneguzzi, 1994). However, high vorticity regions may represent tube-like and sheet-like structures simultaneously. These two structures are essentially different. Tube-like structures show solid body rotation in the core region, while sheet-like structures show same magnitude of strain and rotation. For the case with a strong mean shear like the flow near wall or center of free shear flows, employment of high vorticity or enstrophy regions fail to represent coherent vortical structure.

To avoid this contradiction, Tanaka and Kida (1993) have used $\nabla^2 p$ to represent streamwise vortices in homogeneous shear flows. $\nabla^2 p$ corresponds to two times of second invariant Q of velocity gradient tensor. By using positive Q regions, Tanahashi and Miyauchi (1995) have shown that tube-like fine scale eddies appear in the transitional stage of mixing layers. Detail analysis of these methods have been reported by Jeong and Hussain (1995) and they have proposed a new identification method: λ_2-definition. All of these identification methods, however, can educe only 'regions' of coherent structures under each definition. The regions specified by these methods depend on thresholds of variables that were employed in the definition such as $|\omega|$, Q and λ_2. In this study, a new identification scheme was developed to identify the coherent fine scale structure in turbulence. In this scheme, local flow patterns are directly used to find out the coherent fine scale structure in turbulence. In the case of incompressible flows, local flow patterns are classified into four categories: stable focus-stretching, unstable focus-compressing, stable node-saddle-saddle and unstable node-saddle-saddle (Chong *et al.*, 1990). These patterns are easily determined by evaluating the second and third invariant ($Q = -(S_{ij}S_{ij} - W_{ij}W_{ij})/2$ and $R = -(S_{ij}S_{jk}S_{ki} + 3W_{ij}W_{jk}S_{ki})/3$) of characteristic equations for velocity gradient tensor $A_{ij} = S_{ij} + W_{ij}$, where $S_{ij} = (\partial u_i/\partial x_j + \partial u_j/\partial x_i)/2$ and $W_{ij} = (\partial u_i/\partial x_j - \partial u_j/\partial x_i)/2$ is symmetric and asymmetric parts of A_{ij}. The identification scheme consists of the following steps:

- Evaluation of Q at each collocation point from the results of DNS.
- Probability of existence of positive maximals of Q near the collocation points is evaluated at each collocation point from Q distribution. The case that a maximal of Q coincides with a collocation point is very rare, so it is necessary to define probability on collocation points.
- Collocation points with high possibility of existence are selected to survey actual maximals of Q. Locations of Q maximals are determined by applying a three dimensional cubic spline interpolation to DNS data.
- At the maximal second invariant point, a horizontal plane perpendicular to the vorticity vector is defined and a cylindrical coordinate system setting the maximal point as the origin is considered. The velocity vectors are transformed on this coordinate and mean azimuthal velocity is calculated.
- Point that has minimum variance of azimuthal velocity is surveyed near the maximal point. In this process, a new cylindrical coordinate system is always defined around a newly searched point.
- Statistical properties are calculated around the point.

Figure 1 shows contour surfaces of normalized second invariant ($Q^* =$

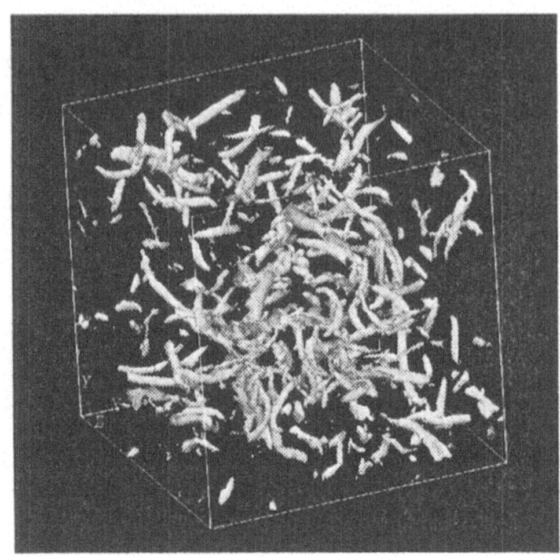

Figure 1. Contour surfaces of normalized second invariant of the velocity gradient tensor ($Q* = 0.03$) for the case of $Re_\lambda = 87.9$. Visualized region is 1/27 of the whole calculation domain. Second invariant is normalized by Kolmogorov microscale and r.m.s. of fluctuating velocity.

0.03) for the case of $Re_\lambda = 87.9$. Here, an asterisk denotes normalization by Kolmogorov microscale (η) and r.m.s. of velocity fluctuation (u_{rms}). This normalization is due to the results described below. Tube-like structures are randomly oriented in homogeneous isotropic turbulence as shown by previous studies by She *et al.* (1990), Vincent and Meneguzzi (1991) and Jimenez *et al.* (1993). Note that positive Q regions are corresponding to solid body rotations where rotation rate ($= W_{ij}W_{ij}$) exceed strain rate ($= S_{ij}S_{ij}$). By the above described identification scheme, points on the axis of tube-like structure can be specified. As shown in Fig.1, tube-like structures in homogeneous isotropic turbulence show strong three dimensional characters. Three-dimensional features in the axial direction are investigated by following additional steps:

- From the point that determined by above procedure, the investigated point is moved in the axial direction with short distance ds. ds is parallel to the vorticity vector $\boldsymbol{\omega}$.
- Near the newly investigated point, a point that have minimum variance of azimuthal velocity is searched by the same procedure described above.
- After the calculation of statistical properties, above steps are repeated until minimum variance point can not be found.

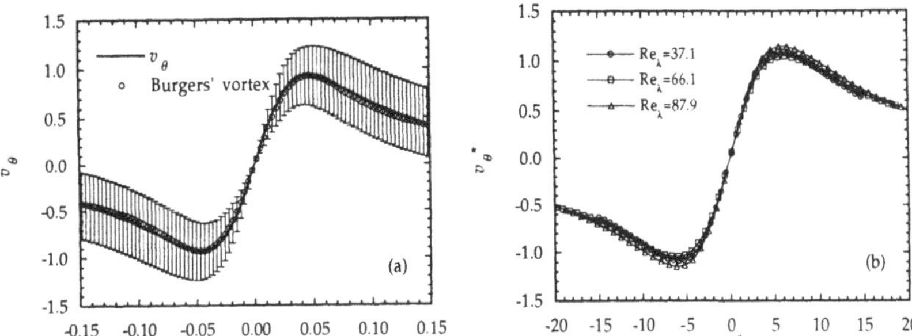

Figure 2. Mean azimuthal velocity profile of the coherent fine scale structure in homogeneous isotropic turbulence. (a): mean azimuthal velocity profile without any scaling($Re_\lambda = 87.9$). Symbols: velocity profile of a Burgers' vortex. Error bars denote variances of azimuthal velocity. (b): normalized by Kolmogorov microscale and r.m.s. of fluctuating velocity.

3. Characteristics of Coherent Fine Scale Structures

3.1. STATISTICS OF 2-D CUTS OF THE TUBE-LIKE STRUCTURES

Figure 2(a) shows a mean azimuthal velocity profile of fine scale structures for the case of $Re_\lambda = 87.9$. The average was taken over 921 samples. Symbols in Fig.2(a) show an azimuthal velocity profile of a Burgers' vortex. Mean azimuthal velocity profile of the coherent fine scale structures shows a good agreement with that of a Burgers' vortex. In Table 2, characteristic length scales of turbulence and mean diameters of the coherent fine scale structures are listed for three different Reynolds numbers, where length scales are normalized by Kolmogorov microscale. In this study, radius of the coherent fine scale structures is determined by a distance between the center and location where the mean azimuthal velocity reaches the maximum value. Taylor microscale and dissipation maximum scale nondimensionalized by the Kolmogorov microscale become large with the increase of Reynolds number. However, mean diameters of the coherent fine scale structures are about 12 times of Kolmogorov microscale and does not depend on turbulent Reynolds number.

In Fig.2(b), mean azimuthal velocity profiles normalized by Kolmogorov microscale and r.m.s. of velocity fluctuation are shown for three different Reynolds number cases. Mean azimuthal velocity profiles for different Reynolds number coincide very well. The maximums of mean azimuthal velocity profiles are close to u_{rms}. These results show that mean diameter and intensity of the coherent fine scale structures can be scaled by Kolmogorov microscale and r.m.s. of fluctuating velocity in these Reynolds number range. These coherent fine scale structures have high and thin energy dissipation regions around them and these regions contribute to total energy dissipation significantly (Tanahashi *et al.*, 1996).

TABLE 2. Characteristic length scales and mean diameter of the coherent fine scale structure. l: integral length scale, λ: Taylor microscale, $l_{D.M.}$: dissipation maximum scale, D_m: mean diameter of the coherent fine scale structure, V: investigated volume, n: number of detected coherent fine scale structure. Listed length scales are normalized by Kolmogorov microscale.

ID	Re_λ	l^*	λ^*	$l^*_{D.M.}$	D^*_m	V	n
2	37.1	52.6	12.0	24.7	11.8	$30.8l^3$	195
3	66.1	140	15.9	29.2	12.7	$27.9l^3$	733
4	87.9	147	18.4	34.9	12.3	$29.3l^3$	921

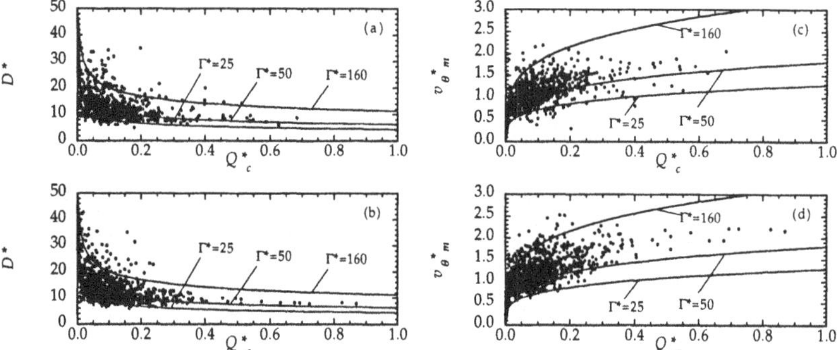

Figure 3. Scatter plots of diameters and maximum azimuthal velocity of the coherent fine scale structure in homogeneous isotropic turbulence. (a) and (c) $Re_\lambda = 66.1$, (b) and (d)$Re_\lambda = 87.9$.

Figures 3(a) and 3(b) shows relation between diameter and second invariant at the center of each coherent fine scale structure. Diameters and second invariants are also normalized by Kolmogorov microscale and r.m.s. of velocity fluctuation. Distributions of diameter as a function of second invariant show good agreement with each other. Variance of the diameter is relatively large for low second invariant and small for large second invariant. Asymptotic value of diameter for large second invariant is $6 \sim 8$ times of Kolmogorov microscale, which is also a lower limit of those diameters for all range of second invariant. These behaviors coincide with the prediction based on Burgers' vortex model (Tanahashi *et al.*, 1997a). Maximums of mean azimuthal velocity are plotted in Figs.3(c) and 3(d) with respect to the second invariant. For all cases, maximum values reach $2.0 \sim 2.5$ times of u_{rms}. Asymptotic behaviors are also observed for the intensity of the azimuthal velocity. The asymptotic values depend on Reynolds number. This Reynolds number dependence has been investigated by Tanahashi *et al.* (1997a). These coherent fine scale structures produce large velocity dif-

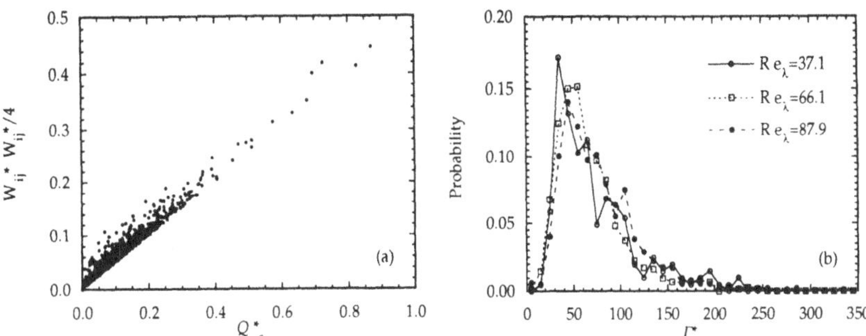

Figure 4. (a): Relation between second invariant and enstrophy at the center of the coherent fine scale structure ($Re_\lambda = 87.9$). (b):Probability of circulation of the coherent fine scale structure. Circulation is normalized by Kolmogorov microscale and r.m.s. fluctuating velocity.

ferences (about 5 times of u_{rms}) in a very small spatial region (about 8η). This character suggest that the coherent fine scale structures are directly related to the intermittency of the velocity fields in small scales.

Figure 4(a) shows relations between enstrophy and second invariant at the center of coherent fine scale structure for the case of $Re_\lambda = 87.9$. The coherent fine scale structures with large azimuthal velocity and small radius correspond to the high enstrophy regions that have been reported by previous studies ((She *et al.*, 1990; Vincent and Meneguzzi, 1991; Jimenez *et al.*, 1993)). Jimenez *et al.* (1993) have reported that high vorticity regions in statistically steady homogeneous turbulence have a radius of about 4 times of Kolmogorov microscale. Their result coincide very well with present asymptotic values ($3 \sim 4\eta$) as was shown in Figs.3(a) and 3(b). The high vorticity regions in turbulence are frequently called as 'worms' (Yamamoto and Hosokawa, 1988) or 'sinews' (Moffatt *et al.*, 1994). Present results suggest that 'worms' or 'sinews' reflect one aspect of the coherent fine scale structures in turbulence. Figure 4(b) shows probability density functions of Γ^* for three Reynolds number cases. In spite of Reynolds number dependence of the intense coherent fine scale structures (Tanahashi *et al.*, 1997a), pdfs of Γ^* constructed by all structures coincide very well and show a peak at $\Gamma^* \approx 50$.

3.2. STRUCTURE IN THE AXIAL DIRECTION OF COHERENT FINE SCALE STRUCTURES

Length of coherent fine scale structures can be defined by searching in the direction of the axis of the coherent fine scale structures. Figure 5(a) shows relation between length of axis and second invariant of typical coherent fine scale structure for the case of $Re_\lambda = 37.2$. From the relation between Q and s, one can determine the length of the coherent fine scale structures.

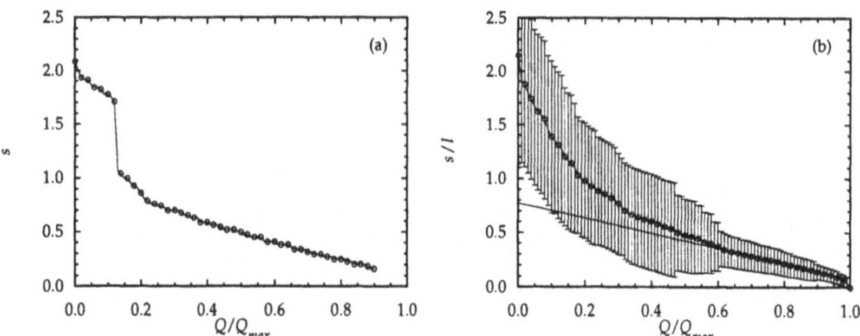

Figure 5. Length of coherent fine scale structures ($Re_\lambda = 37.2$). (a): a typical coherent fine scale structure. (b):Mean length of 70 samples

Figure 5(b) shows mean $Q - s$ relation averaged over 70 coherent fine scale structures. Variance of length is relatively small for Large $Q(Q/Q_{max} > 0.6)$ and is large for small Q. Length of coherent fine scale structures can be defined by a linear extrapolation by using gradient of s for $0.6 < Q/Q_{max} < 0.9$. Solid line in Fig.5(b) shows a result of the former definition. The mean length of coherent fine scale structure is 1.54 times of integral length scale.

Figure 6 shows mean velocity distributions along the axis of the coherent fine scale structures. Averaging was conducted for the coherent fine scale structure with negative third invariant of the velocity gradient tensor because negative third invariant represents the coherent fine scale structure with stretching and most of the coherent fine scale structures have negative third invariant (Tanahashi *et al.*, 1997a). Figure 6 shows that the coherent fine scale structures have strong azimuthal velocity around the axis and velocity component along the axis is accelerated with the distance from the center.

4. CONCLUSIONS

In this study, coherent fine scale structures in homogeneous isotropic turbulence are identified and scaling law of coherent fine scale structure in turbulent flow was investigated. To identify the coherent fine scale structure, a new identification scheme that uses local flow patterns to determine the axis of fine scale structure was employed. The detected coherent fine scale structures of turbulence, which show tube-like features, can be scaled by Kolmogorov microscale and r.m.s. of velocity fluctuation. Mean diameter of the structures is close to 12 times of Kolmogorov microscale and maximum of mean azimuthal velocity is close to r.m.s. of velocity fluctuation. Mean azimuthal velocity profiles can be approximated by a Burgers' vortex. Diameters of the coherent fine scale structures show asymptotic behaviors with the increase of second invariant and reaches about $6 \sim 8$

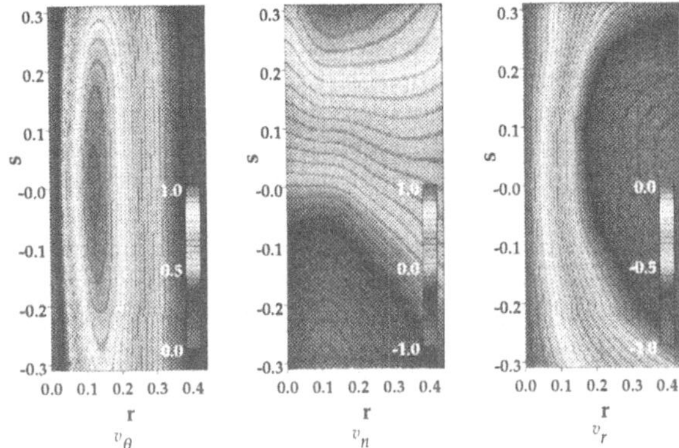

Figure 6. Mean three-dimensional flow pattern of coherent fine scale structures ($Re_\lambda = 37.2$).

times of Kolmogorov microscale. The asymptotic values is a lower limit of diameters of the coherent fine scale structures. These characteristics do not depend upon Reynolds number. The mean length of the coherent fine scale structures is order of integral length scale.

Our findings imply that turbulence is composed of fine scale structures which is universal. Our resent works (Tanahashi *et al.*, 1997a; Tanahashi *et al.*, 1997b) also show the evidence of a universal coherent fine scale structure in turbulence.

Acknowledgments

This work is partially supported by The Kawakami Memorial Foundation.

References

Chong, M.S., Perry A.E. and Cantwell, B.J. (1990) A general classification of three-dimensional flow field, *Phys. Fluids*, Vol. A2, pp. 765–777.

Comte-Bellot, G. and Corrsin, S. (1971) Simple eulerian time correlation of full- and narrow-band signals in grid-generated 'isotropic' turbulence, *J. Fluid Mech.*, **Vol. 48**, pp. 273–337.

Yamamoto, K. and Hosokawa, I. (1988) A decaying isotropic turbulence pursued by the spectral method, *J. Phys. Soc. Japan*, Vol. 57, pp.1532–1535.

Jeong, J. and Hussain, F. (1995) On the identification of a vortex, *J. Fluid Mech.*, Vol. 285, pp.69–94.

Jimenez,J., Wray, A.A., Saffman,P.G. and Rogallo, R.S. (1993) The structure of intense vorticity in isotropic turbulence, *J. Fluid Mech.*, Vol. 255, pp. 65–90.

Kerr, R.M.V. (1985) Higher-order derivative correlations and the alignment of small-scale structures in isotropic numerical turbulence, *J. Fluid Mech.*, Vol. 153, pp. 31–58.

Lundgren, T.S. (1982) Strained spiral vortex model for turbulent fine structure, *Phys. Fluids*, Vol. 25, pp.2193–2203.

Moffatt, H. K., Kida, S. and Ohkitani, K. (1994) Stretched vortices – the sinews of turbulence; large-Reynolds -number asymptotics, *J. Fluid Mech.*, Vol. 259, pp.241–264.

Pullin, D.I. and Saffman, P.G. (1993) On the Lundgren-Townsend model of turbulent fine scales, *Phys. Fluids*, Vol.A5, pp. 126–145.

Ruetsch, G.R. and Maxey, M.R. (1991) Small-scale features of vorticity and passive scalar fields in homogeneous isotropic turbulence, *Phys. Fluids*, Vol. A3, pp. 1587–1597.

Ruetsch, G.R. and Maxey, M.R. (1992) The evolution of small-scale structures in homogeneous isotropic turbulence, *Phys. Fluids*, Vol. A4, pp. 2747–2760.

Saffman, P.G. and Pullin, D.I. (1994) Anisotropy of the Lundgren-Townsend model of fine-scale turbulence, *Phys. Fluids*, Vol. A6, pp.802–807.

She, Z.-S., Jackson, E. and Orszag, S.A. (1990) Intermittent vortex structures in homogeneous isotropic turbulence, *Nature*, Vol344, pp.226–228.

Tanahashi, M. and Miyauchi, T. (1995) Small scale eddies in turbulent mixing layer, *Proc. 10th Symposium on Turbulent Shear Flows*, Vol. 1, pp. P-79–P-84.

Tanahashi, M., Miyauchi, T. and Yoshida, T. (1996) Characteristics of small scale vortices related to turbulent energy dissipation, *Proc. 9th International Symposium on Transport Phenomena*, Vol. 2, pp.1256–1261.

Tanahashi, M., Miyauchi, T. and Ikeda, J. (1997a) Scaling law of coherent fine scale structure in homogeneous isotropic turbulence, *to be appeared in 11th Symposium on Turbulent Shear Flows.*

Tanahashi, M., Miyauchi, T. and Matsuoka, K. (1997b) Coherent fine scale structures in turbulent mixing layers, *to be appeared in 2th International Symposium on Turbulence, Heat and Mass Transfer.*

Tanaka, M. and Kida, S. (1995) Characterization of vortex tubes and sheets, *Phys. Fluids*, Vol. A5, pp. 2079–2082, 1993.

Tennekes, H. (1968) Simple model for the small-scale structure of turbulence, *Phys. Fluids*, Vol. 11, pp.669–761.

Townsend, A.A. (1951) On the fine-scale structure of turbulence, *Proc. R. Soc. Lond.*, Vol. A208, pp.534–542.

Vincent, A. and Meneguzzi, M. (1991) The spatial structure and statistical properties of homogeneous turbulence, *J. Fluid Mech.*, Vol. 225, pp.1–20.

Vincent, A. and Meneguzzi, M. (1994) The dynamics of vorticity tubes in homogeneous turbulence, *J. Fluid Mech.*, Vol. 258, pp. 245–254.

Van Atta, C. W. and Antonia, R. A. (1980) Reynolds number dependence of skewness and flatness factors of turbulent velocity derivatives, *Phys. Fluids*, Vol. 23, pp. 252–257.

EVOLUTION OF VORTICAL STRUCTURE IN ISOTROPIC TURBULENCE

I. HOSOKAWA
University of Electro-Communications
Chofu, Tokyo 182, Japan

AND

K. YAMAMOTO
National Aerospace Laboratory
Chofu, Tokyo 182, Japan

Abstract. Vortical structure in decaying isotropic turbulence is studied by direct numerical simulation. Enstrophy increases with time at least exponentially in accordance with the wind-up of vortex sheets, and arrives at its peak after about 10 (initial large-eddy) turnover times for all small kinematic viscosities. The peak value increases almost in inverse proportion to kinematic viscosity, suggesting a singularity at a finite time of enstrophy in the inviscid limit. At the fully-developed state, while we find many tubes with strong vorticity created, we recognize that the background vorticity field is never random but so coherent that it involves many tubes with relatively low vorticity cores, sometimes much thicker and longer than the former tubes. The similitude of all these tubes with the Burgers vortices may be acknowledged only in a certain average sense.

Decaying isotropic turbulence is useful for looking into how the flow develops from the initial random field to the organized turbulent field. We executed direct numerical simulation (DNS) of this turbulence on the $128^3 \sim 512^3$ grid by the same alias-free Fourier spectral method (using the decomposed solenoidal velocity field with 4π-cyclic boundaries and the Runge-Kutta-Gill scheme for time integration) and with the same initial energy spectrum $E(k) = Ck^4 \exp(-2k^2)$ (k: wave number, $C = (16/3)(2/\pi)^{1/2}$ as we did before [1, 2]. The calculation was performed on the NAL numerical wind tunnel, a distributed memory-parallel computer with 128 vector-type processor elements. It achieved 113.8 GFLOPS for FFT and 90.3 GFLOPS for this turbulence scheme. The operation team was awarded the 1994 Gordon Bell Prize of the IEEE Computer Society. We present here some aspects predicted by our calculation on the issue of singularity of enstrophy in the inviscid limit and on the issue of coherent structure in isotropic turbulence. The other aspect will be published elsewhere [3].

Figure 1a shows the evolution of total dissipation ε for kinematic viscosity $\nu = 0.002$, 0.001 and 0.0005. Time t is normalized by turnover time of the initial eddy with $k = 1$. We can see here a remarkable fact that the peak of ε appears at almost the same time $t \simeq 10$. In Fig.1b is shown the trend of the peak values ε_ν against ν. Even though these values are influenced by errors, it is reasonable to think that we have a finite value of ε_ν for ν_0, which is easily predicted by the quadratic fit:

$$\varepsilon_\nu = 0.072336 - 247.21\nu + 5016.7\nu^2. \tag{1}$$

The value of ε_0 is indicated by the circle in Fig.1a. Hence we know that however small value of ε (for small ν) the initial random field started with, its peak value would be very close to that for $\nu = 0.0005$. The existence of ε_0 substantiates the basic ansatz of fully-developed isotropic turbulence by Kolmogorov [4] for the decaying case.

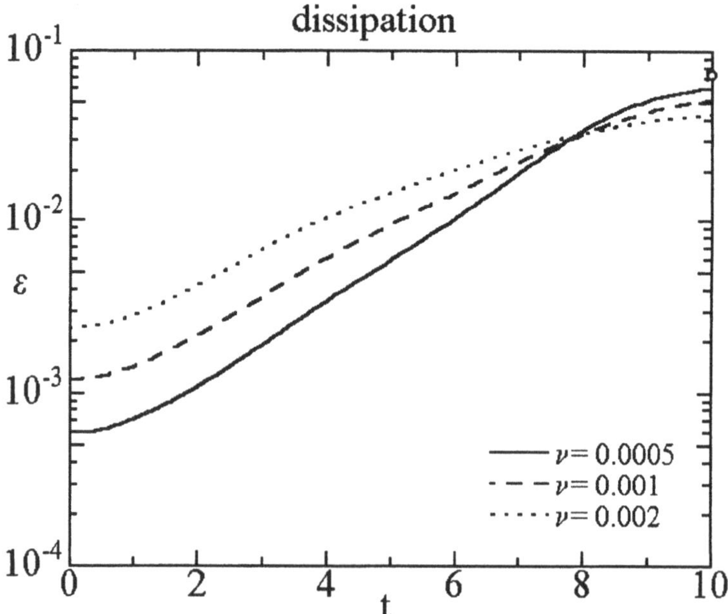

Figure 1a. Dissipation vs. time for various ν. The circle indicates the predicted maximum dissipation in the inviscid limit.

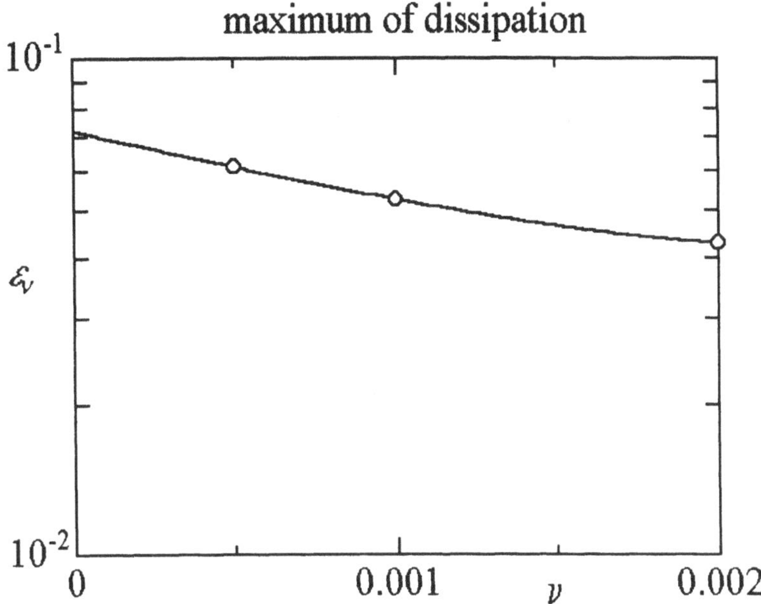

Figure 1b. Maximums of dissipation against various ν. The solid line is a quadratic fit to the DNS data indicated by circles, the formula of which is shown as eq.(1).

Let us reconsider the fact from the viewpoint of enstrophy:

$$\Omega = \varepsilon/(2\nu) \tag{2}$$

in Fig. 2. The theoretical initial value is $\Omega = \int k^2 E(k) \mathrm{d}k$, which is equal to 5/8 for the given energy spectrum. We can see that the period of exponential development increases as ν decreases, and the peak value of $\log \Omega$ increases in proportion to $\log \varepsilon_0 - \log(2\nu)$ as $\nu \to 0$ by virtue of eq.(1) and eq.(2). Then, enstrophy would be singular at about $t = 10$ in the inviscid limit. This fact must have a relation with the blow-up problem of solution of the Euler equation for an initial random field [5-7].Our consideration apparently supports the possibility of blow-up of enstrophy at a finite time [8, 9] also for our initial condition. Since viscosity, however small, always suppresses the enstrophy increase, the purely non-viscous development of enstrophy should not be less rapid than that in our predicted limit solution.

Now we go into the next issue. As Herring and Kerr stated [7], we can see many sheet-like blobs with high vorticity formed in the flow in the initial period of time. They tend to wind up to make tubes with much higher vorticity. At $t = 5$, sheet-like blobs and their wind-up coexist in the flow. We present Fig.3 which visualizes the vorticity contour of level 2.1 in the subbox of scale $L/8$ (L: main scale) with the largest local average of vorticity at this time for the case of $\nu = 0.0005$; level 1 is the global

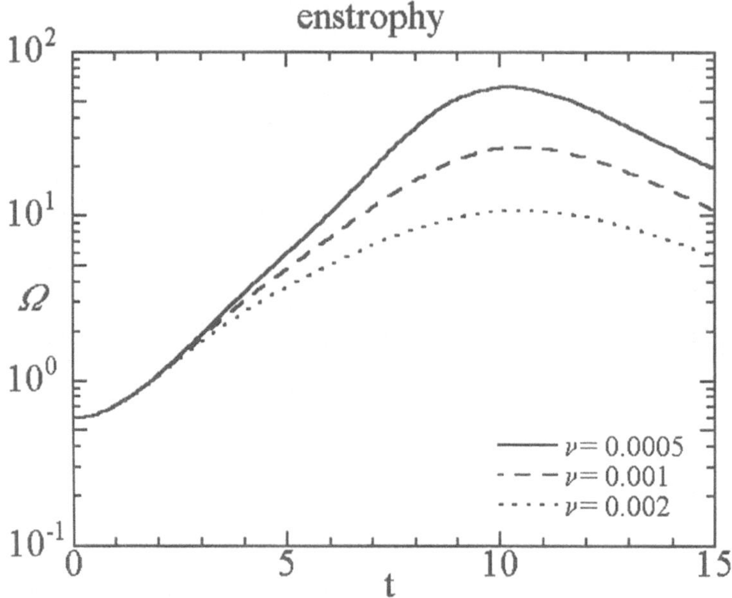

Figure 2. Enstrophy vs. time for various ν.

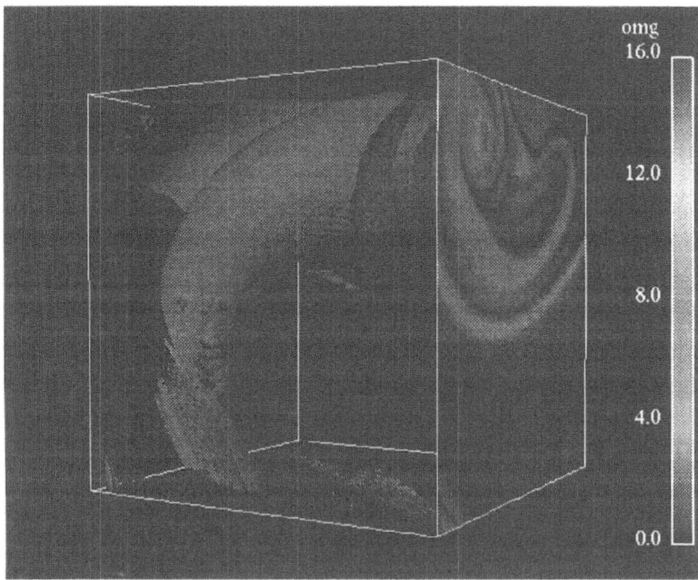

Figure 3. Contour of vorticity magnitude at level 2.1 in the subbox of scale $L/8$ with the largest local average vorticity at $t = 5$ for the case of $\nu = 0.0005$. The sectional contours in the right face of the subbox are measured by the color scale on the right side. (See color plates.)

average value at the respective time. The vorticity field in the right face of the subbox is measured by the color scale put at the right end. We can see here a remarkable wind-up of sheet which has a tube with high vorticity at the center. This big-scale tube is considered to be a typical premordial form of the well-known vortex tube seen at the fully-developed state of turbulence, which was dubbed "worm" [1, 2, 10]. On the other hand, there is no wind-up yet but only thin blobs with sheet-like high vorticity inside in the subbox of scale $L/8$ with the smallest local average of vorticity at the same time. The tube formation in our turbulence seems still difficult to fully explain, even though some interesting analyses [11, 12] have been done. Our simulation well supports the self-similar model of Brachet et al. [11] up to the formation of a sheet-like vorticity aligned with the eigenvector of strain tensor for the intermediate eigenvalue. The sheet, as soon as it evolves from a thin blob, appears to wind up basically by the Kelvin-Helmholz instability to make a tube, but under a complex strain caused by neighboring tubes and/or sheets. Figure 3 substantially reminds us of Lundgren's theory of strained spiral vortex [13, 14].

The flow reaches the fully-developed state at $t \simeq 10$, where enstrophy becomes maximum as was seen in Fig.2. Figure 4a visualizes the vorticity contours of level 3 in the subbox of scale $L/8$ with the largest local average of vorticity at this time for the case of $\nu = 0.001$. The contours embody many bent or twisted worms, well-known coherent structures in isotropic turbulence [15]. But what is in the background occupying the overwhelmingly large volume? Figure 4b visualizes the vorticity contours of level 1 in the same subbox. There we can see many tubes of various sizes, some of which fully cover a worm like a sleeping bag and others form an independent tube with not-so-high vorticity ($<$ level 3) at the core. We may call the latter "soft worms," distinguishing from the conventional (hard) worm with high vorticity at the core. Hence it is difficult to think that the background vorticity field is plainly random. Our result indicates that everywhere are coherent structures of vorticity without big voids.

We show in Fig.5 the visualization of the vorticity contours of level 0.2 in the subbox of scale $L/8$ with the smallest local average of vorticity in the same flow at the same time. Here in this subbox is no vorticity far higher than level 1. As is obvious from the color contours in the faces of the subbox, here are parts of very soft worms with relatively high vorticity at the cores, but much (several to ten times) thicker and longer than usual worms. We assured that all vorticity vectors around the cores of these soft worms are aligned with their axial directions. Note that a big soft worm often accompanies spiral vortex sheets around itself. Then it is natural to think that the tube formation exists in all regions, with vorticities lower as well as higher than the global average. In the former regions local Kol-

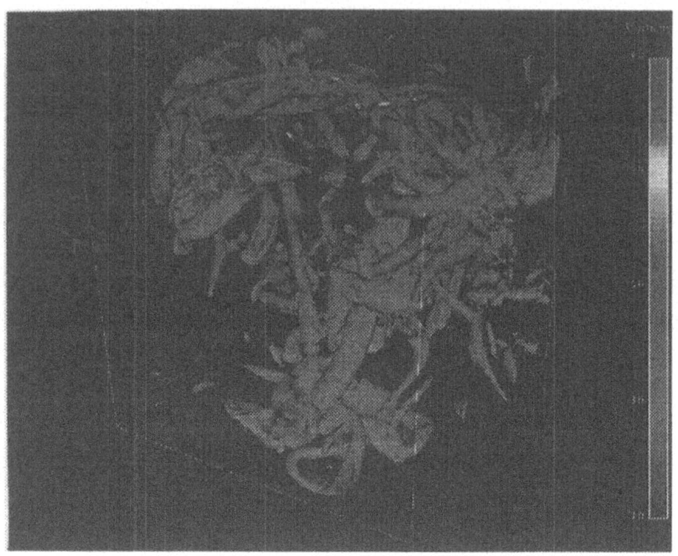

Figure 4.a. Contour of vorticity magnitude at level 3 in the subbox of scale $L/8$ with the largest local average vorticity at $t = 10$ for the case of $\nu = 0.001$. The sectional contours in each face of the subbox are measured by the color scale on the right side. (Transcription from Ref.17. See color plates.)

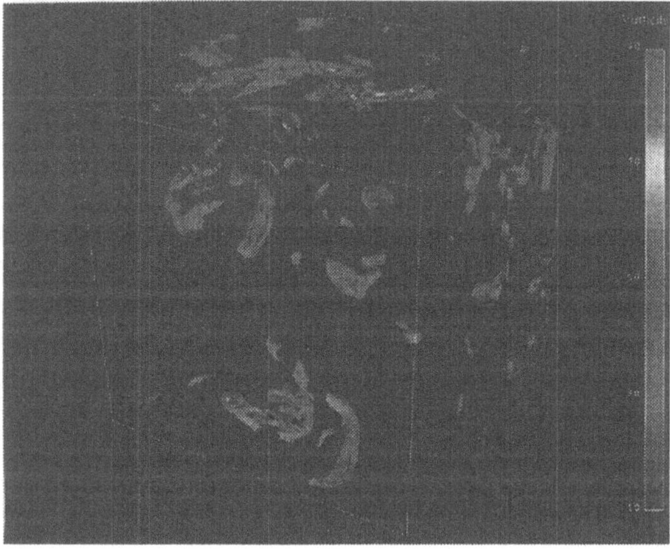

Figure 4.b. Contour of vorticity magnitude at level 1 in the subbox of scale $L/8$ with the largest local average vorticity at $t = 10$ for the case of $\nu = 0.001$. The sectional contours in each face of the subbox are measured by the color scale on the right side. (Transcription from Ref.17. See color plates.)

mogorov length as well as local integral scale are larger, so that the size of worms there must naturally be bigger. The statistics of worms investigated by Jimenez et al. [10] is imperfect in not including these soft worms with core vorticity under the average. If they are totally covered, the statistics will be considerably changed, since the vorticity magnitude of a half region of the volume is under the average. We think that the definitions of "hard" and "soft" are rather qualitative. It looks natural that there is a continuous spectrum of worms from hard to soft; the spectrum should tell the distributions of radius, length and vorticity magnitude in the center of a worm. It is a challenging task in future to fix this total spectrum somehow. Many small worms and spots seen in Fig.4a are considered to be the core parts of softer worms. Such soft worms, occupying eventually a considerable volume of space, are considered to contribute to the revelation of scale-similar or multifractal structure of turbulence [16] in low vorticity regions.

We investigated the cross-sectional vortical structures of both hard and soft worms, but failed fo identify them as the Burgers vortices at all. To begin with axial strain is never uniform in the cross-section of a worm but is often partly negative. All worms are, more or less, non-axisymmetric and distorted. However, there can be found a stable signature of the Burgers vortex, only if we look at these tubes in the circumferential average, as follows. In Fig.6a we plot the circulation $\Gamma(r)$ along circles of various radii r around the central axis in a certain cross-section of a typical hard worm for the case of $\nu = 0.0005$, which is in a high-vorticity region similar to Fig.4a. $\Gamma(r)$ is easily calculated from the vorticity field in the cross-section by Stokes' theorem. This $\Gamma(r)$ is smooth, matching very well with the circulation of the Burgers vortex:

$$\Gamma(r) = \Gamma(\infty)[1 - \exp(-ar^2/4\nu)] \tag{3}$$

if we take $\nu = 0.0005, \Gamma(\infty) = 0.415$ and $a = 1.15$, which is near the real value of the axial strain averaged over the circle area of radius r at $r = 0.06$. But we note that the average axial strain increases monotonically, starting from about 0.3 at $r = 0$; of course, this indicates how non-uniform in the cross-section the axial strain is. A similar matching for a typical soft worm taken from a low-vorticity region similar to Fig.5 can be seen in Fig.6b, where $\Gamma(\infty) = 1.30$ and $a = 0.09$. This value of a is close to the average axial strain at $r = 0.29$, but in this case the average axial strain is never monotonic with respect to r, varying so wildly as to get negative for r less than 0.05 and again for $r = 0.18 \sim 0.27$. Thus, a picture of the local Burgers vortices in turbulence may be retained only with these facts in mind. In other words, it is a tremendously bold idealization of worms after all various perturbations to them have been overlooked. In this idealization, however, the full spectrum of worms above-mentioned will be very relevant, because

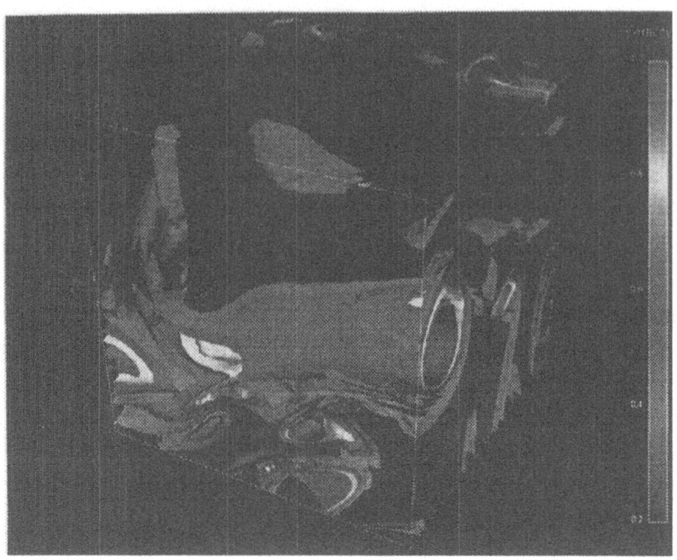

Figure 5. Contour of vorticity magnitude at level 0.2 in the subbox of scale $L/8$ with the smallest local average vorticity at $t = 10$ for the case of $\nu = 0.001$. The sectional contours in each face of the subbox are measured by the color scale on the right side. (Transcription from Ref.17. See color plates.)

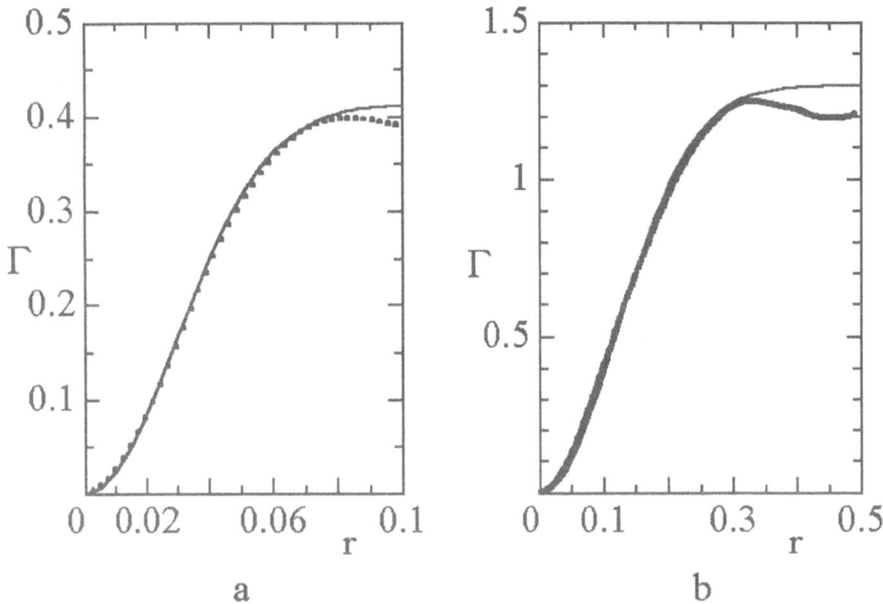

Figure 6. Circulation Γ along the circle of radius r around the central axis in a typical cross-section of a hard worm(a) and that of a soft worm(b) for the case of $\nu = 0.0005$. (Transcription from Ref.17.)

it is to designate how strong and what sizes of Burgers vortices distribute in what a way.

We thank S. Oide, T. Satoh and H. Yamaguchi for their computational assistance.

References

1. K. Yamamoto and I. Hosokawa, A Decaying Isotropic Turbulence Pursued by the Spectral Method, *J. Phys. Soc. Jpn.* **57**, 1532 (1988).
2. I. Hosokawa and K. Yamamoto, Intermittent structure of dissipation in isotropic turbulence viewed from a direct simulation, in *Turbulence and Coherent Structures* eds. O. Metais and M. Lesieur (Kluwer Acad. Publ., Dordrecht, 1991), p.177.
3. S. Oide, I. Hosokawa and K. Yamamoto, Direct Numerical Simulation of Decaying Isotropic Turbulence with a Passive Scalar, *Nagare* **16**, 259 (1997).
4. U. Frisch, *Turbulence* (Cambridge, 1995), p. 84.
5. A. Pumir and E. Siggia, Collapse solutions to the 3-D Euler equations, *Phys. Fluids* **A 2**, 220 (1990).
6. O. N. Boratav, R. B. Pelz and N. J. Zabusky, Reconnection in orthogonally interacting vortex tubes: Direct numerical simulations and quantifications, *Phys. Fluids* **A 4**, 581 (1992).
7. J. R. Herring and R. M. Kerr, Development of enstrophy and spectra in numerical turbulence, *Phys. Fluids* **A 5**, 2792 (1993).
8. R. M. Kerr, Evidence for a singularity of the three dimensional, incompressible Euler equation, *Phys. Fluids* **A 5**, 1725 (1993).
9. O. N. Boratav and R. B. Pelz, On the local topology evolution of a high-symmetry flow, *Phys. Fluids* **7**, 1712 (1995).
10. J. Jimenez, A. A. Wray, P. G. Saffman and R. S. Rogallo, The structure of intense vorticity in isotropic turbulence, *J. Fluid Mech.* **255**, 65 (1993).
11. M. E. Brachet, M. Meneguzzi, V. Vincent, H. Politano and P. L. Sulem, Numerical evidence of smooth self-similar dynamics and possibility of subsequent collapse for three-dimensional ideal flows, *Phys. Fluids* **A 4**, 2845 (1992).
12. T. Passot, H. Politano, P. L. Sulem, J. R. Angilella and M. Meneguzzi, Instability of strained vortex layers and vortex tube formation in homogeneous turbulence, *J. Fluid Mech.* **282**, 313 (1995).
13. T. S. Lundgren, Strained spiral vortex model for turbulent fine structure, *Phys. Fluids* **25**, 2193 (1982).
14. T. S. Lundgren, A small-scale turbulence model, *Phys. Fluids* **A 5**, 1472 (1993).
15. Z.-S. She, E. Jackson and S. A. Orszag, Intermittent vortex structures in homogeneous turbulence, *Nature* **344**, 226 (1990).
16. I. Hosokawa, S. Oide and K. Yamamoto, Isotropic Turbulence: Important Difference between True Dissipation Rate and Its One-Dimensional Surrogate, *Phys. Rev. Lett.* **77**, 4548 (1996).
17. I. Hosokawa, S. Oide and K. Yamamoto, Existence and Significance of 'Soft Worms' in Isotropic Turbulence, *J. Phys. Soc. Jan.* **66**, 2961 (1997).

FRACTAL AND SPIRAL ORGANISED STRUCTURES : SPECTRA AND DIFFUSION

J.R. ANGILELLA AND J.C. VASSILICOS

DAMTP
University of Cambridge, Silver Street, CAMBRIDGE, CB3 9EW, U.K.

1. Introduction

Objects displaying a well-defined Kolmogorov capacity, like fractals or spirals, have attracted the interest of mathematicians and physicists in the last decades, mainly because of their occurence in a wide variety of situations. For example, scalar fields rolled-up in a spiral way (Gilbert [2]), or displaying a fractal interface (Nicolleau [6]), are very common features of fluid flows. It has been recently shown that they might have remarkable physical properties (Vassilicos [7], Flohr & Vassilicos [1], Gurbatov & Crighton [3]), due to the space-filling feature of the distribution of their gradients. One of the most remarkable features of these convoluted structures is their spectral signature. Moffatt [5] and Gilbert [2] showed that the energy spectrum of a rolled-up patch of scalar differs from the spectrum of an isolated patch of scalar, and that the exponent of their spectrum depends on the geometrical characteristics of the spiral. Vassilicos and Hunt [8] generalized these results to both fractals and spirals by showing that algebraic spirals have a well-defined Kolmogorov capacity (like fractals), and that a scalar field characterised by sharp interfaces with well-defined Kolmogorov capacity has an energy spectrum which depends on that capacity. One of the main results of these authors is that the space-filling feature of the on-off scalar field makes it more singular. In the present paper we exhibit cases where space-filling properties have a regularising effect, leading to remarkable diffusive properties.

We investigate a one-dimensional field $u_0(x)$ defined as a sum of N Dirac delta functions (also referred to as pulses) located at positions x_i :

$$u_0(x) = \sum_{i=1}^{N} m_i \delta(x - x_i), \tag{1}$$

where m_i denotes the integral of the pulse located at $x = x_i$. Points (x_i) will hereinafter be referred to as discontinuity points. We consider two generic cases :

- non-alternating case : $m_i = m$ for all i,
- alternating case : $m_i = (-1)^{i+1}\, m$ for all i.

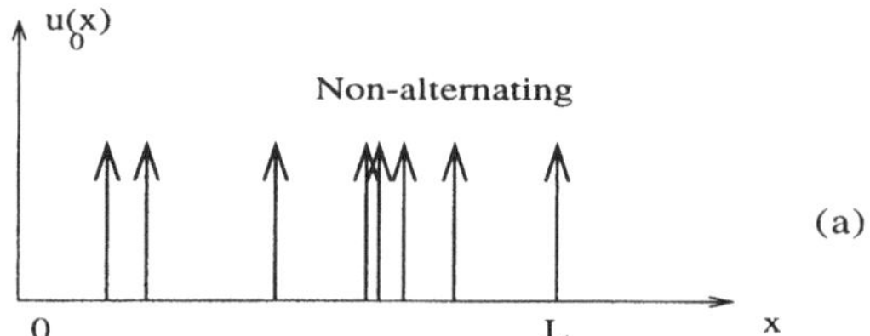

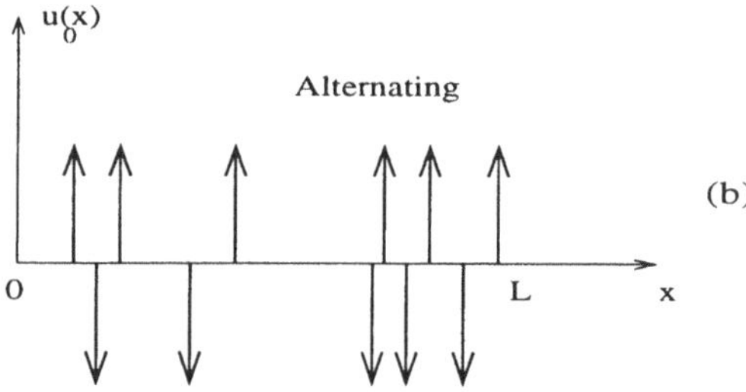

Figure 1. Sketch of the field $u_0(x)$. Vertical arrows denote δ functions.

We assume that the points (x_i) display a well-defined Kolmogorov capacity $D \in]0, 1[$, i.e. the minimal number of segments ("boxes") of size r required to cover all the points (x_i) reads

$$N_{boxes}(r) \simeq N \; for \; r < \eta \quad , \tag{2}$$

$$N_{boxes}(r) \simeq N \left(\frac{r}{\eta}\right)^{-D} \simeq \left(\frac{r}{L}\right)^{-D} \; for \; \eta < r < L \quad , \tag{3}$$

$$N_{boxes}(r) \simeq 1 \; for \; L < r, \tag{4}$$

where N is the total number of points, η is the smallest separation between points and L is the overall length scale of the structure. In general, η is fixed by some physical constraint (e.g. viscous cut-off) and L is prescribed by geometrical boundaries. Note that the total number of points is such that $N \sim (\eta/L)^{-D}$, so that it depends on η/L when D is fixed.

This paper is devoted to the analysis of the diffusive properties of the field (1), and to the comparison between alternating and non-alternating cases. In the next section we calculate the energy spectrum of the field $u_0(x)$. We then use this spectrum in section 3 to investigate the effect of molecular diffusion on $u_0(x)$.

2. The energy spectrum of fractal and spiral pulse fields

The energy spectrum $E_0(k)$ of u_0 in the alternating case has already been calculated (Vassilicos & Hunt [8]) and scales like k^D for $1/L \ll k \ll 1/\eta$. This result is valid for both fractals and spirals. Here we calculate this spectrum in the non-alternating case by extending the method of these authors. First we calculate the spectrum of the velocity potential $\psi(x) = \int_0^x u_0(x)dx$ from its structure function $< \delta\psi^2(r) >$, where $< . >$ denotes space average. Then we obtain the spectrum of u by multiplying the spectrum of ψ by k^2. For a given length r we have

$$\psi(x+r) - \psi(x) = m.q(x,r),$$

where $q(x,r)$ is the number of discontinuity points lying in the segment $[x, x+r]$, so that :

$$< \delta\psi^2(r) >= m^2 \frac{1}{L} \int_{-\infty}^{+\infty} q(x,r)^2 dx.$$

In the case of *homogeneous* fractals [4] the above integral can be estimated by $\int_{-\infty}^{+\infty} q(x,r)^2 dx \simeq \sum_{boxes} r \times \bar{q}(r)^2 \simeq r \times \bar{q}(r)^2 N_{boxes}(r)$, where $\bar{q}(r)$ is the number of points lying in each box of the minimal coverage of the structure with boxes of size r. We obtain

$$< \delta\psi^2(r) > \simeq m^2 N^2 \left(\frac{r}{L}\right)^{1+D}, \quad \eta \ll r \ll L.$$

Similarly, the structure function of order $p \geq 2$, in the limit $r \ll L$, is $< \delta\psi^p(r) > \simeq m^p N^p \left(\frac{r}{L}\right)^{1+(p-1)D}$. This result shows that the space-filling properties of the discontinuity points influence the structure function in such a way that they tend to make the field more autocorrelated. This is a remarkable difference with the alternating case, where space-filling properties are known to make the field less autocorrelated. For $D = 0$ we recover $< \delta\psi^2(r) > \sim r$, as expected for an isolated discontinuity. From the structure function calculated above we conclude that the energy spectrum of ψ

scales like k^{-D-2}, and therefore the spectrum of non-alterning delta functions displayed in a fractal manner scales like

$$E_0(k) \sim \frac{m^2}{L} N^2 (kL)^{-D}, \quad 1/L \ll k \ll 1/\eta,$$

and is flat elsewhere. This result is quite different from the alternating case (Vassilicos & Hunt [8]), where $E_0(k) \sim k^D$ in the "fractal range", i.e. the wavenumber range over which space-filling properties influence the spectrum. It shows that a fractal structure made of accumulated non-alternating delta functions is less singular than an isolated delta function, whereas a structure made of accumulated delta functions with alternated signs is more singular than an isolated delta function.

In the spiral case, we cannot use the homogeneity assumption, but the location of discontinuity points is known (e.g. $x_i = Li^{-\alpha}$, where $\alpha > 0$ determines the Kolmogorov capacity of the spiral [8]) and enables to integrate q^2, and therefore to calculate the structure function of ψ. We obtain the energy spectrum

$$E_0(k) \simeq \frac{m^2}{L} N^{2+\alpha} (kL)^{-1}, \quad 1/x_N \ll k \ll 1/\eta,$$

and $E_0(k)$ is flat elsewhere. We observe that the energy spectrum is modified by the space-filling structure, but its exponent does not depend on D, in contrast with the fractal non-alternating case, and the alternating case. It is the *range* over which the spectrum is proportional to k^{-1} ("spiral" range) that depends on D, and the larger the D, the larger the spiral range.

3. Trapping and accelerated diffusion

We use the spectra calculated above to investigate the energy decay of $u(x,t)$, when submitted to a diffusion process :

$$u(x,0) = u_0(x), \tag{5}$$

$$\frac{\partial}{\partial t} u(x,t) = \nu \frac{\partial^2}{\partial x^2} u(x,t), \tag{6}$$

where ν is the diffusion coefficient. Our initial field $u_0(x)$ is the sum of delta-functions defined in (1). Equation (6) implies that the energy spectrum of u at time t is

$$E(k,t) = E(k,0) e^{-2\nu k^2 t},$$

and the total energy is calculated by $E(t) = \int_0^{+\infty} E(k,t) dk$. For $t \ll \eta^2/\nu$, integration of the energy spectrum leads to a $t^{-1/2}$ energy decay in all cases, corresponding to the diffusion of isolated pulses. We now calculate the energy decay for $\eta^2/\nu \ll t \ll L^2/\nu$.

3.1. HOMOGENEOUS FRACTALS

In the case of homogeneous fractals, integration of $E(k,t)$ leads to

Non-alternating case :

$$\frac{E(t)}{E_\eta} \sim \left(\frac{\nu t}{\eta^2}\right)^{\frac{D-1}{2}} + O\left((\eta/L)^{1-D}\right), \; for \; \frac{\eta^2}{\nu} \ll t \ll L^2/\nu, \tag{7}$$

Alternating case :

$$\frac{E(t)}{E_\eta} \sim \left(\frac{\nu t}{\eta^2}\right)^{-\frac{D+1}{2}} + O\left((\eta/L)^{1+D}\right), \; for \; \frac{\eta^2}{\nu} \ll t \ll L^2/\nu, \tag{8}$$

where

$$E_\eta = \int_0^{1/\eta} E_0(k)dk$$

approximates the energy of the signal at $t = \eta^2/\nu$. One can see that the topology of the signals has huge effects on their energy decay, and that this effect is quite different in the alternating and non-alternating cases. In the alternating case the larger the Kolmogorov capacity the faster the energy decay, whereas in the non-alternating case, the larger the Kolmogorov capacity the slower the energy decay. This slow-down in energy decay we call diffusion trapping.

Note that the two relations lead to $E(t) \sim t^{-1/2}$ for $D = 0$, as this case corresponds to the decay of an isolated pulse. The predicted energy decay has been checked numerically by computing the energy of $u(x,t)$ defined as a sum of gaussian functions with a thickness that grows like $(\nu t)^{1/2}$. We use 2^{18} grid points and the number of pulses is of order 100. Results are depicted in figures 2 and 3. We now investigate the autocorrelation length-scale $\mathcal{L}(t)$ which is given by the weighted average

$$\mathcal{L}(t) = \frac{1}{E(t)} \int_{1/L}^{+\infty} \frac{1}{k} E(k,t)dk. \tag{9}$$

For short times ($t \ll \eta^2/\nu$) one can easily check that $\mathcal{L}(t) \sim (\nu t)^{1/2}$ in both the alternating and non-alternating cases. For longer times ($\eta^2/\nu \ll t \ll L^2/\nu$), we proceed like for the calculation of energy and neglect the

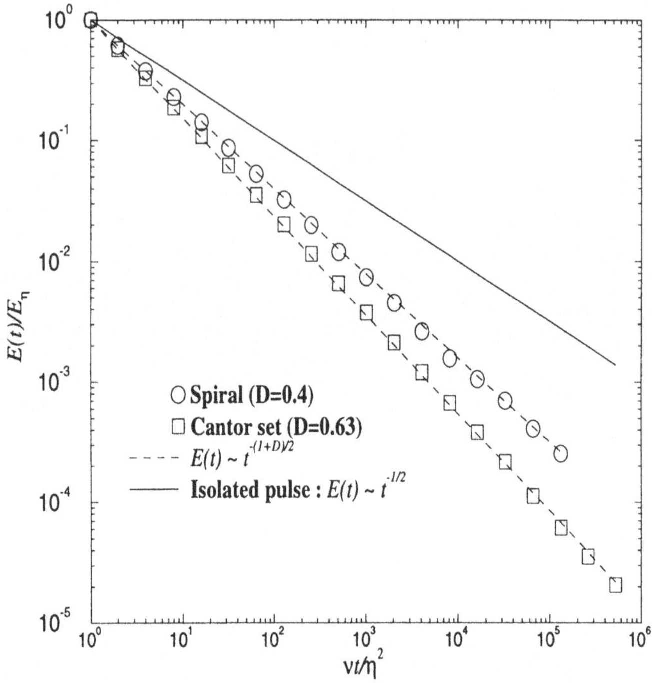

Figure 2. Energy decay of $u(x,t)$ in the alternating case, for fractals and spirals. The energy decay is accelerated.

terms corresponding to $k \gg 1/\eta$ in integral (9), since they are exponentially small. In the non-alternating case this leads to :

$$\mathcal{L}(t) \sim L\left(\frac{\nu t}{L^2}\right)^{\frac{1-D}{2}} + O\left(L\left(\frac{\nu t}{L^2}\right)^{\frac{1}{2}}\right), \quad \frac{\eta^2}{\nu} \ll t \ll L^2/\nu. \qquad (10)$$

A similar calculation in the alternating case leads to $\mathcal{L}(t) \sim \sqrt{\nu t}$ at leading order. Hence, the autocorrelation length-scale is unaffected by the geometry of the structure in the alternating case, whereas in the non-alternating it grows faster than in the non-fractal situation. This difference can be explained by noticing that two pulses with same sign (and same strength) rapidly diffuse into a single pulse, and that this process affects the autocorrelation length-scale, whereas two pulses with opposite sign (and same strength in absolute value) never merge into a single pulse, so that the correlation length-scale is only affected by the increase of the thickness of individual pulses (as $(\nu t)^{1/2}$).

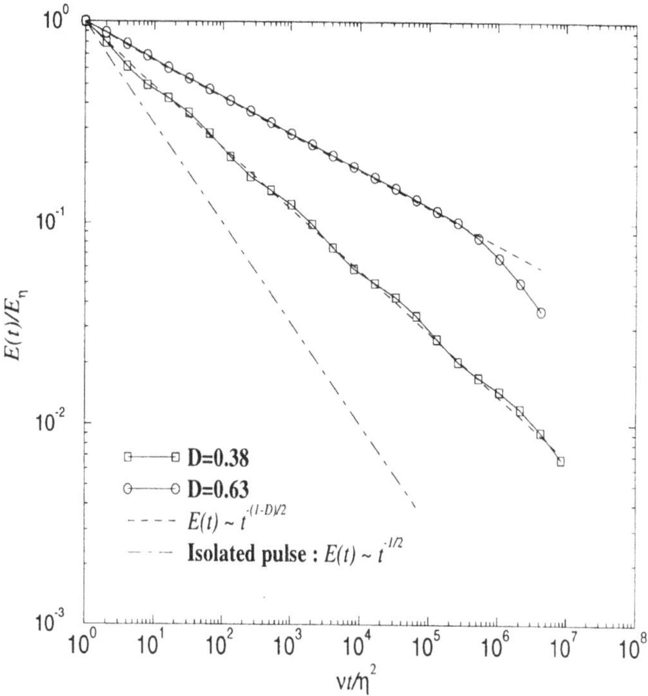

Figure 3. Energy decay of $u(x,t)$, for homogeneous fractals, in the non-alternating case. The energy decay is delayed.

3.2. SPIRALS

In the spiral alternating case the energy decay is similar to the fractal alternating case, since the initial spectrum is the same. In the spiral non alternating case, integration of the energy spectrum calculated above leads to

$$\frac{E(t)}{E_\eta} \sim \frac{\ln\left(\frac{x_N}{\sqrt{\nu t}}\right)}{\ln\left(\frac{x_N}{\eta}\right)} \quad for\ \eta^2/\nu \ll t \ll x_N^2/\nu,$$

and to a $t^{-1/2}$ decay over the time range $x_N^2/\nu \ll t \ll L^2/\nu$. This effect differs from the fractal case in that $E(t)$ is roughly constant ($E(t) \sim t^0$) in the time range $\eta^2/\nu \ll t \ll x_N^2/\nu$, then decays similarly to an isolated pulse ($E(t) \sim t^{-1/2}$) for longer times. This "trapping" of energy between η^2/ν and x_N^2/ν corresponds to the fact that pulses merge very often in the vicinity of x_N, and that this leads to a very weak decay of the local energy $u(x)^2$ in this region. Spiral diffusion trapping is more dramatic than homogeneous fractal diffusion trapping. Figure 4 shows a sketch of the

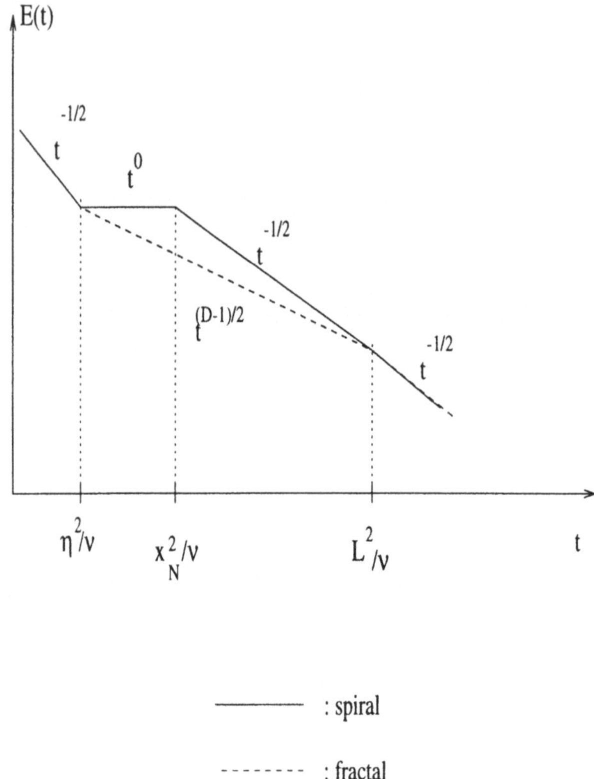

Figure 4. *Sketch of the energy decay of $u(x,t)$, for homogeneous fractals (dashed line) and for spirals (solid line), in the non-alternating case. In the spiral case the energy is "trapped" on the time range $\eta^2/\nu \ll t \ll x_N^2/\nu$, whereas in the homogeneous fractal case diffusion trapping is reflected in the slower self-similar energy decay from η^2/ν to L^2/ν.*

energy decay in the non-alternating fractal and spiral cases. In the fractal case energy is self-similar from η^2/ν to L^2/ν, whereas the spiral structure has an intermediate time scale x_N^2/ν.

The autocorrelation length-scale is also affected by the space-filling properties of the structure. It increases like $t^{1/2}$ for $t \ll \eta^2/\nu$ and for $x_N^2/\nu \ll t$, and increases only logarithmically in the time range $\eta^2/\nu \ll t \ll x_N^2/\nu$.

Figure 5 shows numerical results for a spiral with $D = 0.7$ and $N = 274$.

4. Conclusion

We have analyzed spectral and diffusive properties of fractal and spiral sets of delta functions corresponding to two generic situations. In the case

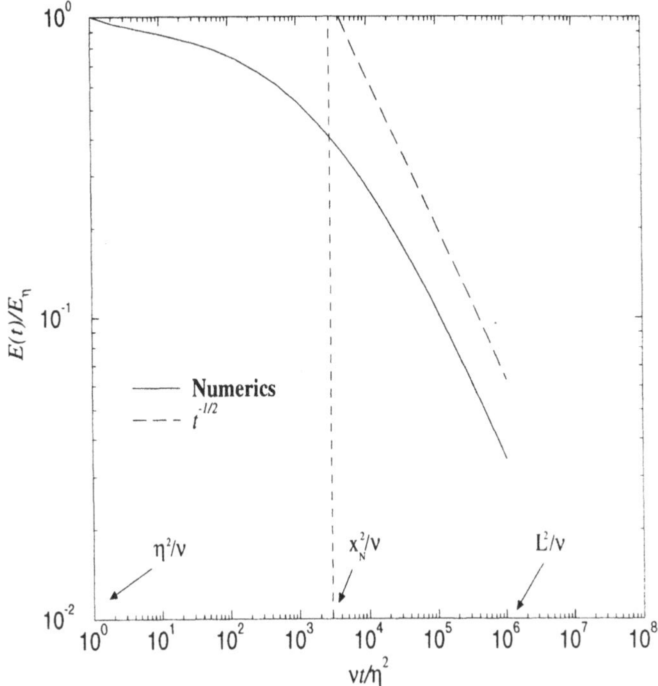

Figure 5. Energy decay of $u(x,t)$ in the spiral case with non-alternated signs.

of a fractal or spiral distribution of delta functions with alternated signs, the space-filling feature of the structure makes it more singular, as already pointed out by Gilbert [2], Moffat [5], and Vassilicos & Hunt [8]. In the case of a fractal or spiral distribution of identical delta functions, we find that the space-filling feature of the structure tends to *regularize* it, i.e. to make it more autocorrelated.

We have investigated the diffusive properties of these structures and observe that their space-filling feature does not always imply accelerated dissipation (Vassilicos [7], Flohr & Vassilicos [1]). In the non-alternating case dissipation is delayed, whereas it is accelerated in the alternating case.

In the non-alternating case, spectral and diffusive properties of fractals and spirals are similar. In the alternating case the homogeneity of the structure (in the sense of Hentschel & Procaccia [4]) is observed to be of major importance and is responsible for quantitative differences between fractals and spirals.

References

1. P. FLOHR and J.C. VASSILICOS. Accelerated scalar dissipation in a vortex. *Journal of Fluid Mechanics*, Accepted 1997.
2. A. GILBERT. Spiral structures and spectra in two-dimensional turbulence. *Journal of Fluid Mechanics*, 193:475–497, 1988.
3. S. N. GURBATOV and D.G. CRIGHTON. The nonlinear decay of complex signals in dissipative media. *CHAOS*, 5(3):524–530, 1995.
4. H.G.E. HENTSCHEL and I. PROCACCIA. The infinite number of dimensions of probabilistic fractals and strange attractors. *Physica D*, 435, 1983.
5. H. K. MOFFAT. Simple topological aspects of turbulent vorticity dynamics. *Proceedings IUTAM Symp. on Turbulence and Chaotic phenomena in Fluids. (ed. Tatsumi)*, page 223, 1984.
6. F. NICOLLEAU. Numerical determination of turbulent fractal dimensions. *Physics of Fluids*, 8(10):2661, 1996.
7. J. C. VASSILICOS. Anomalous diffusion of isolated flow singularities and of fractal or spiral structures. *Physical Review E*, 52(6), December 1995.
8. J. C. VASSILICOS and J.C.R. HUNT. Fractal dimensions and spectra of interfaces with application to turbulence. *Proc. R. Soc. Lond.*, A 435(505), 1991.

V. Geostrophic and Stratified Flows

VORTEX WAKES IN STABLY-STRATIFIED FLUIDS

G.R. SPEDDING

Department of Aerospace Engineering
University of Southern California
Los Angeles, CA 90089-1191

Abstract.

The evolution of an initially-turbulent wake behind a bluff-body is a useful model for a broad range of flows involving the decay of a local turbulent patch in a stable background density gradient. Such flows encompass many atmospheric and oceanic applications, the latter including also the evolution of wakes of submerged bodies.

In the case of the towed-sphere wake, the initially-turbulent motions evolve into a chain of alternate-signed vortices that remain close to the centreline. The characteristic, well-ordered wake is generated regardless of the relative magnitude of the background density gradient (provided it is nonzero), because eventually, buoyancy forces dominate the decaying flow. The wake geometry appears to be very stable, and the basic pattern has a very long persistence time.

When the large scale structures are generated in the presence of weaker background turbulence, then a vortex pattern detection problem inevitably arises as a requirement for identifying the foreground motions and their possible interactions with the ambient. The space of independent parameter combinations is now large, but experiments can help to identify relevant dimensionless groups.

Although the horizontal motions are most well-known, and are the easiest to visualize, the structure in the vertical direction is also important in determining the dynamics, stability and longevity of the wake in both quiet and noisy environments.

1. Turbulence and vortex motions in stratified fluids

1.1. LIMITING LENGTHSCALES

Turbulent motions in the atmosphere and ocean are affected by the stabilising background vertical density gradient over a large range of horizontal scales – from $1 - 10^3$km in the atmosphere, and from metre scales up to 10^2km in the ocean. Turbulence-generating events are also frequently episodic and limited in spatial extent, and the decay and evolution of an initially-turbulent patch in a stratified fluid is a canonical flow with numerous geophysical and engineering applications.

When inertial and buoyancy effects are in balance, the internal Froude number, $F = u/Nl = 1$ and turbulent motions can occur up to length scales limited by the Osmidov scale, which can be written,

$$l_0 = \left(\frac{\epsilon}{N^3}\right)^{\frac{1}{2}} \approx \left(\frac{u^3}{N^3 l}\right)^{\frac{1}{2}}, \tag{1}$$

where ϵ is the kinetic energy dissipation rate, u is a turbulent velocity, l is an integral length scale, and $N = (-g/\rho_0(\partial\rho/\partial z))^{\frac{1}{2}}$ is the buoyancy frequency. Initially, $l < l_0$, and the turbulence can evolve as if unaffected by stratification, but as it does so, l increases and u decreases until the two length scales are equal. Further fluid motions are constrained and become strongly anisoptropic. Nevertheless, the early turbulent motions can create disturbances that later cannot be reversed because insufficient kinetic energy is available. The turbulence continues to evolve into a vertically-layered structure that constitutes a memory of the initial conditions.

Although the maximum length scale argument can be made more sophisticated, for example, to account for the appearance of structures with vertical scales $> l_0$, the basic phenomenology seems well-supported by both numerical and laboratory experiment (*e.g.* Lin & Pao 1979; Browand *et al.* 1987; Liu *et al.* 1987; Staquet 1995; Xu *et al.* 1995; Fincham *et al.* 1996; Kimura & Herring 1996; Majda & Grote 1997).

1.2. VORTEX STRUCTURES

At the same time, the quasi-*2D* fluid motions within each layer become organised into contiguous patches of vertical vorticity that are coherent over very large horizontal distances. Recent years have seen excellent progress in characterising the emergence of these coherent motions from turbulent initial conditions (*e.g.* Voropayev *et al.* 1991; Voropayev & Afanasyev 1994), and in analysing 1-, 2-, 3- & 4-vortex topologies (*e.g.* Voropayev & Afanasyev 1992; Flór *et al.* 1993; Flór & van Heijst 1994, 1996). The idea is that more complex flows can be decomposed into 'atomic' structures whose properties are known in great detail.

Research in the evolution and decay of turbulent wake flows has also been stimulated by application to undersea vehicles and their detection. The pioneering work of Lin & Pao (1979) demonstrated the emergence of the 'pancake vortex' and contained hot film measurements of near-wake velocities for towed and self-propelled spheres and ellipsoidal profiles. That body of work has now been extensively updated by studies that capitalise on improved, quantitative experimental techniques. These include exquisite shadowgraph-based, or derived methods (Sysoeva & Chashechkin 1991), combined dye and streak photography (Lin *et al.* 1992), careful hot-film and innovative fluorescent dye techniques (Hopfinger *et al.* 1991; Chomaz *et al.* 1993; Bonneton *et al.* 1993) and particle-tracking or DPIV methods concentrating on the far wake (Chomaz *et al.* 1993; Spedding *et al.* 1996a,b; Spedding 1997).

1.3. OBJECTIVES

In this paper, the current state of knowledge of the evolution of initially-turbulent sphere wakes in a stably stratified environment will be briefly reviewed, and combined with some new results that are described in greater, but still selective, detail. In particular, the interaction of the wake with a turbulent background is postulated to depend on the vertical variability, and subsequent examples of the vertical distribution of vortex patches serve to emphasise the very different dynamics of stratified flows, as compared with purely *2D* or planar flows. A number of these properties emerge during a transitional, non-equilibrium phase that probably characterises any decaying turbulent flow in a stably-stratified environment.

2. Notes on the experimental methods

2.1. TOWED SPHERES

All experiments were performed using towed spheres suspended by thin, obliquely-mounted steel cables under high tension. Horizontal cable mounts cannot be used because, even when the initial Reynolds numbers are high, the flow induced by the late wake on the stationary cable generates a thick, low Re_θ boundary layer containing vorticity of opposite sign to the ambient mean vorticity, completely disrupting the vortex topology in the centreplane (unpublished, and unintended, results).

2.2. DPIV METHODS

The use of DPIV is becoming commonplace in fluid mechanics, but careful attention must be paid to the practical and theoretical limits in resolution and accuracy. A variant of DPIV, termed Correlation Image Velocimetry (CIV) (Fincham & Spedding 1997) was used to estimate velocity fields on nearly-horizontal isopycnals, or within vertical slices (or slabs) cut by laser. Comparatively modest sensors can be used because the experimental facilities are large, and the timescales in the late-wakes are long. Unless the smallest measurable scales approach the Kolmogorov scale then peak velocities and their gradients will invariably be under-estimated by DPIV methods. In late-wake flows the energy at the small scales where under-sampling can occur is very small, and can be neglected. At early times, sampling errors can be estimated and checked by simultaneous acquisition at different resolutions, and further details can be found in the original references.

When the velocity measurements in a plane are projections onto $\{x, y\}$ of motions on the isopycnal, and $\vec{q} = \{u, v\}$ is the velocity vector in $\{x, y\}$, then it is convenient to separate $\vec{q}$ into components,

$$\left.\begin{array}{l}\omega_z = \nabla \times \vec{q} \\ \Delta_z = \nabla \cdot \vec{q}\end{array}\right\}. \tag{2}$$

ω_z and Δ_z, can be regarded as approximate measures of the vortex and wave motions, respectively in the wake. The approximation becomes increasingly accurate with time as the isopycnal surface becomes horizontal and vertical velocities, $|w| << |\vec{q}|$.

2.3. FROUDE, REYNOLDS NUMBERS

The initial Froude and Reynolds numbers based on the tow speed, U and body diameter, D, are

$$F = 2U/ND, \quad Re = UD/\nu,$$

and are the only dimensionless numbers in the problem for steady motion of a given geometry, far from exterior boundaries. Typically Re and F are both large ($Re = O(10^8 - 10^9), F = O(10^2)$) in undersea vehicle wakes, and while $Re = O(10^{10})$, F can be $O(1)$ due to the large length scales in meteorological flows. From the applied point of view, there is strong incentive to extend to high Re and cover a broad range of F. It is quite simple to achieve high values of F through combinations of high U and small D, but it is harder to reach Re much beyond 5×10^4 in finite-sized laboratory facilities. At present these requirements also exceed available resources for numerical experiments.

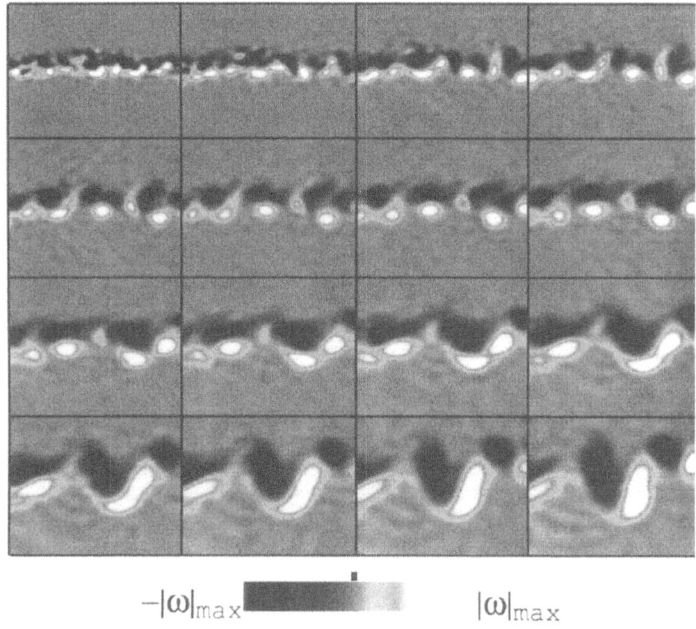

Figure 1. Time series of the vertical vorticity distribution, $\omega_z(x, y, Nt)$. $F = 40$ and $Re = 5400$, and the timesteps are logarithmically spaced from $Nt = 8$ (top left) to $Nt = 350$ (bottom right). The observation box is 72×54 cm, and the sphere passed from right to left through it.

2.4. INTERPRETATION OF SLICES

Interpretation of structure from single, or even multiple *2D* slices though *3D* objects can be misleading, particularly when the orientation of the cut is arbitrary with respect to flow. The late wake in a stably stratified flow however, has very strong anisotropy in the z axis that is aligned with the gravitational vector, g. Cuts that are normal, or parallel to this direction are thus not arbitrarily aligned with respect to the flow physics. Nevertheless, due caution must be exercised in making only statements about vorticity distributions in specified planes, without adding implicit arguments about *3D* structure for which there is no direct evidence.

3. Coherent structures in towed-sphere wakes

Figure 1 is a time series of the experimentally-measured vertical vorticity, $\omega_z(x, y)$. Coherent vortices can be distinguished from the first frame onwards, and they increase in size, and decrease in number primarily by pairing interactions between closest neighbours. $F = 40$ is a moderately high

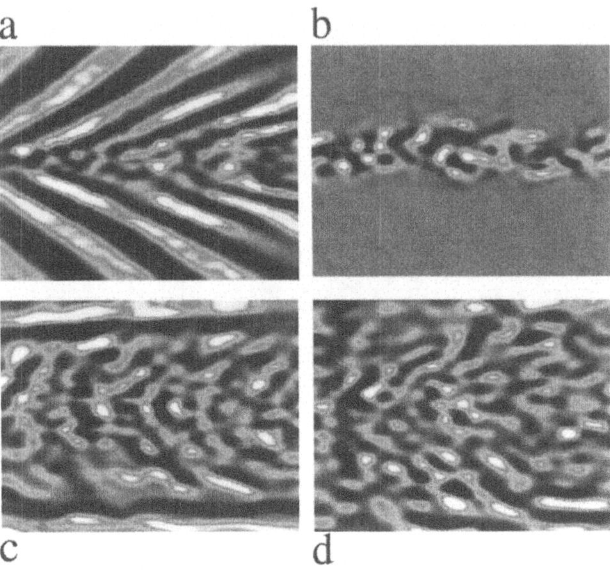

Figure 2. $\Delta_z(x,y)$ for low (a), and high (b-d) Froude number wakes, as measured at the centreplane. (a) $F = 1.2$, $Re = 4.9 \times 10^3$, $Nt = 44$ (b-d) are for $F = 40$, $Re = 5.4 \times 10^3$, and $Nt = 8$, 34 & 73, respectively. These times correspond to frames 1, 3 & 6 in figure 1, so the wave locations with respect to the wake vortices can be verified.

number, when the arguments in the introduction imply a large number of eddy turnover times before buoyancy forces can supress vertical motions at the large scales. Nevertheless, by the end of the top line, at $Nt = 50$, the vortices appear quite well-ordered, in the sense that the spacing is quasi-regular, and there is no other identifiable structure. The pairing and merging interactions between like-signed neighbours can readily be seen on the next two lines, up to $Nt = 220$. Although figure 1 stops at $Nt = 350$, similar processes continue, and while the vortex patches continue to grow in size and decline in number by a combination of neighbour-interactions and viscous diffusion, a recognisable wake structure can be seen up until $Nt \approx 2000$. The vortex-induced motions on the isopycnal correspond to the time-invariant, residual motions discussed by Lighthill (1996). Persistence times are very long, scaling up to 10 days or more in ocean applications. Making the equivalence $x/D = Nt.F/2$, the final frame on figure 1 corresponds to a downstream location, $x/D \approx 7 \times 10^3$. The same phenomenon can be observed over the entire Froude number range $F \in [1, 240]$ for all $Re > 4 \times 10^4$ currently amenable to experiment.

Stably-stratified flows also support internal waves, whose characteristics in the sphere wakes depend strongly on the initial F. Figure 2 shows characteristic examples of two different Froude number regimes. In figure 2a, $F = 1.2$, and the internal waves are primarily lee waves from the body itself. The remaining three panels show three characteristic times in the $F = 40$ wave field. In figure 2b, the divergence field is associated with overturning and strongly *3D* motions that have scales similar to the wake vortices themselves; indeed the distinction between vortex and wave modes is not very useful because the coherent patterns that will eventually characterise both are still emerging. In figure 2c, $Nt = 34$, and the most evident patterns are almost parallel to the wake centreline, and several vortex diameters from it. The wave-like disturbance is a single packet containing several cycles, that moves away from the centreline, and it is characteristic of high-F wakes. When F is high, turbulence timescales are short compared with N^{-1}, and so the numerous intrusions and subsequent recoil of turbulent eddies into the surounding fluid that comprise the overall expansion and contraction of the re-adjusting wake, contribute in ensemble to an oscillatory disturbance, almost parallel to the centreline, with a long coherence length in x. It is very interesting that this occurs after several ($34/2\pi \approx 5$) buoyancy periods. Figures 2c & d also show a third source of internal waves that persists at later times ($Nt = 73$) where wave packets can be observed radiating from the edge of the wake vortices as they advect through the fluid. The pattern of figure 2d and the elapsed time accord well with the figure 7 appearing in Hopfinger *et al.* (1991). Since the vortex spacing is not random, but quite regular, then the wave emissions are not random either, but should have an equivalent Strouhal number that accords with the vortex shedding in the wake, and Bonneton *et al.* (1993) have shown that this is so.

4. Particular and general characteristics of late wakes

One particular consequence of the highly-ordered vortex array in a stratified wake is an unusually high mean wake defect velocity (figure 3a), so that at any given x/D, the mean centreline velocity, U_0 is significantly higher than for the unstratified wake. U_0/U decays at a rate that is indistinguishable, experimentally, from the unstratified turbulent wake ($U_0/U \sim (x/D)^{-2/3}$). This in itself is surprising since there is no particular reason to expect such a similarity at large x/D. The apparent Froude number independence, and the coincidental similarity with turbulent decay rates in homogeneous fluids extends even to turbulence quantities, such as the mean wake-averaged vorticity magnitude, $W^{1/2}$ (figure 3b). Note how the ordinate is rescaled by F, so that when the data are plotted as a function of Nt, the same scaling relations are maintained as for the unstratified wake. Again, although this

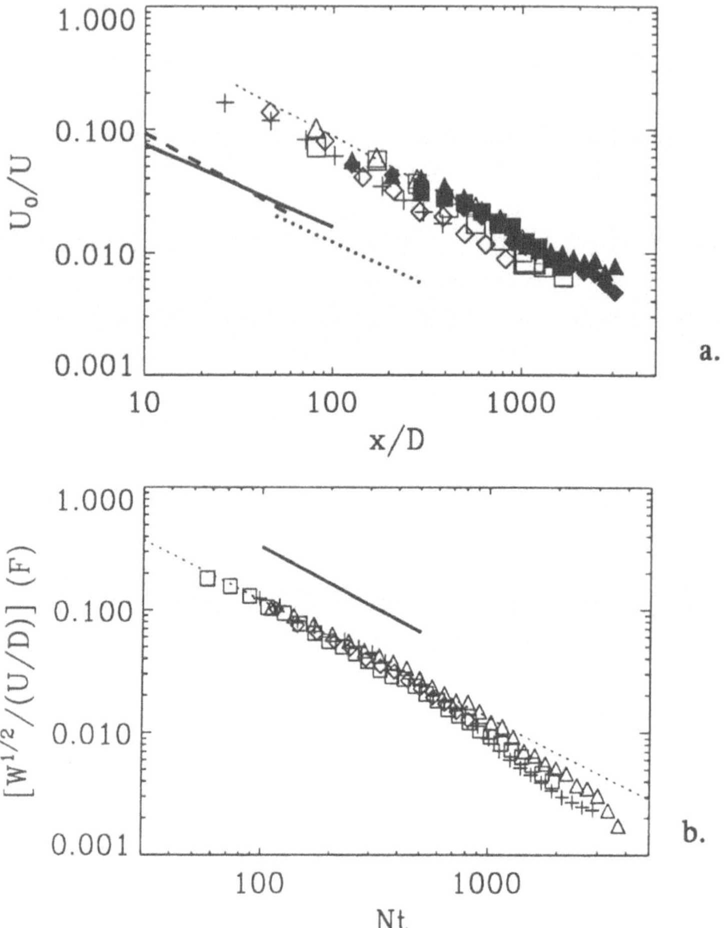

Figure 3. (a) The mean centreline velocity *vs.* downstream distance for a range of F from 1.2 to 15, and $Re = 5$ or 10×10^3. The literature values for the unstratified wakes are given by the solid lines. (b) Rescaled, wake-averaged vorticity magnitude *vs.* Nt for $F \in [1, 8]$ and $Re \approx 5 \times 10^3$. The solid line has arbitrary offset, but has a -1 slope that is found in unstratified, *3D* turbulent wakes. Figures from Spedding *et al.* 1996b, which may be consulted for further details.

seems like a reasonable assumption to make for early times, before the stratification effects are felt in the wake, it is curious to find collapse and similar decay rates out to such large values of $Nt \approx 1000$.

The $U_0/U(x/D)$ curves of figure 3a can be rescaled by N in a similar fashion, and figure 4a shows the result of doing this for $F \in [10, 240]$, where improvements in the experimental hardware and software allowed measurements at smaller values of Nt than before. The scatter is somewhat higher, as reflected in the larger symbol sizes, but the data appear to collapse on two lines, with a break at $Nt \approx 50$. Beyond $Nt = 50 (\equiv Nt_{II})$, the decay rate is approximately -2/3, as previously measured, but for $Nt < Nt_{II}$, the decay rates are significantly lower. The end result is that an initially-unspectacular mean defect velocity will be higher than its unstratified counterpart by the time it reaches Nt_{II}.

If one measures the kinetic energy dissipation rates due to horizontal gradients of horizontal velocities, they do not show the same breakpoint, but a consistent explanation for the decreased initial energy decay rates at the centreplane is that potential energy of displaced fluid is being converted back to kinetic energy as the wake restratifies. In fact, any kind of wake collapse that implies a return of energy towards the centreline must be accompanied by an increase in the local energy density there. If the initial decay is unaffected by stratification then the curves can be extrapolated back to their intersection with the empirical fits for *3D*, $(x/D)^{-2/3}$ decaying wakes. The intersection occurs at different values of x/D (depending on F), but at a constant value of $Nt \approx 2$ for the experimental data in figure 4a, and a general picture of the stratified wake time history can be drawn as in figure 4b.

Two characteristic times, $Nt_I = 2$ and $Nt_{II} = 50$ seem to divide three flow regimes: (i) 3D ($Nt < 2$) – a fully three-dimensional and turbulent decaying flow, (ii) NEQ ($2 < Nt < 50$) – relaxation of displaced fluid back to its equilibrium location with resulting low energy decay rates close to the centreline, and (iii) Q2D ($Nt > 50$) – a quasi-two-dimensional regime where nonlinear pairing interactions that are essentially planar can occur, but vertical displacements are increasingly limited. Comparisons are sometimes made with two-dimensional turbulent flows, but this regime is not at all similar. The empirically measured decay rates (e.g. figure 3) are different from *2D* or planar flows, and the dynamics are strongly affected by vertical gradients of horizontal velocity. Purely *2D* flows have very little in common with the Q2D regime that characterises late time evolution of stratified turbulence.

This kind of flow field evolution (3D $\rightarrow$ NEQ $\rightarrow$ Q2D) is quite general, requiring only an initial turbulent patch with some minimum values of $\{F, Re\}$ (scaling arguments and experimental measurements in Spedding *et al.* (1996a,b) suggest $F_{min} = 4$ & $Re_{min} = 4 \times 10^3$), so a turbulent flow can develop over a range of scales before the stabilising buoyancy forces reorganise the flow field. There is some evidence that the time Nt_I is a common feature of many such flows, but that the value of Nt_{II} is configuration-dependent (Spedding, 1997). The NEQ regime is a transitional one where several concurrent physical processes can be envisioned.

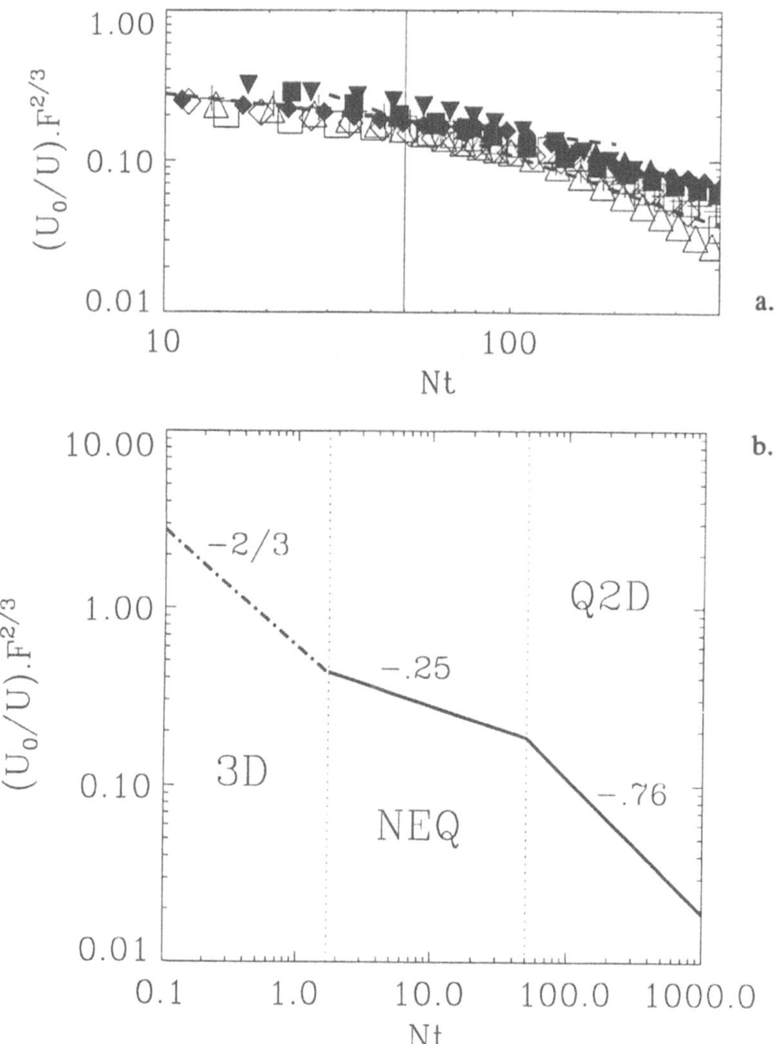

Figure 4. (a) Mean centreline velocity rescaled with stratification parameters for $F \in [10, 240]$. Dashed lines mark mean least-squares fits that intersect at approximately $Nt = 50$. (b) General curve that collapses time evolution of initially turbulent, stratified wake velocity profiles. The solid lines in the non-equilibrium (NEQ) and quasi-*2D* (Q2D) regimes are empirical fits to the data. The dot-dashed line in the 3D regime is presumed to apply at early times, but has not been directly measured. Figures taken from Spedding (1997).

The relaxation or recoiling of wake disturbances leads to emission of internal wave packets. The wave emission at $Nt = 34$ in figure 2c falls in the middle of the postulated NEQ regime, while the waves generated at the leading edge of the wake vortices (figure 2d, $Nt = 73$) are observed right up to its end, or even in the early part of the Q2D regime. The low overall NEQ energy decay rate reflects this balance between production of nearly horizontal motion near the centreline and the generation of wave-like disturbances that transport energy away from it.

In the specific case of the towed-sphere, it is clear that information regarding the initial vortex shedding is preserved through the NEQ phase and that the patterns established during this time dictate the late-wake structure, which can be maintained for very long times. What happens if these conditions are disrupted, say, by the existence of a background shear and/or turbulence?

5. Identification of wake-turbulence interactions

The interaction of the wake flow with a pre-existing, decaying, ambient turbulence field can be studied by making the tow shortly after rake-generated turbulence is created in the same tank. Currently, the initial condition for the wake is fixed as the turbulent field with initial $Re_t = 3000$ at an evolution time, $Nt_t = 300$, $x/M = 33$, where Re_t is based on the horizontal mesh spacing, M, of the vertical grid bars. The characteristics of such a turbulence have been extensively measured (see Fincham *et al.*, 1996). At this time vertical layering is well-established and horizontal and vertical integral length scales, l_h, l_v, are approximately $13D$ and $4D$, respectively. Figure 5 shows the time evolution of the sphere wake that is generated in the turbulent background but is otherwise identical to that shown in figure 1. The disruption of the wake geometry can be seen towards the end of the top line, where large-scale turbulent eddies are sweeping up wake vortices of the same sign. Although small scale remnants can be distinguished by eye up until the end of the third line ($Nt = 215$), they have been displaced from their normal location close to the centreline, and are decreasing in amplitude faster than the background turbulence (initial energy decay rates are approximately proportional to $t^{-4/3}$ in both flows in isolation).

In order to distinguish between a simple superposition of the wake flow on some larger-scale background from the case where there is some nontrivial nonlinear interaction between the two, it is necessary to measure wake quantities that take into account the geometric distortion of the wake centreline by the background. In finding a solution to this pattern detection problem, we also allow more quantitative estimates of the nature of the interactions. Figure 6 illustrates the process. The usual definition of wake-averaged quantities is shown in the top left panel where streamwise-averages, $U_x(y)$, are simply taken on straight lines of constant y. Clearly the same approach will fail for the distorted wake in the top right panel. Even when the vortices themselves can be readily identified, the definition

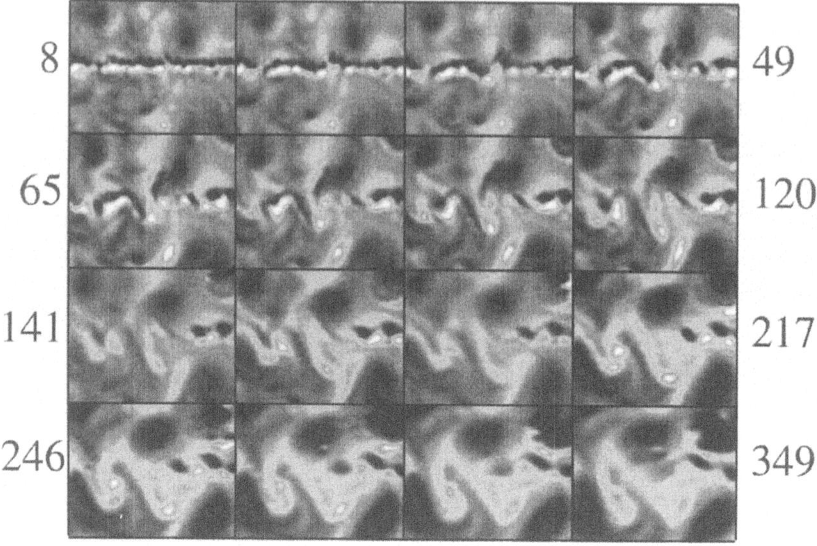

Figure 5. Wake turbulence interaction. The undisturbed wake would be the same as figure 1, although the observation box size has doubled to 146 × 111 cm. The numbers mark dimensionless times, Nt, for the left and right columns of data.

of the wake boundary is not so evident.

The necessary first step is to define a curved centreline, which is done by computing the streamfunction in the appropriate rotated and translated coordinates. The streamfunction is

$$\nabla^2 \psi = -\omega_z, \tag{3}$$

and the following transformations are applied to give the streamfunction in a frame of reference that is steady with respect to the mean wake:

$$\left.\begin{array}{c} \psi \to \psi - \bar{U}_x y + \bar{U}_y x \\ \psi \to \psi + \Omega R^2 \end{array}\right\}, \tag{4}$$

where $\bar{U}_x$ and $\bar{U}_y$ are mean translations, and Ω is a rotation rate about $R = 0$. These quantities can be found from the mean motion of the wake vortices from frame to frame, which is calculated using an automated tracking routine. All non-closed streamlines are now followed and the tangential velocity component is integrated along their length. The streamline with the highest value is denoted the centre streamline, ψ_0. In line 3 of figure 6, transects taken at regular intervals normal to ψ_0 are used to interpolate

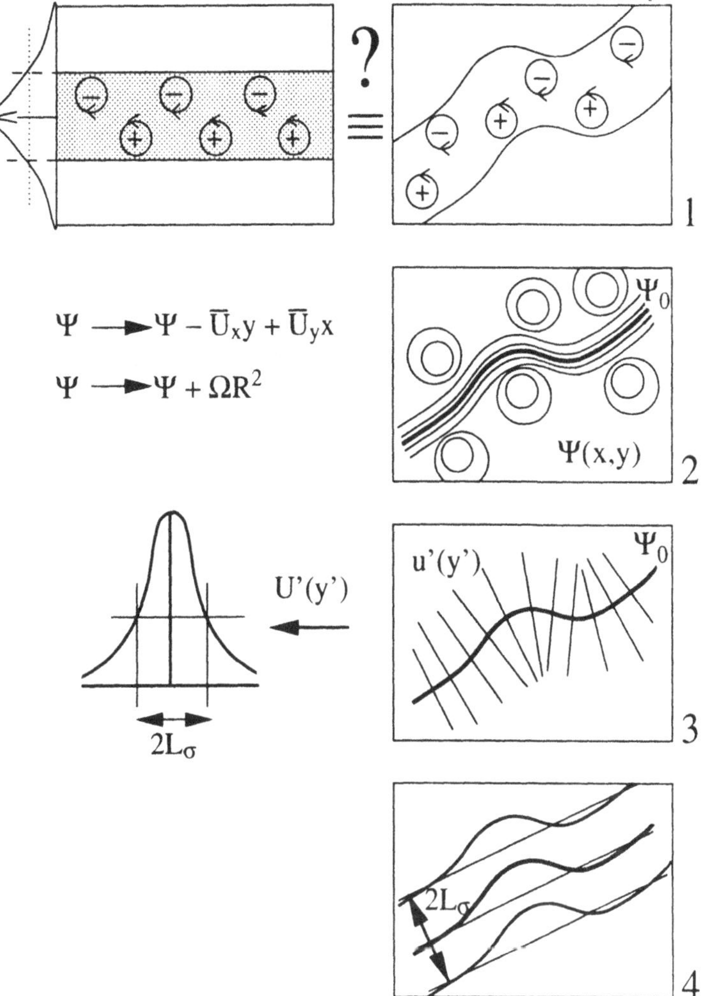

Figure 6. An algorithm for following curved wake profiles in a noisy background.

local profiles of normal velocity, $u'(y')$. The sum of these gives a mean normal profile, $U'(y')$, whose half width can be found and is denoted L_σ. The wake boundary is defined as the curved path that is L_σ removed from ψ_0, measured perpendicular to a linear fit of $\psi_0(x, y)$. During the procedure, normal and averaged velocity profiles are automatically accumulated, as are the mean centreline velocity and the mean wake width.

Figure 7 shows a time series of mean wake profiles collected in this way for the experiment in figure 5. The velocity profile is normalised by the local defect magnitude and so is always 1 at the $y' = 0$ location. Although the wake and turbulence kinetic energy decay at similar rates ($\sim t^{-4/3}$) in isolation, now the wake defect decays faster than the surrounding turbu-

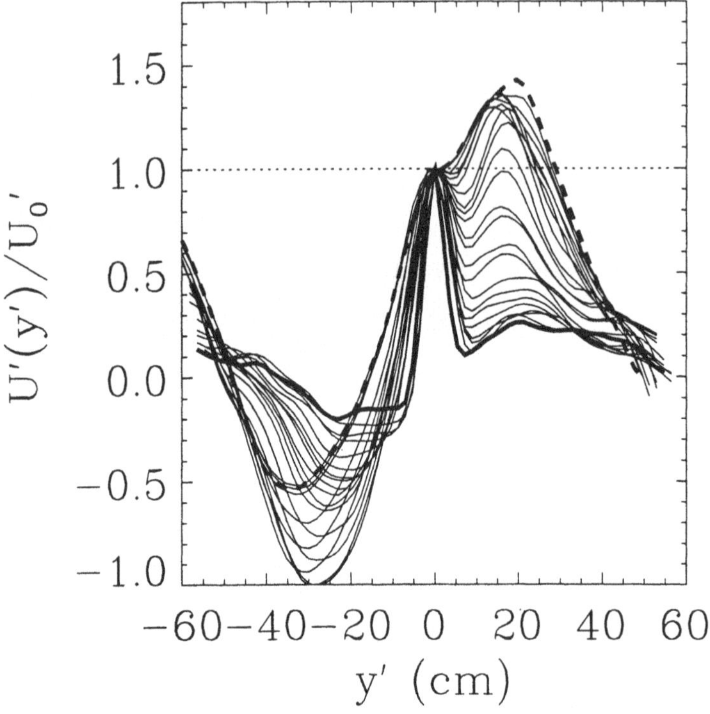

Figure 7. Normalised wake velocity profiles that diminish in amplitude with respect to the background turbulence. The first timestep is at $Nt = 8$ (thick solid line) and the last is at $Nt = 217$ (thick dashed line). The experiment is the same one shown in figure 5, and so corresponds to the first three lines in that figure.

lence, and so is indistinguishable by the last timestep, appearing only as a weak inflection in the otherwise arbitrary profile that arises from averaging normal profiles through the background turbulence. By this measure, a wake extinction time, $Nt_x \approx 200$ can be estimated for $F = 40$, $Re = 5400$. A similar experiment for $F = 4$, $Re = 6200$ showed the wake profiles to be comparatively unaffected, with $Nt_x > 1000$. It is tempting to offer an explanation based only on F, but it is not possible to maintain Re and modify F without also modifying the ratio of wake:turbulence initial length scales. Further experiments have also demonstrated a clear Reynolds number effect, where low Re wakes can be quickly disrupted.

The question arises as to which dimensionless groups determine Nt_x? The number of combinations of independent parameters is large, with separate, and changing, scales of length and time for both the wake and the background turbulence. Some physical arguments can help to focus ideas somewhat. For example, identifying a Reynolds number ratio would be rea-

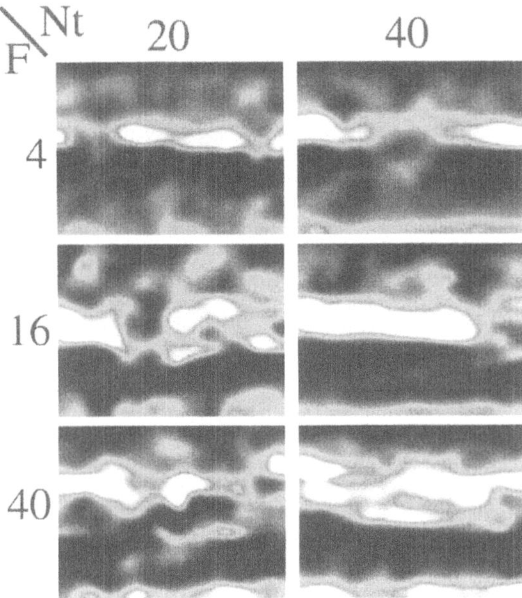

Figure 8. The horizontal vorticity distribution, $\omega_y(x, z)$ in the vertical centreplane ($y = 0$), for the $\{F, Re\}$ pairs, $\{4, 1.3 \times 10^3\}$, $\{16, 5.2 \times 10^3\}$ & $\{40, 1.3 \times 10^4\}$ at $Nt = 20$ & 40. The field of view is approximately 21 × 15 cm in x and z, and the sphere passed from left to right.

sonable if the main disturbance mechanism were essentially a *2D* process with vortex straining and merging on a plane. On the other hand, an explanation that involves F implicitly invokes arguments involving vertical scales and buoyancy forces. It is unsurprising that both Re and F seem to be important in determining Nt_x, and work is continuing on this problem.

6. Vertical structure

The vertical wake structure has been alluded to in discussion of multiple layers, propagation of internal waves, confinement effects (natural and artificial) and finally, in the wake-turbulence interaction problem. Actual data are just now beginning to emerge, and a sample is given in figure 8, where the component of vorticity normal to the vertical centreplane, $\omega_y(x, z)$, is shown for three different values of F (and Re), and at two comparatively early times in the wake formation, $Nt = 20$ & 40; they mark intermediate and late times during the NEQ regime.

In the low F, low Re case, the early wake is quite thin in the vertical, but because it meanders horizontally, the centre moves through the plane of observation. This is primarily an effect of low *Re* and is one reason why low *Re* ($< 3 \times 10^3$) wakes do not have the same scaling behaviour as those at higher *Re*. For $F = 16$, the initial wake is thicker, and more disorganised. At $Nt = 40$, it has begun to relaminarise, but the elongated vortex structures are not horizontal. There is some overlap between consecutive vortex patches of the same sign, and the wake remains thick. In the lower row, for $F = 40$, the initial and final wakes are thicker still, and there is more evidence for a more disjoint vertical structure.

In the two higher Froude number wakes ($F = 16$ & 40), $|\omega_y|_{max}$ is approximately 1/2 that of $|\omega_z|_{max}$ at these times (the two frames at $F = 40$ correspond roughly to frames 2 and 4 of figure 1), but while $\langle \epsilon_z \rangle$ (wake-averaged dissipation due to gradients of velocity components normal to z) decreases from $Nt = 20$ to $Nt = 40$, $\langle \epsilon_y \rangle$ increases during the same period to about $1.5\langle \epsilon_z \rangle$. Although the contribution of $\langle \epsilon_x \rangle$ to the total dissipation will not be so large (because of the absence of the mean, in-plane wake flow), it is evident that vertical shearing motions are significant contributors to the dynamics of the decaying wake.

It was conjectured in the introduction that high *Re*, high *F* wakes should leave a vertically-banded structure that contains information about the initial generating conditions (note that the horizontal structure does so too, *c.f.* figure 1), and the evidence in figure 8 supports the idea, although the geometry of the iso-vorticity surfaces is not simple. The increased complexity of the high-*F* wake is also consistent with the *F*-dependence of the wake-turbulence interaction experiments.

7. Concluding remarks

Initially-turbulent wakes of towed spheres can be used as the most simple reference case for wake turbulence of undersea objects, and also as a general exemplar of the long time evolution of a decaying turbulent patch in a stabilising density field. The non-equilibrium (NEQ) regime that marks the transition between initially-*3D* turbulent flow, and long-time dynamics that are dominated by buoyancy forces, exhibits fascinating and subtle dynamics that underlie the late-wake formation, and that may have quite broad application to geophysical flows.

Acknowledgements Many thanks to Prof. F.K. Browand for his continued support and helpful discussions, and to Robert Bell and Jared Ball for assistance in the production and analysis of the experimental data. The support provided by ONR Grant #'s N00014-92-J-1062 & N00014-96-1-1001, administered by Dr. Pat Purtell, is most gratefully acknowledged.

References

Bonneton P., Chomaz J.M. and Hopfinger E.J. (1993) Internal waves produced by the turbulent wake of a sphere moving horizontally in a stratified fluid. *J. Fluid Mech.*, **254**, 23–40.

Browand F.K., Guyomar D., and Yoon S.C. (1987) The behaviour of a turbulent front in a stratified fluid: experiments with an oscillating grid. *J. Geophys. Res.*, **92**, 5329–5341.

Chomaz J.M., Bonneton P., Butet A., and Hopfinger E.J. (1993) Vertical diffusion in the far wake of a sphere moving in a stratified fluid. *Phys. Fluids*, **5**, 2799–2806.

Chomaz J.M., Bonneton P., and Hopfinger E.J.(1993) The structure of the near wake of a sphere moving horizontally in a stratified fluid. *J. Fluid Mech.*, **254**, 1–21.

Fincham A.M., Maxworthy T., and Spedding G.R. (1996) Energy dissipation and vortex structure in freely-decaying, stratified grid turbulence. *Dyn. Atmos. Ocean*, **23**, 155–169.

Flór J.B., Govers W.S.S., van Heijst G.J.F., and van Sluis R. (1993) Formation of a tripolar vortex in a stratified fluid. *Appl. Sci. Res.*, **51**, 405–409.

Flór J.B. and van Heijst G.J.F. (1994) Experimental study of dipolar vortex structures in a stratified fluid. *J. Fluid Mech.*, **279**, 101–134.

Flór J.B. and van Heijst G.J.F. (1996) Stable and unstable monopolar vortices in a stratified fluid. *J. Fluid Mech.*, **311**, 257–287.

Hopfinger E.J., Flór J.B., Chomaz J.M. and Bonneton P. (1991) Internal waves generated by a moving sphere and its wake in a stratified fluid. *Exp. Fluids*, **11**, 255–261.

Kimura Y. and Herring J.R. (1996) Diffusion in stably stratified turbulence. *J. Fluid Mech.*, **328**, 253–269.

Lighthill M.J. (1996) Internal waves and related initial-value problems. *Dyn. Atmos. Ocean*, **23**, 3–17.

Lin J.T. and Pao Y.H. (1979) Wakes in stratified fluids: a review. *Ann. Rev. Fluid Mech.*, **11**, 317–338.

Lin Q., Lindberg W.R., Boyer D.L., and Fernando H.J.S. (1992) Stratified flow past a sphere. *J. Fluid Mech.*, **240**, 315–354.

Liu Y.N., Maxworthy T., and Spedding G.R. (1987) Collapse of a turbulent front in stratified fluid 1. Nominally two-dimensional evolution in a narrow tank. *J. Geophys. Res.*, **92**, 5427–5433.

Majda A.J. and Grote M.J. (1997) Model dynamics and vertical collapse in decaying strongly stratified flows. *Phys. Fluids.* (in press).

Spedding G.R. (1997) The evolution of initially-turbulent bluff-body wakes at high internal Froude number. *J. Fluid Mech.*, **337**, 283–301.

Spedding G.R., Browand F.K., and Fincham A.M. (1996a) The long-time evolution of the initially-turbulent wake of a sphere in a stable stratification. *Dyn. Atmos. Ocean*, **23**, 171–182.

Spedding G.R., Browand F.K., and Fincham A.M. (1996b) Turbulence, similarity scaling and vortex geometry in the wake of a sphere in a stably-stratified fluid. *J. Fluid Mech.*, **314**, 53–103.

Staquet C. (1995) Two-dimensional secondary instabilities in a strongly stratified shear layer. *J. Fluid Mech.*, **296**, 73–126.

Sysoeva E.Y. and Chashechkin Y.D. (1991) Vortex systems in the stratified wake of a sphere. *Izv. Akad. Nauk SSSR, Mekh. Zhidk. Gaza*, **4**, 82–90.

Voropayev S.I. and Afanasyev Y.D. (1992) Two-dimensional vortex-dipole interactions in a stratified fluid. *J. Fluid Mech.*, **236**, 665–689.

Voropayev S.I. and Afanasyev Y.D. (1994) *Vortex Structures in a Stratified Fluid.* Chapman & Hall.

Voropayev S.I., Afanasyev Y.D., and Filippov I.A. (1991) Horizontal jets and vortex dipoles in a stratified fluid. *J. Fluid Mech.*, **227**, 543–566.

Xu Y., Fernando H.J.S., and Boyer D.L. (1995) Turbulent wakes of stratified flow past a cylinder. *Phys. Fluids*, **9**, 2243–2255.

BAROCLINIC TRIPOLAR GEOSTROPHIC VORTICES

Formation and subsequent evolution

XAVIER J. CARTON AND STÉPHANIE M. CORRÉARD
SHOM/CMO
13 rue du Chatellier, 29275 Brest, France

Abstract

In the Bay of Biscay, large anticyclonic vortices (called swoddies) have been observed to form surface-intensified tripoles. Here we idealize this process in a two-layer quasi-geostrophic model. First, we compute the growth rate of elliptic disturbances on circular baroclinic vortices. Such perturbed vortices, either with a continuous or a piecewise-constant potential vorticity profile, are then used as initial conditions for nonlinear simulations in a numerical quasi-geostrophic model at high Reynolds number. Finite-amplitude evolutions yield baroclinic dipoles in the case of strong instability, but also uniformly rotating states, such as surface-intensified or arch-shaped tripoles and orthogonal elliptical vortices, when the instability is moderate. These baroclinic tripoles are novel stationary solutions of the stratified geostrophic dynamics. Their long-term evolution is naturally stable, though an asymmetric breaking into a monopole and a dipole can be induced by increased dissipation or by beta-effect.

1. Introduction

In-situ measurements and satellite observations have shown the formation of large anticyclonic vortices (swoddies) from the slope water current over the Cape Ferret Canyon in the Bay of Biscay (Pingree & LeCann, 1992). Soon after their formation, these swoddies acquired a tripolar structure, with an elliptical anticyclone at the center surrounded on each side by two elongated cyclones. Tripole formation is a well-known evolution of unstable geostrophic vortices in 2D flows (Hopfinger & van Heijst, 1993): when elliptically perturbed, strongly unstable circular vortices form two dipoles drifting in opposite directions, while less unstable vortices form a steadily rotating tripole, with a nonlinear vorticity-streamfunction relation (Carton & Legras, 1994). If the vorticity distribution is piecewise-constant instead of continuous, it has also been shown that tripoles are stationary solutions of the two-dimensional Euler equation (Polvani & Carton, 1990).

In stratified flows, unstable, purely baroclinic vortices do not form steady states when elliptically perturbed, but transform into counter-rotating, pulsating ellipses (Carton & Mc Williams, 1989 and 1996). To idealize swoddy evolutions in a two-layer fluid, we study unstable circular vortices, for which barotropic and baroclinic instabilities compete. The very novel result of this study is the formation of steady and robust baroclinic tripoles. After presenting the mathematical and numerical models (section 2), we show that unstable baroclinic vortices with continuous profiles can form tripoles (section 3). We restrict the problem to piecewise-constant vortices, for which the parameter space is smaller and classify the various nonlinear regimes (section 4). In section 5, the tripolar end-states of these nonlinear simulations are proved to be stationary and naturally long-lived vortex solutions of the stratified quasi-geostrophic dynamics.

2. The mathematical and physical models

Our model is based on the two-layer quasi-geostrophic equations, which express the conservation of layerwise potential vorticity in the unforced, inviscid limit:

$$\frac{dQ_j}{dt} = \partial_t Q_j + J(\psi_j, Q_j) = 0 \tag{1}$$

where the two-dimensional advecting velocity is $\vec{u}_j = \vec{k} \wedge \vec{\nabla}\psi_j$, $J(a,b) = \partial_x(a)\partial_y(b) - \partial_x(b)\partial_y(a)$ is the Jacobian operator, and the three-dimensional potential vorticity is

$$Q_j = \nabla^2\psi_j + \frac{f_0^2}{g'H_j}(\psi_k - \psi_j), \quad k = 3 - j. \tag{2}$$

Here $j = 1, 2$ is the (upper, lower) layer index; $g' = g(\rho_2 - \rho_1/\rho_0)$ is the reduced gravity (ρ_j is the density in layer j and ρ_0 its vertical average); f_0 is the Coriolis parameter; H_j is the thickness of layer j and $H = H_1 + H_2$. The internal radius of deformation is given by $R_d^2 = g'H_1H_2/f_0^2H$. These equations are parity-invariant.

The numerical model relies on a pseudo-spectral projection and truncation of the quasi-geostrophic equations on a biperiodic square grid ($L = 4\pi$), with 256^2 nodes. The time-step is controlled by the Courant-Friedrichs-Lewy stability criterion. Numerically, a traditional biharmonic viscosity is used to remove the small-scale noise: the zero on the right-hand side of Eq. 1 is replaced by $\nu_4\nabla^6\psi_j$, with $\nu_4 = 5\,10^{-9}$.

The initial conditions consist in a circular baroclinic vortex, either with a continuous (section 3) or with a piecewise-constant (section 4) radial profile of potential vorticity. These vortices are surface intensified, and the volume integral of their potential vorticity vanishes, so that they are isolated, as are ocean vortices (Morel, 1995). For swoddies, the radius of maximum velocity R lies at 25 km; it is normalized to unity in the model. In-situ data also gives values of 15 to 22.5 km for the first internal radius of deformation R_d in the Bay of Biscay. The dimensionless ratio R/R_d thus varies between 1.66 and 1.11. Considering that most of the swoddy energy was found above the 700m depth, with a noticeable decrease around 350m, we initialize our model with equal layer thicknesses $H_1 = H_2 = 350m$.

3. Instability of baroclinic vortices with a continuous vorticity profile

In-situ data of azimuthal velocities (Pingree & LeCann, ibid) suggest that a circular vortex idealizing a swoddy can be defined by the following relative vorticity distribution:

$$\bar{\zeta}_1 = \nabla^2 \bar{\psi}_1 = \zeta_0 (1 - n\, r^n/2)\, exp(-r^n), \quad \bar{\zeta}_2 = \alpha \bar{\zeta}_1 \tag{3}$$

where $n = 2$ or $n = 3$. Note that this implies that the initial potential vorticity distribution varies with stratification, contrary to section 4.
First, we solve the linear instability problem for this vortex. A normal mode perturbation is added to the mean flow:

$$\psi_j = \bar{\psi}_j + \psi'_j, \quad \psi'_j = \phi_j(r)\, exp[i\ell(\theta - ct)], \quad |\phi_j| / |\bar{\psi}_j| << 1 \tag{4}$$

We linearize the quasi-geostrophic equations (1-2) around the mean flow (3), and compute the growth rate $\ell Im(c)$ of elliptical disturbances ($\ell = 2$) by means of a numerical eigenvalue solver. On Fig. 1, we plot the growth rate with respect to the mean vertical shear α and to the Froude number $F = (H_1 H_2 / H^2)(R/R_d)^2$ which represents the intensity of stratification. Here we have chosen $n = 3$ so that the instability is both barotropic and baroclinic (Carton & McWilliams, 1989), yet with a stronger baroclinic contribution as shown by the increase of the growth rate with the Froude number and with the vertical shear of the mean vortex.
For nonlinear simulations in the spectral code, α is set to 0.5 from oceanic data. The parameter space is thus reduced to $(n,\ R/R_d)$. For $n = 2$, the mean vortex is marginally unstable for $R/R_d = 1.11$, and forms an elliptical vortex for $R/R_d = 1.66$. For $n = 3$ and $R/R_d = 1.66$ and 1.11, the resulting tripole is like-signed in the upper and lower layers (see the time series of potential vorticity maps in both layers on Fig. 2). Once formed, the tripolar structure rotates with little deformation (an occasional filamentation is

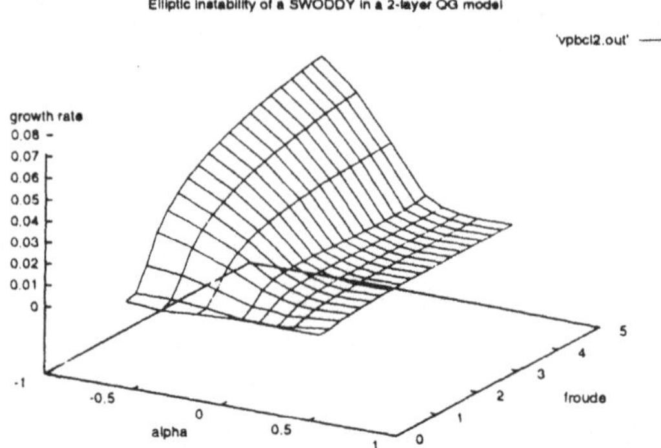

Figure 1. Growth rate of the elliptic perturbations (normal modes) for the mean vortex defined by Eq. 3,with n=3, and with respect to the vortex baroclinicity α and to the Froude number.

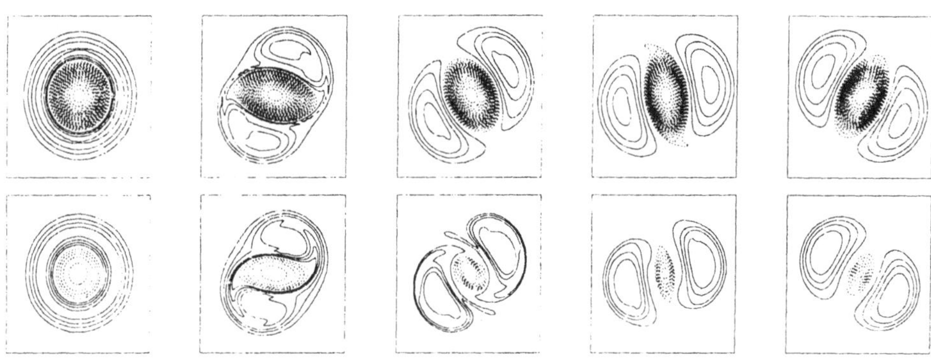

Figure 2. Time series of potential vorticity maps for the instability of the continuous vortex defined in section 3 with $n = 3$ and $R/R_d = 1.66$; the upper (lower) panels represent the upper (lower) layer, at time $t = 0, 80, 160, 240, 480$. Contour intervals are 0.05 (0.03) in the upper (lower) layer. Solid (dashed) lines indicate positive (negative) vorticity.

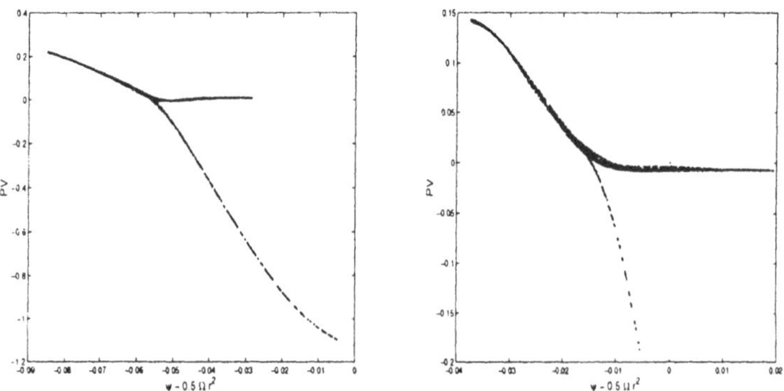

Figure 3. Scatter plots of potential vorticity versus relative (corotating) streamfunction in the upper and lower layers, for the final tripolar state ($\Omega = -4.078\,10^{-2}$).

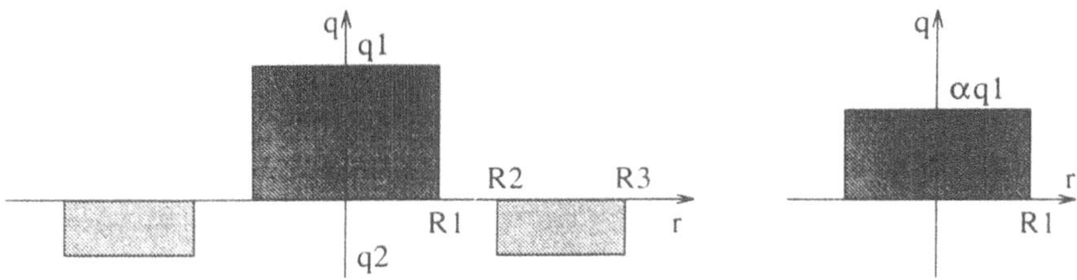

Figure 4. Radial profiles of potential vorticity for the circular vortices considered in section 4; left-hand (right-hand) panel shows the upper (lower) layer.

observed at the saddle points lying at each tip of the elliptical core). The two layerwise vortices rotate at the same rate Ω. In the rotating frame of the reference, the potential vorticity is a simple (quasi-linear) function of the relative streamfunction $\psi_j - (1/2)\Omega r^2$ (see the scatter plots on Fig. 3); this indicates stationarity for the whole tripolar structure. Time series (not shown here) of energy E, angular momentum M and enstrophy Z (squared potential vorticity) reveal that the final baroclinic tripole has equal E and M as the initial circular vortex, but lower Z. This has already been observed in two-dimensional flows for multipolar vortices (Morel and Carton, 1994). Note that for $R/R_d = 1.11$ the tripole becomes asymmetrically unstable after $t \sim 600$, perhaps due to long-term effects of viscosity (see Carton & Legras, 1994).

4. Formation of tripoles from unstable baroclinic piecewise-constant vortices

To idealize the problem even further, we now consider vortices with piecewise-constant potential vorticity profiles (see Fig. 4). The upper layer vortex is similar to that used in (Morel & Carton, 1994) for the formation of barotropic tripoles, now in potential vorticity. The condition for a vanishing volume integral of vorticity is $q_2 = q_1\,(1+\alpha H_2/H_1)\,[R_1^2/(R_2^2 - R_3^2)]$. As stated earlier, the core radius R_1 is set to unity, while velocity data inside swoddies suggest that $R_2 = 1.6R_1, R_3 = 2.6R_1$ to adequately represent the external vorticity ring of these oceanic eddies (note that other values have been tested in numerical simulations, but space lacks here for an extended parametric investigation). Numerically, we have to smooth the boundaries of these vorticity patches to avoid the Gibbs phenomenon in the spectral code; this is achieved by applying iteratively a 4-point smoother.

Two parameters, α and R_1/R_d, are thus left. The nonlinear evolutions of this baroclinic vortex, elliptically perturbed initially ($\ell = 2$) are summarized as follows:

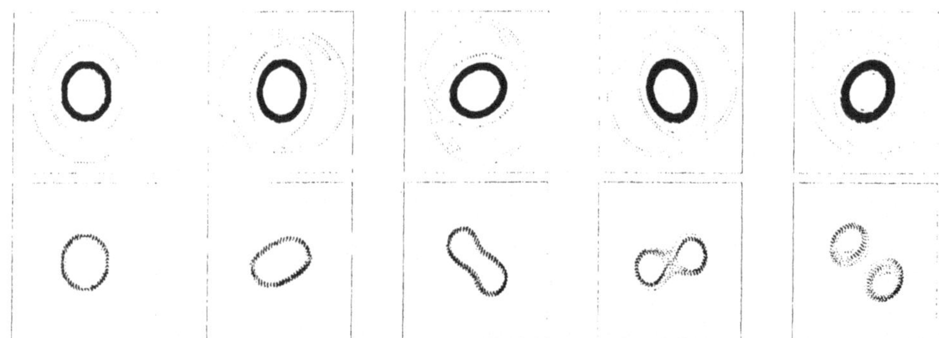

Figure 5. Time-series of potential vorticity showing the formation of an arch-shaped tripole for $R_1/R_d = 1.11, \alpha = -0.2$; the upper (lower) row of figures represents the upper (lower) layer; times shown are $t = 0, 40, 80, 280, 560$.

(A) For small R_1/R_d, baroclinic instability is moderate, so that the upper layer vortex should behave similarly to its two-dimensional counterpart:
- For $\alpha = 1$, the core vortex is like-signed in both layers, the opposite-sign vorticity is concentrated in the external ring in the upper layer, and thus is quite intense; the resulting shear on the upper core leads to a horizontal dipolar breaking.
- When α decreases ($\alpha = 0.5$), the horizontal shear on the upper core is weaker, and a surface-intensified tripole can form at first (see also Carton & Legras, ibid), but the shear is still sufficient to break the tripole asymmetrically into a monopole and a dipole.
- For more surface-intensified cores ($\alpha = 0.2$ or $\alpha = 0$), the upper external ring is sufficiently weak to allow the formation of a steadily rotating surface-intensified tripole. This tripole is a novel stationary state of stratified quasi-geostrophic flows, though it bears strong similarities with two-dimensional tripoles (Morel & Carton, ibid). For $\alpha = 0.2$ the major axis of the elliptical lower core is aligned with that of the upper core.
- For $\alpha = -0.2$, the upper vortex again forms a tripole, but the lower core now has nearly the same potential vorticity as the upper satellites. It thus breaks into two pieces which align under the satellites. The final tripolar configuration has a core in the upper layer and nearly barotropic satellites; we call it arch-shaped (Fig. 5).
- For $\alpha = -0.4, -0.6, -0.8$, the upper satellites are now too weak to induce breaking of the lower core vortex. It simply elongates as an ellipse in the direction of the satellites, i.e. perpendicular to the elliptical upper core. Again this configuration is stationary in rotation.
- Finally, for $\alpha = -1$, the outer ring vanishes in the upper layer ($q_2 = 0$), and we observe a counter-rotating, pulsating baroclinic ellipse at the center,

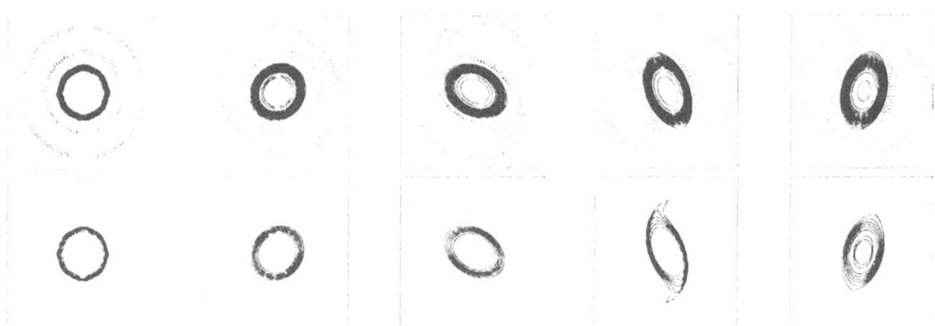

Figure 6. Time-series of potential vorticity showing the formation of a surface-intensified tripole for $R_1/R_d = 1.66, \alpha = 0.2$; the upper (lower) row of figures represents the upper (lower) layer: times shown are $t = 0, 80, 120, 160, 380$.

as in (Carton & McWilliams, 1989).

(B) For large R_1/R_d, the baroclinic effects are now essential:
- For $\alpha = 1, 0.5, 0.2, 0$, the upper layer vortex forms a tripole and never breaks into dipoles; indeed barotropic instability is reduced by increased stratification. The lower layer core becomes elliptical and aligns under the upper core (Fig. 6).
- For $\alpha = -0.2$, the lower core breaks into two pieces which align under the satellites of the upper layer tripole as for $R_1/R_d = 1.11$. The final structure is arch-shaped.
- For $\alpha = -0.4, -0.6, -0.8$, the larger vertical shear of the core and the stronger layer coupling first lead to the breaking of the lower core into two pieces which move outwards to align under the upper satellites. But the shear that these lower satellites induce on the upper core is now sufficient to break it into two pieces also. Each piece pairs with a lower satellite to form a baroclinic dipole which drifts away.
- Finally, for $\alpha = -1$, in the absence of the upper outer ring, a counter-rotating, pulsating baroclinic ellipse is again observed at the center.

5. Stationarity and stability of baroclinic tripoles

We want to prove that the baroclinic tripoles found with the spectral code are indeed stationary (Fig. 7). Since their potential vorticity distribution has been smoothed by the code, we replace it by a constant value (equal to the maximum) inside the outermost vorticity contour which closes on itself.

Then we compute the exact steady states of baroclinic tripoles with identi-

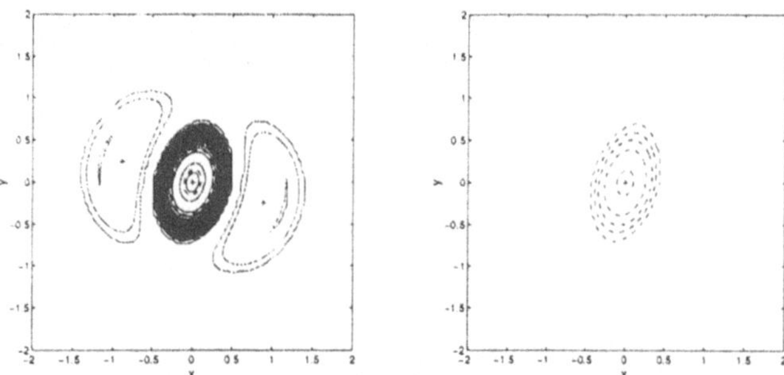

Figure 7. Upper and lower layer potential vorticity distributions of the baroclinic tripole found as an end-state with $R_1/R_d = 1.66, \alpha = 0.2$, at time $t = 380$.

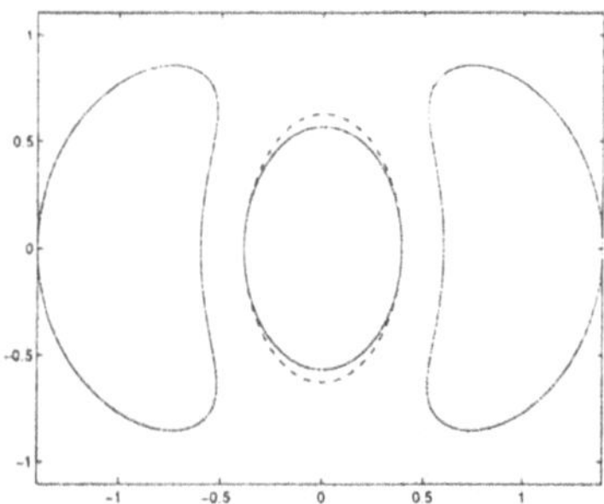

Figure 8. Potential vorticity contours for the exact steady state of a rotating baroclinic tripole with the same size and intensity as that shown on Fig. 7 (upper layer in solid lines, lower layer in dashed lines).

cal characteristics, by means of the adapted (Wu *et al.*, 1984) second-order algorithm. This iterative procedure is initialized with an idealized tripole:
- two circular cores with the same characteristics as those found with the spectral code,
- two circular satellites with the same potential vorticity as the satellites found with the spectral code, but much smaller.
For such small satellites, the interaction between the vorticity patches is weak, and their shape is close to circular. The procedure thus converges rapidly; it is then restarted with larger satellites. When these satellites become too large and elongated, the procedure stops converging.

We can now compare the exact steady state obtained with this iterative algorithm (Fig. 8) with the tripolar end-state of the spectral simulations (Fig. 7). The area integral of potential vorticity in the upper core is 0.79 versus 0.77, in the lower core 0.139 versus 0.137 and in the satellites -0.34 versus -0.35. We can conclude that the two tripoles coincide, and thus that the tripolar end-state of the spectral simulation is an exact steady state.

When extended further in time, the spectral simulations which have formed baroclinic tripoles show a short-term destabilization of these tripoles only when viscosity is increased by a factor 50 or when planetary beta-effect (an ambiant vorticity with a constant gradient along the y axis) is added. Therefore, baroclinic tripoles appear as naturally stable vortex solutions of the two-layer quasi-geostrophic equations.

6. Discussion, conclusions

Our study has shown that steadily rotating vortices, with a baroclinic tripolar shape, can form from the mixed barotropic-baroclinic instability of isolated vortices, either with a continuous or a piecewise-constant radial vorticity profile, once elliptically perturbed (mode $\ell = 2$). These tripoles rotate steadily, are robust and exist for a long time compared to their internal turnover period if viscosity is small and in the absence of beta-effect. Their shape is remarkably similar to that of the exact stationary tripolar solutions of the two-layer quasi-geostrophic equations, obtained by a relaxation method. We also remark that all our trials to form baroclinic quadrupoles (mode $\ell = 3$) with different geometrical characteristics failed, contrary to two-dimensional flows where stable quadrupoles exist (Morel & Carton, ibid). Transient pentapoles (mode $\ell = 4$) appeared, but their four satellites soon merged into two to form a tripole. Baroclinic tripoles can also result from the alignment of baroclinic vortices with constant relative vorticity distributions (Corréard & Carton, 1997), or from the head-on collision of two baroclinic dipoles (Sokolovskyi, 1989). Tripoles appear therefore as relatively unique stable multipoles in stratified geostrophic dynamics. Obviously, more work is still needed to fully understand their dynamics. In that respect, an inviscid model such as the contour surgery algorithm (Dritschel & Saravanan, 199[illegible]) may prove to be a more appropriate tool than spectral codes.

Acknowledgements

Special thanks are due to SHOM and DRET (French Ministry of Defence) for support under the research program "Processus Associés aux Modèles Intermédiaires et Régionaux". Discussions with Yves Morel, David Dritschel and Mikhail Sokolovskyi are gratefully acknowledged. The authors express gratitude to José Da Cruz and to Béatrice F[illegible]rgeau for their important contribution to the model simulations.

References

Carton, X.J. and Legras, B. (1994) The life-cycle of tripoles in two-dimensional incompressible flows, *J. Fluid Mech.*, **Vol. no. 267**, pp.53–82.

Carton, X.J. and Mc Williams, J.C. (1989) Barotropic and baroclinic instabilities of axisymmetric vortices in a quasi-geostrophic model. In: *Mesoscale/ Synoptic Coherent Structures in Geophysical Turbulence*, **Vol. no. 50**, Ed. J.C.J. Nihoul & B.M. Jamart, Elsevier, pp.225–244.

Carton, X.J. and Mc Williams, J.C. (1996) Nonlinear oscillatory evolution of a baroclinically unstable vortex, *Dyn. Atmos. Oceans*, **Vol. no. 24**, pp.207–214.

Corréard, S.M. and X.J. Carton (1997) Vertical alignment of geostrophic vortices: On the influence of the initial distribution of potential vorticity, to appear in *Proceedings of the IUTAM/SIMFLOW Symposium, Lyngby, Denmark*, Kluwer Acad. Publ.

Dritschel, D.G. and Saravanan, R. (1994) Three-dimensional quasi-geostrophic contour dynamics, with an application to stratospheric vortex dynamics, *Q.J.R. Meteorol. Soc.* **Vol. no. 120**, pp.1267–1297.

Hopfinger, E.J. and van Heijst, G.J.F. (1993) Vortices in rotating fluids. *Ann. Rev. Fluid Mech.*, **Vol. no. 25**, pp.241–289.

Morel, Y.G. (1995) Etude des déplacements et de la dynamique des tourbillons géophysiques; application aux meddies. *Thèse de Doctorat de l'Université Joseph Fourier - Grenoble 1*, 155 pp.

Morel, Y.G. and Carton, X.J. (1994) Multipolar vortices in two-dimensional incompressible flows. *J. Fluid Mech.*, **Vol. no. 267**, pp.23–51.

Pingree, R.D. and Le Cann, B. (1992) Three anticyclonic Slope Water Oceanic eDDIES (SWODDIES) in the Southern Bay of Biscay in 1990, *Deep-Sea Res.*, **Vol. no. 39, 7/8**, pp.1147–1175.

Polvani, L.M. and Carton, X.J. (1990) The tripole: a new coherent vortex structure of incompressible two-dimensional flows. *Geophys. Astrophys. Fluid Dyn.*, **Vol. no. 51**, pp.87–102.

Sokolovskyi, M. (1989) O vstrechnom stolknovenii raspredelennykh khetonov. *Doklady Akademii Nauk SSSR*, **Vol. no. 306**, pp.198–202.

Wu, H.M., Overman II, E.A. & Zabusky, N.J. (1984) Steady-state solutions of the Euler equations in two dimensions: Rotating and translating V-States with limiting cases. I. Numerical algorithms and results. *J. Comp. Phys.*, **Vol. no. 53** pp.42–71.

VERTICAL ALIGNMENT OF GEOSTROPHIC VORTICES

On the Influence of the Initial Distribution of Potential Vorticity

STÉPHANIE M. CORRÉARD AND XAVIER J. CARTON
SHOM/CMO
13 rue du Chatellier, 29275 Brest, France

Abstract

In a two-layer quasi-geostrophic numerical model, we study the vertical alignment of geostrophic vortices; we compare its efficiency for RVI conditions (a disk of constant relative vorticity in each layer initially) to that for PVI conditions (a disk of constant potential vorticity in each layer; Polvani, 1991). The disks of radius R are initially separated by a distance d, and R_d is the internal radius of deformation. At small R_d/R and large d/R, a multipolar potential vorticity distribution is associated with these RVI conditions. This distribution is responsible for a much weaker efficiency of the alignment process in the RVI case than in the PVI case: indeed, the formation of horizontal or of vertical dipoles prevents alignment. Even more interesting is the case where alignment does occur: under PVI conditions, the end state is a simple vertical column of potential vorticity, while for RVI vortices it is a steadily rotating baroclinic tripole, a novel stationary feature of stratified geostrophic dynamics.

1. Introduction

High-resolution numerical experiments have evidenced the key role of vortices in the evolution of the energy and enstrophy spectra in stratified geostrophic turbulence (McWilliams, 1989). In particular, the final vertical alignment of like-signed vortices has been associated with the conversion of baroclinic to barotropic energy. The importance of this mechanism was assessed by Polvani (1991): using a two-layer quasi-geostrophic numerical model, initialized with a disk of constant potential vorticity in each layer (PVI conditions), he showed that alignment would occur for weak stratification and initially close vortices (less than 3.3 radii apart). Vertical vortex alignment was also observed in the ocean: in the Azores region, a surface-intensified anticyclone got locked above a deep anticyclone, a meddy, and

both propagated together (Tychensky & Carton, 1997). But in that case, the two vortex cores had a nearly constant relative vorticity distribution (RVI conditions).
The difference between PVI and RVI conditions has already been investigated in the context of baroclinic vortex merger (Verron *et al.*, 1990), to understand the discrepancy between numerical simulations (Polvani *et al.*, 1989) and laboratory experiments (Griffiths & Hopfinger, 1987). They have shown that baroclinic vortex merger is much more sensitive to stratification for RVI than for PVI conditions. Here, our objective is to compare vertical vortex alignment under RVI conditions with the PVI case. We first describe the two-layer quasi-geostrophic equations and numerical model. Then we present the potential vorticity distribution associated with one disk of constant relative vorticity in each layer (RVI conditions). We classify and explain the various nonlinear evolutions obtained with the numerical model when stratification and the initial separation between the vortices are varied. Finally, we show that the aligned end-state under RVI conditions is a baroclinic tripole, a novel stationary feature of two-layer quasi-geostrophic dynamics.

2. Mathematical and physical frameworks of RVI geostrophic vortex alignment

The two-layer quasi-geostrophic equations are obtained under the shallow-water and Boussinesq approximations, assuming also incompressibility, dominant effects of stratification and of global rotation. These equations express the conservation of potential vorticity in unforced, inviscid conditions:

$$\frac{dQ_j}{dt} = \partial_t Q_j + J(\psi_j, Q_j) = 0 \tag{1}$$

where the advecting velocity is $\vec{u}_j = \vec{k} \wedge \vec{\nabla}\psi_j$, $J(a,b) = \partial_x(a)\partial_y(b) - \partial_x(b)\partial_y(a)$ is the Jacobian operator, and the potential vorticity is

$$Q_j = \nabla^2\psi_j + \frac{f_0^2}{g' H_j}(\psi_k - \psi_j), \quad k = 3 - j. \tag{2}$$

Here $j = 1, 2$ is the (upper, lower) layer index; $g' = g(\rho_2 - \rho_1)/\rho_0$ is the reduced gravity (ρ_j is the density in layer j and ρ_0 its vertical average); f_0 is the Coriolis parameter; H_j is the thickness of layer j (here we have kept $H_1 = H_2$). The internal radius of deformation is given by $R_d^2 = g' H_1 H_2 / (f_0^2 (H_1 + H_2))$. These equations are parity-invariant.
The numerical model relies on a pseudo-spectral (Galerkin) projection and truncation of the quasi-geostrophic equations on a biperiodic square grid

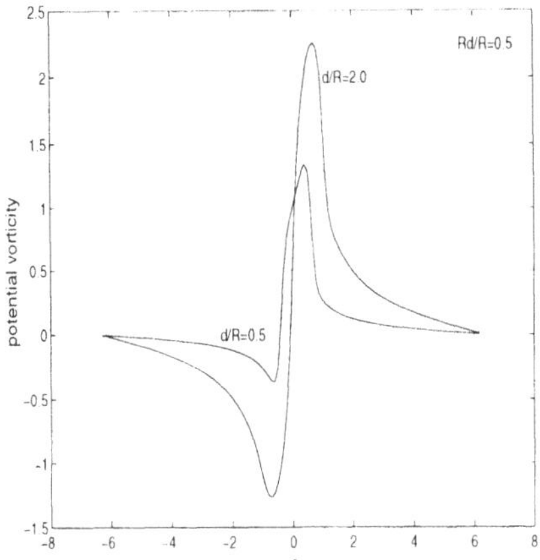

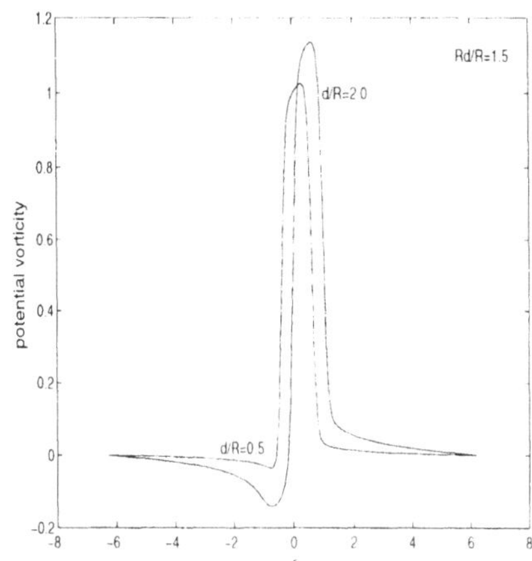

Figure 1. Upper-layer radial potential vorticity profiles associated with RVI conditions for $d/R = 0.5; 2.0$ and $R_d/R = 0.5; 1.5$.

($L = 4\pi$), with 256^2 nodes. The time-step is imposed by the Courant-Friedrichs-Lewy stability criterion. Enstrophy accumulation at small scales is removed by a traditional biharmonic viscosity, for which the coefficient has been kept to a minimum compatible with nonlinear processes such as filamentation (explicitly the zero on the right-hand side of Eq. 1 is replaced by $\nu_4 \nabla^6 \psi_j$, with $\nu_4 = 5\,10^{-9}$).

The initial conditions consist in each layer in one disk of unit relative vorticity and of radius $R = 0.5$, which we call the cores. The vortices are like-signed (cyclonic) and initially separated by a distance d. On Fig. 1, we present the upper-layer radial potential vorticity profiles associated with these initial conditions for $d/R = 0.5; 2.0$ and $R_d/R = 0.5; 1.5$. We see that for small d/R and/or large R_d/R, the potential vorticity distribution resembles that of relative vorticity: one single positive pole, relatively narrow. On the contrary, at small R_d/R and large d/R, the potential vorticity distributions are more intense and more spread out. Moreover, a negative potential vorticity pole (a satellite) accompanies each core; this satellite is due to the vortex stretching of the opposite-layer core (vortex stretching is the second term on the right-hand side of Eq. 2). The situation is then clearly dipolar.

3. Nonlinear evolutions of two RVI vortices in the $(R_d/R, d/R)$ parameter space

By running a series of sixty experiments with the numerical code, we obtained a nonlinear regime diagram for RVI vortex alignment (Fig. 2). In the $(R_d/R, d/R)$ plane, four regimes appear:

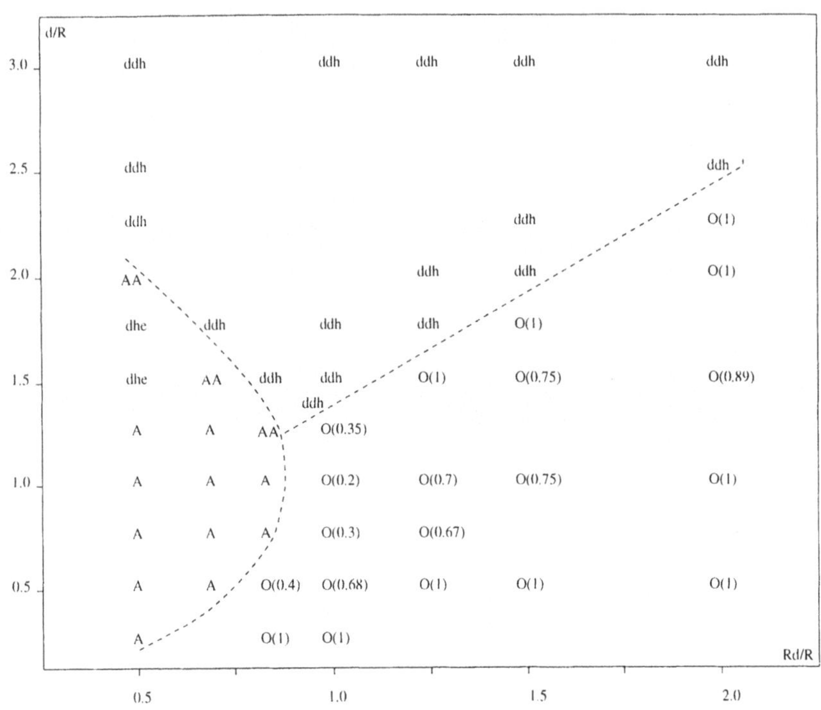

Figure 2. Nonlinear regime diagram in the $R_d/R, d/R$ plane for RVI vortex alignment.

- alignment (A) for close and large vortices,
- formation of horizontal dipoles (ddh) when the vortices are initially well separated,
- an oscillatory co-rotation (O, with $d_{final}/d_{initial}$) when the two vortices are rather small and close,
- hetonic divergence (dhe) for large vortices barely overlapping (hetons are vertical dipoles).
- a subregime of accelerated alignment (AA) was also observed close to the regime of hetonic divergence, but it will not be detailed here.

The variety of nonlinear evolutions in the RVI case contrasts with the simple "alignment versus co-rotation" regime diagram in the PVI case (Fig. 7 of Polvani, 1991); the presence of the satellites whose strengths vary with R_d/R and d/R (cf. previous section) is responsible for this variety in the RVI case. When $d/R \leq 1$, the RVI and PVI regime diagrams are similar. In that parameter range, the influence of the satellites remains limited since their potential vorticity noticeably overlaps with that of the cores. Their influence is further reduced at large R_d/R. On the contrary, for $d/R > 1$, the dipolar effects due to the satellites can fully develop if R_d/R is not too large.

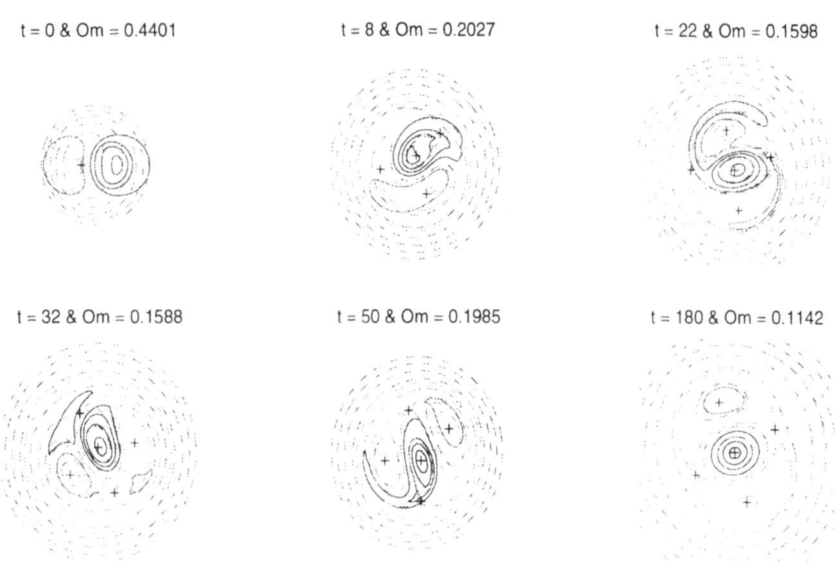

Figure 3. Upper layer potential vorticity (blue: positive core, red: negative satellite) and corotating streamfunction (purple) with separatrices (green), saddle points of the separatrix and vortex centers (crosses) for the alignment of RVI vortices with $R_d/R = 0.5, d/R = 1.0$; times shown are t=0, 8, 22, 32, 50, 180 and $Om = \Omega$.

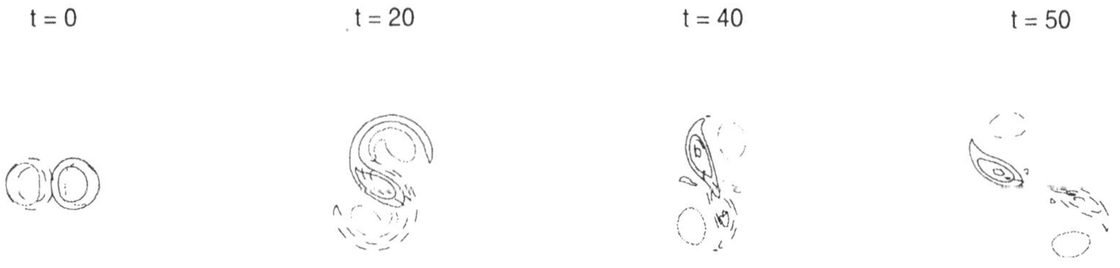

t = 0 t = 20 t = 40 t = 50

Figure 4. Potential vorticity (blue: positive core, red: negative satellite, solid line: upper layer, dashed line: lower layer) for the hetonic divergence of RVI vortices with $R_d/R = 0.5, d/R = 1.5$; times shown are t=0, 20, 40, 50.

Figure 5. Potential vorticity (blue: positive core, red: negative satellite, solid line: upper layer, dashed line: lower layer) for the horizontal dipole formation from two RVI vortices with $R_d/R = 0.5, d/R = 2.5$; times shown are t=0, 10, 20, 30.

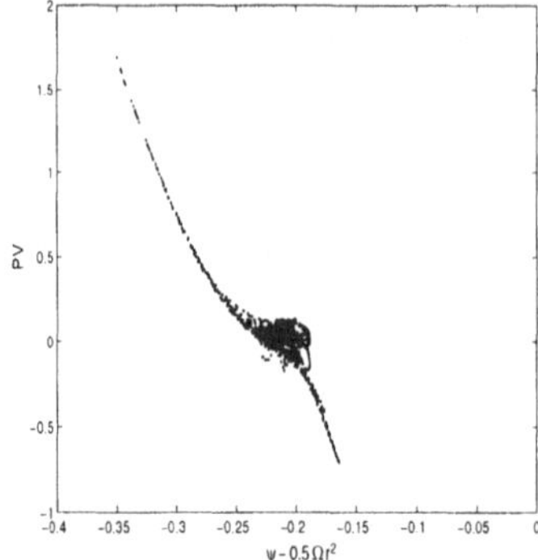

Figure 6. Upper layer potential vorticity versus corotating streamfunction - scatter plot for the final state of the alignment of RVI vortices with $R_d/R = 0.5, d/R = 1.0$.

3.1. ALIGNMENT

Alignment occurs for large and initially close vortices ($R > R_d$, $d \leq 2R$). Since here the process is vertically symmetric, we only present the evolution of potential vorticity and of corotating streamfunction in the upper layer (Fig. 3, for $R_d/R = 0.5, d/R = 1.0$). The corotating streamfunction $\psi_c = \psi - 1/2\,\Omega r^2$ is obtained by substracting the instantaneous rotation of the cores. We observe successive inward and outward motions of the core while the satellite is slowly pushed outwards. These inward (resp. outward) upper core motions result from its interaction with the lower core (resp. satellite). We also note strong deformations with angular modes 1 and 2 on each vortex contour, indicating the intense shear created by the companion vortices. Filamentation occurs when potential vorticity crosses the saddle points of the separatrix. The dominant interaction between the upper and lower cores brings them into an aligned state at the center; the final cores are slightly elliptical and smaller than initially, and the strongly deformed satellites steadily rotate around them. Each potential vorticity pole is then contained within each lobe of the separatrix, indicating stationarity. This stationarity is confirmed by the scatter plot of potential vorticity versus corotating streamfunction, shown on Fig. 6 for the upper layer (the lower layer one is symmetrical).

3.2. HETONIC DIVERGENCE (FORMATION OF VERTICAL DIPOLES)

For initially more distant vortices ($1.5 < d/R < 2.0$) and for small R_d/R, the dominant interaction occurs between each core and the opposite-layer satellite, initially superimposed. The hetonic structure is then very strong. The combined effect of all vorticity poles pushes each satellite ahead of the opposite-layer core in the cyclonic direction (see Fig. 4 for $R_d/R =$

$0.5, d/R = 1.5$). This vertical tilt between core and satellite results in a vertical dipole, which drifts towards the center; this initial stage is similar to that of alignment, with the core vortices reaching the center and shedding filaments. But these cores then undergo a fast outward drift induced by the opposite-layer satellites, now located on their right-hand side. Since the satellites have not been pushed as far aside as during alignment, their vertical coupling with the opposite-layer cores dominates and this outward drift is irreversible; the plot of ψ_c, not shown here, indicates that a saddle-point finally appears at the center between the two cores, splitting ψ_c into two drifting lobes within which the cores are contained.

3.3. HORIZONTAL DIPOLE FORMATION

When the initial d/R is increased again ($d/R > 2.0$), the horizontal dipolar effect prevails over the alignment tendency (Fig. 5 for $R_d/R = 0.5, d/R = 2.5$). Originally, each core pairs with its opposite-layer satellite to drift towards the center, as in the regime of hetonic divergence. But the two hetons collide at the center and the two cores never align, even temporarily. Instead, these cores exchange partner during the collision, from the coupling with an opposite-layer to a same-layer satellite. They finally form horizontal dipoles which drift apart irrevocably.
The shape of the boundary between horizontal dipole formation and other regimes in the $(R_d/R, d/R)$ plane can be understood as follows:
- for $R_d/R < 1$ the vertical coupling dominates, but decreases with increasing R_d/R; horizontal dipoles can thus form for weaker ratios of negative to positive potential vorticities (satellite to core intensities) when R_d/R increases; therefore the regime boundary follows decreasing values of d/R.
- for $R_d/R > 1$, the horizontal interaction dominates. The ratio of negative to positive potential vorticities increases with R_d/R and so does the regime boundary with d/R.

3.4. OSCILLATION AND CO-ROTATION

Everywhere else in the parameter plane, the satellites are negligible, the two cores are elliptical and rotate around the center of the domain; their aspect ratio oscillates in time. For parameters close to the alignment regime, the core separation decreases in time while oscillating, and finally settles at a value slightly smaller than the initial one. This regime does not show significant differences with that described by Polvani (1991).

4. Stationary baroclinic tripoles

The end-state of the alignment process is an elliptical, cyclonic columnar vortex at the center, with one anticyclonic satellite in each layer, the two satellites being symmetrical with respect to the center; these satellites rotate at the same rate as the core major axis (Fig. 7). The whole tripolar structure is thus stationary (see the scatter-plot in Fig. 6). Note that this structure does not have a stationary point-vortex equivalent. Indeed, point vortex satellites create velocities at the center which do not add up to zero, but to equal and opposite values in the upper and lower layers. Such a point-vortex baroclinic tripole would split up into two hetons in finite time. The stationarity of the tripolar end-state of the alignment process is thus due to its finite area.

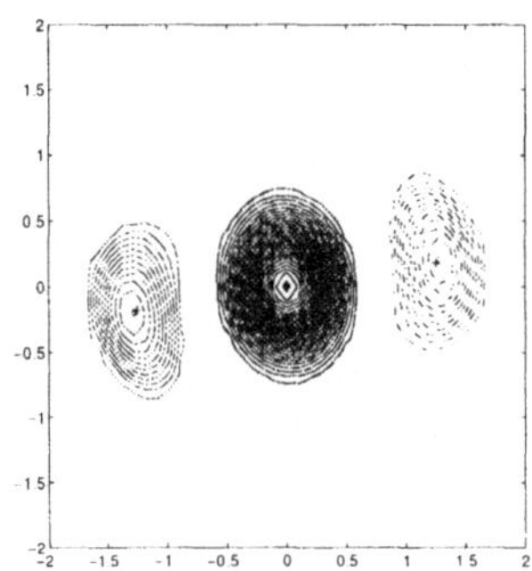

Figure 7. Potential vorticity contours (both layers) for the end-state of the alignment process, with two RVI vortices with $R_d/R = 0.5, d/R = 1.0$.

But we can show that there exists exact tripolar steady states with piecewise-constant potential vorticity (PCPV), identical to the tripolar end-state of Fig. 7.
These exact PCPV steady states are found by means of an iterative relaxation procedure, based on the Wu *et al.* (1984) algorithm. The iteration starts from an initial guess of four small circular vortex patches, two of them at the center and two of them on opposite sides of the core (one per layer). The stationarity condition is that velocity must be tangential to each contour in the rotating frame of reference. Once the procedure has converged for small vortices, the vortex radii are increased until they reach the value of the tripolar end-state of the spectral code. The shape of the exact PCPV tripolar steady state (Fig. 8) coincides with that of the tripolar end-state of the spectral code (Fig. 7). The area integral of potential vorticity for each pole differs by less than 10 % between the two solutions.

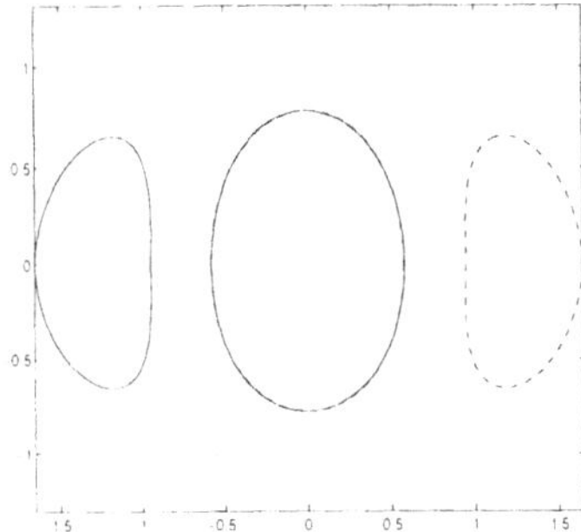

Figure 8. Potential vorticity contours (both layers) for a PCPV baroclinic tripole corresponding to the state of Fig. 7. Solid (resp. dashed) lines indicate the upper (resp. lower) layer.

5. Conclusions

Using a two-layer numerical quasi-geostrophic model, we have shown the essential differences between RVI and PVI conditions for vertical vortex alignment. We have mapped the nonlinear evolutions of the vortex system in the (initial vortex separation versus stratification) parameter plane. For weak or moderate stratifications and sufficient initial separation, dipolar coupling, either horizontal or vertical, prevents the final alignment of the two original vortices in the RVI case. This dipolar evolution is due to the initial existence of opposite-sign satellites near the core vortices. Such an evolution does not occur for PVI alignment (Polvani, 1991). The efficiency of alignment is thus considerably reduced in the RVI conditions. When alignment occurs, a baroclinic tripolar vortex is formed for RVI conditions. This vortex is steadily rotating, possesses a simple vorticity-streamfunction relation and thus constitutes a novel stationary solution of the stratified quasi-geostrophic dynamics. We have noted that this tripole does not have a point-vortex counterpart, whereas two-dimensional tripoles do. Finite-area effects are thus essential. By idealizing this continuous potential vorticity distribution as a piecewise-constant potential vorticity (PCPV) tripole, we have been able to show that steadily rotating, baroclinic PCPV tripoles also exist, and closely resemble the end-states of RVI alignment.

It is of interest to note that steady baroclinic tripoles, with a different spatial distribution of potential vorticity, can form from the mixed barotropic-baroclinic instability of circular vortices in stratified flows (Carton & Corréard, 1997). Long-lived baroclinic tripoles have also been observed in laboratory simulations of baroclinic vortices such as meddies (priv. comm. from A. Fincham, Coriolis Laboratory, Grenoble, France). More work is now necessary to classify all these various forms of baroclinic tripoles.

Acknowledgements

Special thanks are due to SHOM and DRET (French Ministry of Defence) for support under the research program "Processus Associés aux Modèles Intermédiaires et Régionaux". Discussions with Lorenzo Polvani, Sophie Valcke, David Dritschel and Yves Morel are gratefully acknowledged.

References

Carton, X.J. and Corréard, S.M. (1997) Baroclinic tripolar geostrophic vortices: Formation and subsequent evolution. To appear in *Proceedings of the IUTAM/SIMFLOW Symposium, Lyngby, Denmark*, Kluwer Acad. Publ.

Griffiths, R.W. and Hopfinger, E.J. (1987) Coalescing of geostrophic vortices, *J. Fluid Mech.*, **Vol. no. 178**, pp. 73–97.

McWilliams, J.C. (1989) Statistical properties of decaying geostrophic turbulence, *J. Fluid Mech.*, **Vol. no. 198**, pp.199–230.

Polvani, L.M., Zabusky, N.J. and Flierl, G.R. (1989) Two-layer geostrophic vortex dynamics. Part 1: Upper-layer V-States and merger, *J. Fluid Mech.*, **Vol. no. 205**, pp. 215–242.

Polvani, L.M. (1991) Two-layer geostrophic vortex dynamics. Part 2: Alignment and two-layer V-States. *J. Fluid Mech.*, **Vol. no. 225**, pp.241–270.

Tychensky, A. and Carton, X.J. (1997) Hydrological and dynamical characterization of meddies in the Azores region: a paradigm for baroclinic vortex dynamics. To appear in *J. Geophys. Res.*.

Verron, J., Hopfinger, E.J. and McWilliams, J.C. (1990) Sensitivity to initial conditions in the merging of two-layer baroclinic vortices. *Phys. Fluids A*, **Vol. no. 2**, pp.886–889.

Wu, H.M., Overman II, E.A. & Zabusky, N.J. (1984) Steady-state solutions of the Euler equations in two dimensions: Rotating and translating V-States with limiting cases. I. Numerical algorithms and results. *J. Comp. Phys.*, **Vol. no. 53**, pp.42–71.

THE INTERACTION OF A VORTEX WITH A STABLE JET

Quasi-geostrophic numerical modeling of the crossing of an oceanic eddy through a zonal jet

FRÉDÉRIC O. VANDERMEIRSCH
SHOM/CMO and LPO/CNRS-UBO

AND

XAVIER J. CARTON AND YVES G. MOREL
SHOM/CMO
13 rue du Chatellier, BP 426, 29275 Brest Cedex, France

ABSTRACT

We examine the nonlinear interactions between a zonal jet and a vortex (oceanic eddy), using a one-and-a-half-layer quasi-geostrophic numerical model. The eastward jet is defined by two strips of constant and opposite potential vorticity: it is stable in isolation. The anticyclonic eddy is initialized north of the jet by a circular patch of constant potential vorticity. We numerically determine the physical conditions under which the eddy crosses the jet meridionally, or drifts along its northern side. Three groups of dynamic regimes are found: the first one with the eddy staying north of the jet and drifting eastward, the second one with the eddy crossing the jet and then drifting away as a dipole, and the last one with the eddy also crossing the jet, and progressing eastward. Furthermore, an analytical criterion for the crossing of the jet is established, which remarkably compares with the numerical experiments. The sensitivity of these dynamic regimes to environmental parameters is finally examined.

1. Introduction

Ocean vortices are usually long-lived and carry substantial amounts of momentum and tracers away from their region of generation. A striking example of long-lived vortices are meddies, anticyclonic lens eddies, which form on the Iberian shelf from the instability of the Mediterranean water undercurrent. These meddies contain warm and salty water, are intensified around 1000 m depth, have a radius of 30 to 60 km, and an azimuthal velocity reaching 0.25-0.30 $cm.s^{-1}$. Once formed, such meddies

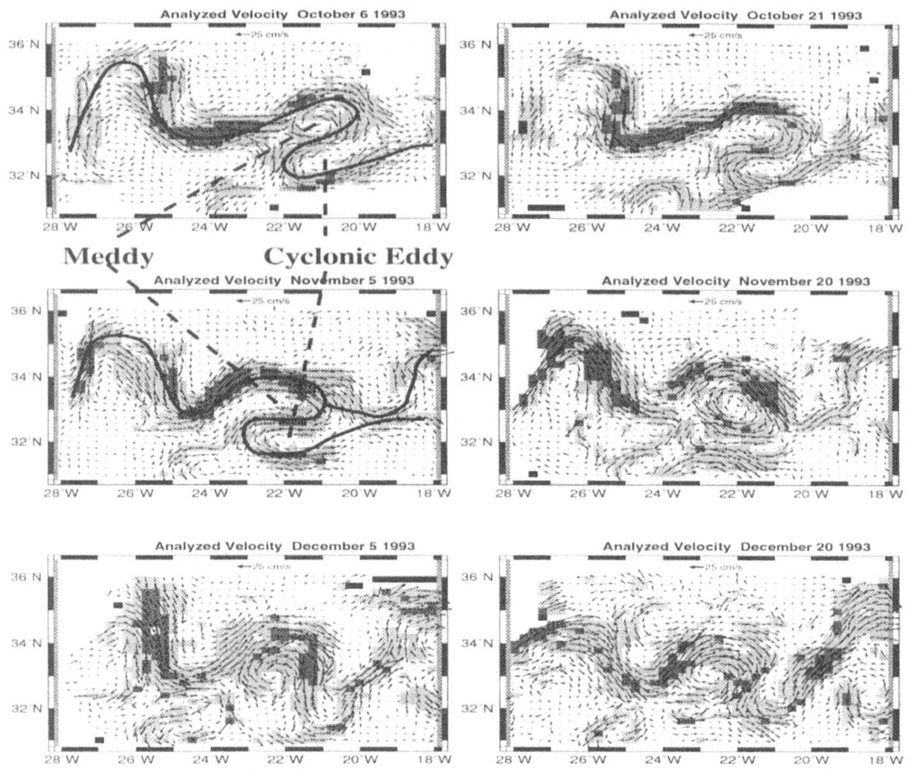

Figure 1. Time evolution of the surface velocities during the SEMAPHORE oceanic cruise (1993-94), computed by objective analysis of surface drifter trajectories measured for three consecutive months.

drift southwestward across the northeastern Atlantic ocean, under the influence of beta-effect (the variation with latitude of the Coriolis force), of mean currents or of bottom topography (Morel, 1995). They can later interact with the general oceanic circulation in the north Atlantic basin. The SEMAPHORE oceanic cruise (1993-94) has observed two such interactions of meddies with the Azores jet. In Fig. 1, we see the surface currents obtained from Lagrangian float trajectories; they show the surface signature of the meddy, and the meddy interaction with the eastward jet, resulting in strong meridional meanders and in cyclonic eddy formation (Tychensky, 1994; Richardson & Tychensky, 1997). This surface cyclone pairs with the meddy and propagates southwestward as a baroclinic (vertical) dipole (at $3.1\ cm.s^{-1}$).

The interaction between a coherent vortex and a zonal jet has already been the subject of several hydrodynamic process studies. Once advected towards the jet by the mean flow or by beta-effect, the vortex generates strong meanders on the mean flow (Stern & Flierl, 1987; Smith & Davis,1989). Until now, the vortex has never been observed to cross the jet in numerical experiments, though twice it came close to doing it (Fig. 14 of Stern & Flierl, ibid; Fig. 18 of Pratt & Stern, 1986; note that this might be attributed in part to their use of contour dynamics without surgery). In section 2, we present the one-and-a-half-layer quasi-geostrophic equations and our numerical model. To fully assess the various possible dynamical regimes in the space of physical parameters, we perform a comprehensive series of numerical experiments with this quasi-geostrophic model scaled on oceanic parameters. These regimes (vortex crossing the jet or not) are detailed and physically explained (section 3). An analytical criterion has been found which determines the crossing of the jet (section 4). This criterion agrees quite nicely with the numerical outcomes. Finally the sensitivity of this process to environmental parameters is touched on.

2. Quasi-geostrophic numerical model

The model used hereafter is based on the quasi-geostrophic approximation (dominance of Earth rotation and of ocean stratification on the nonlinear terms in the horizontal momentum equations, with Boussinesq approximation). The horizontal motion is thus essentially nondivergent and derives from a streamfunction ψ. The variation of the Coriolis parameter with latitude (beta-effect) is included in these equations $f = f_0 + \beta y$ (f_0 the reference Coriolis parameter and β the beta parameter). Vertically, the flow is hydrostatic, and the stratification is idealized: a dynamic layer of finite thickness H (with density ρ) rests on a motionless layer of infinite depth (with density $\Delta\rho + \rho$). In the absence of forcing and of dissipation, the model conserves the total vorticity Q:

$$\frac{dQ}{dt} = \frac{d(q + f_0 + \beta y)}{dt} = \partial_t q + J(\psi, q) + \beta \partial_x \psi = 0 \qquad (1)$$

where the potential vorticity associated with the motion is defined as

$$q = \nabla^2 \psi - (f_0^2/g'H)\, \psi \qquad (2)$$

The reduced gravity is defined as usual $g' = g\Delta\rho/\rho$. This sets a horizontal length scale called the deformation radius ($R_d^2 = g'H/f_0^2$).
The numerical model is based on a finite-difference discretization of the quasi-geostrophic equation in a zonal channel geometry (western inflow, eastern outflow, northern and southern free-slip boundary conditions). The horizontal grid is 151x151, with a dimensionless domain size of 12x12 (in reality 720x720 km). The scaling of the equations is given by a unit velocity in the model representing a 1 $m.s^{-1}$speed in the ocean. In section 3, R_d is

set to 0.50 (in reality 30 km and $\gamma = 1/R_d = 2.0$) while it varies in section 4. The beta-effect will be included only in section 4. The time step is imposed by the Courant-Friedrichs-Lewy stability criterion. The right-hand side of eq. 1 is replaced in the numerical model by a traditional biharmonic viscosity to erase the small-scale noise; the numerical viscosity is constrained to a very small value ($\nu_4 = 10^{-7}$ corresponding to a pseudo-$Re = 1280$) which does not alter the physical outcome of the experiments.

The initial conditions of the model consist in a zonal jet defined by two strips of constant and opposite potential vorticity ($Q_J, -Q_J$) and in a width equal to half a domain size. The vortex is a circular patch of constant potential vorticity Q_T and of radius R_T initially north of the jet at a position x_T, y_T (see Fig. 2).The potential vorticity distribution can be inverted into velocity in the quasi-geostrophic model as:

$$\bar{q}_J(y) = \begin{cases} Q_J & , y > 0 \\ -Q_J & , y < 0 \end{cases} \Longrightarrow \bar{u}_J(y) = Q_J/\gamma \, exp(-\gamma|y|) \tag{3}$$

$$\bar{q}_T(r) = \begin{cases} Q_T , & |r| < R_T \\ 0 , & |r| > R_T \end{cases} \Longrightarrow \bar{u}_T(r) = \begin{cases} Q_T R_T K_1(\gamma R_T) I_1(\gamma r) \\ Q_T R_T I_1(\gamma R_T) K_1(\gamma r) \end{cases} \tag{4}$$

where I_1 and K_1 are modified first-order Bessel functions of the first and the second kind, respectively, $r^2 = x^2 + y^2$ and the following boundary conditions: $\bar{u}_J(y \to \infty) \to 0$ *and* $\bar{u}_T(y \to \infty) \to 0$.

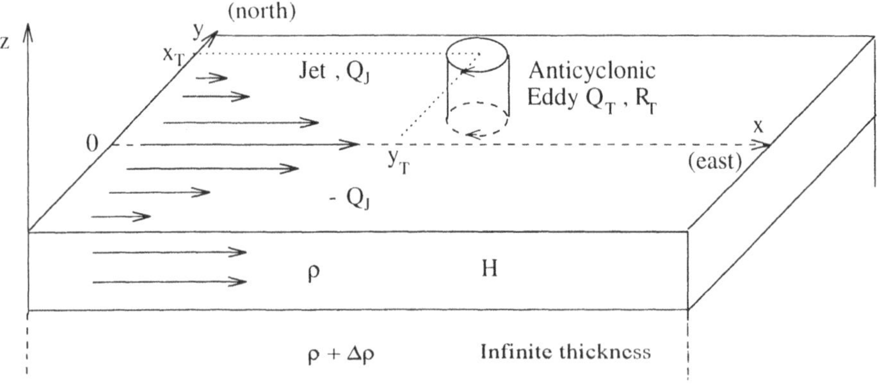

Figure 2. Initial jet-vortex configuration

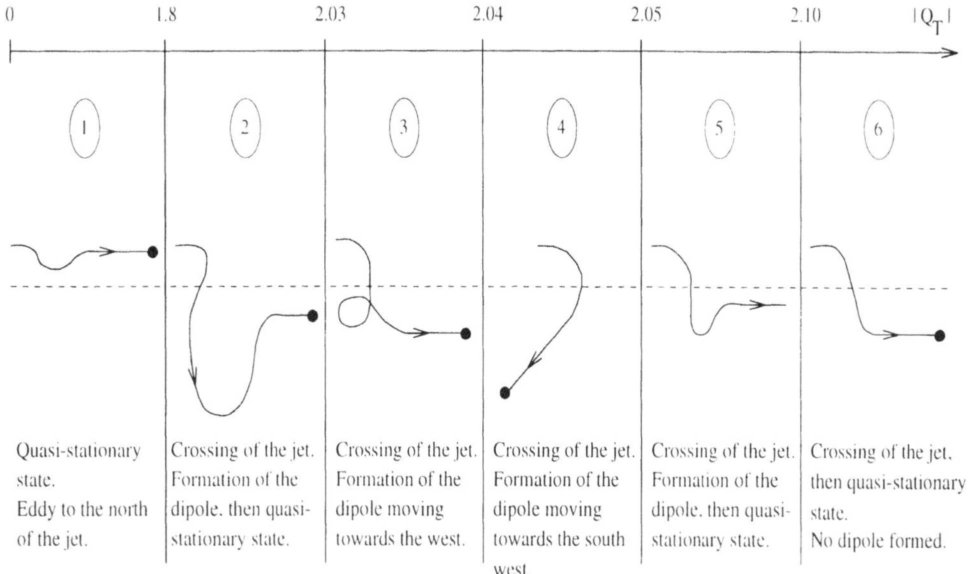

Figure 3. Summary of the six dynamic regimes for the anticyclone initially located north of the jet according to its amplitude Q_T with $Q_J = 0.5.0$, $\gamma = 2.0$. $R_T = 0.30$ and $(x_T, y_T) = (-4, 0.96)$.

3. Nonlinear dynamic regimes

In this section, we have set $\gamma = 2$, $Q_J = 0.5$, $R_T = 0.30$ et $(x_T, y_T) = (-4, 0.96)$ and we have varied Q_T. By symmetry of the quasi-geostrophic equations, we study here only the case of anticyclonic vortices. Such vortices can cross the eastward jet only if they are originally located north of it (Stern & Flierl, 1987). This can be understood by considering the local deformation of the jet velocity field induced by the vortex. When the vortex-jet interaction is sufficiently strong, a cyclonic vortex can be created south of the jet, and the original anticyclone can pair with it to form a dipole. The final evolution is very sensitive to the initial amplitude of the anticyclone Q_T. In practice, six regimes which form three major groups are observed in the nonlinear model (see Fig. 3): for $|Q_T| < 1.8$, the anticyclone drifts eastward north of the jet without crossing it; for $1.8 < |Q_T| < 2.1$, the anticyclone crosses the jet and pairs with the cyclone as a dipole which drifts away; for $|Q_T| > 2.1$, the anticyclone crosses the jet and simply drifts eastward along it without pairing with the cyclone. In both cases of eastward drift, a quasi-stationary state is found in the moving frame of reference. We will now study regimes 1 and 3 in more detail.

Figure 4. Time evolution of potential vorticity maps for regime 1; $t = 0, 30, 120$ and 240, $Q_T = -1.2$, contour interval $= 0.15$. The little square denotes the initial position of the vortex and the thin solid line its trajectory.

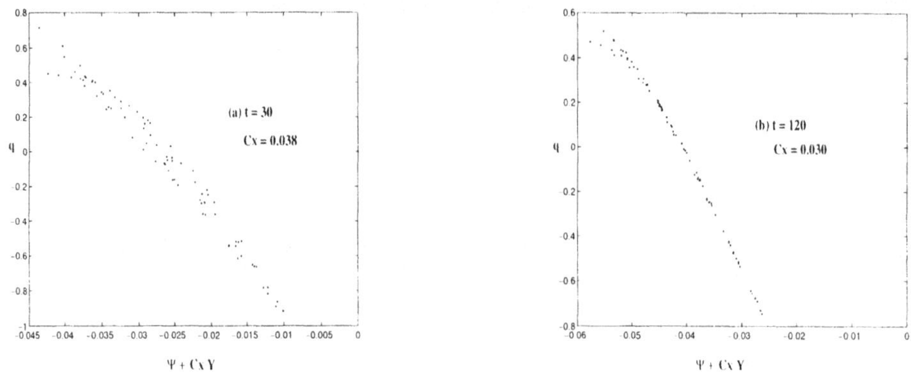

Figure 5. q versus ψ scatter-plots for regime 1;$Q_T = -1.2$. (a) at $t = 30$, dispersion is important for the transient state; (b) at $t = 120$, a quasi-linear relation is found with a zonal velocity of 3.0 $cm.s^{-1}$. A steady state is attained.

3.1. REGIME 1: QUASI-STATIONARY EASTWARD PROPAGATION OF THE VORTEX NORTH OF THE JET

A weak vortex ($|Q_T| < 1.8$) will deform elliptically and propagate eastward at a nearly constant speed under the influence of the zonal jet. Note that this equilibration of the vortex is most easily obtained with small vortices. Fig. 4 shows the time-evolution of the potential vorticity for $Q_T = -1.2$. At $t = 30$, the vortex is in a transient regime. From $t = 60$, the vortex starts drifting zonally to finally reach a quasi-steady state. To check the stationarity of the final vortex state, we show on Fig. 5 the scatter-plot of its potential vorticity versus its translating streamfunction, with a drift speed C_x, C_y given by a minimization of the dispersion. We find $C_x = 3.0\ cm.s^{-1}, C_y/C_x \sim 10^{-2}$. This proves that the vortex motion is mostly zonal. In turn, the jet is deformed by the presence of the vortex, and meanders are induced. A wavelength analysis is performed for these meanders, which shows that among the first 5 modes (scaled by the domain size), mode 2 is prevalent (with wavelength of 360 km).

Figure 6. Time evolution of potential vorticity maps for regime 3: t = 0, 80, 120 and 170, $Q_T = -2.04$, contour interval = 0.14. The little square denotes the initial position of the vortex and the thin solid line its trajectory.

3.2. REGIME 3: VORTEX CROSSING THE JET SOUTHWARD AND FORMING A DIPOLE; COMPARISON WITH OCEANIC DATA

Though in our model the vortex and the jet lie in the same layer, regime 3 nicely illustrates the process observed during the SEMAPHORE oceanic cruise (Fig. 1). During that cruise, the meddy interaction with the jet resulted in a baroclinic dipole propagating southwestward. In our model (see Fig. 6), $Q_T = -2.04$ is sufficient to induce a strong cyclone detachment from the jet. The two vortices pairing as a dipole first propagate westward since the anticyclone lies north of its cyclonic companion (t = 120), then the motion is northward towards the jet axis since the anticyclone is stronger. The shear induced by the jet is then sufficient to break the dipole. Finally (at t=170) both vortices propagate eastward (at 3.1 $cm.s^{-1}$, see the eddy trajectory on Fig. 6).
Very recently, numerical experiments with a two-and-a-half-layer model have demonstrated similar behaviors (i.e the formation of a baroclinic dipole), which can more realistically be compared with the observations at sea. Note that in our numerical model, the absence of beta effect suppresses the permanent southwestward drift (see also Käse & Zenk, 1996).

4. Analytical criterion

In this section we look for an analytical criterion to determine if the vortex crosses the jet axis. Fig. 7 shows the successive steps in that process: (1) the anticyclone is advected eastward by the jet, (2) the anticyclone pushes the jet axis northward upstream and southward downstream, thus creating a cyclonic vortex downstream (and a weaker anticyclone upstream, not shown here). The cyclone and anticyclone thus form a dipole. (3) Due to the angle of the dipole axis, its average motion is southwestward (therefore across the jet axis). Physically, it is reasonable to assume that the absolute value of the maximum velocity due to the anticyclone must be at least equal to the

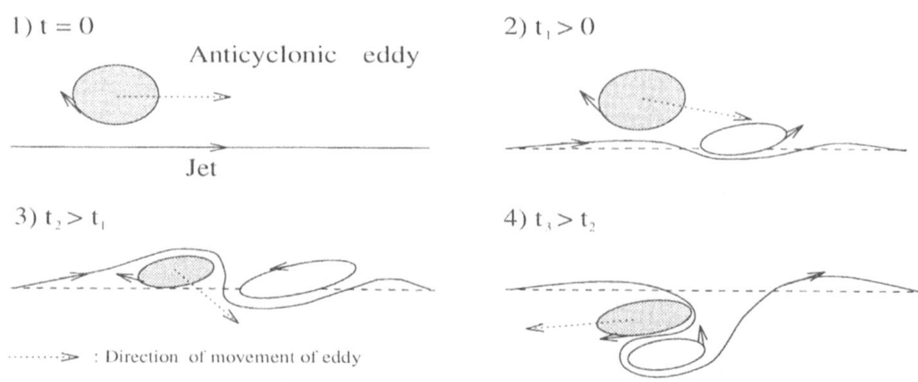

Figure 7. Schematic evolution of a vortex crossing a zonal jet

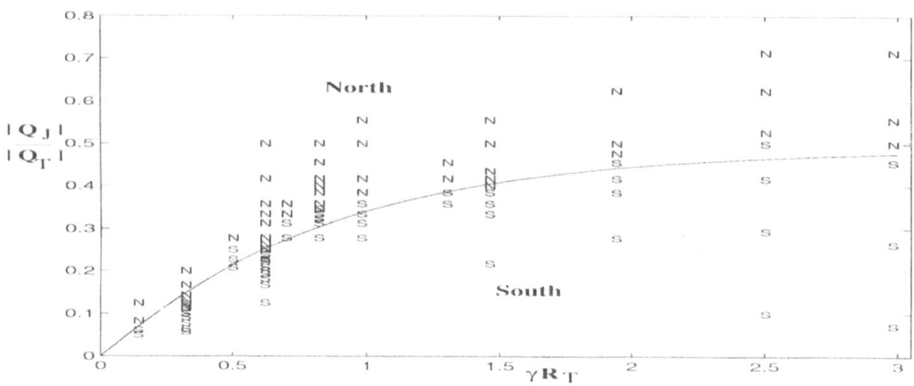

Figure 8. Analytical criterion for an anticyclonic vortex the jet (solid curve) superimposed with the numerical results of the nonlinear model.

velocity on the jet axis for the anticyclone to cross the jet (4). This yields the following analytical relation for the critical value of Q_{Tc}:

$$Q_J/Q_{Tc} = \gamma R_T \; I_1(\gamma R_T) \; K_1(\gamma R_T) \tag{5}$$

(Note that when $\gamma R_T \to \infty$, $\gamma R_T I_1(\gamma R_T) K_1(\gamma R_T) \to 0.5$).

In Fig. 8, we plot this analytical relation in the $(|Q_J/Q_T|, \gamma R_T)$ plane. This curve separates the domains where the vortex does or does not cross the jet (resp. labeled south or north). On this figure, we superimpose the results of the nonlinear quasi-geostrophic model simulations, obtained by varying Q_J, Q_T, R_T and γ. The agreement is remarkably good, with an error reaching at most 15 % on Q_{Tc}. For large γR_T, the vortex must have an intensity at least equal to the potential vorticity jump of the jet $2Q_J$

to cross it. Conversely, a small vortex will need a strong potential vorticity to cross the jet; this confirms our previous observation that small vortices will stay north of the jet, and drift eastward.
This dimensionless curve can now be applied to Gulf-Stream rings ($1.5 < \gamma R_T < 2.5$). Since their potential vorticity is at most $2Q_J$, anticyclonic rings shed north of the Gulf-Stream cannot cross it southward later. They will either translate north of it or be reabsorbed.

We have also investigated the influence of environmental parameters on these dynamic regimes:
- Influence of the initial jet-vortex separation: by increasing this distance, we increase the time needed for the vortex to reach the jet, and hence the viscous decay of the vortex during its approach. To cross the jet, the initial vortex intensity Q_T will therefore have to be slightly larger.
- Influence of the initial vortex intensity on a possible jet-vortex merger: contrary to the observation made by Smith & Davis (1989), a vortex can merge with a stable zonal jet if its intensity is close to the vorticity jump of the jet.
- Influence of the initial vorticity profiles of the vortex and jet: if we replace their piecewise-constant vorticity profiles by continuous profiles with similar characteristics, the physical outcome is not qualitatively altered since the physical mechanisms are identical.
- Influence of the beta-effect: numerically, we observe that beta-effect (the dimensionless value is $\beta = 0.072$) favors the crossing of the jet: this might be attributed to the southwestward drift induced by beta on anticyclonic vortices. Indeed, the vortex is initially substantially accelerated by the beta-effect and the final vortex trajectory is more southward. Still, numerical experiments show that the critical value Q_{Tc} diminishes only slightly. The analytical criterion thus still holds in the presence of the beta-effect.

5. Conclusions

With a one-and-a-half-layer quasi-geostrophic numerical model, we have been able to reproduce the oceanic observations of eddy-jet interaction. By varying the initial intensity of the vortex, we have found very different final trajectories. The ratio of vortex to jet intensities and the ratio of vortex to deformation radii are thus two critical parameters for the southward crossing of the jet by the vortex. For weak vortices, the final trajectory was eastward, north of the jet, and the final state was quasi-stationary in the moving frame of reference.
The analytical criterion that we have established gives a remarkable order of magnitude for the vortex intensity necessary to cross the jet. Moreover,

this analytical criterion states that, for large vortices compared to the deformation radius, the minimum vortex intensity must be at least twice the vorticity jump of the jet for southward crossing. An application to Gulf-Stream vortices show that they can only follow the jet or be reabsorbed.

The conditions for jet-crossing are not extremely dependent on the jet potential vorticity profile, provided it is stable; the piecewise-constant vorticity condition is therefore not necessary. The role of beta-effect is striking: it accelerates the eddy-jet interaction and favors the crossing of the jet by the eddy. The final trajectory of the eddy is also more southward. On the contrary, the beta-effect affects only weakly the critical value Q_{Tc}. The essential parameters for crossing remain Q_J/Q_T and γR_T.
Further work will concentrate on the stratified case with a two-and-a-half-layer model for application to the meddies and Azores current. Preliminary work shows that a similar criterion applies on the dominant baroclinic component of the flow.

Acknowledgements

Special thanks are due to SHOM and DRET (French Ministry of Defence) for support under the research program "Processus Associés aux Modèles Intermédiaires et Régionaux".

References

Bell G. I. and L. J. Pratt, 1992: The interaction of eddy with an unstable jet, *J. Phys. Oceanogr.*, **Vol. no. 22**, pp.1229–1244.
Käse R. H. and W. Zenk, 1996: The Warmwatersphere of the north Atlantic ocean, Edited by W. Krauss. *Gebrüder Borntraeger*, pp.365–395.
Morel Y., 1995: Etude des déplacements et de la dynamique des tourbillons géophysiques, application aux meddies, *Thèse de Doctorat d'Université*. Grenoble I. 155 pp.
Richardson P. L. and A. Tychensky, 1997: Semaphore meddy trajectories. Submitted to *J. Geophys. Res.*.
Smith, D.C. and G. P. Davis, 1989: A numerical study of eddy interaction with an ocean jet. *J. Phys. Oceanogr.*, **Vol. no. 19**, pp.975–986.
Stern, M. E. and G. R. Flierl, 1987: On the interaction of a vortex with a shear flow. *J. Geophys. Res.*, **Vol. no. 92**, C10, pp.10733–10744.
Tychensky A. and X. J. Carton, 1997: Hydrological and dynamical characterization of meddies in the Azores region: a paradigm for baroclinic vortex dynamics. Submitted to *J. Geophys. Res.*.

VI. Topological Aspects

STREAMLINE TOPOLOGY OF AXISYMMETRIC FLOWS

MORTEN BRØNS
Department of Mathematics
Technical University of Denmark
DK-2800 Lyngby, Denmark

1. Introduction

When a fluid velocity field $v(x,t)$ is given, the streamlines at the time instant t_0 are found as trajectories of the system of ordinary differential equations $\dot{x} = v(x,t_0)$. In general, this is a nonlinear system, and qualitative (topological) information on the streamlines may be obtained using tools from the theory of nonlinear dynamics.

Any investigation of a system of nonlinear differential equations starts with a local analysis close to the critical points. The local streamline topology is basically determined by Taylor expansion coefficients of v, and bifurcations of the streamline patterns occur when these coefficients vary through certain degenerate configurations. Several authors have performed such studies, see e.g. Dallmann [6], the review paper by Perry and Chong [13], and the monograph by Bakker [2]. Lugt [11, 12] and Brøns [3] have used the topological viewpoint to describe the flow close to viscous and free interfaces.

Here we examine bifurcations of streamline patterns in axisymmetric, incompressible, viscous flows close to the axis. Our approach is based on coordinate transformations using normal form theory [9], which allows a reduction of the number of nonlinear terms, resulting in systems which are significantly simpler to analyse. Bakker [1] has touched upon this method in an analysis of three-dimensional flows, but to the authors knowledge this approach has not been used systematically in the field before. In section 5.3 we briefly compare the results with experimental and numerical results on the Vogel-Ronneberg flow [7, 10]. Some of the results have been preliminarily reported in [4].

2. Derivation of the streamline equations

We consider axisymmetric flow in cylindrical coordinates (r, θ, z) with velocity components (u, v, w) depending only on r, z. Continuity is satisfied when a stream function $\psi(r, z)$ is introduced [10] such that

$$u = \frac{1}{r}\frac{\partial \psi}{\partial z}, \quad w = -\frac{1}{r}\frac{\partial \psi}{\partial r}. \tag{1}$$

Since the axis is a streamline, we can choose the boundary condition

$$\psi = 0 \ \text{ for } \ r = 0. \tag{2}$$

Due to the rotational symmetry, the flow field must be an even function of r. This makes is convenient to use $\rho = 1/2r^2$ as a radial variable, and the equations for the intersection of the axisymmetric stream-surfaces with the meridional plane (1) become

$$\dot{\rho} = \frac{\partial \psi}{\partial z}, \quad \dot{z} = -\frac{\partial \psi}{\partial \rho}, \tag{3}$$

which is a Hamiltonian system with total energy ψ.

To study the flow close to the origin we introduce the Taylor expansion

$$\psi = \sum_{n+m=0}^{\infty} a_{nm}\rho^n z^m. \tag{4}$$

The boundary condition (2) is satisfied if

$$a_{0m} = 0, \quad m = 0, 1, \ldots \tag{5}$$

Using this, eqns. (3) become

$$\dot{\rho} \;=\; \rho \sum_{n+m=0}^{\infty} (m+1)a_{n+1\, m+1}\rho^n z^m, \tag{6a}$$

$$\dot{z} \;=\; -\sum_{n+m=0}^{\infty} (n+1)a_{n+1\, m}\rho^n z^m, \tag{6b}$$

which we study in detail now.

3. Regular critical points

If $a_{10} = 0$, the origin is a critical point for (6). Linearizing at the origin one obtains the system matrix

$$J = \begin{pmatrix} a_{11} & 0 \\ -2a_{20} & -a_{11} \end{pmatrix}. \tag{7}$$

If $a_{11} \neq 0$, the origin is a hyperbolic saddle with eigenvalues $a_{11}, -a_{11}$. The separatrix corresponding to $-a_{11}$ is the center axis, while the other separatrix is

$$z = -\frac{a_{20}}{a_{11}}\rho + O(\rho^2). \tag{8}$$

In terms of the original radial variable r, it follows that the dividing streamline is orthogonal to the center axis. Depending on the sign of a_{11}, the point is either a point of separation or attachment. See figure 1.

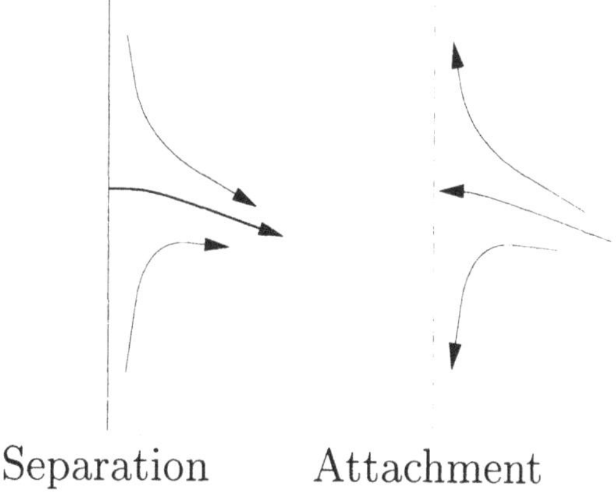

Figure 1. Flow near a regular critical point on the axis shown in physical (r, z) coordinates. Separation occurs when $a_{11} > 0$, attachment when $a_{11} < 0$.

4. Normal form close to a simple degenerate critical point

If $a_{10} = a_{11} = 0$, the origin is a degenerate critical point because 0 is an eigenvalue of J of algebraic multiplicity two. Two sub-cases must be treated individually. If $a_{20} \neq 0$, the geometric multiplicity m_g of the eigenvalue is one, and we denote the degeneration as *simple*. If $a_{20} = 0$, $m_g = 2$, the degeneracy is double. We do not treat this case in general here, but briefly touch upon it in section 6.

Since we want to study the streamlines not only at the exact degeneracy, but with parameters taking values close to this, we introduce two small parameters by

$$\epsilon = a_{10}, \quad \delta = a_{11}. \tag{9}$$

When the linear part of a system of ordinary differential equation is degenerate, higher order terms determine the dynamics close to the origin. We will simplify these terms by transforming the equations to normal form [9, 5]. Rather than giving the full details, we show how the second-order terms of (6) are simplified.

The equations are Hamiltonian, and with a transformation that preserves this property a streamfunction also exists in the new coordinate system. It is convenient to use a canonical transformation since these can easily be found from generating functions [8]. To provide an almost-identity transformation we choose

$$S = \rho y + \sum_{i+j+k+l=3} s_{ijkl}\rho^i y^j \epsilon^k \delta^l \tag{10}$$

which defines a canonical transformation by the relations

$$x = \frac{\partial S}{\partial y}, \quad z = \frac{\partial S}{\partial \rho}. \tag{11}$$

With this choice of S, the transformation is the identity plus terms of second and higher order in both the state variables ρ, z and the small parameters δ, ϵ. Solving eqn. (11) for x, y and keeping terms up to the second order results in

$$\begin{aligned} x &= \rho + 2s_{1200}\rho z + s_{2100}\rho^2 + s_{1101}\rho\delta + s_{1110}\rho\epsilon + 2s_{0210}z\epsilon \\ &\quad + s_{0111}\epsilon\delta + s_{0120}\epsilon^2 + 3s_{0300}z^2 + 2s_{0201}z\delta + s_{0102}\delta^2, \end{aligned} \tag{12a}$$

$$\begin{aligned} y &= z - s_{1200}z^2 - s_{1101}z\delta - s_{1002}\delta^2 - s_{1110}z\epsilon - s_{1011}\epsilon\delta \\ &\quad - s_{1020}\epsilon^2 - 2s_{2100}\rho z - 2s_{2001}\rho\delta - 2s_{2010}\rho\epsilon - 3s_{3000}\rho^2. \end{aligned} \tag{12b}$$

While any choice of the s_{ijkl} is possible, we want x to be radial-like variable and want the transformation to map the axis $\rho = 0$ to $x = 0$. This gives

$$s_{0210} = s_{0111} = s_{0120} = s_{0300} = s_{0201} = s_{0102} = 0. \tag{13}$$

Inserting the transformations in the stream function (4) yields to third order

$$\begin{aligned} \psi &= \epsilon x + a_{20}x^2 + (-2a_{20}s_{2100} + a_{30})\,x^3 + (-4a_{20}s_{1200} + a_{21})\,x^2 y \\ &\quad + (-s_{2100} - 2a_{20}s_{1110})\,x^2\epsilon - 2a_{20}s_{1101}\delta x^2 + a_{12}xy^2 \\ &\quad + \delta xy - 2s_{1200}\epsilon yx - s_{1110}x\epsilon^2 - s_{1101}\delta\epsilon x. \end{aligned} \tag{14}$$

With the choice

$$s_{1200} = \frac{a_{21}}{4a_{20}}, \quad s_{2100} = \frac{a_{30}}{2a_{20}}, \quad s_{1110} = -\frac{a_{30}}{4a_{20}^2}, \quad s_{1101} = 0, \tag{15}$$

a large number of terms vanish, and we get

$$\psi = \epsilon\left(1 + \epsilon\frac{a_{30}}{4a_{20}^2}\right)x + a_{20}x^2 + d_1 xy + a_{12}xy^2, \tag{16}$$

where

$$d_1 = \delta - \frac{a_{21}}{2a_{20}}\epsilon. \tag{17}$$

If $a_{12} \neq 0$ the cubic term in the state variables of (16) is non-degenerate. Assuming this, further simplification is possible. We replace y by $y - d_1/2a_{12}$, replace x by $a_{12}/a_{20}x$ and finally multiply ψ by $a_{20}/2a_{12}^2$ – corresponding to scaling the time in (6) – and finally obtain

$$\psi = -\mu x + \frac{1}{2}x^2 + \frac{1}{2}xy^2, \tag{18}$$

where μ is a new small parameter depending on δ and ϵ. The corresponding differential equations for the streamlines are

$$\dot{x} = xy, \quad \dot{y} = \mu - x - \frac{1}{2}y^2. \tag{19}$$

Note, that since $\rho \geq 0$, we are interested in the half-plane $x \geq 0$ if $a_{20}/a_{12} > 0$, but if $a_{20}/a_{12} < 0$, the relevant case is $x \leq 0$.

If a_{12} is also small, higher-order terms must be included in the normal form until some non-degeneracy condition holds. Proceeding as above, one obtains in the general case

Theorem (Normal form). *Let $a_{10}, \ldots, a_{1\,N-1}$ be small parameters. Assuming the non-degeneracy conditions $a_{20} \neq 0, a_{1N} \neq 0$, a normal form of order N for the system (6) is*

$$\dot{x} = xf'(y), \quad \dot{y} = \mu - x - f(y), \tag{20}$$

$$f(y) = \sum_{n=1}^{N} c_n y^n, \quad c_{N-1} = 0, \quad c_N = \frac{1}{N}. \tag{21}$$

where μ and $c_n, n = 1, \ldots, N-2$ are small parameters. The region of physical significance is $x \geq 0$ if $a_{20}/a_{1N} > 0$, and $x \leq 0$ if $a_{20}/a_{1N} < 0$.

In addition to a simplification of the equations, the number of small parameters is also reduced by the theorem. The number of small parameters is N from the start, but in the normal form only $N-1$ parameters occur. This is obtained by a translation of the origin along the y-axis, and has been denoted *the movement principle* [2, 3]. Mathematically, this is just the basic fact that one may remove the term of the next-highest degree of a univariate polynomial by a suitable choice of the origin.

Before we embark on a study of the equations (20) for specific values of N, some general remarks are in place. The Hamiltonian of the system is

$$H = x(x/2 + f(y) - \mu). \tag{22}$$

Dividing streamlines which start or end at the axis are given by $H = 0$. At the degeneration where $\mu = 0,\ c_n = 0, n = 1, \ldots, N-2$, there is one such streamline given by

$$x = -\frac{2}{N}y^N. \tag{23}$$

Thus, the origin is a *saddle with four hyperbolic sectors* [2]. For N odd, the dividing streamline is symmetric around the origin. For N even, the dividing streamline lies entirely on one side of the axis. In the latter case, the bifurcations will be different depending on which half-plane that is physically significant.

Locating the critical points is the first step in the analysis of (20). To determine their type, the Jacobian matrix

$$J = \begin{pmatrix} f'(y) & xf''(y) \\ -1 & -f'(y) \end{pmatrix} \tag{24}$$

is needed. If $|J| < 0$, the critical point is a saddle (a stagnation point), if $|J| > 0$, it is a center. If $|J| = 0$, the point is degenerate, and bifurcation occurs.

Critical points on the axis satisfy $x = 0, f(y) = \mu$. Here the Jacobian determinant is

$$|J| = -f'(y)^2, \tag{25}$$

showing that bifurcation occurs when $f'(y) = 0$. Critical points off the axis satisfy $f'(y) = 0, x = f(y) - \mu$. Here

$$|J| = xf''(y). \tag{26}$$

Local bifurcations occur when a critical point is degenerate. Global bifurcations occur when saddles are joined by separatrices. For the system (20) there can be no heteroclinic connections involving critical points on and off the axis, since, for a critical point off the axis, $H = 3/2x^2 \neq 0$ and $H = 0$ on the axis. This limits the number of possible bifurcations significantly.

5. Unfolding of degenerate configurations

5.1. NORMAL FORM OF ORDER 2

For $N = 2$, the system (20) is

$$\dot{x} = xy, \quad \dot{y} = \mu - x - \frac{1}{2}y^2. \tag{27}$$

The system has critical points on the axis given by $(x, y) = (0, \pm\sqrt{2\mu})$. These exist only for $\mu \geq 0$. There is also an off-axis critical point $(x, y) =$

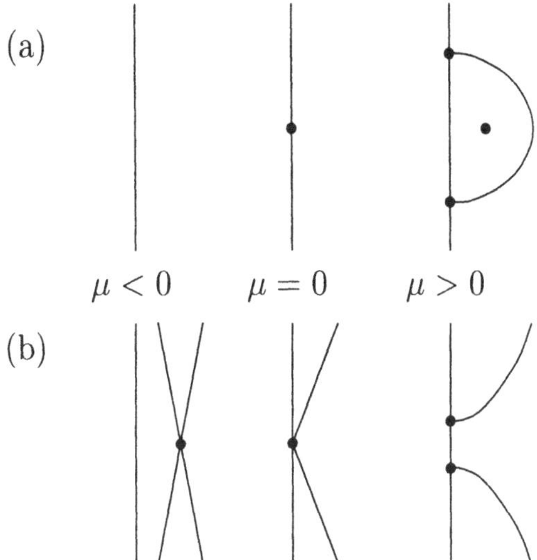

Figure 2. Bifurcation diagram for the second-order normal form (27). (a): Streamlines in physical coordinates for $a_{20}/a_{12} > 0$. (b): Streamlines in physical coordinates for $a_{20}/a_{12} < 0$.

$(\mu, 0)$. Since $|J| = 2\mu$ this point is a saddle for $\mu < 0$ and a center for $\mu > 0$. The corresponding bifurcation diagrams are shown in fig. 2. Physically, case (a) is the creation of a recirculation bubble on the axis. Case (b) will occur when two recirculation bubbles merge.

5.2. NORMAL FORM OF ORDER 3

For $N = 3$, the system (20) becomes

$$\dot{x} = x(c_1 + y^2), \quad \dot{y} = \mu - x - c_1 y - \frac{1}{3}y^3. \tag{28}$$

One easily obtains that bifurcation of critical points on the axis occur when $\mu^2 = -4/9c_1^3$ and bifurcation of critical points off the axis occur when $c_1 = 0$. This results in the bifurcation diagram shown in figure 3. Since N is odd, the same bifurcations occur for both signs of a_{13}. These bifurcations will appear when a small and a large recirculation bubble merge. The analysis shows that this is a two-stage process, where a homoclinic loop is formed before a center and a saddle merge and disappear off the axis.

5.3. NORMAL FORM OF ORDER 4

For $N = 4$, the system (20) becomes

$$\dot{x} = x(c_1 + 2c_2 y + y^3), \quad \dot{y} = \mu - x - c_1 y - c_2 y^2 - \frac{1}{4}y^4. \tag{29}$$

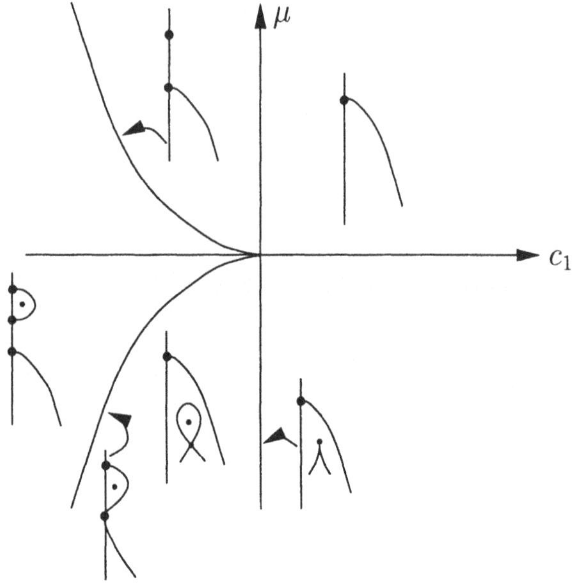

Figure 3. Bifurcation diagram for the third-order normal form (28) for $a_{20}/a_{13} > 0$.

The condition for bifurcation of critical points on the axis at (x, y) is

$$\frac{3}{4}y^4 + c_2y^2 + \mu = 0. \tag{30}$$

It is convenient to use the discriminant of this equation

$$\lambda = c_2^2 - 3\mu \tag{31}$$

as a parameter rather than μ. When $\lambda \geq 0$, bifurcation occurs when

$$c_1^2 = \left(\frac{2}{3}\right)^3 (-c_2 \pm \sqrt{\lambda})(2c_2 \pm \sqrt{\lambda})^2. \tag{32}$$

If $\lambda < 0$, no critical points bifurcate on the axis. Bifurcation off the axis happens when

$$c_1^2 = -4\left(\frac{2}{3}c_2\right)^3. \tag{33}$$

From this, it is not difficult to obtain the bifurcation diagrams. We show in figure 4 the result for $a_{14} > 0$, which describes the possibilities for the creation and merging of two small separation bubbles.

The topologies in the top panel of figure 4 are exactly those that occur in the Vogel-Ronneberg flow [7, 10], where a fluid in a cylindrical vessel

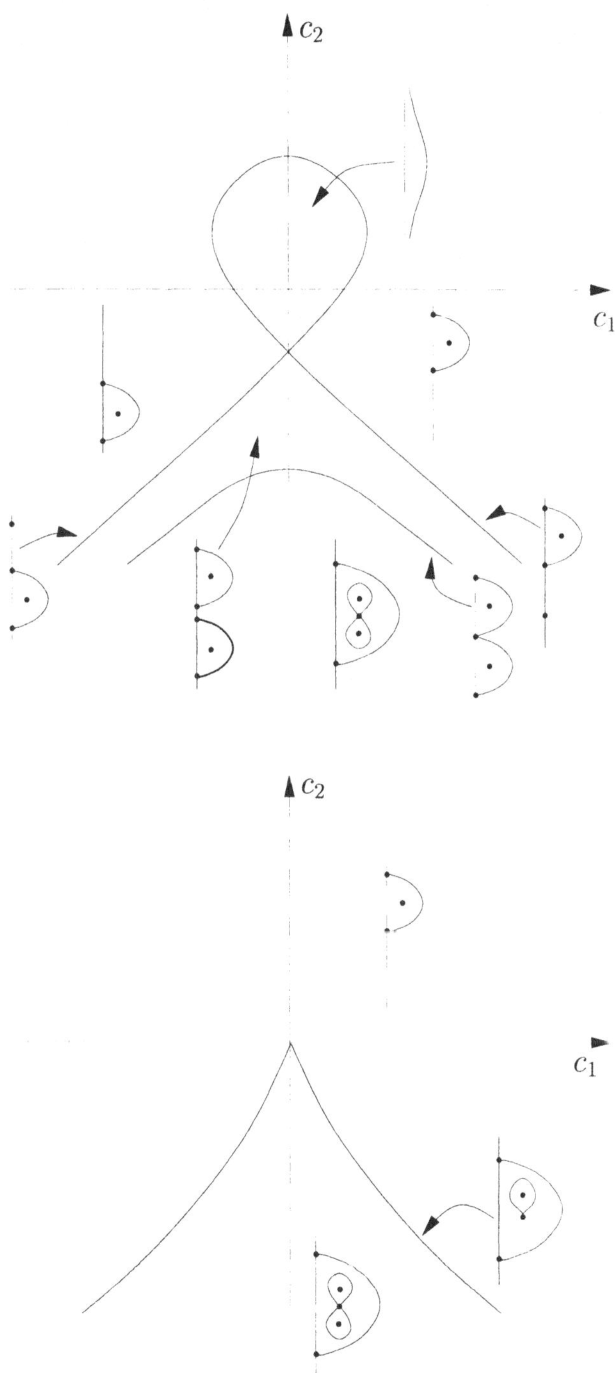

Figure 4. Bifurcation diagrams for the fourth-order normal form (29) for $a_{14} > 0$. Top panel shows $\lambda > 0$, bottom panel is $\lambda < 0$.

is brought in motion by rotating one of the end-walls. The bifurcation diagrams obtained numerically and experimentally show that depending on the aspect ratio, up to two recirculating bubbles can occur which afterwards may merge as shown. The author is not aware of any example of the creation of a recirculating bubble with internal structure as appears in the bottom panel of figure 4.

6. Conclusions and perspectives

The normal form approach to flow topology used in the present paper makes the analysis significantly easier than in the cases where no transformations are used [2]. Also, it will be much clearer what the effects are of further constraints, such as those imposed by the Navier-Stokes equations in a steady flow.

However, if a singularity occurs where the linear part vanishes completely, the normal form transformations used here do no provide any simplifications. More sophisticated normal forms will be needed, and work to exploit such possibilities is in progress.

References

1. P. G. Bakker. On the topology of three-dimensional separations, a guide for classification. In D. Roose, editor, *Continuation and Bifurcations: Numerical Techniques and Applications*, pages 297–318. Klüwer, Dordrecht, 1990.
2. P. G. Bakker. *Bifurcations in Flow Patterns.* Klüver Academic Publishers, Dordrecht, 1991.
3. M. Brøns. Topological fluid dynamics of interfacial flows. *Physics of Fluids*, 6(8):2730–2737, 1994.
4. M. Brøns. Topological fluid mechanics with applications to free surfaces and axisymmetric flows. *Zeitschrift für Angewandte Mathematik und Mechanik (ZAMM)*, 76 suppl. 5:73–74, 1996.
5. S.-N. Chow, C. Li, and D. Wang. *Normal Forms and Bifurcation of Planar Vector Fields.* Cambridge University Press, 1994.
6. U. Dallmann. Three-dimensional vortex structures and vorticity topology. *Fluid Dynamics Research*, 3:183–189, 1988.
7. M. P. Escudier. Observations of the flow produced in a cylindrical container by a rotating endwall. *Experiments in Fluids*, 2:189–196, 1984.
8. H. Goldstein. *Classical Mechanics.* Addison-Wesley, Reading Mass., 1950.
9. J. Guckenheimer and P. Holmes. *Nonlinear Oscillations, Dynamical Systems, and Bifurcations of Vector Fields.* Springer Verlag, New York, 1983.
10. J. M. Lopez. Axisymmetric vortex breakdown part 1. Confined swirling flow. *Journal of Fluid Mechanics*, 221:533–552, 1990.
11. H. J. Lugt. Local properties at a viscous free surface. *Physics of Fluids*, 30:3647–3652, 1987.
12. H. J. Lugt. Oblique vortices on a solid wall and on an interface between two immiscible fluids. *Physics of Fluids A*, 1:1424–1426, 1989.
13. A. E. Perry and M. S. Chong. A description of eddying motions and flow patterns using critical-point concepts. *Annual Review of Fluid Mechanics*, 19:125–155, 1987.

FLOW TOPOLOGY AND TOMOGRAPHY FOR VORTEX IDENTIFICATION IN UNSTEADY AND IN THREE-DIMENSIONAL FLOWS

UWE CH. DALLMANN AND HEINRICH VOLLMERS

DLR - Institute of Fluid Mechanics
D-37073 Göttingen, Germany

AND

WEN-HAN SU

Fluid Mechanics Institute,
Beijing University of Aeronautics and Astronautics,
Beijing 100083, China

Abstract. The characterization of organized structures is discussed with emphasis on changes of the flows' topologies. The changes may be monitored in physical space. However, structural changes also appear in the space of invariants of various vector-gradient tensors. "Vortices" are only one part of a flow's structure to be identified herewith. Jacobian invariants of vector-field mappings onto physical space provide a topological characterization of flow structures and their changes in the physical space and in the vector-field spaces. Several problems of structure identification are highlighted with respect to instantaneous flow fields and within cross-sectional flow patterns of iso-surfaces of scalars or components of vectors. On the other hand, the topological structures of flows (defined by any physical vector quantity of interest) are well-defined by singular flow surfaces in physical space. Their geometric complexity in three-dimensional and/or unsteady flows suggests to monitor especially their local genesis which is accompanied by spatio-temporal vector-field degeneracies. Flow tomography is suggested to provide a means for flow analyses in physical space.

1. Introduction

Spatio-temporal symmetry breaks in flows lead to topological flow changes. Such changes characterize the formation and the breakdown of well-defined organized structures within the various scalar, vector and tensor fields associated with fluid motion. Changes of flow symmetry are the basis in theories of various flow instabilities, of laminar-turbulent transition phenomena, of the formation of separated vortex flow regions and/or of vortex breakdown. Last but not least, there is confidence (also supported by this IUTAM-Symposium) that a modern theory of small-scale turbulence can be derived which is based on characteristic structures of turbulence. Such a theory will be of geometrical and of topological nature. Hence, this approach requires a proper characterization of coherent structures and - most important - their identification via topological changes in a space of proper invariant quantities. Vortices in velocity fields are only one feature resulting from such flow changes.

Dallmann (1983a, 1983b, 1985, 1988) and Dallmann & Gebing (1994) have shown how topological **changes** of three-dimensional flow structures create elementary steady or instantaneous velocity and vorticity structures in an incompressible flow at a wall. The structures evolve through flow parameter variations via so-called unfoldings of local flow degeneracies and global structural bifurcations within the velocity and/or vorticity fields under consideration. It is obvious that spatial structures of any other vector fields will evolve under such variations, too, and that their local genesis is also associated with flow degeneracies at critical points within their vector fields. However, structures of different quantities do not (dis)appear or change simultaneously in time and/or flow-parameter space. Therefore, the search for a complete and proper description of the structures of a flow field (defined by several vector fields) continues. The present paper summarizes our own efforts in this search.

Various structural flow bifurcations within three-dimensional velocity fields close to a wall have also been identified theoretically by Bakker and de Winkel (1990) and the theory of unfoldings of local flow degeneracies has been applied recently by Brøns (see this IUTAM-Symposium Proceedings) to describe axisymmetric structures of vortex breakdown.

Topological aspects of changes within velocity fields have been considered theoretically by Vieillefosse (1982, 1984) and Cantwell (1992) in an attempt to study the dynamics of a (simplified inviscid model) flow solely via the dynamics of invariants of the velocity-gradient tensor. An algorithm for classification of turbulent flow structures employing the second invariant of the velocity-gradient tensor has been proposed by Wray and Hunt (1990). Various tensor data have been taken from experimental and numerical simulations in the topological studies of Perry & Chong (1987), Chong *et al.* (1990a, 1990b), Chen *et al.* (1990), Soria & Cantwell (1993) and others in order to identify any correlations between various tensor invariants.

For two-dimensional vortical flows diagnostic equations have been derived by Kolář (1997) neglecting the effect of unsteady irrational straining while Jeong and Hussain (1995) deal with the identification of three-dimensional vortices but, in addition, also neglect viscous effects. So far certain regions of "organized vortical structures" in only two-dimensional incompressible but "turbulent" flows have been attributed to a criterion based on a local balance between pressure, strain-rate and vorticity by Weiss (1990) and Larcheveque (1993). Here we reconsider our own efforts on the identification of vortex-flow regions via tensor invariants.

Many further investigations have been focused on the identification of instantaneous flow structures within cross-sectional flow patterns of iso-surfaces of scalars or of convenient vector components. In this paper we emphasize some problems encountered with such kind of cross-sectional flow analyses.

2. On the spatio-temporal structures of unsteady flows

Even for two-dimensional flows severe difficulties arise in answering the question: What is a vortex in an unsteady flow? Lugt (1979) points out that:

> "...the definition of a vortex requires knowledge of the flow development in time. If, instead of pathlines, a time-sequence of instantaneous streamline patterns is used, vortices may be detected by the following procedure: All possible inertial frames are checked, to locate regions with closed streamlines. Then, at another instant, the closed streamline regions in their perspective reference frames are examined to see if their centers have moved. When the centers of such regions do not change with time, these regions represent vortices. When they move, these regions are not vortices."

Lugt (1979) considered two-dimensional flows. Within a three-dimensional flow the existence of a vortex does not require closed streamlines in general. Hence, an appropriate definition and identification of a three-dimensional unsteady vortex is even more difficult, apart from the fact that within a coherent flow structure a "vortex" (whatever definition is used for its identification) is only one kinematic feature amongst others (Dallmann 1988; Dallmann & Gebing 1994).

In order to follow any flow's structures in time only instantaneous velocity-field measurements, for instance obtained with Particle Image Velocimetry, provide valuable information. However, phenomena in experiments like phase jitter, intermittency, etc., pose natural limits on direct instantaneous flow-structure comparisons with any numerical simulations. Also, comparing two different unsteady flows of numerical simulations (for instance, performed on different grids) by instantaneous patterns of any quantity gives rise to severe difficulties. In general, one will not be able to trace the same discrete time sequence in two simulations in order to identify all of the instantaneous topological changes within the two simulations. Nevertheless, coherent flow structures can be identified via the identification of topological changes of a flow field in terms of its dependent variables. The physics of flow structures cannot be understood in terms of velocity-field structures only! Simply by inspection of the equations of motion it is obvious that various vector- and tensor-field properties influence the genesis and dynamics of a flow's structures.

A complete instantaneous flow topology consists of the instantaneous topological properties of all its dependent variables. The dynamics of spatial flow structures may be characterized by temporal topological changes within several different vector fields. Unfortunately, all of these topological changes of all the variables do not appear simultaneously in time (Dallmann & Gebing 1993). Hence, the complexity of a description of non-simultaneous, instantaneous topological changes of all the dependent variables and its dependences on variations of flow parameters like Reynolds number, etc. leads to a search for a proper space where one can separate spatial from temporal characteristics in the parameter-dependence of flow structures.

One approach in this direction, extensively applied to vortex or eddy identification, is to expand any flow field quantity $\vec{u}(\vec{x},t)$ of interest in terms of so-called Karhunen-Loeve-eigenmodes $\vec{\sigma}_i(\vec{x})$ multiplied by time-dependent decomposition coefficients $\xi_i(t)$ (see Lumley 1970 and the contributions to this IUTAM-Symposium Proceedings). This method is commonly referred to as "Proper Orthogonal Decomposition". Since the Karhunen-Loeve-eigenmodes $\vec{\sigma}_i(\vec{x})$ depend only on the physical space coordinates their topological structure can change only due to physical parameter changes or resolution changes, whereas the purely time-dependent coefficients $\xi_i(t)$ define a proper phase-space (also dependent on physical and/or numerical modelling) to be analyzed by dynamical-systems methods. Hence, a set of time-independent topologies characterizing a time-dependent flow can be given by the set of Karhunen-Loeve modes of all of the dependent variables. However, other decompositions may provide this simplification as well. Such a decomposition into Karhunen-Loeve modes turnes out to be valuable for validations of numerical flow simulations with respect to necessary spatial resolutions (Dallmann *et al.* 1995a).

3. On topological changes of three-dimensional flow structures in terms of invariants of vector-gradient tensors

In the following we reconsider a topological characterization of the structure of a flow via Jacobian invariants of vector-field mappings onto physical space as proposed by Dallmann (1983a). At first we stress an important fact of topology. Consider any steady or instantaneous velocity field in a plane where a vortex is seen via streamlines turning around one or several nodal points or centers according to fig. 1. By any change of flow control parameters, like Reynolds number, or simply by a change of the observer's reference frame a vortex never (dis)appears by (dis)appearance of only one nodal point (either one node or one focus) or one center. Topological constraints require that structures (dis)appear locally via the (dis)appearance of pairs or clusters of critical points, for instance within a plane vector field through nodal-saddle-point pairs. Hence, local structural flow changes are always accompanied by structural flow degeneracies, i.e. by saddle-nodes, cusps or so-called higher-order critical points. In addition, global bifurcations appear in various topology changing recombinations of (locally created) structures via changes in saddle-saddle-point connections (see fig. 1). Such structural changes also appear in three-dimensional vector fields $\vec{Q}(U,V,W)$ in physical space $\vec{X}(x,y,z)$ and are defined on certain characteristic surfaces in the space of the invariants of the vector-gradient tensor, i.e. of the Jacobian matrix

$$M = \partial(U,V,W)/\partial(x,y,z)$$

which is discussed in detail by Reyn (1964) and with respect to fluid mechanical classifications by Chong *et al.* (1990a, 1990b). In three dimensional flows certain characteristic surfaces in the space of the three Jacobian invariants (P,Q,R) define degeneracies, so-called general non-hyperbolic critical points which are singularities on stream-surfaces, vorticity-surfaces, etc.. Hence, Dallmann (1985, 1988) and Dallmann & Gebing (1994) suggested that a classification of flow structures should be based on unfoldings of such flow-field singularities. Elementary topological flow structures with singular streamsurfaces, singular vorticity surfaces, etc. defined by clusters of critical points evolve from there. Hence, a key issue is the identification of flow degeneracies!

The local structure of a velocity field in physical space can be defined by the invariants (P,Q,R) of the velocity-gradient tensor provided any frame of reference is fixed. Also the invariants of the acceleration-gradient tensor are of interest because its antisymmetric part leads to the vorticity-transport equation and its symmetric part contains information on the pressure gradient field via the invariants of the Hessian of the pressure $\wp$, i.e. of the (pressure gradient)-gradient tensor $\wp_{,ij}$. Parts of this tensor have been used by Jeong and Hussain (1995) to provide a new definition of a vortex which is focused on pressure minima. However, such pressure minima are not always present within vortices (Lugt & Copeland 1994). Hence, only the search for a proper space of invariants to define a flow's structure instead of a flow's "vortex" seems to be meaningful.

Of special interest to define a flow's structure are the surfaces of vanishing Jacobian determinant J which is the third invariant of M . (We recall that two-dimensional flow patterns are defined by the two invariants (p, q) of the corresponding (2×2) matrix M.) The $(J = 0)$-surfaces are naturally occuring singularities in flow fields which can be viewed in the following way:

In a three-dimensional flow we may understand any vector field $\vec{Q}(U, V, W)$ - where (U, V, W) may be components of any vector $\vec{Q}$ of interest - as the mapping $\vec{Q} : \vec{X} \rightarrow \vec{Q}$. Usually one examines the character of a flow field in terms of the $\vec{Q}$-vectors' dependence on spatial coordinates, i.e. one studies streamlines, vorticity lines, etc. in physical space. Equivalently one could analyze the flow field under consideration also in the image space of the vector $\vec{Q}(U, V, W)$. But there may be a set of points in physical space, where the determinant J of the Jacobian matrix (for instance the invariants q or R of the velocity-gradient tensor in two- or three-dimensional flows, respectively) vanishes. These points are singular points in $\vec{X}(x, y, z)$, since it is there, where the area/volume of the image $\vec{Q}$ is zero although the corresponding area/volume of the original neighbourhood $\vec{X}$ is not. These singularities have the important property to be invariant under any small perturbations on the flow field. In the velocity space $\vec{Q}(U, V, W)$ we can identify these singularities $J = 0$ as folded surfaces which may exhibit lines of cusps, so-called ribs, and may intersect themselves to form cusp points on the ribs. The dependence of the flow-field structure on time and/or on parameters like Reynolds number will change these $(J = 0)$-surfaces. For a certain set of parameters and/or at certain instants in time the flow structure in physical space $\vec{X}$ and thereby the $(J = 0)$-surfaces in the vector space $\vec{Q}$ may change their topology. Therefore, we suggest to monitor changes of the $(J = 0)$-surfaces of the dependent vector fields to characterize the dynamics of a flow field.

Another characteristic surface in the space of Jacobian invariants of a vector-gradient tensor separates regions with complex eigenvalues (see Reyn 1964) from those with real eigenvalues. This surface, which we call the $(D = 0)$-surface, provides a further topological characteristic not only for velocity fields (where this surface helps to identify conditional vortex-flow regions as discussed below). Unfortunately we loose such information on a flow's "vortex" structure if we restrict a topological analysis of any general vector to an analysis of its scalar-components because gradient fields like $\nabla(scalar)$-fields have only real eigenvalues. This is, unfortunately, quite often done by looking only at some components of the vorticity vector.

4. On the identification of vortex-flow regions

Cantwell (1978) analyzed topological properties of coherent structures (including eddies) in (p, q)-phase-plane plots of particle trajectories. Dallmann (1983a) and Vollmers *et al.* (1983) made a first attempt to define in the phase-space of the Jacobian invariants (P, Q, R) three-dimensional structures of incompressible $(P \equiv 0)$ flows. They suggested a mathematical definition of a region in physical space where a vortex could be identified under continuous variation of Galileian frames of reference by locating the region in physical space where the velocity-gradient tensor exhibits complex eigenvalues. This definition of a "vortex" was later considered by others. The formation of such a spatial region defines a topological change of a flow, however, not necessarily in physical space for a given frame of reference. If such a region exists then there is always a finite range of possible frames of reference where "vortices" (i.e. foci or centers) will be seen in the velocity field. Such vortices are, nevertheless, only one part of the game of changing topological structures. We have to follow the evolution of a flow structure rather than to identify only "vortices"! Different spatial structures appear in three-dimensional flows which do not exhibit any swirling "vortex motion" at all (Dallmann 1983a, 1983b, 1988; Dallmann & Gebing 1994). The scatter diagrams of Chen *et al.* (1990) and Soria *et al.* (1993), also discussed by Perry & Chong (1992), indicate the importance of flow features which are different to "vortices" in the velocity field. Nevertheless, the problem of "vortex identification" (see, for instance, Jeong and Hussain 1995) is still present even at this IUTAM-Symposium. By inspection of fig. 2 one might conclude that a necessary condition for a vortex to appear in a three-dimensional velocity field should be that the flow pattern within a cross-section cutting a (still to be identified!) "vortex axis" should exhibit nodal character (nodes or foci) on this vortex axis. Unfortunately, for an observer moving with the flow a cross-sectional flow pattern with nodal points (either a node or a focus) appears to be present everywhere in a three-dimensional incompressible flow. This is due to the fact that the local flow structures which can occur in the vicinity of a critical point within an incompressible flow $(P \equiv 0)$ are those displayed in fig. 3, i.e. they all exhibit one sectional flow pattern with either one node or one focus. Hence, a cross-sectional flow pattern analysis alone cannot identify vortices in velocity fields!

On the other hand, changes of the flow's structure can appear in physical space with respect to an observer moving with the fluid particles and this allows to characterize a flow field: The surfaces in physical space where the Jacobian discriminant $D = 27R^2 + (P^3 - 18PQ)R + (4Q^3 - P^2Q^2)$ of M

changes sign separates regions where no vortices at all and where vortices may be detected locally. This provides the complex-eigenvalue criterion, i.e. the $(D > 0)$-criterion for three-dimensional vortex flow regions as suggested by Dallmann (1983a) and Vollmers *et al.* (1983). However, it is also obvious from fig. 2 that a vortex can appear in a velocity field with respect to certain moving observers also as a swirling motion around a node, swirling only at some distance away from the axis. Hence, the local appearance of a focus in a cross-section within the complex eigenvalue region is not necessary for a globally present vortex. A necessary condition for its spatial and/or temporal genesis is a vanishing Jacobian determinant $J = 0$, i.e. three-dimensional vortices may appear as part of a complex structure "downstream" of $R = 0$. For incompressible flows where $P \equiv 0$ it follows that the surfaces $R = 0$ and $Q = 0$ intersect on the surface $D = 0$.

5. On sectional flow pattern analyses

For the purpose of physical modelling comparisons between flow simulations are commonly performed via analyses of isolines or iso-surfaces of various physical quantities taken at certain instants in time. Intuitively one is focused on the identification of any structural changes apparent within certain sets of cross sections, i.e. on planes or surfaces cutting through a flow in three-dimensional space at certain instants in time. Mainly extremal properties within such iso-surface data are of interest and are used for "vortex identification". Such information, however, could be directly provided by the gradient operator. This leads to the identification of critical points, separatrices and singular surfaces of $\nabla(scalar)$, i.e. to topological properties of vector fields. Apart from the above mentioned fact that information about a general vector field is lost by analyzing only the gradient fields of its vector-components, extremal properties of components of three-dimensional vector quantities have limited significance for physical modelling in three-dimensional space. Sectional flow analyses of any vector fields have to be used with great care otherwise the physical models derived herewith could solely depend on the spatial orientations of the set of cross-sections chosen. For instance, different orientations of cross-sections through the iso-pressure and iso-vorticity component surfaces displayed in fig. 4 would clearly identify a different number and very different spatial locations of extrema within such cross-sections simply because of the convoluted nature of the iso-surfaces.

Let us consider incompressible flows around a prolate spheroid (see Dallmann *et al.* 1995b for details and for references). At certain sets of Reynolds number, excentricity of the prolate spheroid and angle of attack streamwise "vortices" appear in simulations (fig. 4). These large-scale vortices are of basic interest since the spatial formation of the "vortex axes" is still unclear. No criterion has been provided sofar for the so-called open separation line. These vortices allow studies not only on the identification of organized structures ("Where are the vortices?") but especially on their structural changes. At further critical sets of parameters flow unsteadiness and laminar-turbulent transition sets in. In addition, laminar and turbulent vortex breakdown is observed. It is of engineering interest to recognize the coincidence of the wall-flow patterns in the time-dependent numerical simulations, in experiments and also in the (quasi-)steady numerical simulations up to the separated flow region including the formation process of the leeside vortices (see fig. 4). In fig. 5 the spatial evolution of the separating vortex is clearly displayed by a quasi-steady simulation of the cross-sectional computer tomography in good agreement with the experimental visualization by laser-light-sheet technique (Su *et al.* 1992, 1993). Dallmann & Schulte-Werning (1990) point out that such an orientation-invariant image of a three-dimensional flow can be obtained by indicating the locations of particles when they pass through a set of cross-sections. Particles which have been uniformly distributed either on or upstream of the body will be mapped into a non-uniform distribution within the cross-sections at subsequent downstream positions. In fig. 5 particles have been released in the vicinity of the front stagnation point (also seen in fig. 4(f)); they follow the near-wall skin-friction pattern before they leave the vicinity of the body at separation lines and critical points of the wall-flow pattern. The particles are then monitored as they cross subsequent cross-sections. As a matter of fact, topological flow tomography patterns could be obtained by releasing particles solely from the critical points of a vector field (Dallmann & Schulte-Werning 1990), hence, by tracing the singular stream-surfaces of a velocity (or any other vector) field.

The information obtained by such means of flow tomography is different from that obtained via cross-sectional streamline plots! The latter are obtained by integrating the direction field defined solely by the velocity components parallel to the chosen cross-sections. In contrast to flow tomography patterns the cross-sectional stream-line patterns may strongly change their topological structures with the cross-section orientation! Let us compare two sequences of steady flow patterns obtained in differently oriented cross-sections according to fig. 6(c). Different patterns and different topological flow changes appear in fig. 6(a) and fig. 6(b) as we apply different cross-sectional scanning through the three-dimensional flow.

Although the here presented example is based on a steady flow solution it represents a generic problem of instantaneous flow analyses within cross-sections. By inspection of fig. 6(a) one might conclude that the leeside vortices behind a prolate spheroid are created within the plane of flow symmetry and above the body while in fig. 6(b) they appear on the body wall! Flow-physics modelling, like identification of the spatial origin of vortices, transport of passive scalars, turbulence properties, etc. along such cross-sectional flow patterns within a three-dimensional vector field is always questionable (and will quite often be simply wrong!). The sequences of the structural changes seen in fig. 6(a) and fig. 6(b) are not at all topologically equivalent, hence, they are not necessarily characteristics of the velocity field, they are influenced by the cross-section orientations!

Acknowledgements

The authors appreciated very much their collaboration with Prof. Hong-Quan Zhang, who provided unsteady numerical flow simulations around the prolate spheroid while staying with us as an Alexander v. Humboldt fellow. A collaboration project between DLR and the Chineese Aeronautical Establishment CAE kindly supported the very fruitful one-year stay of Prof. Wen-Han Su in Göttingen, Germany.

References

Bakker, P.G., de Winkel, M.E.M., (1990), "On the topology of three-dimensional separated flow structures and local solutions of the Navier-Stokes equations." In: Moffat, H.K., Tsinober, A. (eds.) Proc. *IUTAM-Symp. Topological Fluid Mechanics.* Cambridge Univ. Press, Cambridge, pp. 384–394.

Cantwell, B.J., (1978) "Coherent tubulent structures as critical points in unsteady flow." Arch. Mech. Strosow. (Archives of Mechanics), Vol. **31**, pp. 707–721.

Cantwell, B.J., (1992), "Exact solution of a restricted Euler equation for the velocity gradient tensor." Phys. Fluids A, Vol.**4**, No.4, 782-793.

Chen, J.H., Chong, M.S., Soria, J., Sondergaard, R., Perry, A.E.,Rogers, M., Moser, R., Cantwell, B.J., (1990), "A study of the topology of dissipating motions in direct numerical simulations of time developing compressible and incompressible mixing layers." Proc. of the Summer Program, Center for Turbulence Research, Stanford, CA, 141-164.

Chong, M.S., Perry, A.E., Cantwell, B.J., (1990a), "A general classification of three- dimensional flow fields." Phys. Fluids A, Vol.**2**, No.5, 765-777.

Chong, M.S., Perry, A.E., Cantwell, B.J., (1990b), "A general classification of three- dimensional flow fields." In: Moffat, H.K., Tsinober, A. (eds.) Proc. IUTAM-Symp. Topological Fluid Mechanics. Cambridge Univ. Press, Cambridge, 408-420.

Dallmann, U., (1983a), "Topological structures of three-dimensional flow separations." DFVLR-IB 221-82 A07 .

Dallmann, U., (1983b), "Topological structures of three-dimensional vortex flow separation." 16th Fluid and Plasma Dynamics Conf., Danvers, USA, AIAA-83-1735.

Dallmann, U., (1985), "On the formation of three-dimensional vortex flow structures." DFVLR-IB 221-85 A13.

Dallmann, U., (1988),"Three-Dimensional Vortex Structures and Vorticity Topology." Fluid Dynamics Research **3**, North Holland, 183-189.

Dallmann, U., Schulte-Werning, B., (1990), "Topological changes of axisymmetric and non-axisymmetric vortex flows." In: Moffat, H.K., Tsinober, A. (eds.) Proc. IUTAM-Symp. Topological Fluid Mechanics, 1989. Cambridge Univ. Press, Cambridge, 372-383.

Dallmann, U., Gebing, H., (1993), "How to validate unsteady flow simulations? On topologi- cal equivalence of separated flows around spheres and ellipsoids." In: Daiguji, H. (ed.) Proc. 5th International Symposium on Computational Fluid Dynamics, A Collection of Technical Papers (Part II). Sendai, Japan, 19-26.

Dallmann, U., Gebing, H., (1994), "Flow attachment at flow separation lines. On uniqueness problems between wall-flows and off-wall flow fields." Acta Mech. Vol.**4**, Springer-Verlag Wien, 47-56.

Dallmann, U., Herberg, Th., Gebing, H., Su, W.-H., Zhang, H.-Q., (1995a), "Flow field diag- nostics: topological flow changes and spatio-temporal flow structure." 33rd. Aerospace Sciences Meeting and Exhibit, Reno, USA, AIAA 95-0791.

Dallmann, U., Su, W.-H., Zhang, H.-Q., (1995b), "Three-Dimensional Separated Flows around Prolate Spheroids - Numerical Simulations versus Experimental Investigations." In: Hui, W.H., Kwok, Y.-K., Chasnov, J.R. (eds.) First Asian Computational Fluid Dynamics Conference, Proc., The Hong Kong Univ. of Science & Technology, Vol.**3**, 1103-1109.

Jeong, J., Hussain, F., (1995), "On the identification of a vortex." J. Fluid Mech., Vol.**285**, 69- 94.

Kolář, V. (1997), "Diagnostic equations for two-dimensional vortical flows." Acta Mech. Vol.**120**, pp. 227–231.

Larchevêque, M., (1993), "Pressure field, vorticity field, and coherent structures in tow-dimensional incompressible flows." Theor. Comput. Fluid Dynamics **5**, 215-222.

Lugt, H.J., (1979), "The dilemma of defining a vortex." In: Müller, U., Roesner, K.G., Schmidt, B. (eds.) Theoretical and Experimental Fluid Mechanics, Springer, 310-322.

Lugt, H.J., Copeland, G.S., (1994), "On pressure minima in two-dimensional vortex flows." Phys. Fluids A, Vol.**6**, 2230-2232.

Lumley, J.L., (1970), "Stochastic tools in turbulence." Appl. Mathem. and Mech., Academic Press, Vol.**12**, New York and London.

Perry, A.E., Chong, M.S., (1987), "A description of eddying motions and flow patterns using critical-point concepts." Ann. Rev. Fluid Mech., Vol.**19**, 125-155.

Perry, A.E., Chong, M.S., (1993), "Topology of flow patterns in vortex motions and turbulence.": Bonnet, J.P., Hunt, J.C.R., (eds.), Eddy Structure Identification in Free Turbulent Shear Flows. Proceedings IUTAM Symposium, Poitiers, 1992, Kluwer, 339-362.

Reyn, J.W. (1964), "Classification and description of the singular points of a system of three linear differential equations." ZAMP, Vol.**15**, 540-557.

Soria, J., Cantwell, B.J., (1993), "Identification and classification of topological structures in free shear flows." In: Bonnet, J.P., Hunt, J.C.R., (eds.) Eddy Structure Identification in Free Turbulent Shear Flows. Proceedings IUTAM Symposium, Poitiers, 1992, Kluwer, 379-390.

Su, W.H., Tao, B., Xu, L., (1992), "Experimental investigation of three dimensional separated flow over a prolate spheroid." In: Zhuang, F.G. (ed.) Proc. 1st International Conference on Experimental Fluid Mechanics, Chengdu, China, 1991, International Academic Publishers, Beijing, China, 205-210.

Su, W.H., Tao, B., Xu, L., (1993), "Three-dimensional separated flow over a prolate spheroid." AIAA J. Vol.**31**, 2175-2176.

Vieillefosse, P., (1982) "Local interaction between vorticity and shear in a perfect incompressible fluid." J. Phys. (Paris) **43**, 837.

Vieillefosse, P., (1984), "Internal motion of a small element of fluid in an inviscid flow." Physica A **125**, 150.

Vollmers, H., Kreplin, H.-P., Meier, H.U., (1983), "Separation and vortical-type flow around a prolate spheroid - Evaluation of relevant parameters." AGARD Symp. Aerodynamics of Vortical Type Flows in Three Dimensions, Rotterdam, AGARD-CP-342.

Weiss, J., (1990), "The dynamics of enstrophy transfer in two-dimensional hydrodynamics." Physica D **48**, 273-294.

Wray, A.A., Hunt, J.C.R., (1990), "Algorithms for classification of turbulent structures." In: Moffat, H.K., Tsinober, A. (eds.) Proc. IUTAM-Symp. Topological Fluid Mechanics, 1989. Cambridge Univ. Press, Cambridge, 75-104.

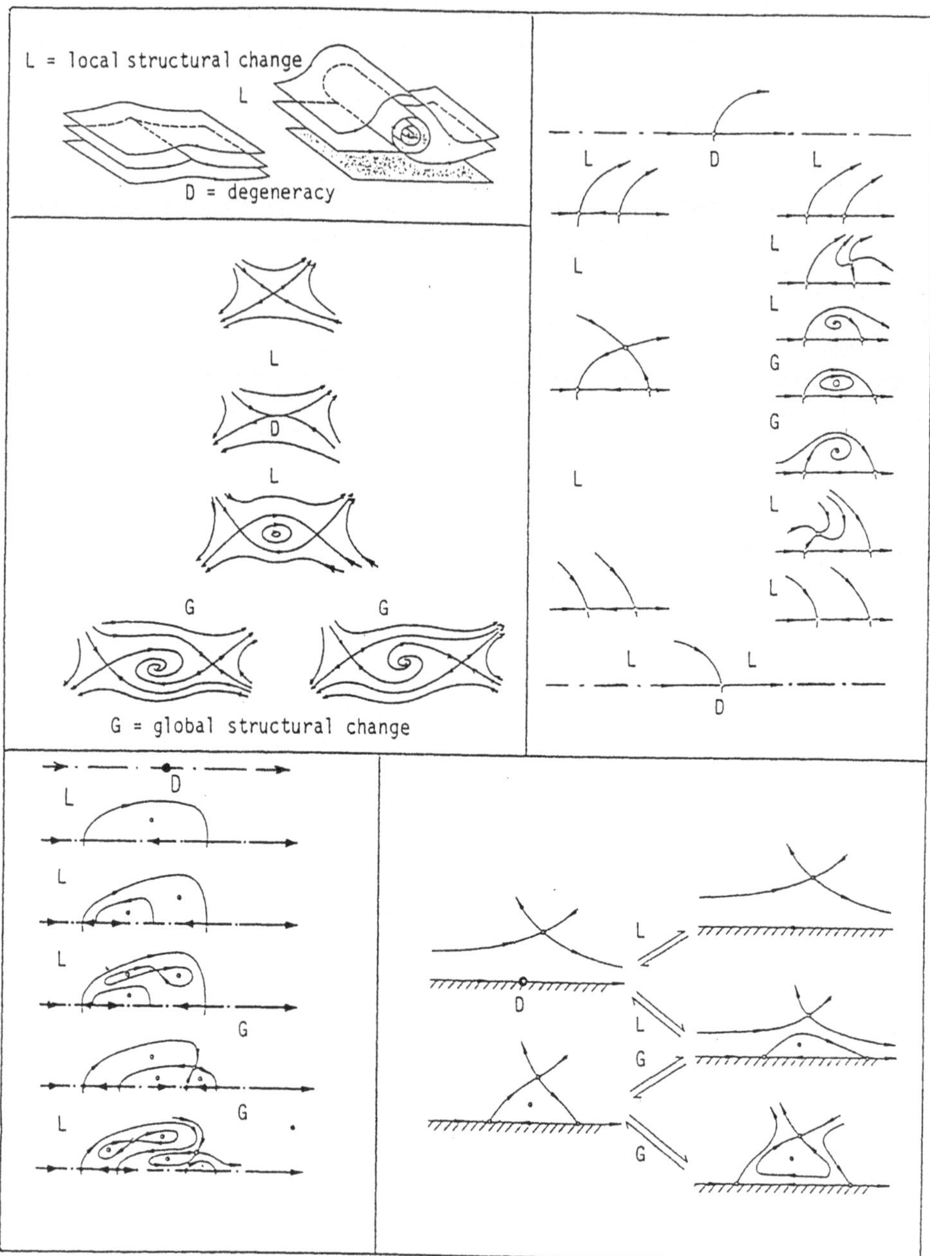

Figure 1. Examples of sectional streamline patterns (trajectories of the velocity components within a cross-section) indicating that local flow degeneracies D precede local structural changes L where "vortex" formation is only one of several features and that global structural changes rearrange vortex flow regions.

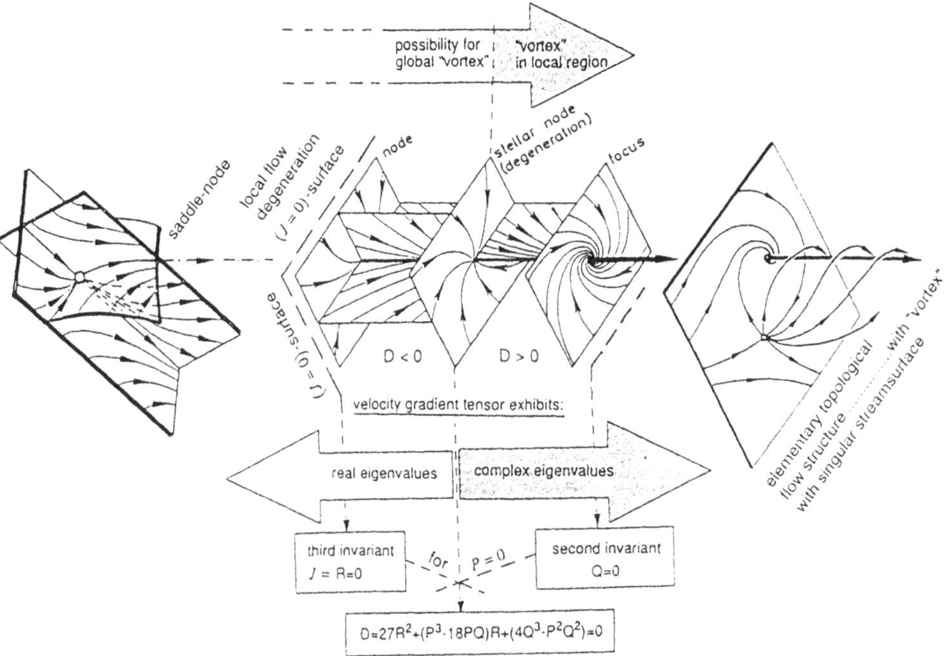

Figure 2. Sketch of the genesis of a "vortex" from a local saddle-node degeneration (degeneracy) within a three-dimensional velocity field w.r.t. different criteria based on the invariants (P, Q, R) of the velocity-gradient tensor.

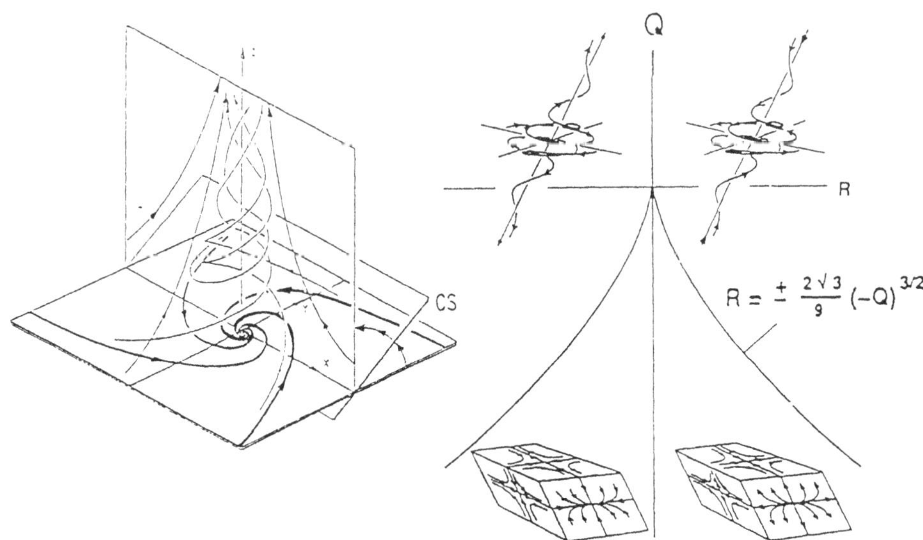

Figure 3. Elementary structures of incompressible flows ($P \equiv 0$) in the (Q, R)-space of invariants of the velocity-gradient tensor (Soria & Cantwell 1993). A nodal-point (node or focus) pattern necessarily changes into a saddle-point pattern by a change of orientation of the cross-section (CS) .

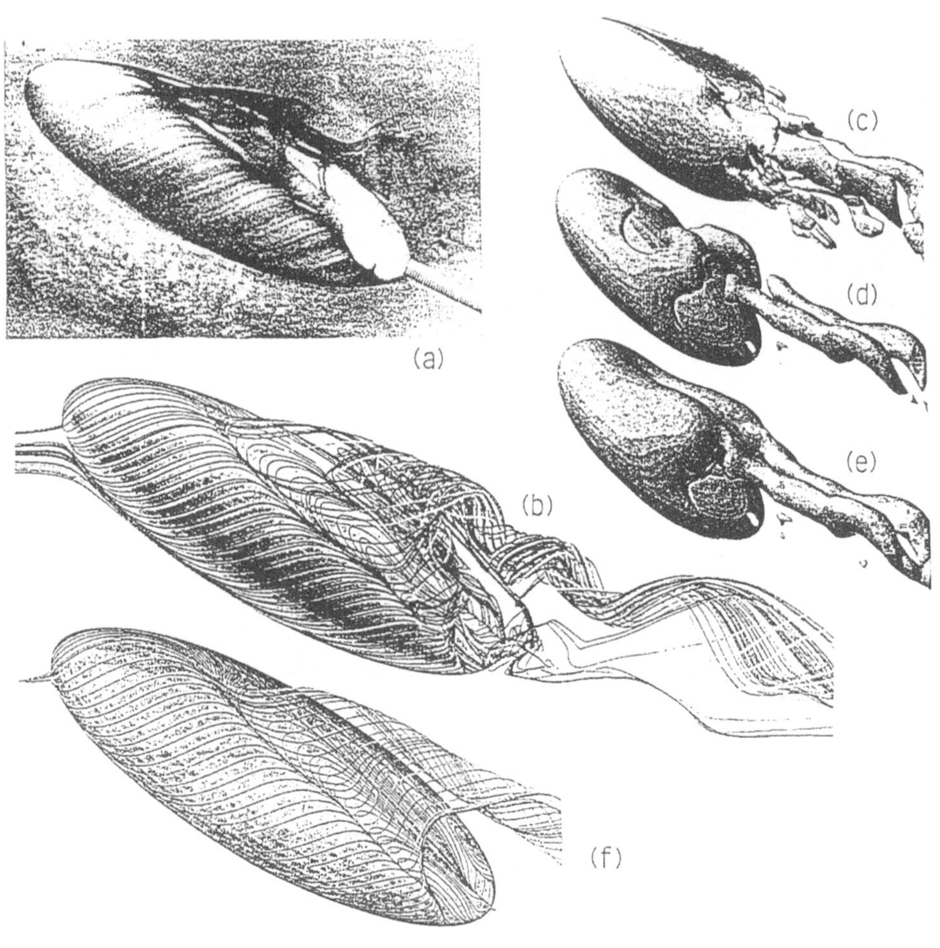

Figure 4. Flow around a prolate spheroid at $Re_a = 1.32 \times 10^4$, $a/b = 3, \alpha = 30^o$.
(a) Water-tunnel flow visualization by Su *et al.* (1992, 1993) and
(b) - (e) unsteady numerical simulation by Zhang (1995, private com.):
(b) instantaneous streamlines,
(c) iso-contour surface of vorticity component ($\omega_x = const.$),
(d) (e) iso-contour surfaces of pressure (two values)
(f) quasi-steady numerical simulation indicating near-wall streamlines (started close to the front stagnation point) and formation of separated leeside vortex caused by open boundary-layer separation.

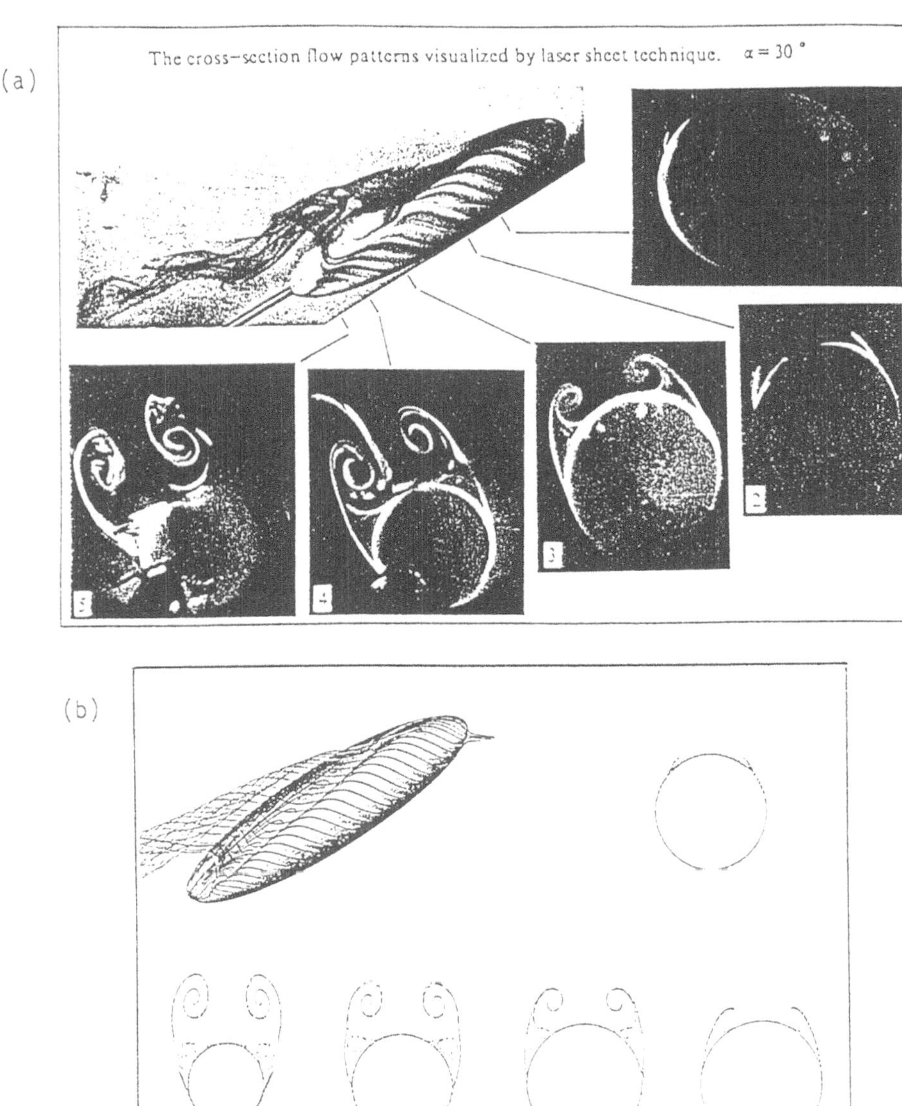

Figure 5. Flow Tomography. Cross-sectional flow patterns visualized at $Re_a = 10000, a/b = 4, \alpha = 30^o$, by:
(a) laser-sheet technique and
(b) quasi-steady numerical simulation

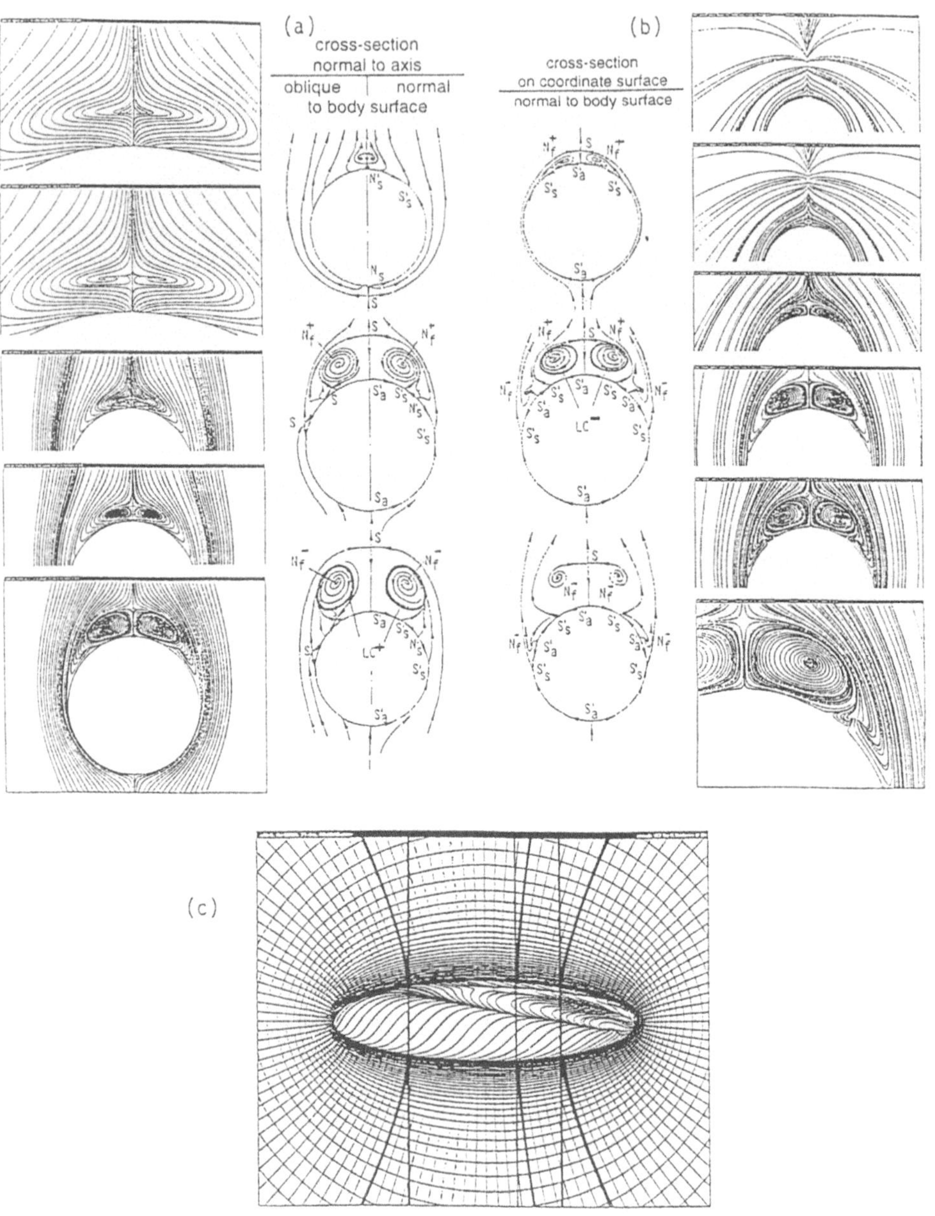

Figure 6. Sequences of cross-sectional streamline patterns on cross-sections which are (a) normal to body axis and (b) parallel to coordinate surfaces at the positions indicated in (c) of flow around a prolate spheroid at $Re_a = 10000, a/b = 4, \alpha = 25^o$.

COHERENT STRUCTURES IN FLUIDS ARE TOPOLOGICAL TORSION DEFECTS

R. M. KIEHN
Physics Department, University of Houston
Houston, TX, 77004 *USA*

Abstract: Cartan's theory of a global 1-form of Action on a projective variety permits the algebraic evaluation of certain useful geometric and topological objects which can be singular. The projective algebraic methods therefore lend themselves to the development of a theory of coherent structures and defects in which the concept of translational shear dislocations and rotational shear disclinations can be put on equal footing. The topological methods not only lead to a precise definition of coherent structures in fluids, but also produce a non-statistical test for thermodynamic irreversibility on a symplectic manifold of dimension 4, and therefore yield a necessary criteria for turbulence.

1. Introduction

The objective is to devise non-statistical theoretical methods that will describe the one key feature of a turbulent flow that everyone agrees upon, the feature of irreversibility, and then to show that topological torsion defects in such irreversible regimes have long lived observable consequences that permit the defects to be defined as coherent structures. The intuitive suggestion is that starting from arbitrary initial conditions on 4 dimensional variety $\{x,y,z,t\}$, an irreversible process will decay to one of its (non unique) "stationary states", a long-lived self-organized or coherent state. The mathematical suggestion is that irreversible processes occur on symplectic manifolds of Pfaff dimension 4 (or topological class) [1], and conformally decay or are attracted into closed sets of measure zero and Pfaff dimension 3. On the 4 dimensional manifold, the anholonomic differential (non-statistical) fluctuations in the classic kinematic formulas, which lead to irreversibility, disappear on the sub-manifolds of measure zero (the long-lived coherent structure).

Most of these ideas are based primarily upon the calculus of variations as extended by Cartan's theory of differential forms [2], and secondly upon Cartan's concept of the Repere Mobile on a projective manifold [3]. A particular topological feature of the Cartan method that has been ignored by the hydrodynamic community is the concept of the Torsion Current [4], perhaps because the idea involves non-Riemannian manifolds and their implicit non-uniqueness of solutions. It will be demonstrated below that when the Torsion Current has a non-zero space-time divergence, then the associated dynamical system is irreversible in a thermodynamic sense, and is of topological dimension 4. The associated dynamical system decays to "coherent" states where the divergence of the Torsion Current vanishes, yielding a conservation law for the evolution of the resulting (coherent) structure [5].

It is this Cartan idea of spaces with torsion [6] that is the major theme of this article. It is known that in a space with an affine connection it is possible to have torsion defects produced by shears of translation [7]. In this article, it is emphasized that in a fluid the dominant torsion defect is not induced by translational shears, but instead is induced by rotational shears, and their attendant accelerations. Such rotational torsion defects (disclinations) do not occur in affinely connected manifolds, but are latent in projective manifolds. Affine translational shears preserve parallelism; rotational projective shears do not. In hydrodynamics, such topological torsion defects are representatives of deRham period integrals, and are generated by Harmonic vector fields. As Harmonic vector fields do not produce any contributions to the RHS of the Navier-Stokes equations, no matter how large the kinematic viscosity, they do not induce dissipation. They are topological limit sets which will produce the visible wakes or coherent structures often seen in experiments [8]. Often these wakes, as coherent structures or topological defects, will appear as tangential discontinuities that, like minimal surface soap films, are globally stabilized. These concepts have been reported elsewhere [9].

In part 2, some historical background and motivation is provided for the present stage of the theory. In part 3, Cartan's Magic Formula (from the calculus of variations) will be used to describe topological evolution, and to develop a thermodynamic criteria for irreversibility. Due to space limitations, part 4, the constructive development of Cartan's Repere Mobile on a projective domain, the demonstration of a new equation of differential geometric structure involving rotational torsion 2-forms, and its application to disclination defects, rotational shears and coherent structures in hydrodynamics, will be presented elsewhere. (See http://www.uh.edu/~rkiehn)

2. Some Historical Motivation

It is important to understand the motivation behind this article. It started in 1974, when, using Cartan's techniques of exterior calculus [10], it was suggested (on intuitive grounds) to examine evolutionary systems that satisfied the equation:

$$i(\mathbf{V})d\mathrm{A} = \Gamma\mathrm{A} + d\Theta, \tag{1}$$

rather than the classic Cartan-Hamilton (extremal) equation:

$$i(\mathbf{V})d\mathrm{A} = 0. \tag{2}$$

These ideas were extended and compared to the projective features of the conformal group. Later, more detailed applications to hydrodynamics were made that led to a derivation of the Navier-Stokes equations on a 4-D space-time setting [11]. In 1975, Cartan's methods of differential topology were applied to the theory of period integrals, with the introduction of a novel 3-dimensional period integral, the integral of quantized spin [12]. The 3-dimensional spin integral is distinct from, but related to, the 3-dimensional period integral of Topological Torsion, which forms the basis of the current work.

In 1977 it was determined that irreversibility could be associated with continuous topological evolution, which, although not deterministically predictive, was deterministically retrodictive on the space of exterior forms (covariant anti-symmetric tensor fields) [13]. A natural logical arrow of time is built into the set of differentiable, but not homeomorphic, maps. It was also suggested about that time that the transition to turbulence must involve the failure of the Frobenius integrability theorem, but the details were not clear. It was argued that the streamline state of a fluid implied that the Frobenius condition, $A\hat{}dA = 0$, was satisfied; as the turbulent state was the antithesis of the streamline state, the Frobenius condition must fail in the turbulent regime. The key idea, however, was that the failure of the Frobenius theorem implied the necessity of including the topic of torsion into the analysis. In the current mathematics literature these ideas have migrated into what are called the Chern-Simons forms.

For a vector field that fails the Frobenius condition, the associated dynamical system can not be planar; the space curve has (Frenet) torsion and a helical signature. In 1979 it was determined that parity symmetry breaking and time reversal symmetry breaking could occur in macroscopic electromagnetic systems, but the Pfaff dimension had to be 3 or greater [14] (a necessary condition for the failure of the Frobenius theorem). Moreover, such electromagnetic systems can support a new form of propagating discontinuities defined as a Torsion wave. In fact, it was determined that helical or torsional electromagnetic waves propagate with different speeds in different directions (a result verified by experiments in dual polarized ring laser systems)! The Torsional waves could not be represented as functions of a single variable (scalar longitudinal waves), or even an ordered pair of variables (complex transverse polarizable waves), but were irreducibly 4 dimensional [15]. The basis of the Torsional waves was the division algebra of quaternions. Such waves can also appear in fluids, but they have been little studied.

In 1986, while in Rio de Janeiro, the author became aware of what are now known as Falaco Solitons [16]. They are easily produced - easily observed - long lived topological defects, obviously involving rotational shears, in a dynamical fluid system. Observations of these long lived topological defects gave credence to the theory of topological defects in hydrodynamic systems. These defects are not to be associated with affine translational shears, as the Falaco effect is dominated by rotational shears. The two 2D surface defects whose Snell projections produce the black spots on the floor of the swimming pool are connected by a 1D string defect that is not visible in the photograph unless dye is injected into the water. The 1D string connects the vertices of the two dimensional surface dimples, and globally stabilizes the coherent structure. Helical torsion waves will propagate along the guiding center furnished by the invisible string connecting the surface defects, much in the fashion of whistlers along the earth's magnetic field lines. These topological defects will last for more than 15 minutes in a still pool of water. For more details and pictures, see [17]

In 1990, using the ideas of Pfaff reduction, some exact solutions to the Navier Stokes equations were obtained in a rotating frame of reference. The extraordinary feature of such solutions is that they replicated certain features of the Falaco Solitons, and exhibited topological phase changes as certain flow coefficients were varied. In one example, bifurcation into a Torsion bubble was produced as the mean flow speed parameter was increased beyond a critical value; the bifurcation took place at constant vorticity! These results are not too widely known, but should be of interest to those at this conference who study coherent structures in rotating systems. The results offer another alternative to the problem that goes by the name of "Vortex bursting" in the hydrodynamics literature. It is suggested herein that this phenomena has nothing to do with vorticity per se, but is an exhibition of one coherent structure of topological torsion transforming into another [18]. In thermodynamics, this event would be called a phase transition.

The concept that a 1-form of Action for a fluid system, when constrained with anholonomic differential fluctuations, would lead to a derivation of the Navier-Stokes equations was presented at the 1992 SECTAM conference in Tennessee [4]. The idea was to define a hydrodynamic action as the 1-form constructed from a classical Lagrange Action, but with possibly non-holonomic differential fluctuations $(d\mathbf{r} - \mathbf{v}dt) \neq 0$ included as constraints on the kinematics. In the following equation, the coefficients, $\mathbf{p}$, are to be considered as Lagrange multipliers.

$$\mathrm{A} = L(\mathbf{r},\mathbf{v},t)dt + \mathbf{p} \circ (d\mathbf{r} - \mathbf{v}dt) \tag{3}$$

If all of the variables are independent, the domain of definition is 10 dimensional, $\{\mathbf{r},t,\mathbf{v},\mathbf{p}\}$. For the 10 dimensional velocity vector $\mathbf{V} = \{\mathbf{v},1,\mathbf{a},\mathbf{f}\}$, the virtual work 1-form becomes

$$\begin{aligned} W &= i(\mathbf{V})d\mathrm{A} \\ &= (\mathbf{f} - \partial L/\partial\mathbf{r}) \circ (d\mathbf{r} - \mathbf{v}dt) - (\mathbf{p} - \partial L/\partial\mathbf{v}) \circ (d\mathbf{v} - \mathbf{a}dt) \neq 0 \end{aligned} \tag{4}$$

The fundamental result is that if the system under consideration is without differential fluctuations ($(d\mathbf{r} - \mathbf{v}dt) \Rightarrow 0$, $(d\mathbf{v} - \mathbf{a}dt) \Rightarrow 0$), then the virtual work must vanish. But this can happen only on a manifold of odd Pfaff dimension! In contrast, if the system is a symplectic system of even Pfaff dimension, then the virtual work 1-form can never vanish. The key feature is that if the Pfaff dimension is even, then differential fluctuations are to be expected, and these lead to dissipation. The result implies that the evolution is described only imperfectly by a single parameter group of a dynamical system on the symplectic space. In the SECTAM reference explicit expressions were given for a Navier-Stokes system, for which the criteria of irreversibility required that

$$\nu\, curl\mathbf{v} \circ curl\, curl\mathbf{v} \neq 0. \tag{5}$$

Now it is known that a Lagrange system constrained by non-holonomic differential kinematic fluctuations leads to a non-compact symplectic manifold of dimension 2n+2. This (thermodynamic) manifold will not admit unique extremal vector fields that will leave the action integral stationary as a relative integral invariant (the virtual work must vanish for extremal fields, which is impossible on the symplectic manifold). There do exist non-extremal vector fields on the symplectic manifold that leave the Action integral invariant, but they are non unique and are dependent upon initial conditions that may require closed additions to be imposed on the Action 1-form. In modern language, the vector fields that produce stationary states (Bernoulli-Casimir functions) in a symplectic system are not gauge invariant. However, it has been observed that there does exist a unique, gauge independent, vector field on the symplectic manifold that would leave the Action integral a conformal, but not a stationary, invariant; this unique vector field, the Torsion vector field, will satisfy the thermodynamic criteria of irreversibility defined below.

3. Differential Topology - Pfaff Dimension

The basic tool for studying topological evolution is Cartan's magic formula [19], in which it is presumed that a physical (hydrodynamic) system can be described adequately by a 1-form of Action, A, and that a physical process can be represented by a contravariant vector field, **V**, which can be used to represent a dynamical system or a flow:

$$\begin{aligned} \mathrm{L}_{(\mathbf{V})}\int A &= \int \mathrm{L}_{(\mathbf{V})}A = \int\{i(\mathbf{V})dA + d(i(\mathbf{V})A)\} \\ &= \int\{W + d(U)\} \;\; = \int Q. \end{aligned} \tag{6}$$

The base manifold will be the 4-dimensional variety $\{x,y,z,t\}$ of engineering practice, but no metrical features are presumed a priori. In fact, the defect analysis is based upon a projective space in which concept of length has been abrogated away. If the RHS of equation 6 is zero, the $\int A$ is said to be an integral invariant of the evolution generated by **V**.

From the point of view of differential topology, the key idea is that the Pfaff dimension, or class [20], of the 1-form of Action specifies topological properties of the system. Given the Action 1-form, A, the Pfaff sequence, $\{\mathrm{A}, d\mathrm{A}, \mathrm{A}\hat{}d\mathrm{A}, d\mathrm{A}\hat{}d\mathrm{A}, \ldots\}$ will terminate at an integer number of terms $\leq$ the number of dimensions of the domain of definition. On a 2n+2=4 dimensional domain, the top Pfaffian, $d\mathrm{A}\hat{}d\mathrm{A}$, will define a volume element with a density function whose singular zero set (if it exists) reduces the symplectic domain to a contact manifold of dimension 2n+1=3. This (defect) contact manifold supports a unique extremal field that leaves the Action integral "stationary", and leads to the Hamiltonian conservative representation for the Euler flow in hydrodynamics. The irreversible regime will be on an irreducible symplectic manifold of Pfaff dimension 4, where $d\mathrm{A}\hat{}d\mathrm{A} \neq 0$. Topological defects (or coherent structures) appear as singularities of lesser Pfaff (topological) dimension, $d\mathrm{A}\hat{}d\mathrm{A} = 0$.

Classical hydrodynamic processes can be represented by certain nested categories of vector fields, **V**. Recall that in order to be Extremal, the process, **V**, must satisfy the equation

$$\textit{Extremal}--(\textit{unique Hamiltonian}): \qquad i(\mathbf{V})d\mathrm{A} = 0; \tag{7}$$

in order to be Hamiltonian the process must satisfy the equation

$$\textit{Bernouilli}--\textit{Casimir}--\textit{Hamiltonian}: \qquad i(\mathbf{V})d\mathrm{A} = d\Theta; \tag{8}$$

in order to be Symplectic, the process must satisfy the equation

$$\textit{Helmholtz}--\textit{Symplectic}: \qquad di(\mathbf{V})d\mathrm{A} = 0. \tag{9}$$

Extremal processes cannot exist on the non-singular symplectic domain, because a non-degenerate anti-symmetric matrix (the coefficients of the 2-form dA) does not have null eigenvectors on space of even dimensions . Although unique extremal stationary states do not exist on the domain of Pffaf dimension 4, there can exist evolutionary invariant Bernoulli-Casimir functions, Θ, that generate non-extremal, "stationary" states. Such Bernoulli processes can correspond to energy dissipative symplectic processes, but they, as well as all symplectic processes, are reversible in the thermodynamic sense described below. The mechanical energy need not be constant, but the Bernoulli-Casimir function(s), Θ, are evolutionary invariant(s), and may be used to describe non-unique stationary state(s).

The equations, above, that define several familiar categories of processes, are in effect constraints on the topological evolution of any physical system represented by an Action 1-form, A. The Pfaff dimension of the 1-form of virtual work, $W = i(\mathbf{V})dA$, is 2 or less for the three categories. The extremal constraint of equation 7 can be used to generate the Euler equations of hydrodynamics for a incompressible fluid. The Bernoulli-Casimir constraint of equation 8 can be used to generate the equations for a barotropic compressible fluid. The Helmholtz constraint of equation 9 can be used to generate the equations for a Stokes flow. All such processes are thermodynamically reversible. None of these constraints above will generate the Navier-Stokes equations, which require that the topological dimension of the 1-form of virtual work must be greater than 2.

A crucial idea is the recognition that irreversible processes must on domains of Pfaff dimension which support Topological Torsion, $A\hat{}dA \neq 0$, with its attendant properties of non-uniqueness, envelopes, regressions, and projectivized tangent bundles. Such domains are of Pfaff dimension 3 or greater. Moreover, as described below, it would appear that thermodynamic irreversibility must support a non-zero Topological Parity 4-form, $dA\hat{}dA \neq 0$. Such domains are of Pfaff dimension 4 or greater. The existence of Topological Torsion leads to the realization that the classical constraints of kinematic perfection,

$$\Delta\mathbf{x} = (d\mathbf{r} - \mathbf{v}dt) \Rightarrow 0, \textit{ and } \Delta\mathbf{v} = (d\mathbf{v} - \mathbf{a}dt) \Rightarrow 0, \tag{10}$$

put severe restrictions on the topology of the evolutionary process, restrictions that need not be realized in nature. Indeed, such constraints of null anholonomic differential fluctuations, when substituted into the Action 1-form given by equation 3, would lead to the conclusion that the maximum Pfaff dimension of such constrained systems is 2. Such domains cannot support non-zero topological torsion which require topological domains of Pfaff dimension 3 or more. Therefore, constraints such as $\Delta\mathbf{x} = 0$ and $\Delta\mathbf{v} = 0$ are not realized during the irreversible phase of a process; moreover, such differential fluctuations can cause the lifetime of a "stationary state" to be finite. Anholonomic differential fluctuations may be viewed as multiple parameter topological replacements for Langevin noise.

Although there does not exist a unique gauge independent stationary state on the symplectic manifold of Pfaff dimension 4, remarkably there does exist a unique vector field on the symplectic domain, with components that are generated by the 3-form $A\hat{}dA$. This unique (to within a factor) vector field is defined as the Torsion Current, **T**, and satisfies (on the 2n+2=4 dimensional manifold) the equation,

$$i(\mathbf{T})dx\wedge dy\wedge dz\wedge dt = \mathrm{A}\wedge d\mathrm{A} \tag{11}$$

This (four component) vector field, **T**, has a non-zero divergence almost everywhere, for if the divergence is zero, then the 4-form $d\mathrm{A}\wedge d\mathrm{A}$ vanishes, and the domain is no longer a symplectic manifold! The Torsion vector, **T**, can be used to generate a dynamical system that will decay to the stationary states ($div_4(\mathbf{T}) \Rightarrow 0$) starting from arbitrary initial conditions. These processes are irreversible in the thermodynamic sense. It is remarkable that this unique evolutionary vector field, **T**, is completely determined (to within a factor) by the physical system itself; e.g., the components of the 1-form, A, determine the components of the Torsion vector.

To understand what is meant by thermodynamic irreversibility, realize that Cartan's magic formula of topological evolution is equivalent to the first law of thermodynamics.

$$\mathrm{L}_{(\mathbf{v})}\mathrm{A} = i(\mathbf{V})d\mathrm{A} + d(i(\mathbf{V})\mathrm{A}) = W + dU = Q. \tag{12}$$

A is the "Action" 1-form that describes the hydrodynamic system. **V** is the vector field that defines the evolutionary process. W is the 1-form of (virtual) work. Q is the 1-form of heat. From classical thermodynamics, a process is irreversible when the heat 1-form Q does not admit an integrating factor. From the Frobenius theorem, the lack of an integrating factor implies that $Q\wedge dQ \neq 0$. Hence a simple test may be made for any process, **V**, relative to a physical system described by an Action 1-form, A:

If $\mathrm{L}_{(\mathbf{v})}\mathrm{A}\wedge\mathrm{L}_{(\mathbf{v})}d\mathrm{A} \neq 0$ *then the process is irreversible.*

This topological definition implies that the three categories (above) of symplectic, Hamiltonian or extremal processes, $\subset$ **S**, are reversible (*as* $\mathrm{L}_{(\mathbf{S})}d\mathrm{A} = dQ = 0$). However, for evolution in the direction of the Torsion vector, **T**, direct computation demonstrates that the fundamental equations lead to a conformal evolutionary process, a process which is thermodynamically irreversible:

$$\mathrm{L}_{(\mathbf{T})}\mathrm{A} = \sigma\mathrm{A} \quad \textit{and} \quad i(\mathbf{T})\mathrm{A} = 0, \tag{13}$$

such that

$$\mathrm{L}_{(\mathbf{T})}\mathrm{A}\wedge\mathrm{L}_{(\mathbf{T})}d\mathrm{A} = Q\wedge dQ = \sigma^2\mathrm{A}\wedge d\mathrm{A} \neq 0. \tag{14}$$

Turbulent flows must have a component along the Torsion vector to be irreversible ($\sigma \neq 0$). A coherent structure is the end result of an irreversible decay process that forms a set of measure zero, $d\mathrm{A}\wedge d\mathrm{A} = 0$, on space time, but such that the integral over a closed 3-dimensional hypersurface, $\iiint_{closed}\mathrm{A}\wedge d\mathrm{A} \neq 0$, is a relative integral invariant for the remainder of the evolution. In such domains,

$$\mathrm{L}_{(\mathbf{v})}\iiint_z \mathrm{A}\wedge d\mathrm{A} = \iiint_z\{i(\mathbf{V})(d\mathrm{A}\wedge d\mathrm{A}) + d(i(\mathbf{V})(\mathrm{A}\wedge d\mathrm{A}))\} = 0 + 0, \tag{15}$$

if $dA\hat{}dA \Rightarrow 0$, hence the closed integral is an evolutionary, although deformable, invariant.

For a hydrodynamic system, consider the Action 1-form defined by the equation

$$A = \mathbf{v} \circ d\mathbf{x} - (\mathbf{v} \circ \mathbf{v}/2 + \int dP/\rho + \lambda\, div\ \mathbf{v})dt, \tag{16}$$

with a topological (non-Hamiltonian) constraint involving non-holonomic fluctuations in the kinematic velocity field:

$$i(\mathbf{V})dA = \upsilon\, curl\ curl\mathbf{v} \circ (d\mathbf{r} - \mathbf{v}dt). \tag{17}$$

Substitution of the Action 1-form, A, into the constraint yields the Navier-Stokes equations as the equations of constrained topological evolution [4]. By direct evaluation of equation 11, the Torsion vector has 4 space time components $\{with\ \ h = \mathbf{v} \circ curl\mathbf{v}\}$:

$$\mathbf{T} = \{h\mathbf{v} - (\mathbf{v} \circ \mathbf{v}/2)curl\mathbf{v} - \nu\, curl\, curl\mathbf{v};\ h\}, \tag{18}$$

and a 4 divergence given by the expression:

$$div_4\mathbf{T} = -2\nu\ curl\mathbf{v} \circ curl\, curl\mathbf{v} = -2\sigma. \tag{19}$$

When the 4 divergence, -2σ, does not vanish, it follows from equation 14 that the flow, $\mathbf{v}$, is thermodynamically irreversible. Such irreversible solutions to the viscous Navier-Stokes equations must generate lines of vorticity that have non-zero helicity, and can exist only on domains where the Action 1-form is of Pfaff dimension 4.

Consider sets of measure zero (topological Torsion defects) on the space of 4 dimensions such that $div_4\mathbf{T} = 0$. Such domains are at most of Pfaff dimension 3 (relative to the given 1-form of Action) and define a coherent structure. Note that these defect domains (in a Navier-Stokes fluid) do not require that the viscosity coefficient vanish, $\nu \neq 0$, and yet they support thermodynamically reversible processes. Such domains usually evolve in a deformable manner that preserves both the topological property of Pfaff dimension 3 and the topological Torsion integral defined in equation (15). It is important to realize that the evolutionary invariant integral, which gives the defect structure its "coherence", does not necessarily preserve the helicity of the velocity field, defined as $h = \mathbf{v} \circ curl\mathbf{v}$. The helicity of the velocity field, h, is only the fourth component of a covariant Topological Torsion tensor of rank 3. A common feature of such coherent structures in a Navier-Stokes fluid is that the vorticity field satisfies the integrability criteria of Frobenius; e.g., as a three vector field, the vorticity vector must be proportional to a gradient. It follows that the velocity field may have helicity, but the vorticity field does not.

It would appear that the concept of two dimensional turbulence is paradoxical, for it requires four dimensions to support an irreversible flow according to the definitions given above. It should be remarked that the definition of irreversibility, $Q\hat{}dQ \neq 0$, implies that there are two topological classes of irreversibility. Either $dQ\hat{}dQ = 0$, implying that the "heat current" does not stop or start in the interior, or $dQ\hat{}dQ \neq 0$, implying internal sources of heat current (pinch points).

Similar results will hold for coherent structures created in plasmas. From the electromagnetic 1-form of Action, defined in terms of the vector and scalar potentials as,

$$A = \Sigma_{k=1}^{3} A_k(x,y,z,t)dx^k - \phi(x,y,z,t)dt, \tag{20}$$

the topological torsion 3-form, $A \hat{} dA$, induces the torsion current

$$\mathbf{T} = \{(\mathbf{E} \times \mathbf{A} + \mathbf{B}\phi); \mathbf{A} \circ \mathbf{B}\} \equiv \{\mathbf{S}, h\}. \tag{21}$$

If $div_4\mathbf{T} = -2\,\mathbf{E} \circ \mathbf{B} \neq 0$, the electromagnetic 1-form defines a domain of Pfaff dimension 4. Such domains cannot support transverse waves. Evolutionary processes (currents) that are proportional to the Torsion current are thermodynamically irreversible, if $\mathbf{E} \circ \mathbf{B} \neq 0$. Electromagnetic coherent structures are evolutionary deformable domains of Pfaff dimension 3, where $\mathbf{E} \circ \mathbf{B} = 0$. The conformal dissipation function, $\mathbf{E} \circ \mathbf{B}$, is the electromagnetic analogue of the Navier-Stokes function, $\nu\, curl\mathbf{v} \circ curl\, curl\mathbf{v}$.

Epilogue

It is a rare thing to attend a conference where on one day a new theoretical prediction is made, and then on the following day of the conference experimental evidence is presented to support the abstract theory. During the presentation of the material described above on May 27 of the SIMFLO conference, it was stated that in an irreversible turbulent flow there should exist a 4 dimensional defect of topological torsion. For a Navier-Stokes fluid, the signature of such a defect would be a curve of vorticity in the form of a twisted helix, and the basic requirement for the existence of the 4 dimensional symplectic manifold is given by the condition, $curl\mathbf{v} \circ curl\,curl\mathbf{v} \neq 0$. The following day Kuibin and Okulov presented experimental observations with a detailed analysis of a dynamical helical curve of vorticity in a swirling fluid. On the following day, they determined that their independent analysis supported the idea that $curl\mathbf{v} \circ curl\,curl\mathbf{v} \neq 0$, thereby giving credence to the abstract theory of Topological Torsion defects presented above.

References

The entire article in expanded form with hot linked references in the form of pdf files can be found at

http://www.uh.edu/~rkiehn/pd2/pd2home.pdf

1. Zhitomirskii, M. (1992) *Typical Singularities of Differential 1-forms and Pfaffian Equations*. Translations of Mathematical Monographs, AMS Providence . See also, Kiehn, R.M. (1990) Topological Torsion, Pfaff Dimension and Coherent Structures, in H. K. Moffatt and A. Tsinober, editors, *Topological Fluid Mechanics*, Cambridge University Press, p. 225.

2. Giaquinta, M. and Hildebrandt, S. (1995) *Calculus of Variations*, Springer Verlag, Vol 1 p.398

3. Cartan, E. (1937) *La Theorie des Spaces a Connexion Projective*, Hermann, Paris.

4. Kiehn, R. M. (1992) Topological Defects, Coherent Structures and Turbulence in Terms of Cartan's Theory of Differential Topology, in B. N. Antar, R. Engels, A.A. Prinaris and T. H. Moulden, Editors, *Developments in Theoretical and Applied Mathematics*, Proceedings of the SECTAM XVI conference, The University of Tennessee Space Institute, Tullahoma, TN 37388 USA, p. III.IV.2

5. Kiehn, R. M. (1997) When does a dynamical system represent an Irreversible Process, SIAM Snowbird May

1997 poster. Also see http://www.uh.edu/~rkiehn/pdf/siam.pdf

6. Brillouin, L. (1964) *Tensors in Mechanics and Elasticity*, Academic Press, N.Y., p. 93. In 1938, Brillouin wrote: "If one does not admit the symmetry of the (connection) coefficients, Γ_{ijk}, one obtains the twisted spaces of Cartan, spaces which scarcely have been used in physics to the present, but which seem to be called to an important role."

7. Kondo, K. (1962) Unifying Study of Basic Problems in Engineering and Physical Sciences by means of Geometry in RAAG memoirs Vol 3 Tokyo. See also, F.R.N. Nabarro (1987) *The Theory of Crystal Dislocations*, Dover, p. 562.

8. Kiehn, R. M. (1993) Instability Patterns, Wakes and Topological Limit Sets, in J.P.Bonnet and M.N. Glauser, (eds) *Eddy Structure Identification in Free Turbulent Shear Flows*, Kluwer Academic Publishers, p. 363

9. Kiehn, R. M. (1995) Hydrodynamic Wakes and Minimal Surfaces with Fractal Boundaries, in J.M. Redondo, O. Metais, (eds) *Mixing in Geophysical Flows*, CIMNE, Barcelona p.52

10. Kiehn, R. M. (1974) Extensions of Hamilton's Principle to include Dissipative Systems, J. Math Phys. **5**, 9

11. Kiehn, R. M. (1975) Intrinsic Hydrodynamics with Applications to Space-Time Fluids, Int. J. of Engng Sci. Vol 13, pp 941-949

12. Kiehn, R. M. (1977) Periods on manfolds, quantization and gauge, J. of Math Phys **18**, no. 4, p. 614

13. Kiehn, R. M. (1976) Retrodictive Determinism, Int. J. of Eng. Sci. **14**, p. 749

14. Schultz, A., Kiehn, R. M., Post, E. J., and Roberds, R. B., (1979) Lifting of the four fold EM degeneracy and PT asymmetry, Phys Lett 74A, p. 384.

15. Kiehn, R. M., Kiehn, G. P., and Roberds, R. B. (1991) Parity and Time-reversal Symmetry Breaking, Singular Solutions, Phys Rev A, **43**, p. 5665

16. The Falaco Effect as a topological defect was first noticed by the present author in the swimming pool of an old MIT friend, during a visit in Rio de Janeiro, at the time of Halley's comet, March 1986. The concept was presented at the Austin Meeting of Dynamic Days in Austin, January 1987, and caused some interest among the resident topologists. The easily reproduced experiment added to the credence of topological defects in fluids. It is now perceived that this topological phenomena is universal, and will appear at all levels from the microscopic to the galactic.

17. See references [18], [4] and http://www.uh.edu/~rkiehn/pd2/pd2homep.htm

18. Kiehn, R. M., (1991) Compact Dissipative Flow Structures with Topological Coherence Embedded in Eulerian Environments, in R.Z. Sagdeev, U. Frisch, F. Hussain, S. S. Moiseev and N. S. Erokhin, (eds) *Non-linear Dynamics of Structures*, World Scientific Press, Singapore p.139-164.

19. Marsden, J.E. and Riatu, T. S. (1994) *Introduction to Mechanics and Symmetry*, Springer-Verlag, p.122

20. Libermann, P. and Charles-Michel, M., (1986) *Symplectic Geometry and Analytical Mechanics*, Riedel -Kluwer, p. 284

THE VORTEX CONCEPT AND ITS IDENTIFICATION IN TURBULENT BOUNDARY LAYER FLOWS

L. M. PORTELA
Mechanical Engineering Department, Stanford University
Stanford, CA 94305, USA

Abstract. The inadequacy of the use of point-concepts and dynamic-concepts to define a vortex is discussed. An objective and formal kinematic definition of a vortex that corresponds to the basic notion of a swirling motion around a central set of points is proposed. Some results of the application of algorithms based on the proposed definition to the DNS of a flat plate TBL at $Re_\theta = 670$ are presented.

1. Introduction

Vortices are important in many types of flows. In particular, unsteady vortices play a central role in turbulence. However, we have lacked a formal accepted definition, and the issue has generated significant controversy. The basic notion usually associated with a vortex, and the one we consider, is that of a swirling motion around a central set of points.

Essentially, the attempts at defining a vortex that are found in the literature can be classified into three categories: (1) definitions that attempt to convey the visual idea of a swirling pattern; (2) definitions based on local-interpretations (i.e., based on Taylor-expansion ideas) of point-concepts; (3) definitions based on dynamic arguments (i.e., based on the Navier-Stokes equation) for simplified situations.

The problem with definitions of the first type is that, besides not being proper definitions from a logical point of view, they involve a subjective visual judgment; therefore, it is not possible to use them in a formal algorithm of identification that can be implemented in a computer code. An example is the definition proposed by Robinson (1991): "a vortex exists when instantaneous streamlines mapped onto a plane normal to the vortex core exhibit a roughly circular or spiral pattern, when viewed from a frame of reference moving with the center of the vortex core".

The problem with the second type of definitions is that a vortex is a set-concept (a concept that is expressed in terms of conditions imposed on a set of points, not in terms of conditions imposed on a point); in general local-interpretations of point-concepts cannot be transposed to set-concepts. Examples of the second type of definition are: (a) a region of high vorticity; (b) a connected turbulent fluid mass with instantaneous phase-correlated vorticity over its spatial extent (Hussain 1986); (c) a maximal connected spatial region with[1] $N_k > 1$ (for incompressible flows $N_k > 1$ is equivalent to $q > 0$) (Melander and Hussain 1993); (d) a region with complex eigenvalues of $\nabla\vec{V}$ (Chong, Perry and Cantwell 1990) (e) a connected region with two negative eigenvalues of $\tilde{S}^2 + \tilde{\Omega}^2$ (Jeong and Hussain 1995); (f) a region of low pressure with $q > 0$ (Hunt, Wray and Moin 1988). Definitions (a) through (d) are based on kinematic local-interpretations, whereas definitions (e) and (f) are based on a mixture of kinematic local-interpretations with dynamic arguments. In general all the above definitions do not correspond to a vortex; they can include flow patterns other than vortices, and they can fail to identify vortices. For example, definitions (a) and (b) are vorticity-based and can identify regions of high vorticity without any swirling motion (e.g., internal shear layers in a TBL) as a vortex. Definitions (c) and (d) are based on the local-interpretation of $\nabla\vec{V}$ as giving the local streamline pattern. However, as shown by Portela (1997), they can have the same type of problems as vorticity-based definitions; among other examples, Portela (1997) shows that definitions (c) and (d) would identify a laminar flat plate boundary layer as a vortex. Definition (e) mixes kinematic with dynamic arguments, but it also has the same type of problems as vorticity-based definitions, and it also defines a flat plate boundary layer as a vortex (for 2D incompressible flows, definitions (c), (d) and (e) are equivalent). Also, definition (e) can fail to identify vortices with a very simple geometry; for example, Portela (1997) shows that it can fail to identify a Burgers vortex, depending on the parameters of the Burgers vortex. Definition (f) also mixes kinematic and dynamic arguments, and, among other problems, it can also fail to identify vortices with very simple geometries.

The problem with the third type of definitions is that a vortex is either a geometric or a kinematic concept, and not a dynamic concept. A kinematic concept should stand on its own, without recourse to dynamic arguments. Otherwise, apart from the more fundamental problem of mixing completely different concepts, we would need a different vortex definition

[1] $N_k \equiv \|\tilde{\Omega}\|/\|\tilde{S}\|$; $\|\tilde{S}\| \equiv [(\tilde{S}.\tilde{S}^T)]^{1/2}$; $\|\tilde{\Omega}\| \equiv [(\tilde{\Omega}.\tilde{\Omega}^T)]^{1/2}$; $\tilde{S} \equiv \frac{1}{2}(\nabla\vec{V} + \nabla\vec{V}^T)$; $\tilde{\Omega} \equiv \frac{1}{2}(\nabla\vec{V} - \nabla\vec{V}^T)$; q is the second invariant of $\nabla\vec{V}$.

for each type of dynamic situation. For example, dynamic arguments valid for a steady incompressible flow with constant viscosity are not necessarily valid for unsteady flows, or compressible flows, or non-Newtonian flows, or for the Reynolds averaged Navier-Stokes equation, etc. An example of the third category of definition is: elongated regions of low pressure (Robinson 1991). The problem with this criterion is that Robinson's argument for the use of low pressure is based on the very simplistic situation of circular streamlines, who have a minimum of pressure at its center, essentially due to symmetry considerations. Only in very simple situations does the center of a vortex correspond to a minimum of pressure, or a vortex to a low pressure region. Portela (1997) derived the exact conditions for a vortex center to correspond to a minimum of pressure in 2D incompressible flows with constant viscosity; even for this very simple case, a general correspondence exists only in the limit of a steady vortex with an infinite Reynolds number. Definitions (e) and (f) above mix kinematic point-concepts with dynamic pressure-based arguments for incompressible flows with constant viscosity. As discussed above a vortex is a kinematic set-concept, and in general these arguments are not valid, even for very simple situations. The center of a Burgers vortex can correspond to a maximum of pressure or to a minimum of pressure, depending on the parameters. Hence, any pressure-based criterion can fail to identify a Burgers vortex. These conceptual considerations have important practical consequences; for example, Portela (1997) showed that the criterion used by Robinson (1991) to identify vortices in the flat plate TBL - a fluctuating pressure less than minus four times the wall shear stress - fails to identify more than two-thirds of the streamwise vortices.

We propose an objective and formal mathematical definition that corresponds to the notion of a swirling motion around a central set of points. The definition is completely general, and deals with the three major aspects of the vortex concept: definition of the notion of a maximal region with its points swirling around a minimal connected central set of points, the nucleus; extension of the concept of swirling to a 3D situation, by defining swirling around an appropriate axial-set; choice of a proper velocity field, by defining the notion of a reference frame attached to the swirling motion.

The definitions we propose use basic concepts of set theory, topology, and differential geometry. Here we present the basic ideas behind the definitions. The formal definitions, and a more detailed and complete presentation, are found in Portela (1997).

2. Proposed Definitions

We propose a new formalism, based on the Jordan curve theorem, that we call "Jordan structures". A Jordan curve separates the open Euclidean

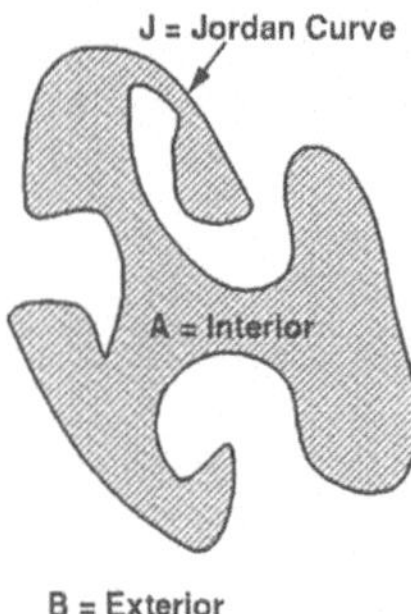

Figure 1. Jordan region.

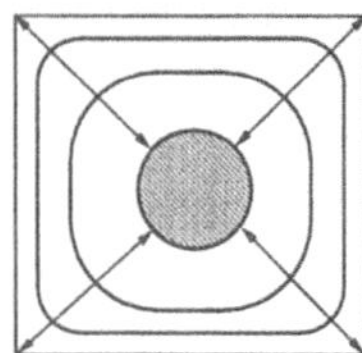

Figure 2. Contraction-expansion from a rectangle to a circle through a series of convex Jordan regions.

plane into two disjoint open connected sets: the interior and the exterior (figure 1). The formalism allows us to make precise the idea of minimal central sets surrounded by maximal sets, with a continuous expansion-contraction from one to the other, with each set of the expansion-contraction series satisfying a certain condition. For example, in figure 2 is shown a continuous expansion-contraction from a rectangle to a circle through a series of convex Jordan regions.

The idea of swirling, both for closed and open streamlines, is formalized using the differential-geometry concept of winding-angle. The winding-angle is a measure of the rotation of the points of a streamline with respect to a point not belonging to the streamline. The concept is illustrated in figure 3, where $\alpha(\vec{Q}, \vec{P}_1, \vec{P}_2)$ is the winding-angle with respect to $\vec{Q}$ when moving along the streamline from $\vec{P}_1$ to $\vec{P}_2$. The winding-angle is used to define a "vortical Jordan-structure region"; this is a Jordan region where the streamlines of all the points of its boundary, moving either forward or backward or both, do not enter its interior, and have a certain amount of kinematic swirl with respect to all the interior points (figure 4). A natural choice for the amount of kinematic swirl is a winding-angle equal to 2π, but that is not necessary.

A vortex nucleus is defined as a minimal set that can be reached through a continuous contraction of vortical Jordan-structure regions; and a vor-

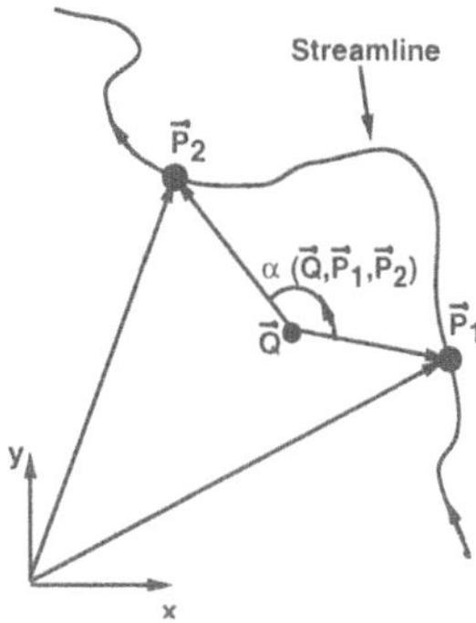

Figure 3. Winding-angle.

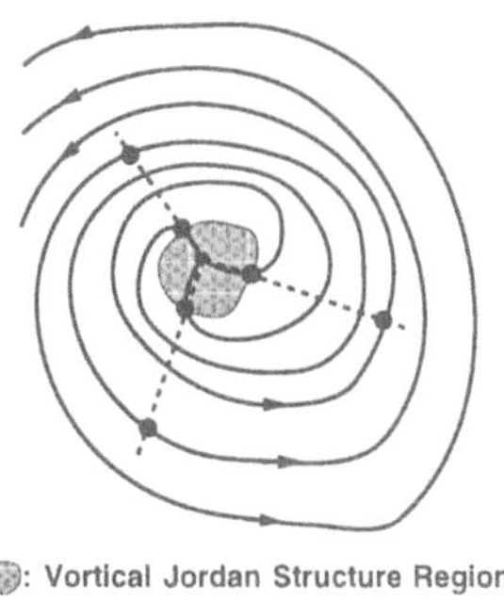

Figure 4. Vortical Jordan-structure region with the required winding-angle equal to 2π.

tex is defined as a maximal set, all of whose points belong to a vortical Jordan-structure region that can reach a vortex nucleus through a continuous contraction of vortical Jordan-structure regions. It can be easily shown (Portela 1997) that there exists a one-to-one correspondence between a vortex and a vortex nucleus. A vortex nucleus does not need to be a point, but if it is a point we call it a vortex center. The definition is illustrated in figure 5 for a case in which the required winding-angle was chosen as 2π and the nucleus is a point. Figure 6 show schematically the streamline pattern due to two co-rotating Oseen vortices. All the streamlines are closed streamlines, except for the curve L, with the shape of an "horizontal eight". L separates the open Euclidean plane into three open disjoint sets, two of them bounded: A_1 and A_2. A_1 and A_2 are closed vortices (i.e., with closed streamlines), and P_1 and P_2 are their centers. E is a vortical-closed Jordan region (i.e., its boundary, S_E, is a closed streamline). The union of A_1, A_2 and L is a vortex nucleus; it is the minimal set that E can reach through a continuous contraction of vortical-closed Jordan regions.

The previous definitions deal with a *given* 2D vector field, not necessarily

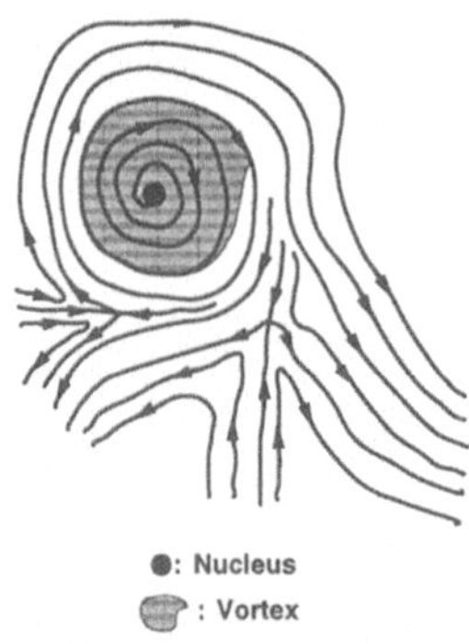

Figure 5. Nucleus and vortex, with the required winding angle equal to 2π.

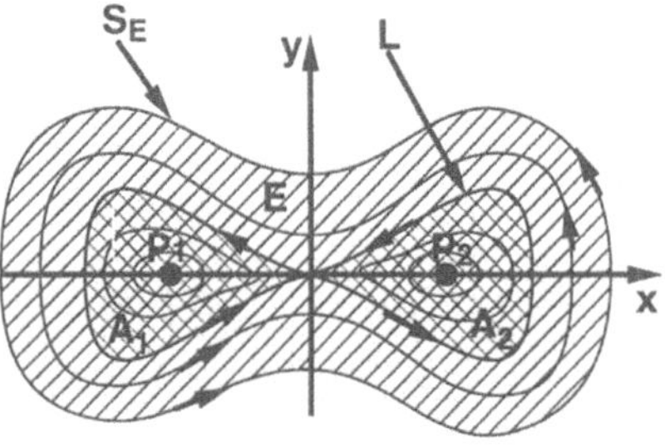

Figure 6. Two co-rotating vortices.

a velocity field; from their perspective it is immaterial how the vector field was obtained. This might be an adequate approach if one is interested in the streamline patterns of a vector field at a particular time. In the case of a velocity field with streamline patterns that can be changing with time, we formalize the notion of a reference frame attached to the swirling motion by using a transformation of coordinates and a transformation of velocities that produce a constant vortex, or a constant vortex nucleus; we call them an absolute vortex and an absolute vortex nucleus. Clearly, a change in the position of the frame of reference will produce the same amount of change in the position of the points that are mapped onto the absolute vortex; it can be easily shown (Portela 1997) that an absolute vortex, or an absolute vortex nucleus, is invariant under a change of the frame of reference, in the same sense that a velocity field and a tensor field are invariant under a change of the frame of reference. If there exists a simple transformation that produces an absolute vortex, or an absolute vortex nucleus (e.g., an uniform translation), then an absolute vortex can have simple and important dynamic interpretations.

The generalization of these ideas to a 3D vector field involves the defi-

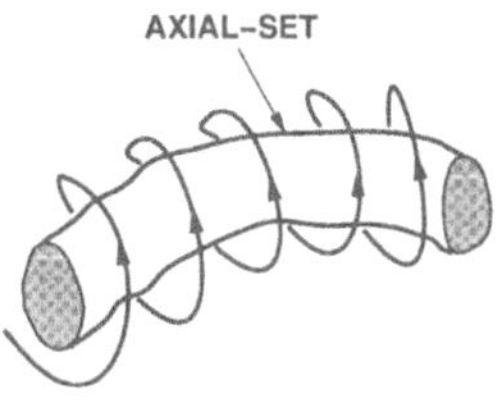

Figure 7. Axial-set.

nition of a 3D curve, a vortex axis, such that at any point of the curve the projection of the 3D velocity field on a plane perpendicular to the curve forms a 2D velocity field that has a vortex center coinciding with the curve (i.e., the vortex axis is the locus of vortex centers defined on planes perpendicular to the vortex axis). If the vortex nuclei are not centers it is possible to define an "axial-set" instead of a curve (figure 7), but the details are more involved and can be found in Portela (1997).

3. Flat Plate TBL

A method of identification based on algorithms using the proposed definitions was developed; the details are found in Portela (1997). This resulted in software that allows the identification of the vortices and their decomposition into different regions. The software was applied to the DNS of the flat plate TBL at $Re_\theta = 670$, using Spalart's code (1988).

The planar-streamwise vortices are defined using the projection of the fluctuating velocity field, $\vec{V}'$ ($\vec{V}' = \vec{V} - \overline{\vec{V}}$), on a zy plane, and the planar-spanwise vortices are defined using the projection of $\vec{V}'$ on a xy plane (figure 8). If the Taylor hypothesis is valid, i.e., if $\partial\vec{V}/\partial t = -\nabla\vec{V}.\overline{\vec{V}}(\vec{P})$, then the planar-streamwise and planar-spanwise vortices are absolute vortices of $\vec{V}(\vec{P})$ under a transformation of coordinates $\vec{P}^* \equiv \vec{P} - \overline{\vec{V}}(\vec{P})t$ and a transformation of velocities $\vec{V}^*(\vec{P}^*) \equiv \vec{V}(\vec{P},t) - \overline{\vec{V}}(\vec{P})$; i.e., they can be obtained using the projection of $\vec{V}^*(\vec{P}^*)$ on their respective planes, which by definition does not change with time. However, the validity of the hypothesis was not checked, and they should be simply considered as planar-vortices of the fluctuating velocity field. A typical zy plane and a typical xy plane are shown in figures 9 and 10, where the uncertainty grid rectangle denotes a rectangle that the program guarantees to contain a vortex center (or a vortex nucleus). For all the vortices the required amount of winding-angle was chosen as 2π. The vortex quasi-closed region is the region where the streamlines of the vortex are quasi-closed (i.e, after one complete turn the

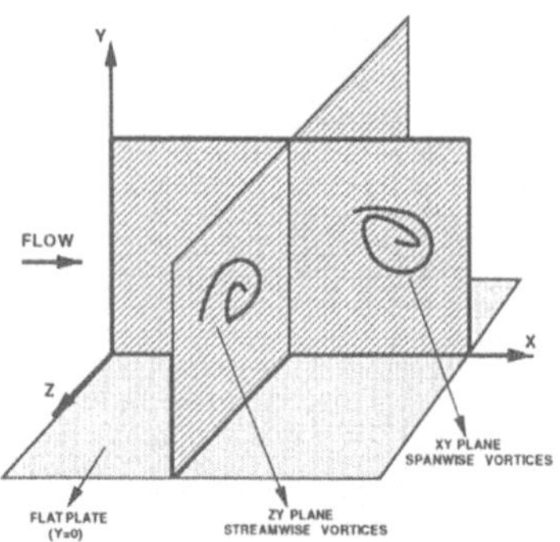

Figure 8. Planar-streamwise and planar-spanwise velocity field projection.

distance to the original position is small[2]). Consecutive planes can be used to obtain a good approximation of the axes of the vortices (a detailed description of the definition and procedure are found in Portela 1997); this is shown in figure 11.

4. Conclusions

A vortex is a kinematic set concept. In general point-concept definitions and definitions using dynamic arguments are not adequate; they can include flow patterns other than vortices, and they can fail to identify vortices. The definitions that appear in the literature, attempting to convey the visual idea of a swirling pattern, are not proper definitions from a logical point of view, and involve a subjective visual judgment; therefore, it is not possible to use them in a formal algorithm of identification that can be implemented in a computer code.

We propose an objective and formal mathematical definition that corresponds to the notion of a swirling motion around a central set of points. The definition deals with the three major aspects of the vortex concept: definition of the notion of a maximal region with its points swirling around a minimal connected central set of points, the nucleus; extension of the concept of swirling to a 3D situation, by defining swirling around an appropriate axial-set; choice of a proper velocity field, by defining the notion of a reference frame attached to the swirling motion.

A method of identification based on algorithms using the proposed definition was developed, This resulted in software that allows the identification

[2]The discussion of several possible criteria is found in Portela (1997). The criterion used in figures 9 and 10 is that the initial and final points of the streamline-segment are contained within a grid rectangle which has roughly the size of the smallest motions that can be resolved by the numerical simulation.

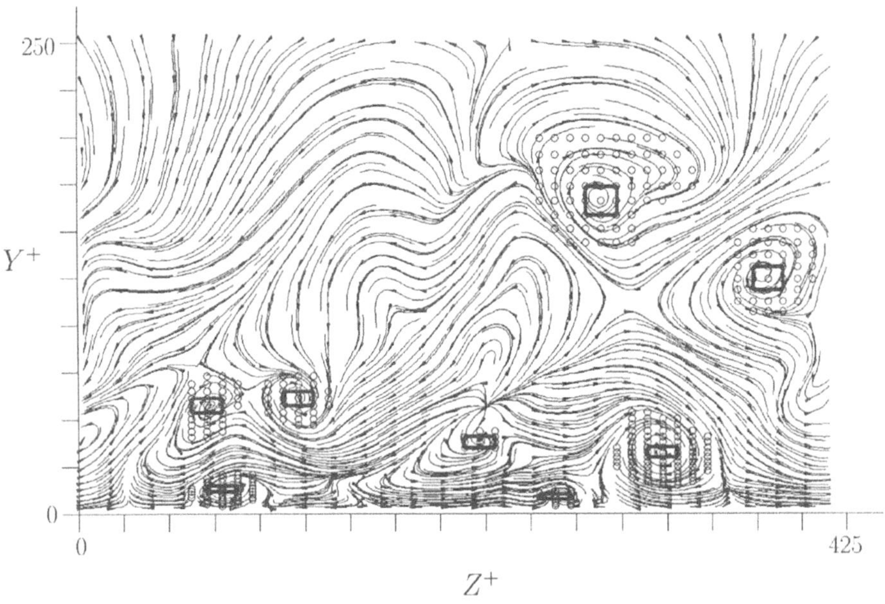

Figure 9. Planar-streamwise vortices.

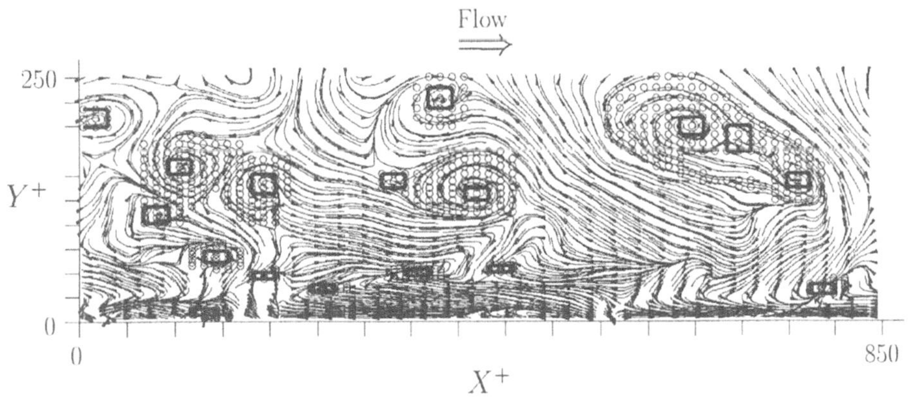

Figure 10. Planar-spanwise vortices.

of the vortices and their decomposition into different regions. The software was applied to the DNS of the flat plate TBL at $Re_\theta = 670$.

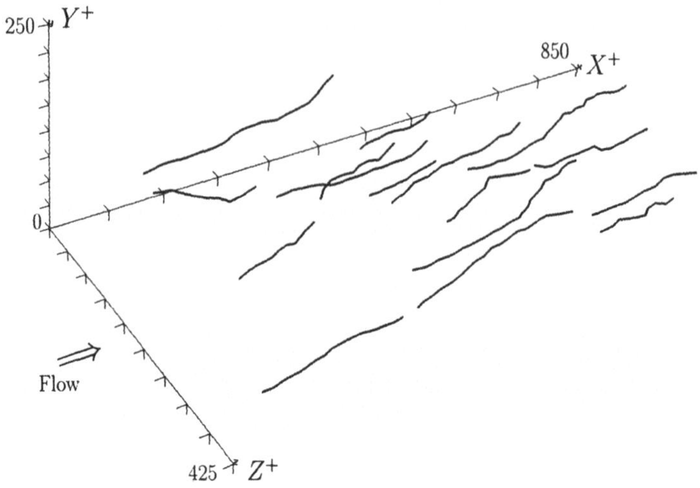

Figure 11. Axes of the quasi-streamwise vortices.

The author gratefully acknowledges the guidance provided by Prof. Bradshaw and Prof. Kline during the course of this work. Prof. Bradshaw revised the manuscript and made numerous suggestions that were incorporated by the author. The author is grateful to Dr. Philippe Spalart who kindly provided the DNS code.

5. References

CHONG, M. S., PERRY, A. E. & CANTWELL, B. J. (1990) A general classification of three-dimensional flow fields. *Phys. Fluids.* **A 2**, 765.

HUNT, J. C. R., WRAY, A. A. & MOIN, P. (1988) Eddies, stream, and convergence zones in turbulent flows. *Center for Turbulence Research Report CTR-S88.* **A 2**, p. 193.

HUSSAIN, F. (1986) Coherent structures and turbulence. *J. Fluid Mech.* **173**, 303.

JEONG, J. & HUSSAIN, F. (1995) On the identification of a vortex. *J. Fluid Mech.* **285**, 69.

MELANDER, M. V. & HUSSAIN, F. (1993) Polarized vorticity dynamics on a vortex column. *Phys. Fluids.* **A 5**, 1992.

PORTELA, L. M. (1997) *Identification and Characterization of Vortices in the Turbulent Boundary Layer.* Ph.D. dissertation, Mech. Eng. Dept., Stanford University.

ROBINSON, S. K. (1991) *The Kinematics of Turbulent Boundary Layer Structure.* NASA Technical Memorandum 103859.

SPALART, P. R. (1988) Direct simulation of a turbulent boundary layer up to $Re_\theta = 1410$. *J. Fluid Mech.* **187**, 61.

VII. Experimental Techniques

COHERENT STRUCTURES IDENTIFICATION IN SEPARATED AND FREE MIXING LAYERS USING HOT WIRES RAKE

S. AUBRUN, H. HA MINH AND H. BOISSON
Institut de Mécanique des Fluides de Toulouse
Av. du Pr. Camille Soula, 31400 Toulouse, France

P. CARLES
Université Pierre et Marie Curie,
Laboratoire de Modelisation en Mécanique,
4 place Jussieu, 75252 PARIS Cedex 05

AND

J.COULOMB
Centre d'Essais Aéronautique de Toulouse (C.E.A.T.),
23, av H. Guillaumet,31056 TOULOUSE cedex

1. Introduction

The study of unsteady phenomena in separated flows is of current interest owing to existence of strong instabilities developing large scale vortical structures in the separated zone (aero-acoustic strain in materials, perturbation in combustion chambers). The physical nature of these structures is not well known yet and, up to now, although some numerical approaches are able to compute unsteady flow with vortical structures, they need experimental data banks to check predicted features of these coherent structures (CS's). Moreover, the flow behind a backward facing step is usually practiced as a numerical test configuration. We have taken account of these considerations to decide that experimental approach seems to be the best way to analyze CS's generated by a separated flow. Similar approaches have been already elaborated in the case of mixing layers and wakes (Bonnet et al. (1992), Hussain et al. (1993)). The challenge is to extend these methods to a more complex case due to the high rate of curvature of the mixing layer. We intend to compare the behavior of CS's developed in a separated flow and in a plane mixing layer.

2. Context

The flow behind a backward facing step belongs to the class of separated flows. Its main advantage comes from the separation at a fixed point imposed by geometry cancels an additional instability. The separation creates a recirculation zone and a high transversal gradient of longitudinal velocity between the external flow and the reverse flow. This kind of description is strongly based on time mean approach and is not representative of the real flow : the relative turbulence intensity is more than 100% in the recirculation "bubble", the instantaneous reattachment position have strong variations around its time mean position, and some coherent motions have been visualized in the shear layer. So, it became necessary to study this flow as an unsteady phenomenon. Particularly, we decided to investigate the development of coherent motions in the shear layer. Visualizations revealed the development of vortical structures in the shear layer and their transit towards the reattachment zone. It seems that their trajectory is globally confined in the upper part of the shear layer , i.e. between the external flow and the zero velocity line. Several authors compared the shear layer in first sections with a plane mixing layer between two streams. Actually, some of them deduced from measurements that the growth of vorticity thickness is linear, as the plane mixing layer one, in first sections (Troutt et al. 1984). On the other hand, Castro & Haque (1987) proved that several paradoxical results are published, and particularly, about the linearity of the growth. Anyway, it is always very difficult to strictly compare some results while initial conditions, configurations , or even aspect ratios are not similar. According to these previous considerations, we choose to examine the shape of CS's which develop in the shear layer behind a backward facing step and achieve a qualitative comparison with the plane mixing layer . It is evident that strong differences exist :
- the plane mixing layer possess a high turbulent zone bounded by two laminar streams and free of the time mean transversal velocity.
- the shear layer produced by separation point possesses a high turbulent zone bounded by a laminar upper part and a turbulent lower part. A low-frequency flapping motion exists, and due to reattachment the layer is incurved towards the wall and the transverse velocity exists. In these conditions, CS's risk to be rather less well-defined than mixing layer ones. Their instantaneous trajectory is deviated by the flapping motion and the low turbulent part of shear layer partially destroys the coherent motion. For brevity in the following, the term separated mixing layer will be used to designate the shear layer produced by the separation point of the backward facing step.

3. Experimental exigencies

By definition, CS's are considered as an accumulation of spatially phase-correlated vorticity (Hussain). Various studies prove that their main features, although globally deterministic, present significant random discrepancies. It is the case, for instance, of their exact shape, spatial extent and trajectory. It follows that with a one-point measurement apparatus, it is not possible to precisely locate the CS relatively to the probe. Multipoint measurements overcome this difficulty since one captures all the spatial extent in a given section of the CS at each time step. Moreover, we can build a detection signal of CS's transit based on its whole spatio-temporal extent, instead of relying on any external signal which could create a false detection. However, this introduces a scale of analysis and a space resolution which can occult some of characteristic scales and thus, some of physical properties of these structures.

4. Experimental apparatus

4.1. THE BACKWARD FACING STEP

The rearward facing step model is fixed in the middle of the wind-tunnel S10 of C.E.A. Toulouse. The test volume is 1 m wide by 2.2 m height by 2 m long. The plate where initial boundary layer is developing is 0.45 m, the reattachment plate is 1.2 m and the step height is 65 mm. The aspect ratio is more than 10 and so, ensures that we simulate a flow in an infinitely high tunnel. Reference velocity is 40 m/s and the turbulence level is 0.7% in the free stream (Reynolds number based on step height is about 170000, Mach number 0.12). Some preliminary Laser Doppler Velocimetry measurements gave access to mean velocity fields and help to check flow bidimensionality as well as to find reattachment length (Xr = 5.7 H), boundary layer thickness ($\delta/H = 0.13$) at separation point, and where to place the rakes in the flow. Then, an experimental study was developed using 16 single hot wires or 8 X hot wires rakes in the mean-gradient direction behind the backward facing-step (figures 1 and 2). The instantaneous streamwise and transversal velocity measurements for a sampling frequency about 12 kHz allowed a good definition of the CS's transit in this area. In addition, measurements with a lower sampling frequency and long duration enable more precise spectral analysis. In this study, we present results based on measurements performed with a rake of 8 X hot wires located at X = 1.2 H. Intervals between wires are $0.24\delta_\omega = 3.2$ mm and the rake covers all the separated mixing layer extend in the transverse direction ($1.8\delta_\omega = 25mm$ where δ_ω is the vorticity thickness in the studied section). The power density spectrum of velocity signal (figure 3) exhibited a concentration of energy around a

Strouhal number of $St = \frac{f\delta_\omega}{U_m} = 0.25$ (i.e. 372 Hz) in the separated mixing layer. We associate this bump to the CS's transit. The shape of this spectrum confirms that natural CS's keep random characteristics and that a spectral analysis or a conditional method based on an external signal detection are not sufficient.

4.2. THE FREE MIXING LAYER

C.E.A.T of Poitiers supplied us with a data bank about a subsonic plane turbulent mixing layer. External velocities are 42.8 m/s and 25.2 m/s. The longitudinal position of the 12 X-hot wires rake is fixed at 600 mm from the trailing edge of the splitting-plate and the vorticity thickness δ_ω is 27.6 mm. The rake extent is greater than twice δ_ω and interval between probes is 6 mm. The Strouhal number is about $St = 0.3$ (i.e. 358 Hz). More detailed informations are given in Bonnet et al. publications.

5. Definition

A time and space dependent signal F can be decomposed as follows :

$$F(t,y) = \overline{F}(y) + f(t,y) = \langle F(t,y)\rangle + f'(t,y) = \overline{F}(y) + \tilde{f}(t,y) + f'(t,y)$$

where $\overline{F}$ is the signal's time average and $\langle F\rangle$ its coherent part. $\tilde{f}$ is deduced from $\overline{F}$ and $\langle F\rangle$ and represents the coherent perturbation around time average. f and f' are the remaining parts through averaging. f' is considered as a completely incoherent signal. For a discrete time series of simultaneous F signal at N transverse locations separated by Δy, during T with a time step Δt, we define:

$$F(t,y) = F(i{*}\Delta t, j{*}\Delta y) = F_{i,j} \quad \text{with} \quad i = 1,2,...,T/\Delta t \quad \text{and} \quad j = 1,...,N$$

6. Dectection of coherent structures with vorticity-based conditional sampling

This procedure was based on the instantaneous spanwise vorticity (using the Taylor hypothesis in the flow direction and finite difference schemes for evaluating derivative terms) and gave us a time display of the vorticity along the transversal length of the rake. We applied a numerical filter (LP without phase difference) to visualize high vorticity areas. Even if this investigation was not the panacea, vorticity peaks were, by definition, the best CS detectors and could clearly be identified from contour plots. Practically, we used a method very close to the vorticity-based conditional sampling technique developed by Hayakawa : From a discrete time series of simultaneous U and V signals at N locations separated by Δy, during T with a time step Δt, we define the instantaneous vorticity :

$$\Omega_{z i,j} = -\frac{1}{U_c}\frac{V_{i+1,j} - V_{i-1,j}}{2 * \Delta t} - \frac{U_{i,j+1} - U_{i,j-1}}{2 * \Delta y}$$

Uc is the average convection velocity of CS's in the longitudinal direction (Uc = 0.5*Umax in our study). We obtained a spatio-temporal map of vorticity field (figure 4). To isolate CS features from the complete motion, we imposed a threshold (Th) on vorticity to select areas where their magnitude was strong. Positions (y_c, t_c) of maximum amplitude in these areas were supposed to be CSs centers :

$$\Omega(t_c, y_c) > Th \quad AND \quad \Omega(t_c, y_c) = \mathrm{Max}_{local}(\Omega(t, y))$$

Then, we obtained a spatio-temporal matrix I that indicate center positions :

$$\text{if} \quad I_{i,j} = 1, \quad t_c = i\Delta t \quad \text{and} \quad y_c = j\Delta y$$

$$\text{else} \quad I_{i,j} = 0$$

The application of phase average on instantaneous signals allowed to extract the coherent motion $\langle F \rangle$ and the pure random motion f'. Furthermore, the coherent motion was assumed to be the CS motion. We chose not to class as the same event structures centered on different transverse positions. These CS's could indeed have distinct features. So, to determine the mean CS centered on a fixed transverse position, we aligned original unfiltered realizations of all transverse positions with respect to each center of reference position and then, we applied ensemble-average on U and V. For instance, to educe the mean CS centered on probe 4, the detection signal was reduced to vector $I_{i,4}$. For τ time delay with respect to the structure center, the phase-average operator was :

$$\langle F(t, y) \rangle_4 = \frac{1}{N}\sum_{t=t_1}^{t_N} F(t + \tau, y) \quad \text{with} \quad \{t_1, ..., t_N\} = \{I_{i,4} = 1\}$$

The "realignment"procedure proposed by Hayakawa was also implemented : phase-average vorticity $\langle \Omega \rangle$ is used to correct the centers of CS's which have already found. This correction should reduce possible time delays between phase average CS and each detected realization. Finally, the selection of events during the detection stage is not based on any structure size criteria. Nevertheless, some structures are completely missed by this technique : long CS could go undetected because their peak vorticity value was under the prescribed threshold. Even, a small structure creating a high but short vorticity peak could be smoothed out by filtering. So, a natural selection of CS's morphology is imposed by using a vorticity-based conditional sampling.

7. Results and discussion

7.1. INSTANTANEOUS BEHAVIOR OF FLOWS

Figure 6 presents vector fields of low-pass filtered instantaneous velocities in a convected frame. For the free mixing layer, we obtain the convected frame by subtracting the mean convection velocity from instantaneous longitudinal velocity. For the separated mixing layer, it is obtained by subtracting the convection velocity but also a transversal velocity. If we do not subtract this term, we do not see CS but only some downward events. In the aim to improve the understanding, the temporal axis is adimensioned with convection velocity Uc and becomes a space axis. In both cases, some vortical structures are regularly detected. In the free mixing layer, we can almost always distinguish vortical events (figures 6c and 6d) but, in the second case, random events or non-vortical ones (6b) can be present during a relatively long time. The coherent motion seems to be regularly destroyed. Although these fields bring to the fore CSs in these kinds of flows, this approach is not sufficiently rigorous to inform us about morphology of CS's. On the other hand , it confirms the use of vorticity-based conditional method.

7.2. EDUCTION BASED ON VORTICITY

On figure 5, we compare the number of educed CS's (fd/f_0, where fd is the number of events per second and f_0 the predominant frequency in the layer) as a function of the transverse position (y) to the time-mean vorticity profile. It appears that, in both cases, these distributions are similar.Indeed, Vortical events are centered on positions where the gradient of velocity is high. But, the use of a detection stage based on a threshold emphasizes this phenomenon since some locations have yet a high level of vorticity, even if they are no additional vortical events. It means that a weak CS would be educed in the center of the shear layer, and not in sides. We present (figure 7) the shape of mean CS centered on reference wires for the separated mixing layer and the free mixing layer. As expected, the vorticity peak was well defined and centered on the trigger point. To check that these vorticity peaks were truly due to vortical structures but not only to transverse gradient, we present coherent streamwise and transverse velocity fields $\tilde{u}$ and $\tilde{v}$.Indeed, locations of over and under velocities are representative : the transverse gradient is optimal for $\tilde{u}$, and the streamwise one for $\tilde{v}$. For the separated mixing layer, the distributions of velocity are less close to an

elliptical and untilted vortical CS. It seems that this phase averaged CS presents a natural tilt. The conditional CS in the separated mixing layer present a greater temporal and spatial extents than one of free mixing layer. Two complementary reasons can be proposed :
- CS's in the separated mixing layer are actually larger than in the free mixing layer. The spread of the time-mean vorticity profile and the high level of turbulent intensity entail an accelerated diffusion.
- CS's in the separated mixing layer are more diversified and thus, the conditional average process educes some CS's which present too much distinct characteristics. The last two plots of each figure depict the shape of mean structure in a convected frame. Indeed we remark that, in the separated mixing layer, the conditional CS have a tilt. To obtain these fields, we subtracted the average onvection velocity of structures Uc from the streamwise coherent velocity $\langle U \rangle$. It is interesting to notice that if we subtract the experimental convection velocity, the structure is not centered on the vorticity peak. On the other hand, if we subtract from the time-mean velocity of the reference position to the phase averaged velocity, the structure is well centered on the vorticity peak where it was educed. This means that averaged structures are convected at their center velocity. Indeed, the global convection velocity includes all structures and so, behave as a spatio-temporal average of the real phenomenon.

8. Conclusion

Although the mixing layer produced by a separation point is more complex than a free mixing layer due to turbulent intensity, adverse pressure gradient, curvature, the CS's educed in a section near the separation point have a similar shape than those of a free mixing layer. This result would not be true in sections close to reattachment. The vorticity-based conditional approach, in the both configurations, enables us to better apprehend the most predominant shape of CS's and check that vortical structures transit in the shear layer. On the other hand, by definition, this approach imposes a particular vortical pattern and eliminate all weak or non-vortical coherent motions. So, it is required to apply less subjective methods as the Proper Orthogonal Decomposition which extracts the most energetic modes from the global information. Nevertheless, we obtain information about the decomposition between coherent motion and incoherent background in a separated flow. These results could improve the semi- deterministic approach in two manners : they can give information to theoreticians to conceive new models or revise old ones, and can be a good data bank to validate some simulations of this reference unsteady flow.

Acknowledgements The author gratefully acknowledge Dr. D. Faghani for fruitful discussions and suggestions. I would like also to thank J.P. Bonnet and J. Delville who advised and designed the hot wires rakes. They also kindly provided us data banks of the plane mixing layer. The measurements were supported by the C.E.A.T. (Centre d'Essais Aéronautique de Toulouse).

References

BELLIN S. (1991), "Etude experimentale des structures cohérentes d'une couche de mélange plane turbulente de fluide incompressible",Thesis of Fluid Mechanics. C.E.A.T. POITIERS

BELLIN S., DELVILLE J., VINCENDEAU E., GAREM J.H., BONNET J.P. (1992), "Largr scale structure characterization in a 2D mixing layer by Pseudo Flow Visualisation and Delocalised Conditional Sampling ", Proceedings of the IUTAM Symposium - POITIERS October 1992, Kluwer Academic Publishers.

CASTRO I.P. & HAQUE A. (1986), "The structure of a turbulent shear layer bounding a separation region", *J. Fluid Mech.* **179**, 439–468.

HAYAKAWA M. (1993), "Vorticity-based conditional sampling for identification of large scale vortical structures in turbulent shear flows", Proceedings of the IUTAM Symposium - POITIERS October 1992, Kluwer Academic Publishers.

HUSSAIN A.K.M.F. & HAYAKAWA M. (1987), "Eduction of large scale organized structures in turbulent plane wake", *J. Fluid Mech.* **180**, 193–229.

TROUTT T.R. SCHEELKE B. & NORMAN T.R. (1984), "Organized structures in a reattaching separated flow field", *J. Fluid Mech.* **143**, 413–427.

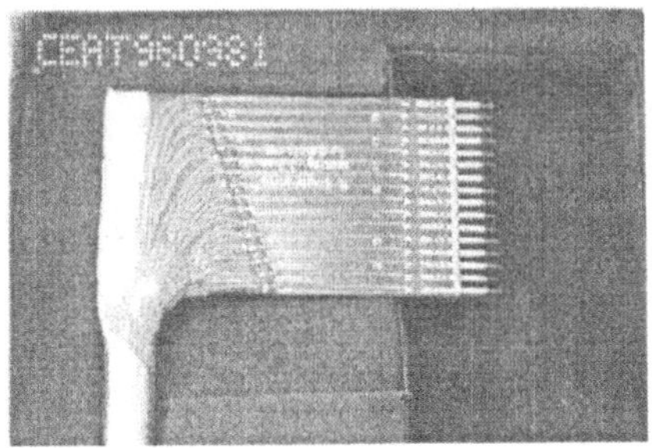

figure 1 : 16 single hot wires rake

figure 2 : hot wire rake "in situ", on a bent support

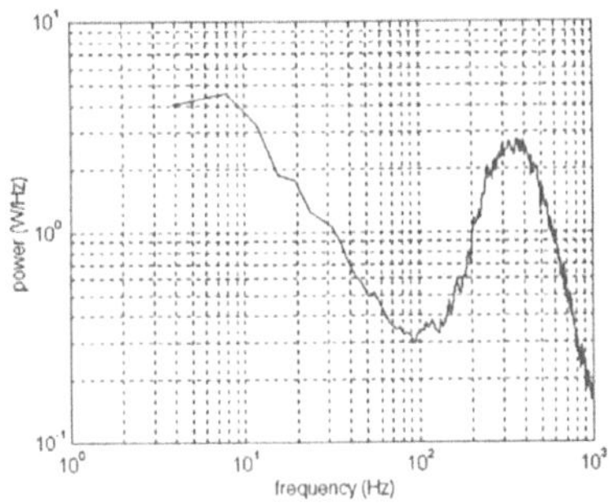

figure 3 : Power spectral density of a velocity signal in the shear layer

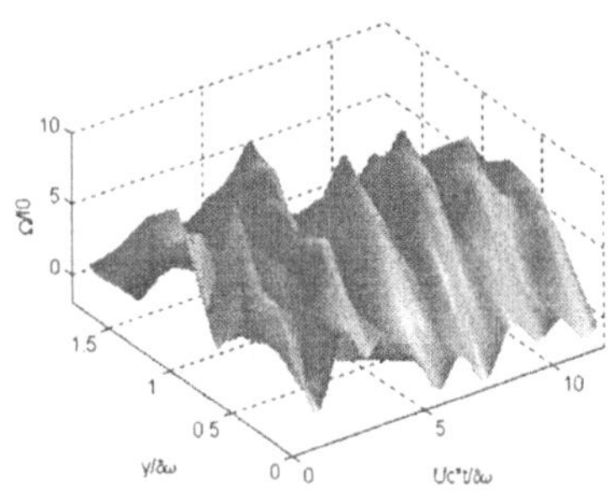

figure 4 : vorticity field

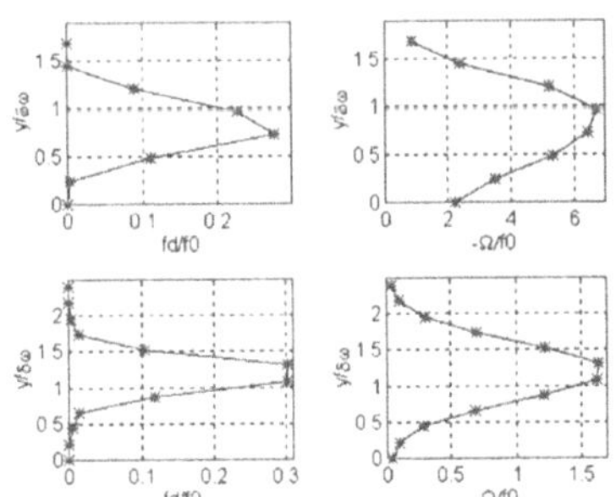

figure 5

figure 5 : top, backward facing step; bottom, plane mixing layer. Left, profile of number of detections per second fd; right, profile of time-mean vorticity $\overline{\Omega}_z$. F_0 is the dominant frequency in layers

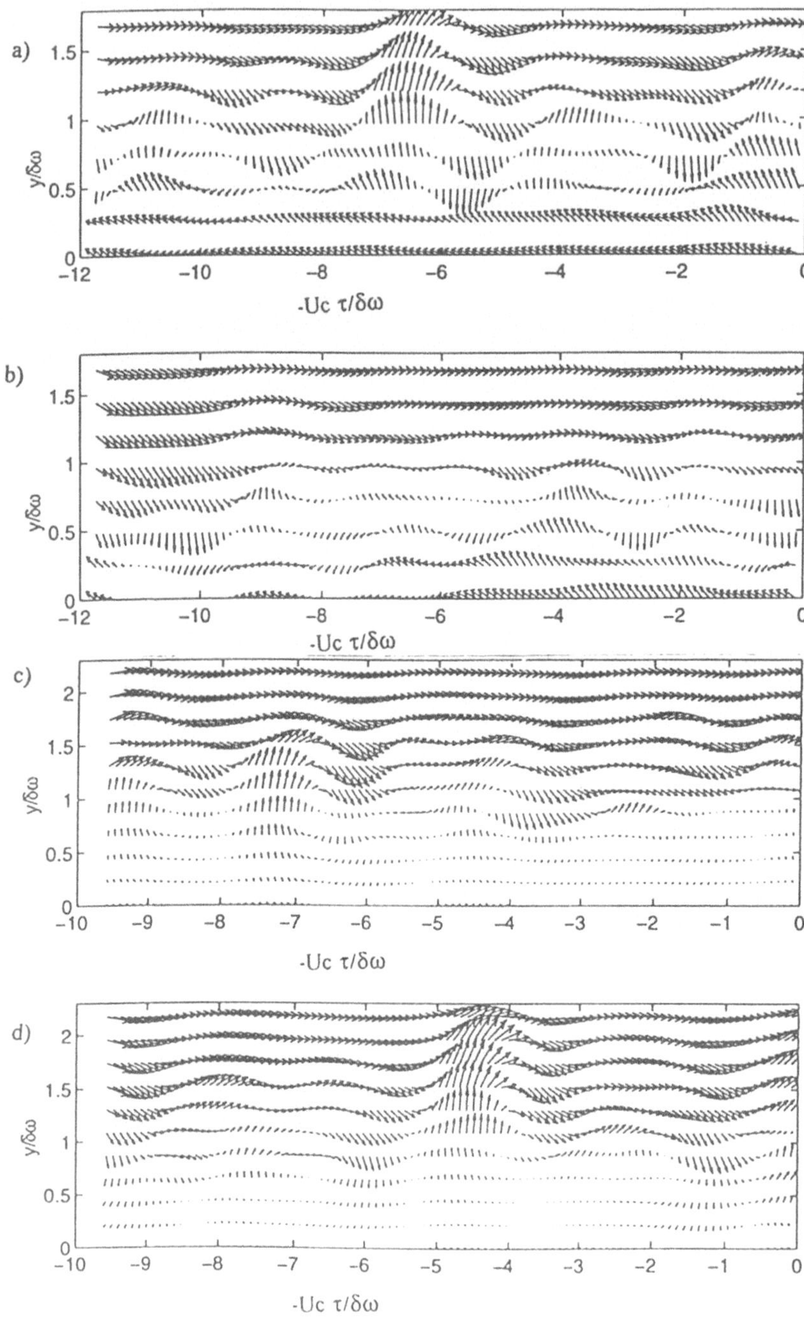

figure 6 : vector fields of instantaneous velocities in a convected frame
a) and b) shear layer; c) and d) plane mixing layer

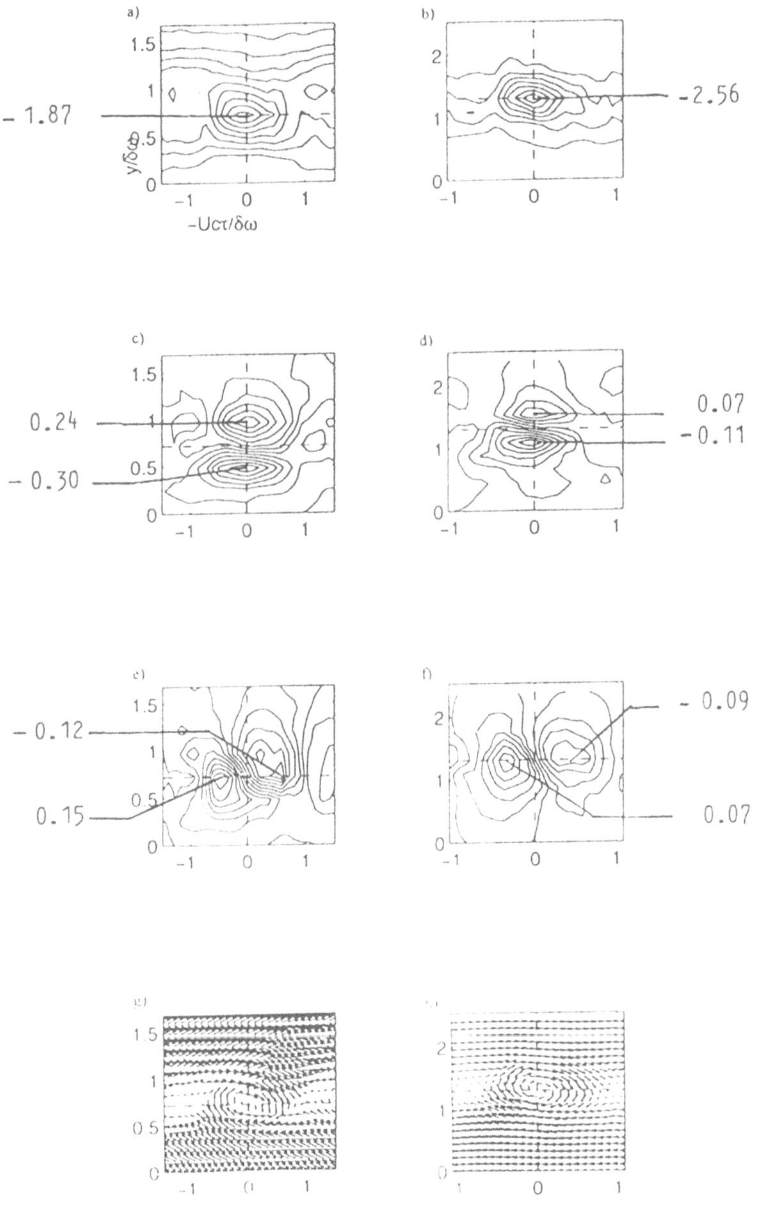

Figure 7 : morphology of phase averaged vortical structures. Left, coherent structure of backward facing step centered on the fourth wire; right, coherent structure of mixing layer centered on the seventh wire.

a) and e) $\dfrac{\langle \Omega_z(\tau, y)\rangle}{\left(\partial \bar{U}/\partial y\right)_{\max}}$. b) and f) $\dfrac{\tilde{u}(\tau, y)}{Um}$. c) and g) $\dfrac{\tilde{v}(\tau, y)}{Um}$

d) and h) vector field of structure.

3D MEASUREMENT OF VORTEX STRUCTURES IN STRATIFIED FLUID FLOWS

A. M. FINCHAM
Laboratoire des Ecoulements Géophysiques et Industriels /Coriolis
UJF-CNRS-INPG, B. P. 53
38041 Grenoble, Cedex 9, France

1. Introduction

Imaging Velocimetry (IV) has become one of the most powerful laboratory measurement techniques for fluid flows. This non-intrusive technique allows instantaneous measurement of 2 components of velocity in a plane and the planar derivatives can usually be evaluated at the resolution of the 2D data grid. With sufficient care, very accurate fields of the component of vorticity perpendicular to the plane can be measured in time using rather modest equipment. The progression from single point hot-wire or LDA measurements is significant, but in the interpretation of complex 3D vortex structures fluid dynamicists still rely on Direct Numerical Simulations (DNS) using planar IV data as experimental support, much as they would use velocity spectra taken with a hot-wire to verify a turbulence model. Just as good flow visualisation can provide insight into general 3D flow structure, individual slices of IV data taken in perpendicular planes, can allow educated speculation as to the true 3D vortex structure.

The extension of planar imaging methods to three dimensions requires moderately sophisticated hardware that for most laboratories is just becoming affordable. Among the many methods, the predominant three are, slice and dice, stereo and holography, see Grant, 1997, for a review. Each has its own benefits and drawbacks but by far the simplest to implement is the slice and dice approach. In the slice and dice method, multiple parallel planes are rapidly acquired, then combined to form a data cube. The main drawback of this approach is the inability to resolve the out-of-plane velocity component without a corresponding orthogonal set of dices. This out-of-plane component can be approximated by integration of the continuity equation away from some known plane, such as a rigid boundary (Utami et al.,1990). As errors are accumulative with the integration process and small errors in the planar velocity fields result in much larger errors in its divergence, in most cases, except when the region of interest is in the close proximity of the known plane the results for the third velocity component are typically just a good approximation.

In the experiments described here, the extreme anisotropy present in late time stratified fluid flows where, $w \cong 0$, is exploited, allowing full 3D vortex reconstruction without recourse to the equation of continuity. This paper proceeds as follows, after motivating the study with a discussion of some properties of stratified flows in section 2, section 3 concentrates on basic planar IV techniques and some of their limitations. Section 4 describes the experiments and reconstruction process in detail and the results are discussed in section 5.

2. Vortex Structures in Late-time Stratified Flows

2.1. COLLAPSE TO THE QUASI-2D STATE

A vertical density gradient characterised by the Brunt-Vaisala or buoyancy frequency, $N = \left(-\frac{g}{\rho}\frac{\partial\rho}{\partial z}\right)^{\frac{1}{2}}$, where ρ is the fluid density, will provide a restoring body force on fluid particles vertically displaced from their equilibrium position. The effects of this force on an initially isotropic disturbance is to quickly diminish the vertical component of velocity along with the vertical growth, while enhancing horizontal development. During this collapse process, the horizontal motion can be seen to organise itself into coherent quasi-2D structures containing vertical vorticity. After several buoyancy periods the vertical velocity component w, exists predominantly as internal gravity wave motions, which radiate energy and horizontal vorticity (Lighthill 1996) away from the source. Once formed these structures persist for long times, due mainly to the inhibition of the 3D turbulent dissipation mechanism, but also to their ability to organise into larger quasi-stable structures with their neighbours.

2.2. TURBULENCE AND VORTEX TOPOLOGY

The decay of initially isotropic turbulence in a stably stratified fluid has been extensively studied (Hopfinger, 1987, Maxworthy et al., 1987, Metais & Herring, 1989, Yap & van Atta, 1993, Fincham et al., 1994, 1996), the late time regime (which is not easily accessible through direct numerical methods) has been described by Fincham et al. to consist of a three dimensional densely packed sea of quasi-2D vortices. These vortices undergo 2D horizontal non-linear interactions, but interact vertically through viscous mechanisms. Though quasi-2D in appearance, they showed that at higher Reynolds numbers, the horizontal component of the vorticity vector contains an order of magnitude more enstrophy that that contained in the vertical component, and is responsible for most of the viscous dissipation of kinetic energy. The resulting vorticity field has a complex 3D topology despite the apparent quasi-2D nature of these flows.

A simple topological vortex model that enforces the connection of vortex filaments into loops was proposed to explain both the persistence and interactions of these 3D structures, see Fig. 1. This necessity for vortex filaments to connect (Saffman, 1992 Chap. 1) explains why DNS of 2D turbulence often exhibit strong isolated vortices that are not observed in late time stratified flow experiments. Unless the fluid is relatively shallow or the flow is 2 dimensionalized (by strong rotation for example), the vortex filaments associated with each vortex must either connect to some adjacent structure or to the boundaries, making the occurrence of an individual isolated vortex impossible.

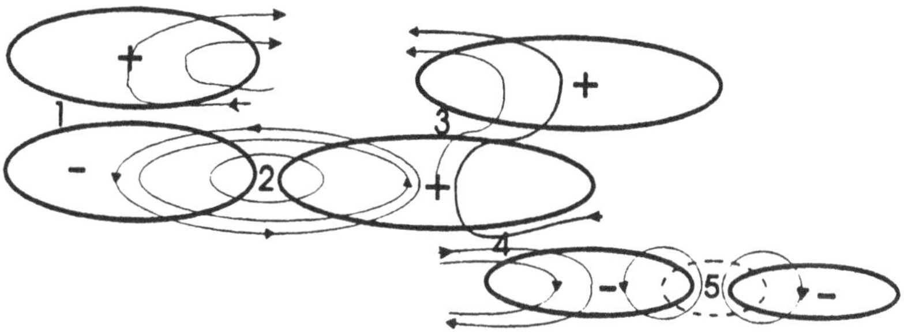

Fig. 1 Cross-sectional schematic diagram of possible vortex line connections between structures. The plus and minus signs indicate the direction of the vertical component of vorticity, ω_z. The type 2 connection can be identified as a vortex dipole, type 3 is commonly observed; type 5 indicated that two like-signed vortices require a third vortex of opposite ω_z in order to be interconnected on the same level. Connections of type 1 and 4 are not commonly realised and may be unstable. (from Fincham et al., 1996)

The quasi-2D appearance of this late time *turbulent* state exhibits a wealth of structures, but is dominated by the presence of multipolar structures, of which dipoles are the most visually evident. It is the purpose of this study to verify this vortex model in-so-as it applies to the stratified vortex dipole (connection type 2 in Fig. 1), and to demonstrate the ability of Imaging Velocity techniques to fully resolve 3D vortex structures.

3. Planar Imaging

3.1. PARTICLE IMAGING VELOCIMETRY

Most IV techniques rely on some sort of image correlation to perform pattern matching between consecutively exposed tracer images. PIV (Adrian, 1991) which relies on optically auto-correlating doubly exposed negatives by analysis of the Youngs Fringe patterns produced by interrogation with a small laser, has given way to DPIV (Adrian, 1991, Willert & Gharib, 1991) where two digitally recorded images are cross-correlated numerically. Beyond the elimination of the photographic process, the removal of any directional ambiguity as well as an improved signal to noise ratio makes the cross-correlation between two separate images preferable to the auto-correlation of a single multiply exposed image.

DPIV typically divides each image into smaller pattern boxes containing several particle images, theses boxes are correlated with the same region in the second image and the peak of the spatial correlation function determines the most likely displacement of the group of particles. Spatial resolution is limited by the resolution of the image sensor and the minimum size of the interrogation box. Classical DPIV algorithms suffer both an increase in the signal to noise ratio and a decreased dynamic range when small boxes are used, making more expensive high resolution image sensors desirable.

3.2. CORRELATION IMAGING VELOCIMETRY (CIV)

All the data presented here has been processed using an independently developed, highly optimised, Correlation Imaging Velocimetry (CIV) system, details of the algorithms and optimisation procedures can be found in Fincham & Spedding, 1997. A basic philosophy, "no compromises between absolute accuracy and computational time", along with a total de-coupling of the search location in image #2 from the original tracer location in image #1 which tolerates the use of very small pattern boxes increasing spatial resolution while preserving the dynamic range of the measurement, this allows better resolvement of strongly vortical flows and provides for a more cost effective use of the available equipment. Continuing optimisation efforts, involving accurately simulated images superposed onto analytical or numerically generated flow solutions, closely coupled with real experiments, has given a thorough understanding of the spectral distribution of errors and the experimentally adjustable parameters that effect them.

3.3. DATA IN THE IMAGE PLANE

As each box of pixels measured the *mean* velocity of the fluid tracers appearing in that part of the image, the camera may be zoomed in or out to measure smaller or larger scales respectively. When zooming out, scales smaller than the pattern box will be averaged (an effective low pass filter of the enstrophy spectrum) but the structures associated with smaller wave-number will still be measured accurately. Zooming in allows resolution of the smallest most dissipative scales typically associated with the strongest vorticity. Unfortunately due to constraints placed on the time interval between the 2 images by the largest most energetic scales which in a 3D flow are removing tracers from the light sheet, (Fincham & Spedding, 1997) at high Reynolds numbers it is not possible to preserve cubic fluid elements with a suitably thin light sheet and the velocity must be averaged through the thickness of the sheet. Nonetheless, as sub grid scales are taken care of by real viscosity, the measurements need not approach the Kolmogorov scale, and the measurable range of useful scales can surpass that of current DNS methods. Utilising two standard cameras synchronised in time with different magnifications, close to 3 orders of scales can be simultaneously resolved (Fincham et al., 1994, 1996, Spedding et al., 1996).

3.3.1. *Filtering*

Although DPIV type methods apparently result in data spaced on a regular grid, in fact, the true location of each velocity vector should be half way between the original location of the particles in image #1 and their final location in image #2. Failure to make this correction will result in a systematic bias in the computation of the spatial derivatives. After this convective correction, the now irregular grid of data must be interpolated back onto a regular grid to ease the post processing procedures. This re-interpolation is done using a thin shell smoothing spline (Spedding & Rignot, 1993) that is not confined to go through every data point. A by-product of the interpolation process is that the spatial derivatives can be computed directly from the spline coefficients. Grid scale fluctuations caused in part by particles entering and leaving the light sheet can cause large spikes in the gradient fields. In order to suppress them a 4 th order butterworth spectral filter, the same size as the grid mesh, is applied before computation of the spatial derivatives. Contrary to most DNS data sets these velocity grids are not periodic. Spectral filtering non-periodic data fields will cause wrap around errors without proper data truncation. In order to minimise these errors, while retaining all of the hard earned resolution, a special spectral flip-filter was developed. The flip-filter flips the data in all directions making a larger (3 x 3) grid to which the filter is applied. Wrap-around errors, are now limited to the boundaries of the extended grid, after the inverse FFT only the central region corresponding to the original data is retained.

3.3.2. *Computable Quantities*

In addition to the velocity field, each planar measurement provides several other quantities of fluid-mechanical significance, the component of vorticity perpendicular to the plane, the in plane divergence and the viscous dissipation of energy due to the velocity gradients within the plane, can all be evaluated at the grid resolution.

4. 3D Experiments

4.1. THE DIPOLE

The experiments were performed in the 13 m diameter rotating tank Coriolis located in Grenoble. Though the dipole discussed here has no significant background rotation, the use of this facility is justified as it provides for the creation of a large (1 m diameter) dipole. It is necessary to produce a dipole with a very long turn-over time to enable completion of the 3D scan before it can move significantly. To keep the Reynolds number significant a large slow dipole is best.

The dipole was generated from the collapse of a weakly turbulent or laminar jet of density matched fluid originating from a horizontal tube of diameter 1.7 cm placed at the mid-depth (33 cm) of the working fluid (water). The fluid was linearly stratified with common salt and had a buoyancy frequency N=0.12 rad/sec. The jet undergoes stratified collapse, and for appropriate injection parameters an almost symmetric stratified dipole is formed downstream. The experimental parameters are as follows; the velocity of the jet was 4 cm/sec, the time of injection was 45 sec, and the distance from the jet exit to the beginning of the measurement volume was 180 cm.

For the dipole discussed here, the Reynolds number based on the jet diameter of 680 is quite small and the jet is still laminar at the exit, as might be expected this laminar injection process proved to give better symmetry to the dipole. At the time of measurement the dipole had a maximum centreline velocity of 0.4 cm/sec and the distance between the poles was 40 cm giving it a horizontal Reynolds number of about 1600, which is significantly larger than that at the jet exit. The weakly dissipative state coupled with an enhanced horizontal growth typical to these stratified structures tends to conserve the horizontal Reynolds number. The Reynolds number based on the vertical scale is of order 400. The data from some preliminary experiments done in a smaller tank, where measurements were made independently in horizontal and vertical planes was used to determine at what stage of development the assumption $w \cong 0$ was valid.

4.2. VOLUME SCANNING SYSTEM

The fluid was seeded with 700 micron diameter polystyrene beads that were carefully prepared by cooking and consecutive density separations, to have a flat distribution of densities matching that of the salt water stratification. This laborious process ensured that there were equal number densities of particles at each depth. Coherent light originating from a 5 watt Argon laser was passed into an optical fibre, the other end of which was attached to a small optical assembly that moved horizontally on a linear bearing traverse. Directly beneath the traverse in the water, was placed a large 45° inclined mirror. The motion of the optical assembly, that basically consisted of a small oscillating mirror, was controlled by a stepper motor coupled to a PC through a programable Anahiem Automation driver box. The oscillating mirror created a vertical sheet of laser light that passed through a thin glass plate held parallel to and just touching the surface of the water above the 45° mirror. In this way, horizontal motion of the optical assembly above the water, translated directly into vertical motion of the laser sheet within the fluid.

Synchronisation between the laser sheet motion and the camera acquisition is achieved through a self calibration scheme that accurately measures the appropriate acceleration times for any specified motion and optimises slice acquisition timing. This 3D-scan module interfaced directly to our online acquisition and processing software package CIVIT (a GUI that implements the algorithms of section 3.2). Synchronous, asynchronous and continuous mode scanning are all possible, continuous mode provided the desired results for these experiments and will be briefly discussed.

The Volume scan process proceeds as follows, an initial scan is made with continuous image acquisition to memory, the light sheet is quickly returned to the starting position, after the appropriate time interval the scan process is repeated to acquire the second image in each pair. This ensures a constant time interval between corresponding slices in each scan. Using a standard resolution digital output camera with 768 x 484 pixels (Pulnix TM-9701), an area of 165 x 120 cm was imaged. The camera output 30 images per second, allowing for a maximum of 30 slices per second. As the maximum velocity associated with the dipole in the measurement area is 0.4 cm/sec and horizontal velocity vectors were to be evaluated as the average of each 2x2 cm square region in the plane, it is reasonable to allow up to 3 seconds between consecutive scans. The actual inter-scan time is determined by the pixel displacement that results in minimal error. Provided the inter scan time is shorter than 3 sec, additional time may be used to gather more slices, or as in the case described here, integrate the camera signal in time (longer exposure) providing better signal quality. As each slice measured may have its own characteristic velocity, it is not possible to minimise errors in all slices with just 2 scans, a third or fourth scan of each volume was made to better resolve the weakly energetic planes. This approach allows proper resolvement of the weakly vortical motions above and below the dipole. This multiple scan process is automatically repeated at predetermined time intervals.

25 slices each containing 74 x 50 velocity vectors were obtained for each volume scan event. This data cube represents an actual fluid volume of 160 x 120 x 20 cm, which is sufficient to vertically contain the dipole while the horizontal extent provides leeway for the deviations of the dipoles trajectory caused by weak background motions. The y axis is parallel to the axis of the jet, with z vertically upward.

4.3. VORTICITY FIELD RECONSTRUCTION

On completion of a scan sequence, the images are quickly archived on disk freeing memory for the next volume-scan event. Immediately online processing of spatially matched image pairs commences. Vector fields of each plane are available on screen for the experimentalist to make any necessary timing changes before the next volume scan event. Meanwhile a slightly larger computer processes the images to higher resolution, each slice is treated as described in section 3. The resulting vector fields are combined into a 3D cube that is fit with a 3D 4 th order tensor product spline, yielding the dominant vertical derivatives of horizontal velocity. With the assumption of $w \cong 0$, all components of the deformation tensor are fully resolved in time. Volume rendering of interesting quantities is then done using appropriate 3D rendering software. Due to the slow time evolution of these late time stratified flows, this 3D reconstruction process can parallel the volume acquisition.

5. Results and Discussion

The purpose of this paper is to demonstrate the ability of current IV methods to accurately measure 3D vortex structures, providing a solid base for the development and validation of conceptual 3D vortex models. In addition, evidence is provided in support of the usefulness of the simple vortex filament model proposed by Fincham et al., to explain the vortex dynamics associated with strongly stratified fluid flows. To this end, the discussion will focus on the topological aspects of the measured vorticity fields. Detailed analysis, including the evolution of this dipolar structure in time and its *eventual* decay will be left for a future publication.

Here we take a single time step and focus on the 3D vorticity field. Fig. 2 shows the vertical component of vorticity ω_z associated with measurement planes perpendicular to the classical dipole axis, the individual colour maps have been re-normalised to better show the structure in each slice.

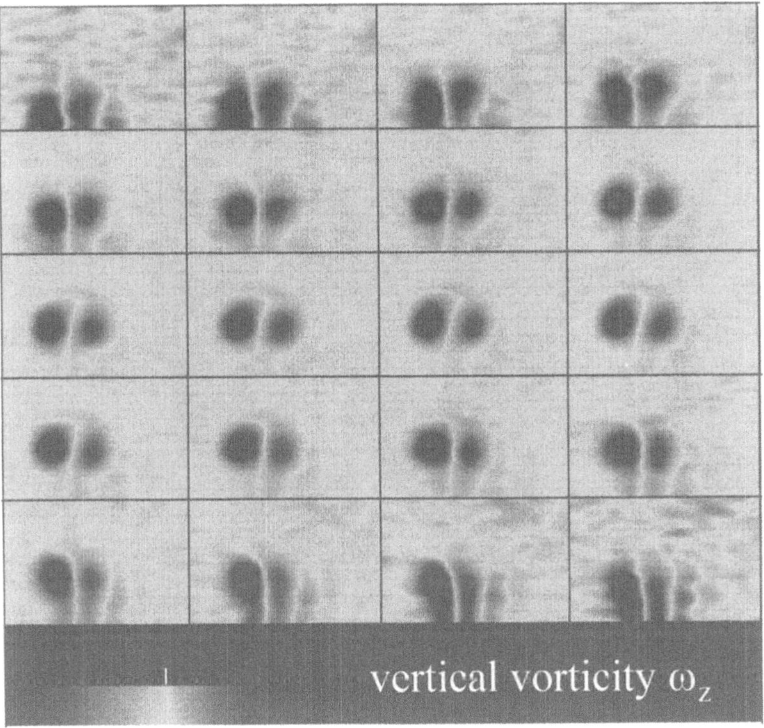

Fig. 2 The vertical component of vorticity ω_z for 20 of 25 horizontal slices through the dipole. Panel 1 in the plane at z=3.

It is clear that the (vertically) central core of the dipole has propagated slightly ahead, giving the vertical profile of ω_z a somewhat banana shape. This deformation by the dipolar self propagation mechanism has previously been observed in the experiments of Flor & Van-Heijst, 1994, and is to be expected, if the velocity field approaches zero both above and below the dipole. Compilation of these slices into a cube allows visualisation of the actual 3D structure of ω_z and is shown in Fig. 3. Here the vertical scale has been expanded by a factor of 5, to better show the topology. The banana shape of ω_z is evident, and the horizontal spacing between the poles can be seen to change with z. This dipole is not perfectly symmetric but serves for the purposes of this discussion.

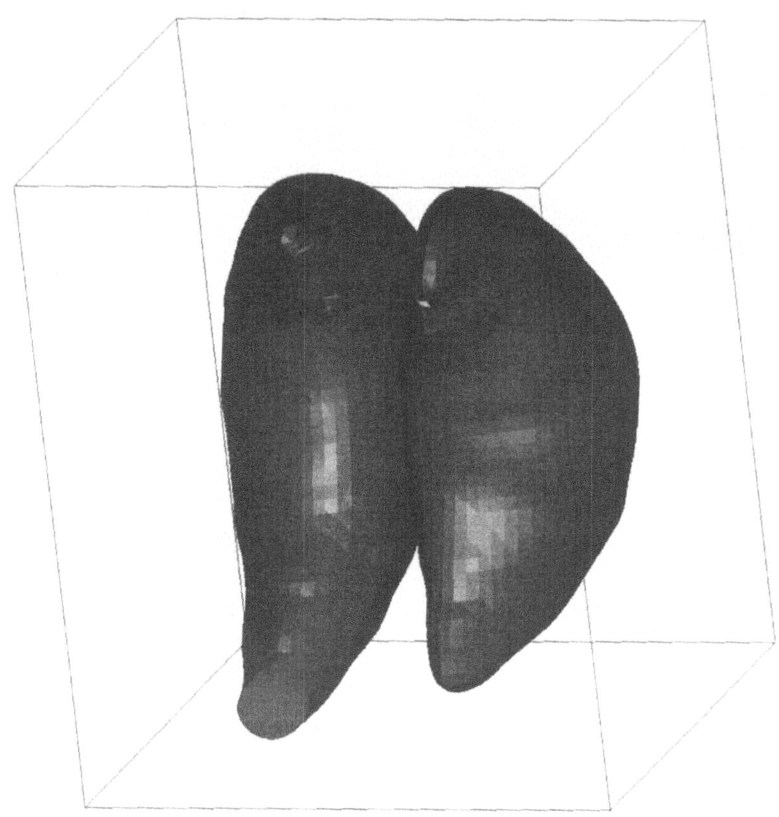

Fig. 3 An iso-surface of the vertical component of vorticity in the region of the dipole. The vertical scale has been expanded by a factor of 5. The dipole is propagating into the page. The curvature evident, both above and below, is caused by strong vertical shearing ($\partial u/\partial z$ and $\partial v/\partial z$) bending the initially vertical vortex lines horizontally.

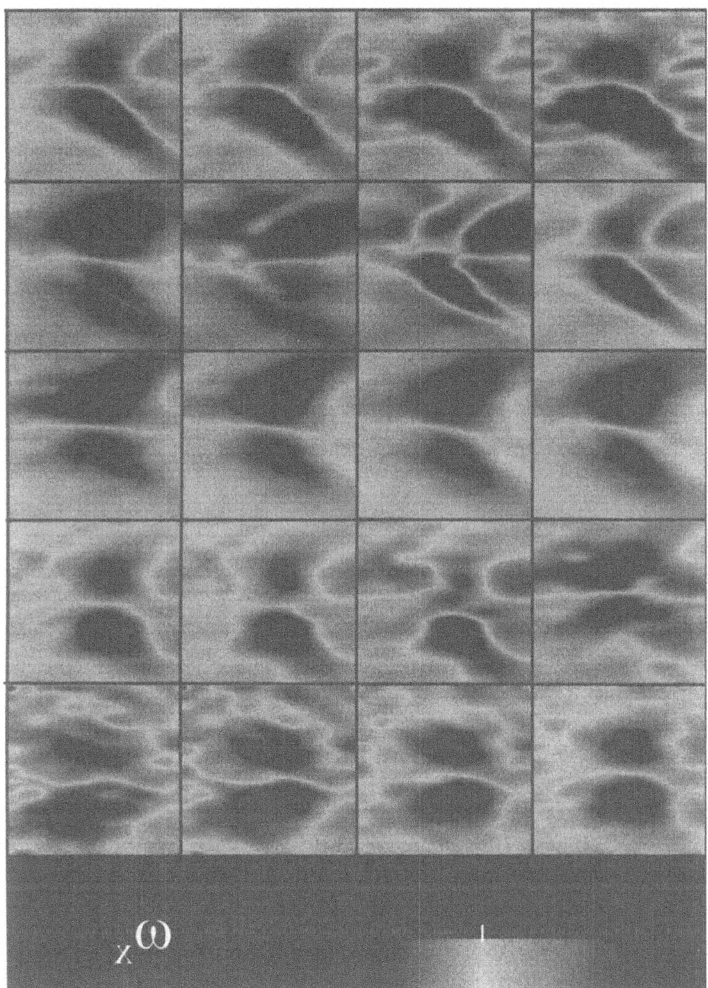

Fig. 4 The horizontal component of vorticity ω_x for a sequence of slices traversing the dipole. Again the vertical scale has been expanded

After full reconstruction, vertical slices taken parallel to the x and y axis reveal the respective horizontal components of vorticity, Figs. 4 and 5. These slices represent a region traversing the dipole, the centre of which is represented in panel 10 of Fig. 4 and panel 11 of Fig. 5. In this central region where slices traverse the core, a complex reorganisation is evident in ω_y (Fig 5.), where the polarity of the quadrapolar type structure is changing. Similar reorganisation can be seen in panels 6 and 14 of Fig. 4, corresponding to the slices aligned with the vertical axis of the two poles.

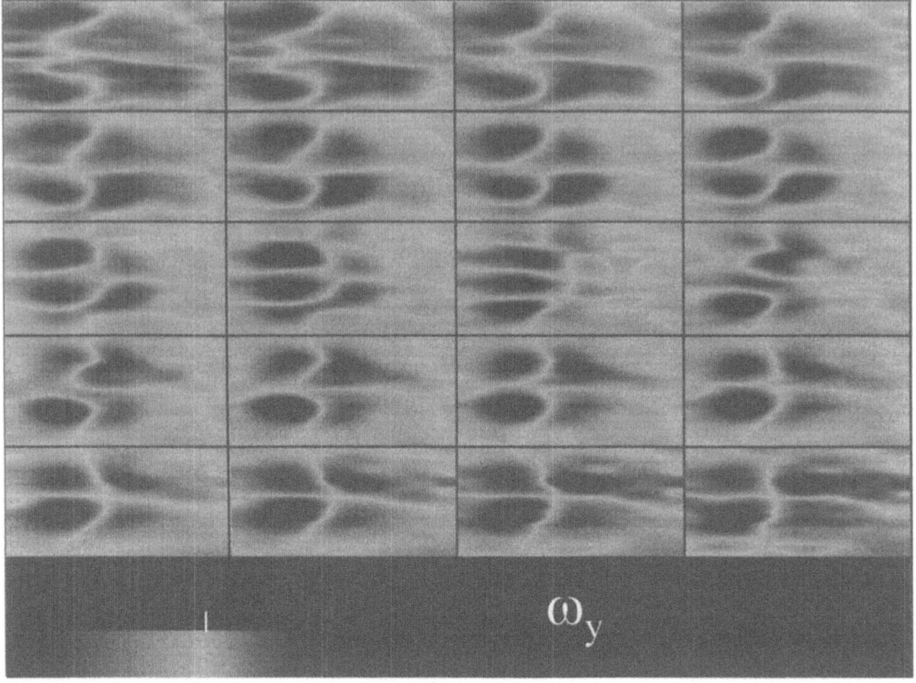

Fig. 5 Horizontal vorticity ω_y traversing the volume.

As mentioned in section 2.2, these horizontal vorticity components associated with pure shearing motions are relatively strong and often determine the direction of the vorticity vector.

An isosurface of the full 3 component enstrophy field is shown in figure 6. It is clearly seen than the enstrophy field takes the shape of a distorted vertically orientated torus like structure, through the centre of which flows the fluid with the largest velocity. The striking similarity with a classical vortex ring creates a small dilemma when one realises that there is no relevant vertical component of velocity. It is the baroclinic torque provided by the background stratification that provides the *sink* for the initially-vertical vortex filaments that are being bent horizontal by the strong vertical shearing motions $\partial u/\partial z$ and $\partial v/\partial z$. This consistency with the previously mentioned vortex model need only be verified by examination of the actual vortex lines themselves.

Fig. 7 shows the evolution of a number of vortex lines originating in a diagonal plane intersecting the centre of both poles, along with an isosurface of ω_z, the colour of the lines is representative of their divergence.

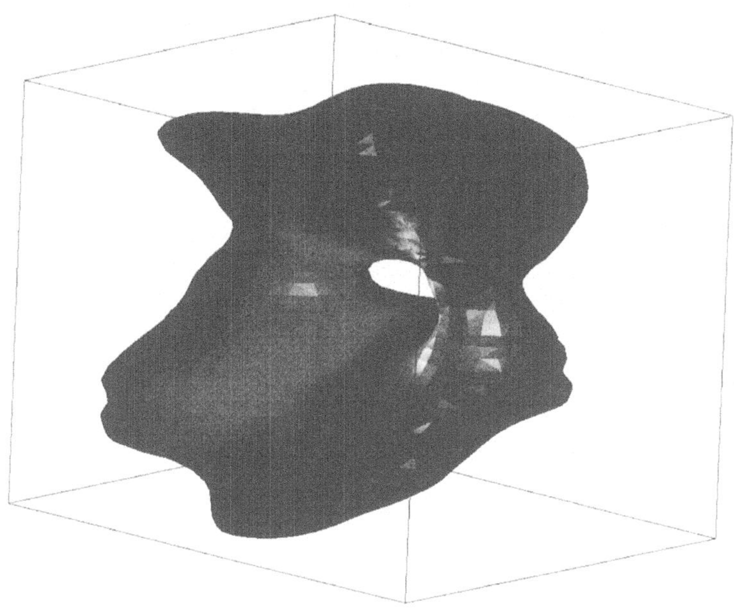

Fig. 6 Isosurface of the 3 component enstrophy field. The dipole is propagating into the page.

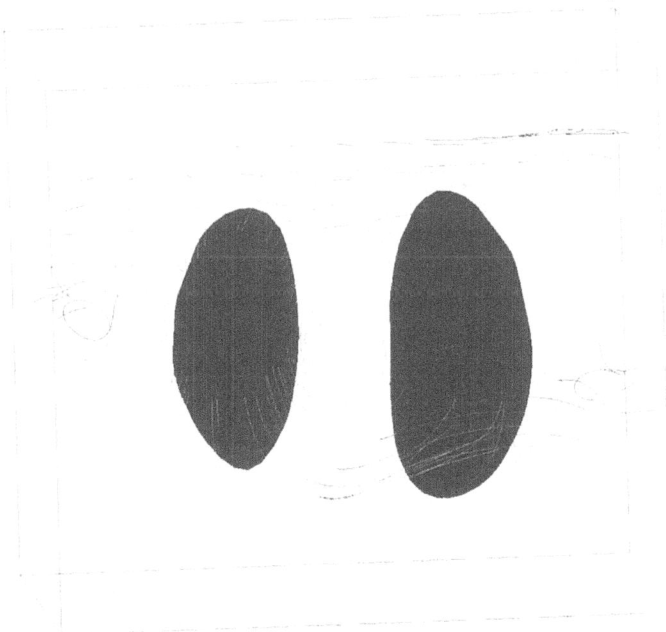

Fig. 7 Vortex lines originating from a diagonal plane traversing the dipole.

Excellent agreement with Fig. 1 is seen, as the vortex lines attempt to close on themselves, connecting the two poles through the strongly sheared regions above and below. Although Helmholtz's theorem is not strictly valid in an inviscid stratified fluid, due to the presence of a baroclinic torque that may make and break vortex lines. At each instance in time, it is expected that vortex filaments will tend to form closed loops, making their measurement and visualisation a useful diagnostic tool for understanding the flow topology.

The fact that the vortex model of section 2.2, was conceived during the analysis of large numbers of horizontal and vertical slices acquired in independent ensembles of the same stratified turbulence experiment, does a lot to demonstrate the value of simple planar IV techniques. On the other hand, the fact that a single time-step of 3D data is sufficient to verify the existence of such structures while providing much additional topological information, only shows the limitations of determining 3D vortex structures from un-correlated 2D slices.

6. Summary and Conclusions

Three dimensional measurement of stratified vortex structures has been achieved for the first time. The measurements have validated the simple vortex filament model of Fincham et al., and provided additional insight into the structure and behaviour of these strongly stratified fluid flows. This has been achieved through the exploitation of the inherent anisotropy associated with these flows, by means of a robust planar CIV technique, combined with a relatively simple volume scanning system.

One of the general conclusions to be made with reference to 3D measurements by slicing, is that it is almost always better to get the Reynolds number from the length scale if possible. Simple scaling can show that these measurements would not have been feasible at this Reynolds number in a smaller tank, unless a higher speed camera and desirably, a pulse laser were available, such technology will eventually become affordable and the technique described here is well positioned to assimilate these predictable advances in technology.

These experiments would not have been possible without the continued assistance of Dr. Henri Didelle and Mr. René Carcel at the Coriolis laboratory who have been instrumental in the development of the complex infrastructure needed for such measurements. They have also participated in the many insightful discussions with Dr. Dominique Renouard that have kept the momentum for this project positive.

References

Adrian, R. J. (1991) Particle-imaging techniques for experimental fluid mechanics. *Annu. Rev. Fluid Mech.* **23**, 261-304

Fincham, A. M. (1994) The structure of decaying turbulence in a stably stratified fluid, using a novel DPIV technique. *PhD Thesis*, Department of Aerospace Engineering, University of Southern California

Fincham, A. M., Maxworthy, T. & Spedding, G. R. (1996) Energy dissipation and vortex structure in freely decaying, stratified grid turbulence. *Dyn. Atmos. Oceans*, **23**, 155-169.

Fincham, A. M. & Spedding, G. R. (1997) Low-cost, high resolution DPIV for measurement in turbulent fluid flows. *Exp. in Fluids*, (in press).

Flor, J.B. & van Heijst, G. J. F. (1994) An experimental study on dipolar structures in a stratified fluid. *J. Fluid Mech.*, **279**, 101-133.

Grant, I. (1997) Particle image velocimetry: a review. *Proc. Instn Mech, Engrs.*, **211**, C, 55-76

Hopfinger, E. J. (1987) Turbulence in stratified fluids: a review. *J. Geophys. Res.*, **92**, 5287-5303.

Lighthill, M. J. (1996) Internal waves and related initial-value problems. *Dyn. Atmos. Oceans*, **23**, 1-20

Maxworthy, T., Caperan, P. & Spedding, G. R. (1987) The kinematics of quasi-2D freely decaying turbulence in a stratified fluid. Third International Symp. On Stratified flows, Pasadena, CA.

Metais, O. & Herring, J. R. (1989) Numerical simulations of freely evolving turbulence in a stratified fluid. *J. Fluid Mech.*, **202**, 117-148.

Saffman, P. G. (1992) Vortex Dynamics. Cambridge University Press.

Spedding, G. R. & Rignot, E. J. M. (1993) Performance analysis and application of grid interpolation techniques for fluid flows. *Exp. in Fluids* **15**, 417-430.

Spedding, G. R., Browand, F. K. & Fincham, A. M. (1996) The long-time evolution of the initially-turbulent wake of a sphere in a stable stratification. *Dyn. Atmos. Oceans*, **23**, 171-182.

Spedding, G. R., Browand, F. K. & Fincham, A. M. (1996) Turbulence, similarity scaling and vortex geometry in the wake of a towed sphere in a stably stratified fluid. *J. Fluid Mech.*, **314**, 53-103.

Utami, T., Blackwelder, R. F. & Ueno, T. 1990: Flow visualisation with image processing of three-dimensional features of coherent structures in an open channel flow. In: Near wall turbulence. (Ed. Kline, S. J.) 289-305, Hemisphere, NY

Willert, C. E. & Gharib, M. (1991) Digital particle image velocimetry. *Exp. Fluids* **10**, 181-193

Yap, C. T. & van Atta, C. W. (1993) Experimental studies of the development of quasi-2-dimensional turbulence in a stably stratified fluid. *Dyn. Atmos. Oceans*, **19**, 289-323.

DNS OF A TURBULENT CHANNEL FLOW TO GUIDE VORTICITY MEASUREMENTS IN THE WALL REGION

P.G. ESPOSITO
INSEAN
Istituto Nazionale per Studi ed Esperienze di Architettura Navale
Via di Vallerano 139 - 00128 Roma - Italy

T. ZHOU AND R.A. ANTONIA
Department of Mechanical Engineering
The University of Newcastle
NSW 2308 - Australia

AND

P. ORLANDI
Dipartimento di Meccanica e Aeronautica
Università degli Studi di Roma "La Sapienza"
Via Eudossiana, 18 - 00184 Roma - Italy

Abstract.
In previous work, the performance of a four hot-wire transverse vorticity probe was tested by comparing measurements in a fully developed turbulent channel flow with corresponding data obtained from direct numerical simulations (DNS) of the same flow and nominally the same experimental conditions. Relative to the DNS data, the probe performed satisfactorily in the outer region, but significantly underestimated the rms vorticity in the wall region. The present work attempts to numerically simulate the probe when it is located in the wall region. The results indicate that the performance of the vorticity probe, when it measures ω_2, should improve as the spanwise dimension of the probe is decreased.

1. Introduction

The experimental determination of vorticity has been a challenging problem for many reseachers because it plays a very important role in describing both kinematical and dynamical aspects of turbulence. Many measurements have been made of the mean square vorticity or enstrophy $\langle\omega^2\rangle$ or one or more of its components in several turbulent flows (e.g. Wallace and Foss (1995)). One way of testing the performance of a vorticity probe is to carry out measurements in decaying grid (or shearless) turbulence (e.g. Zhu et al. (1997)). In this flow, the mean energy dissipation rate $\langle\epsilon\rangle$ and also $\langle\omega^2\rangle$ are known to relatively good accuracy from measurements of the decay rate of the turbulent kinetic energy. Satisfactory performance in shearless turbulence does not necessarily guarantee that the probe will perform adequately in turbulent shear flows, especially near walls where the effect of a large mean velocity gradient is likely to be important. In sheared turbulence, it is possible to test the probe performance by measuring the terms in the vorticity form of the momentum equation. This approach was followed by Antonia and Rajagopalan (1990) to test a one-component vorticity probe in the far-wake of a cylinder. More recently, the same approach and the same probe design were applied to a fully developed turbulent channel flow (Zhou *et al.* (1997)). It was found that the probe performed adequately in the outer region ($x_2^+ \geq 40$, where x_2 is the wall-normal direction and the superscript + denotes normalization by the kinematic viscosity ν and the friction velocity U_τ) of the channel but poorly in the wall region ($x_2^+ < 40$). In particular, the measured rms vorticities were consistently smaller than the published DNS values (e.g. Kim et al., 1987). The main objective of the present work is to simulate the probe using a DNS database for this flow. This approach allows the effect of the separation between hot wires to be readily assessed and can therefore guide possible improvements to the probe dimensions.

2. Measurement of vorticity and experimental conditions

The components ω_2 and ω_3 of the vorticity vector were measured separately using a four hot-wire vorticity probe sketched in Figure 1 consisting of a pair of parallel wires c and d straddling wires a and b of an X-wire. The quantities ω_2 and ω_3 are evaluated as

$$\omega_2 = \frac{\Delta u_1}{\Delta x_3} - \frac{\Delta u_3}{\Delta x_1} = \frac{\Delta u_1}{\Delta x_3} + \overline{U}_1^{-1}\frac{\Delta u_3}{\Delta t} \quad (1)$$

$$\omega_3 = \frac{\Delta u_2}{\Delta x_1} - \frac{\Delta u_1}{\Delta x_2} = -\overline{U}_1^{-1}\frac{\Delta u_2}{\Delta t} - \frac{\Delta u_1}{\Delta x_2} \quad (2)$$

where $\overline{U}_1$ is the local longitudinal mean velocity; Δu_1 is the difference between the longitudinal velocity fluctuations from two parallel hot wires which are separated in either x_2 or x_3 directions; Δu_2 (or Δu_3) represents the difference between values of u_2 (or u_3) at the same point but separated

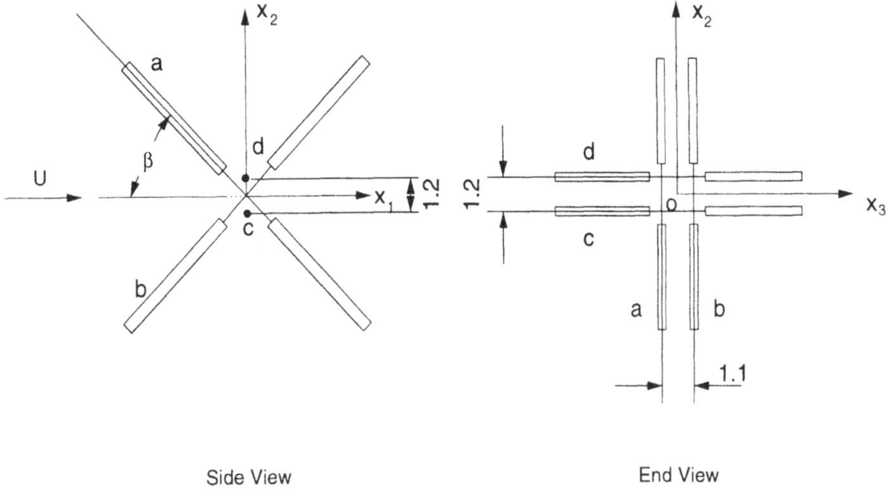

Figure 1. Sketch of the actual one-component vorticity probe.

by a time interval Δt (or may be considered as separated in space by $\Delta x_1 = \overline{U}_1^{-1}\Delta t$). Δu_2 (or Δu_3) was obtained by a central difference scheme to minimize phase distortion. The separations Δx_2 and Δx_3 were chosen to be about 1.2 mm and 1.1 mm, corresponding to about 10 and 11 wall units respectively. The measurements were made in a fully developed turbulent channel flow at a Reynolds number of $Re = 3300$ ($Re = U_0 h/\nu$, h is the channel half-width and U_0 is the mean velocity at the centreline, $U_0 = 2.5$ m/s for this experiment). Further experimental details can be found in Zhou *et al.* (1997).

3. DNS

The simulation was carried out using a second-order finite difference scheme which conserves energy in the inviscid limit. This scheme does not differ significantly from the fractional step method developed by Kim and Moin (1985) for incompressible laminar flows. A finite difference scheme was used instead of a pseudospectral method since it allows an arbitrary coordinate transformation in the normal direction. With a transformation based on the hyperbolic tangent and 150 grid points, the first grid point was at $x_2^+ = 0.4$. Five points lie within 5 wall units from the wall. At the centre of the channel, the resolution was about 3 wall units. One aim of the present simulation was to gain some insight into the poor performance of the vorticity probe close to the wall, while in a previous study it was proved that the probe behaves well around the centre of the channel. For this reason, the simulation was performed in a relatively narrow channel with a spanwise dimension of

$\pi/2$, which is equivalent to about 280 wall units. By using 64 cells, the resolution was about $\Delta x_3^+ \approx 4$. In the streamwise direction, the resolution was doubled. By using a length equal to 4π and 256 cells, the resolution was $\Delta x_1^+ \approx 8$. The resolution at $x_2^+ = 20$ in the normal direction was $\Delta x_2^+ \approx 1.4$.

4. Performance checks

To mimic the vorticity probe on the numerical simulation temporal velocity signals at five locations in the channel were stored for the successive vorticity calculation. Velocity information in enough points around the selected position was stored to allow the evaluation of the vorticity with different grids spacing. In fact, in this study the aim is not to perfectly reproduce the vorticity probes described in Figure 1, but to investigate how the vorticity values are sensitive to the gap between the hot-wires used to evaluate the velocity gradients in each term of the vorticity components. Moreover since it was found difficult to measure the ω_1 streamwise component, we limited our study to identifying the most important contributions to the vorticity components ω_2 and ω_3 given in Eqs. (1) and (2). We define P_1, P_3 and P_5 as "numerical vorticity probes" respectively evaluated with Δx_i, $3\Delta x_i$ and $5\Delta x_i$. This is equivalent to a vorticity probe with $\Delta x_1^+ \approx 8, 24, 40$, $\Delta x_2^+ \approx 1.4, 4, 7$ and $\Delta x_3^+ \approx 4, 12, 20$. In our finite difference scheme, the velocities are located at the centre of the cell faces and the vorticity on the sides of the cell. Standard centred schemes allow in a very simple way to evaluate the vorticity and is is therefore not necessary to show a sketch of the grid. We describe the performance of these probes at $x_2^+ = 20$ so as to understand why the experimental vorticity probe underestimated some quantities near the wall. Figures 2–3 show ω_2' and ω_3', the rms values of ω_2 and ω_3 compared with the numerical results by Kim *et al.* (1987) and with the experimental data by Zhou *et al.* (1997). ω_2' (Figure 2) is strongly affected by the resolution of the probe. P_3 and P_5 yielded values of ω_2' 14% and 30% smaller than that obtained with P_1 and data of Kim *et al.* (1987). The experimental data of Zhou et al. (1997) show quite good agreement with data from P_5. ω_3' (Figure 3) is less sensitive to the spatial resolution of the vorticity probes. In fact, P_3 and P_5 register reductions of 3% and 6% only.

The discrepancies in the evaluation of the rms vorticity fluctuations can be better understood by examining the time history of the vorticity components. Figure 4(a) shows a time record of 20 nondimensional units, corresponding to 146 wall units, for ω_2. P_3 and P_5 smooth the high peaks detected by P_1. However it could be argued that a further increase in the resolution of the probe could lead to higher peaks. This could be verified

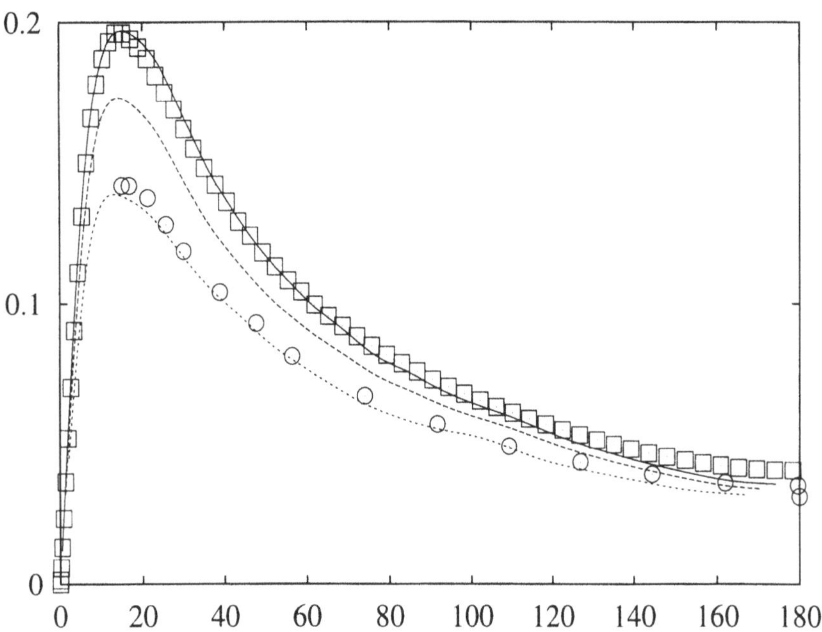

Figure 2. Distribution of ω_2'. Normalization is by wall coordinates. —— P_1; — — P_3; - - - - - P_5; □ DNS by Kim *et al.* (1987); ○ experiments by Zhou *et al.* (1997).

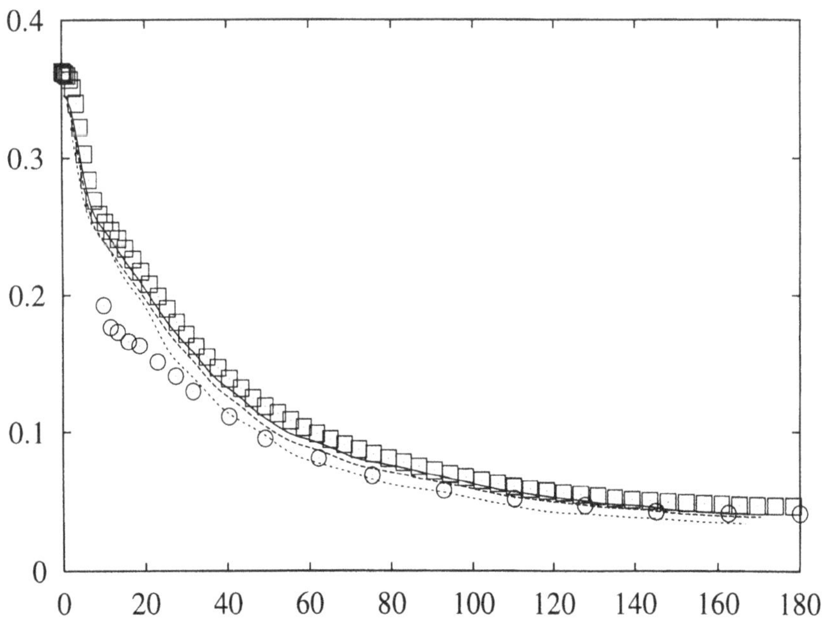

Figure 3. Distribution of ω_3'. Normalization is by wall coordinates.. —— P_1; — — P_3; - - - - - P_5; □ DNS by Kim *et al.* (1987); ○ experiments by Zhou *et al.* (1997).

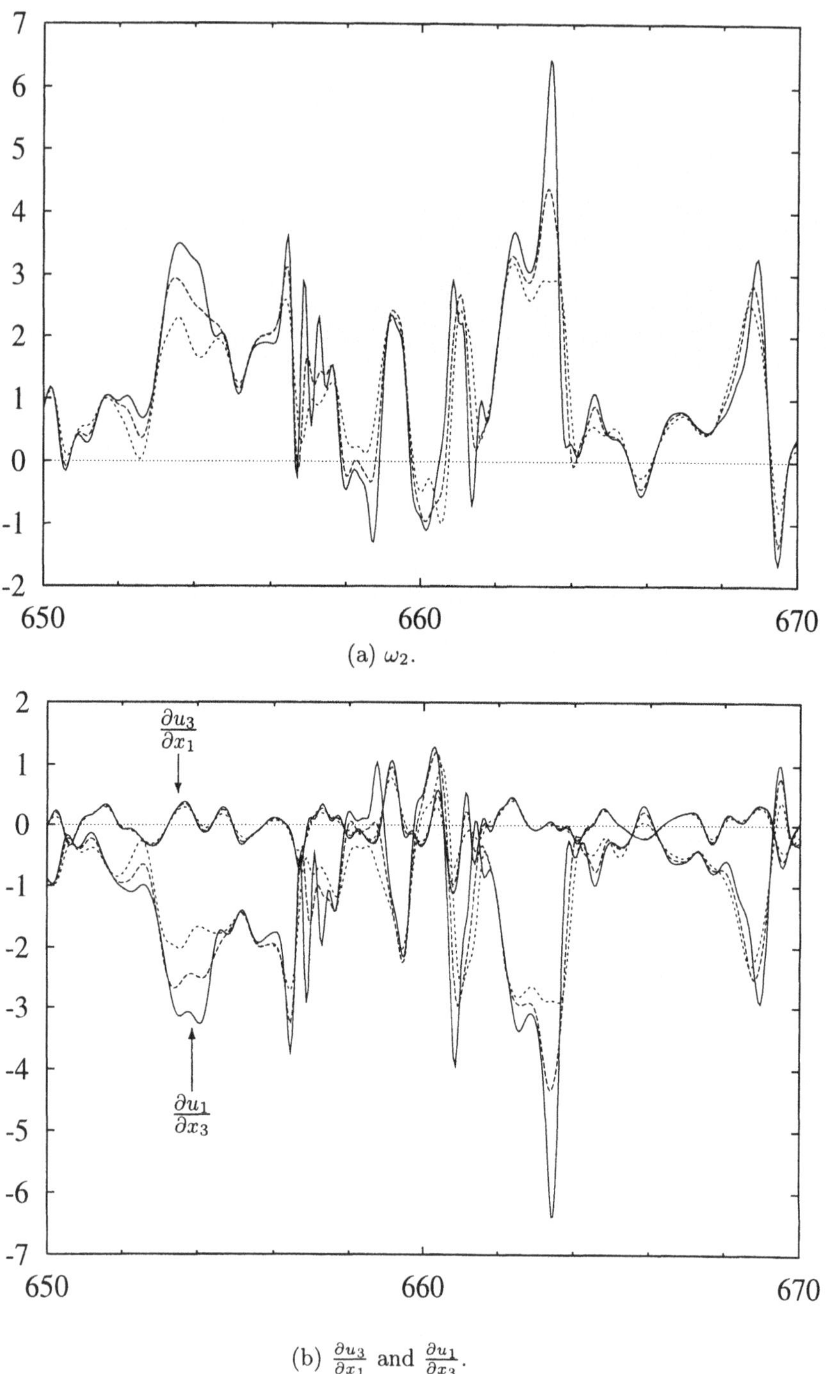

Figure 4. Time histories of ω_2 and its derivatives: —— P_1; — — P_3; - - - - - P_5.

only by a finer simulation, but we believe that vorticity peaks should not grow significantly since the rms values are in good agreement with the data obtained by Kim *et al.* (1987) with a spectral code.

Figure 4(b) shows the time histories of the velocity derivatives that make up ω_2. It appears that the streamwise derivative $\partial u_3/\partial x_1$ provides a small contribution to the overall value of ω_2. As shown in Eqs. (1) and (2), streamwise derivatives are evaluated by Taylor's hypothesis in the experiment. The DNS results suggest that the vorticity components should not be affected by the use of Taylor's hypothesis. On the other hand, the most important term in the evaluation of ω_2 is $\Delta u_1/\Delta x_3$. Figures 5(a)–(b) show the time histories for ω_3 and its component derivatives. ω_3 seems less sensitive to the resolution of the probe. In fact, the vorticity peaks are well resolved even with P_5. Also for ω_3, the most relevant contribution comes from the derivative in the wall normal direction, whereas the contribution from the streamwise derivative is small.

Figures 6(a)–(b) show vorticity spectra in terms of frequency. The plots denote that the main features of the spectra are well captured by the P_3 and P_5 probes. However, Figures 6(c)–(d) show that the probe size causes an attenuation that is nearly constant at low wavenumbers. At intermediate wavenumbers the attenuation is stronger and reaches its maximum around $k \approx 4$. Even for the spectra the effect of probe size is more evident in the ω_2 component than ω_3. In fact the reduction of the ω_2 spectrum is around 0.85 at low wavenumbers for P_3 and 0.7 for P_5, while the ω_3 spectrum has an attenuation of not more than 0.9 for both P_3 and P_5

5. Conclusions

The DNS results imply that the actual vorticity probe (in the ω_2 mode) cannot yield accurate results in the wall region especially because the spanwise resolution is small and the rms value of $\partial u_1/\partial x_3$ is consequently underestimated. The DNS rms vorticity distributions show that the actual probe behaves like P_5 close to the wall. The smaller resolution of P_3 and P_5 result in a spectral attenuation, whose magnitude is practically independent of the wavenumber at low and high wavenumbers with a peak around $k \approx 4$.

6. Acknowledgements

PGE has been partially supported by *Ministero dei Trasporti e della Navigazione*, in the frame of INSEAN research plan 1991-93.

RAA is grateful for the continuing support of the Australian Research Council.

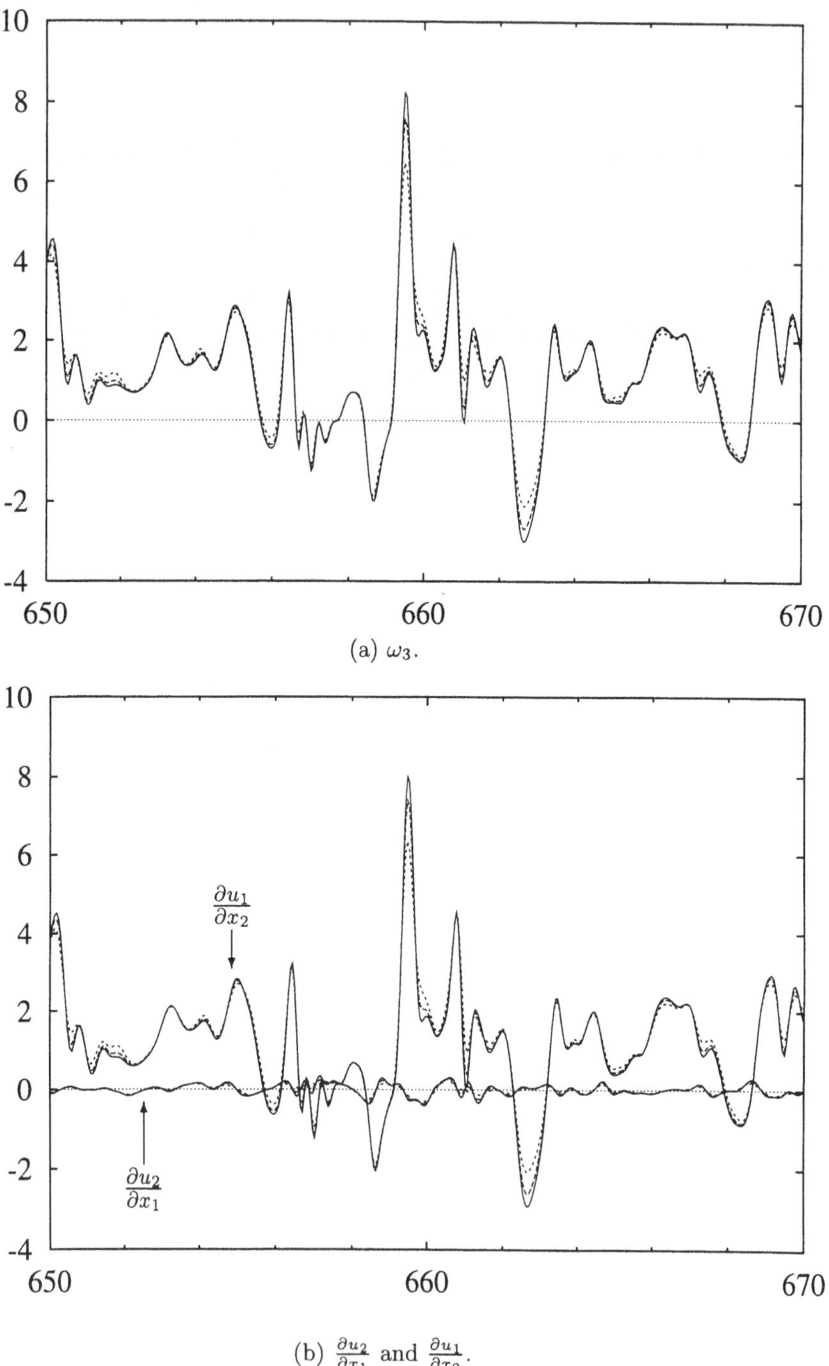

(a) ω_3.

(b) $\frac{\partial u_2}{\partial x_1}$ and $\frac{\partial u_1}{\partial x_2}$.

Figure 5. Time histories of ω_3 and its derivatives: —— P_1; — — P_3; - - - - - P_5.

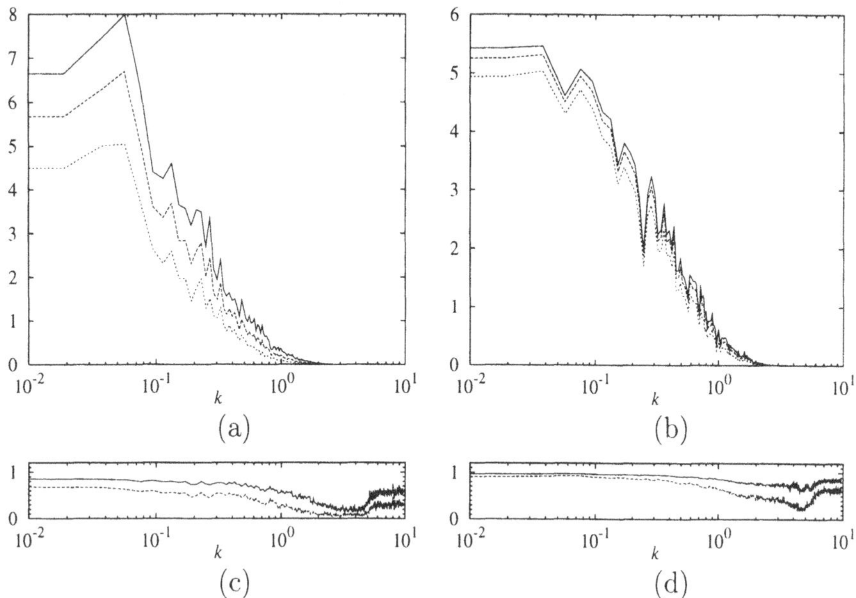

Figure 6. Top: power spectra of ω_2 (a) and ω_3 (b): —— P_1; — — P_3; - - - - - P_5. Bottom: spectrum attenuation factors for ω_2 (c) and ω_3 (d): —— P_3; — — P_5.

References

Kim, J., Moin, P. (1985) Application of a fractional-step method to incompressible Navier-Stokes equations, *J. Comp. Phys.* **59**, pp.219–238

Kim, J., Moin, P., Moser, R. (1987) Turbulence statistics in a fully developed channel flow at low Reynolds number, *J. Fluid Mech.* **177**, pp.133–166

Antonia, R. A., Rajagopalan, S. (1990) Performance of a lateral vorticity probe in a turbulent wake, *Expts. in Fluids* **9**, pp. 118–120

Wallace, J. M., Foss, J. F. (1995) The measurement of vorticity in turbulent flows, *Ann. Rev. Fluid Mech.* **27**, pp. 469-514.

Zhou, T., Antonia, R. A., Zhu, Y., Orlandi, P., Esposito, P.G. (1997) Performance of a transverse vorticity probe in a turbulent channel flow, to appear in *Expts. in Fluids*

Zhu, Y., Zhou, T., Antonia, R. A. (1997) Vorticity measurements in grid turbulence, *Proc. Eleventh Turbulent Shear Flow Conference*, Vol. 1, pp. 10-16 to 10-20

QUANTITATIVE PLANAR IMAGING OF LARGE STRUCTURES DEVELOPED THROUGH THE PRECESSION OF A JET

D.S. Nobes, G.J.R. Newbold
Department of Mechanical Engineering
Z.T. Alwahabi
Department of Chemical Engineering
University of Adelaide
Adelaide, Australia, 5005

1. Abstract

Two dimensional imaging by a Mie scattering technique is used to examine a complex three dimensional mixture field. The flow is generated by a mechanically rotated inclined jet, which is a well defined member of the family of precessing jet flows. This preliminary study of the effects of velocity (Reynolds number) and precession frequency (Strouhal number) on global characteristics of the mixing field, using phase conditional averaging of the concentration field. Instantaneous images show large structures of the order of the exit diameter to develop close to the jet exit. An underlying organisation of structures which are an order of magnitude larger than the exit diameter is also apparent. The phase averaged mixture field also shows the underlying organisation in the flow indicating consistent generation of these structures. The investigation shows that the position of the underlying structures and the development of the large scale turbulence are functions of the initial conditions.

2. Introduction

The Fluidic Precessing Jet (FPJ) has been the ongoing subject of investigation at the University of Adelaide since 1985 (Nathan, 1988). The FPJ nozzle is based on a naturally occurring flow instability which enhances large scale turbulence. The details of the nozzle geometry and of the flow within the nozzle can be found elsewhere (Nathan *et al*, 1997a). The origin of the jet which leaves the FPJ nozzle lies within the nozzle chamber where the time dependent flow field is instantaneously asymmetric at all times. This causes the emerging jet to be deflected relative to the geometric centreline of the nozzle and to precess about that centreline. Jet precession is the rotation of a characteristic jet vector about an axis other than its own. The mixing characteristics of the FPJ nozzle are beneficial to combustion performance and have resulted in it being adopted commercially as a gas burner, under the name 'GYRO THERM', in rotary cement, lime and alumina kilns in Australia and North America. The nozzle offers

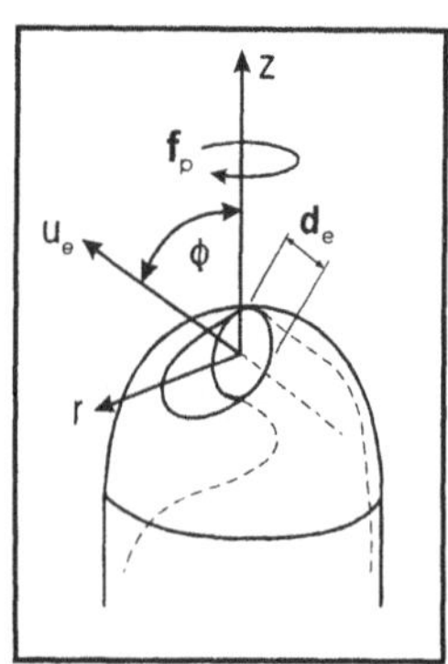

Figure 1 The exit tip of a Mechanical Precessing Jet nozzle

substantial reductions in NOx emissions, while maintaining low CO and reducing fuel consumption (Nathan *et al,* 1991, 1995,1996).

The development and application of the FPJ nozzle has been accompanied by a fundamental investigation of the effects of precession on the mixing characteristics of a jet based on studies of an analogous Mechanical Precessing Jet (MPJ) nozzle. For this investigation the flow is generated by a mechanically rotated, circular cross section jet which is inclined at an angle from the rotational axis. Figure 1 shows a schematic of the exit tip of the MPJ. The parameters which determine the mixing field, ie. the jet exit diameter (d_e), jet exit velocity (u_e), rotational frequency of precession (f_p) and jet deflection angle (ϕ) can each be selected independently for experiments with the MPJ. From these parameters, two dimensionless groups can be shown to characterise the exit conditions: the Strouhal Number ($St_p = f_p d_e / u_e$) and the Reynolds Number ($Re = u_e d_e / \nu$).

Velocity and pressure measurements of the cold flow mixing characteristics of the MPJ nozzle, whose emerging jet is inclined at an angle of 45^0 from the axis of rotation (Schneider *et al,* 1996a,b), have highlighted the significance of the dimensionless frequency (Strouhal number) of the precession. When the nozzle is rotated slowly, at a `low' Strouhal number (say, $St_p < 0.002$), the path of the jet deviates indiscernibly from its initial inclination and locally the jet resembles a fully pulsed jet (cf. Bremhorst *et al.* 1990). As the frequency is increased, pressure fields which are induced in the region of the nozzle tip cause the jet to be deflected both radially and azimuthally. At the higher Strouhal numbers the jet assumes a helical path with an overall diameter which is an order of magnitude larger than that of the nozzle itself. In this `high' Strouhal number flow regime (say, $St_p > 0.01$) mixing on the scale of the helix diameter takes place. There is a rapid decrease in both axial velocity and shear stress and the original jet completely loses its identity within some ten exit diameters from the nozzle. Recent experiments by Nathan *et al*, (1997b) with non-reacting flows have shown that `high' Strouhal number jet precession causes the energy in the turbulence spectrum to be concentrated in the larger scales in that region of the jet flow which corresponds to the base of a flame. This is at the expense of the finer scales.

While past investigations have established the ground work for investigating and describing the topology of the flow field, a new study to achieve, simultaneously, better visualisation and measurement of the mixture fraction field, is needed to determine the effects of the unreacting mixture field on the combustion. The present study is a first step in this direction.

3. Experimental Arrangement and Technique

The experimental technique outlined here allows the concentration of a conserved scalar to be measured over a two-dimensional plane. This allows statistical investigation of the concentration field concurrently with visualisation of the mixing field. The technique of imaging a flow by scattering light from small particles is generally known as Mie Scattering, Lorenz-Mie Scattering or Marker Nephelometry. A detailed description of the theory and experimental use of the technique is given by Becker (1977). The arrangement of equipment used for the present two-dimensional imaging system is shown in Figure 2. Air is introduced at ambient temperature into the nozzle which has a smooth contraction inlet designed to produce a 'top-hat' velocity profile at the exit. The nozzle is orientated vertically into quiescent air under an extraction hood with a very low extraction velocity. Care has been taken to minimise laboratory drafts. Laser pulses of <4nsec at 532 nm from an Nd:YAG laser are expanded by a cylindrical lens to form a thin (<0.25 mm) light sheet. The light sheet reflects from small seeded oil droplets of approximately 0.6μm which are introduced through the nozzle and the scattered signal light is captured by a slow scan, cooled CCD camera having a two-dimensional array of 576x384 pixels. The pixels in the array accumulate a charge which is linearly proportional to the intensity of the scattering signal. Once the image has been collected the array is read out and converted to a digital format, each pixel having 12-bit resolution. The image is transferred to the control computer via a GPIB interface which is also used for communication with and control of the detector. The slow scanning of the array ensures low noise and requires a low framing rate of approximately 12 frames per minute. Images are stored directly to the hard drive of a PC.

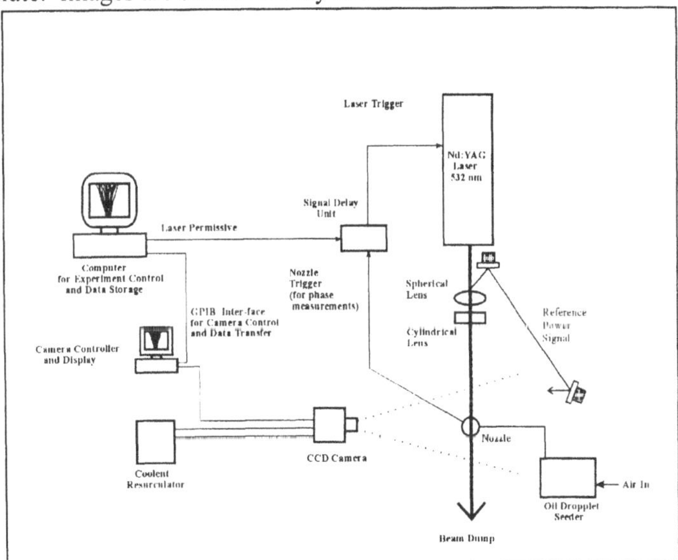

Figure 2 Arrangement of experimental equipment.

Sequences of between 600 and 1200 images have been collected for each experimental condition. For the reduction of data and the calculation of an ensembled averaged or a phase-averaged field, each image is corrected with a background subtraction of noise, a correction for pulse-to-pulse power variation in the laser power and a correction for spatial variations of intensity in the light sheet. To provide accurate calibration a reference signal from unmixed, seeded jet fluid in the potential core is taken for each image (C_0). Another signal for pure ambient fluid is also taken (C_∞). It is assumed that the collected intensity is linearly proportional to the concentration of the seeded fluid (C). Statistical processing of the corrected images has been performed to determine the mean mixture field. The mean mixture fraction $(\overline{C})$ is defined as:

$$\overline{C} = \frac{C - C_\infty}{C_0 - C_\infty} \times 100 \quad ; \qquad 0 \leq \overline{C} \leq 100.$$

The quantitative nature of the results is tested for a simple round jet by comparison with data from four other researchers who used a range of single point measuring techniques. The relevant techniques and experimental conditions of the other researchers is summarised in Table 1.

Table 1 Experimental conditions studied and techniques used by other researchers.

Author / Date	Jet/co-flow	Reynolds Number	Method
Becker *et al* (1967b)	air/air	54,000	Mie Scattering
Birch *et al* (1978)	nat.gas/air	16,000	Raman
Lockwood *et al* (1980)	air/air	50,000	Thermometry
Dowling and Dimotakis	C_2H_4/N_2	5,000	Rayleigh
(1990)	C_3H_6/Ar	16,000	
Present Data	air/air	16,400	Mie Scattering

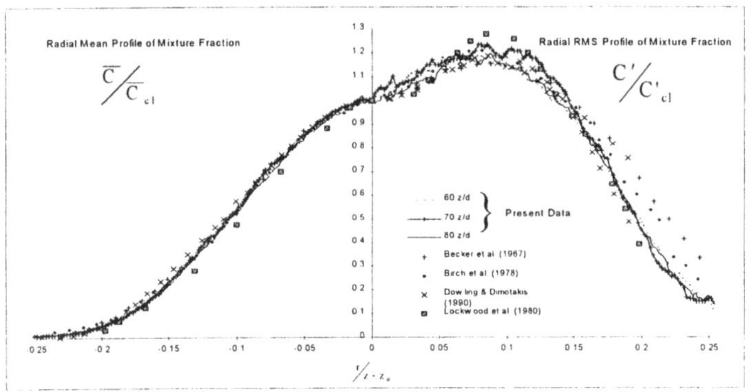

Figure 3 Comparison published data of the similarity of the mean $(\overline{C})$ and fluctuating (C') concentrations of the conserved scalar in the far field with.

Normalised, radial profiles of the mean, $\overline{C}/\overline{C}_{cl}$ and rms, C'/C'_{cl} concentration are shown in Figure 3. For the present simple jet radial profiles at the axial locations z/d = 60, 70 and 80 are given. The results show good agreement with the above- referenced data. This comparison supports the view that the data collection and reduction methodologies yield acceptable quantitative results.

The effect of jet precession on the mixing field is evident from comparisons over the range of combinations of Re and St_p for the MPJ. The particular MPJ used in the experiments has an exit diameter of ϕ3mm and deflection angle of 45°. These parameters remain unchanged in the present experiments. Images of the mixing field of a simple jet of diameter ϕ3mm are also presented for comparison. The conditions of each experiment are given in Table 2.

Table 2 Experimental conditions for the presented results

d_e (mm)	u_e (m/s)	f (Hz)	Re	St_p
3	94	0	16,400	0
3	114	80	20,000	0.0021
3	57	40	10,000	0.0021
3	57	100	10,000	0.005

4. Results and Discussion

4.1 The Simple Jet

An advantage of the two-dimensional imaging technique over the single point techniques is its ability to capture many points simultaneously in time so providing structural information about the flow. All of the images presented here have the same colour map representing nozzle fluid concentrations from 1-30%. Black regions represent nozzle fluid concentration >30%, and white, <1%. All axes are in millimeters.

An instantaneous, corrected image of the mixture field of the simple jet is shown in Figure 4. Note the sharp interface between ambient fluid and mixed fluid. This demonstrates that little ambient fluid is entrained into the flow. A stepwise decrease in concentration along the axis of the flow is also apparent. This appears to be associated with large scale motions in the shear layer of the jet flow and supports the findings of other researchers (Dahm and Dimotakis, 1987; Papanicolau and List, 1988). Fine-scale structures are evident throughout the flow.

In Figure 5 the mean two-dimensional concentration field of the jet shows that the mixture field is symmetrical about the axis of the nozzle. A potential core of nozzle fluid can readily be identified immediately downstream from the nozzle exit, along with the rapid 1/z decay in concentration as the flow proceeds downstream. Note the constant spreading angle of the flow. There is no evidence of structure in the mean mixture field, other than that in the jet itself.

4.2 The MPJ

A critical difference between the data presented for the simple jet and those for the MPJ is that for the MPJ the data is collected conditionally. A signal from a transducer attached to the rotating nozzle is used as a trigger to specify an azimuthal nozzle position. This allows images to be collected only when the nozzle is pointing in the required direction. The general shape of the flow field presented in the images is similar for all cases; this is a consequence of the phase conditional sampling method. A phase angle of 0° is used for the presented data; i.e. the jet is exiting in the plane of the light sheet. Thus for the images presented (Figures 6-9), the z-r plane (cylindrical coordinates) is shown with the jet entering at the bottom of the picture and deflecting to the left from the vertical axis about which the jet rotates. The developing jet is deflected in the azimuthal direction out of the thin light sheet. This is evidenced by the sharp cutoff of high concentration fluid and the presence of ambient fluid crossing the centre line of the flow field. Further downstream, equally spaced on either side of the axis of rotation, are the remnants of the rapidly decaying jet, the first structure at $\theta = 180°$ from the exit (on the right of image) and the second at $\theta = 360°$ (on the left of the image). The image suggests the light sheet has cut through the jet which has been deflected to form a "helix", with high concentration at the center of the structures. Beyond these points no features which are reminiscent of the original jet remain.

Figure 6 shows an instantaneous image with Re = 20,000 and $St_p = 0.0021$. The first structure is divided into two, which is a common feature in the instantaneous images. The pattern in the surrounding fluid suggests that the two objects are counter-rotating vortex tubes which draw ambient fluid into the centre of the structure. The second structure is not as well defined although there are remnants of the vortex tubes. The surrounding flow field shows smaller structures which are of the magnitude of the exit diameter, and unmixed ambient fluid which penetrates deep into the mixture field.

The regular location and shape of the large structures (the vortex tubes) can be seen in the phase averaged image shown in Figure 7. The "inner tube" of the first structure, closest to the spinning axis, has a higher mean concentration than the outer, a feature common to all Re / St_p conditions studied. The regular appearance of these phase averaged structures supports the notion that the exiting jet is deflected into a 'helix'. At the top of the image a return to a semblance of symmetry in the concentration gradients shows that the mixing field is becoming self similar and that the identity of the initial jet is being lost.

The effect of Reynolds number on the flow field can be seen by comparison of Figure 7 with Figure 8. In Figure 8 the Reynolds number has been halved to 10,000 while the Strouhal number has been held constant. The flow fields are remarkably similar. The axial and radial positions of the first and second structures are almost identical. The concentration of nozzle fluid is found in the first of the structures in each of the flows is 14-15%, and in each of the second structures is 8-9%. This indicates that the mixture field is not strongly dependent on the Reynolds number, in agreement with the velocity field of Schneider *et al.* (1996a).

Figure 8 and Figure 9 show the effect of Strouhal number on the mixing field. For the low Strouhal number case (Figure 8), $St_p = 0.0021$, the large structures are axially and radially further from both the jet origin and the axis of rotation than in the high Strouhal number case (Figure 9), $St_p = 0.005$. The concentration of original jet fluid within the structures is much higher in the high Strouhal number case than in the low Strouhal number case, indicating that the mixing rate has been reduced. The general character of the mixing field is largely unchanged. Increasing the Strouhal number has the effect of tightening the helix and increasing the concentration of jet fluid within the structures.

5. Conclusions

The planar, phase averaged Mie scattering technique provides new insights, both qualitative and quantitative, into the mixture field of a strongly 3-dimensional flow. The effect of precession on the flow of a jet is to cause the jet to be deformed into a helix.

The effect of Reynolds number on the mixture field is weak but the Strouhal number has a major influence on both the shape of the 'helix' and the concentration of jet fluid in the 'helix'. It is apparent that precession can be used to provide a measure of control over the characteristics of the mixture field. For a reacting / combusting system this suggests that the mixing characteristics in the region of flame stabilisation could be "tailored" by precession to achieve desirable flame characteristics.

6. Acknowledgments

The authors are members of the Combustion Research Group in the Faculty of Engineering of the University of Adelaide led by Professor R.E. Luxton and Dr. G.J. Nathan of the Department of Mechanical Engineering and by Associate Professor K.D. King and Dr. D-K Zang of the Department of Chemical Engineering. The Group interacts strongly on all projects and the present authors acknowledge the many inputs from other members of the team. Also acknowledged is the support of the Australian Research Council through both direct and collaborative research grants, and of the industrial collaborators in the program, Fuel and Combustion Technology International.

7. References

Becker, H.A., Hottal, H.C. and Williams, G.C. 1967(a) On the Light-scattering Technique for the Study of Turbulence and Mixing, *J. Fluid Mech.* Vol 30 pp 259-284

Becker, H.A. 1977. Mixing, Concentration Fluctions, and Marker Nephelometry. *Studies in Convection*, Vol2, editied by B.E. Launder, Academic Press, New York, pp. 45-139

Birch, A.D., Brown, D.R., Dodson, M.G. and Thomas, J.R., 'The turbulent concentration field of a methane jet' *J. Fluid Mech.* Vol 88 pp431-449 (1978)

Bremhorst, K. and Hollis, P.G (1990) "Velocity Field of an Axisymmetric Pulsed, Subsonic Air Jet", *AIAA Journal*, 28, (12), 2043-2049.

Dahm, W.J.A, and Dimotakis, P.E. (1987). "Measurements of Entrainment and Mixing in Turbulent Jets", *AIAA Journal*, vol. 25(9), pp. 1216-1223.

Dowling, D.R. and Dimotakis, P.E., 'Similarity of the Concentration Field of Gas-Phase Turbulent Jets' *J. Fluid Mech*, Vol 218, pp. 109-141 (1990)

Lockwood, F.C. and Moneib, H.A. 1980. Fluctuating temperature measurements in a heated round free jet *Comb. Sci. Tech.* Vol22, 63-81.

Nathan, G.J., (1988) "The Enhanced Mixing Burner", PhDThesis, Dept. Mech. Eng., the University of Adelaide"

Nathan, G.J., Manias, C.G. and Luxton, R.E., (1991) "Potential Increases in the Efficiency of a Rotary Kiln using an Enhanced Mixing Burner", IEAust Int Mech Eng Congress, (July), Sydney, Aust., Vol 4, pp 58-61.

Nathan, G.J. and Manias, C.G., (1995) "The Role of Process and Flame Interaction in Reducing NOx Emissions", Combustion and Emissions Control, pp 309-318, ISBN 0 90 259 7493.

Nathan, G.J., Turns, S.R. and Bandaru, R.V. (1996), "The Influence of Jet Precession on NO_x Emissions and Radiation from Turbulent Flames", *Combust. Sci. and Technol.*, 112, 211-230.

Nathan, G.J., Hill, S.J and Luxton, R.E., (1997a), "An Axisymmetric Fluidic Nozzle to Generate Jet Precession for Enhanced Large Scale Mixing", *J. Fluid Mech.*, accepted.

Nathan, G.J., Nobes, D.S., Mi, J., Schneider, G.M., Newbold, G.J.R., Alwahabi, Z.T. Luxton, R.E. and King, K.D. (1997b) "Exploring the Relationship between Mixing, Radiation and NO_x Emissions from Natural Gas Flames". Invited Lecture to be presented, *Combustion and Emissions Control*, June, Bath, UK.

Papantoniou, D., and List, E.J. (1989). "Large-scale Structure in the Far Field of Buoyant Jets", *J. Fluid Mech.*, vol. 209, pp 151-190.

Schneider, G.M., Froud, D., Syred, N., Nathan, G.J. and Luxton, R.E. (1996a) "Velocity Measurements in a Precessing Jet Flow using a Three Dimensional LDA System", Experiments in Fluids, (in press)

Schneider, G.M., Nathan, G.J., Luxton, R.E., Hooper, J.D. and Musgrove, A.R., (1996b), "Velocity and Reynolds Stresses in a Precessing Jet", Experiments in Fluids, (in press).

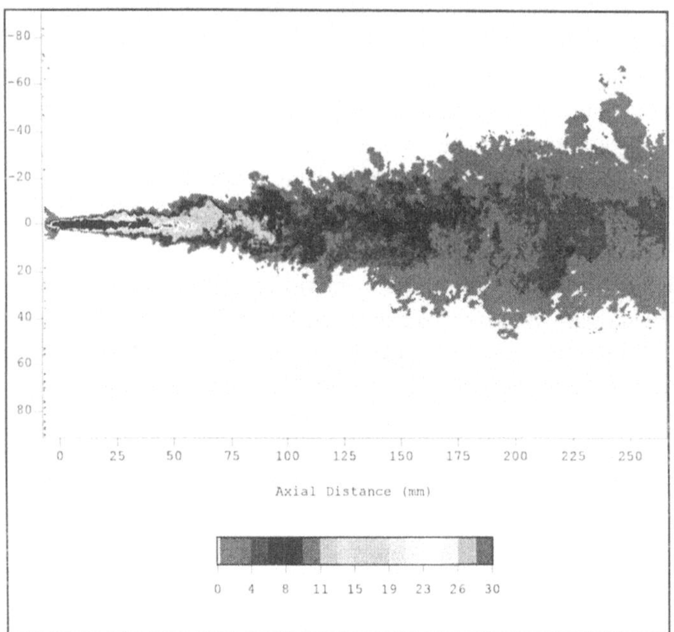

Figure 4 An instantaneous image of the simple round turbulent jet. Re=16,400. (All of the images presented here have the same colour map representing nozzle fluid concentrations of 1-30%. Black region represent nozzle fluid concentration >30% and white <1%. All axis are in mm.)

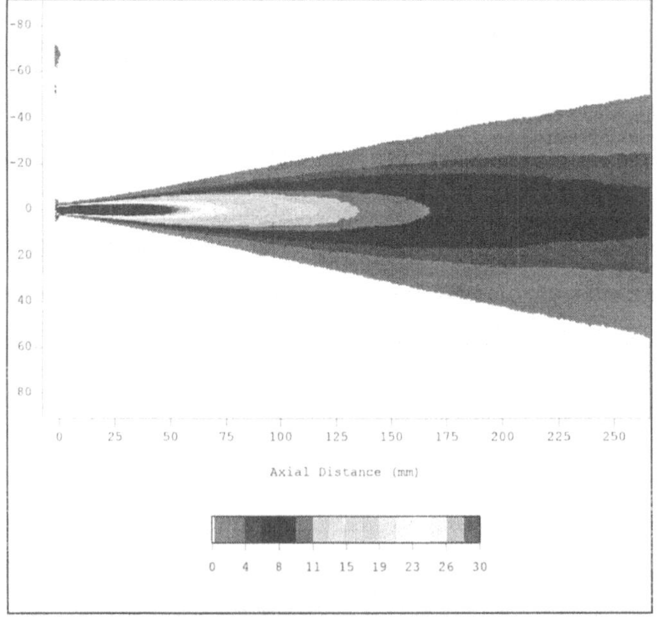

Figure 5 The mean concentration field of a simple round turbulent jet. Re= 16,400.

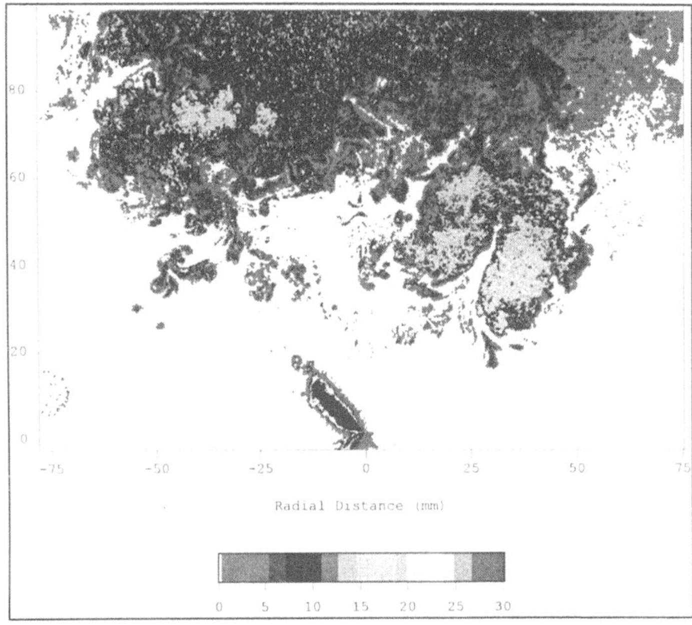

Figure 6 An instantaneous image of an MPJ flow. Re = 20,000, $St_p = 0.0021$

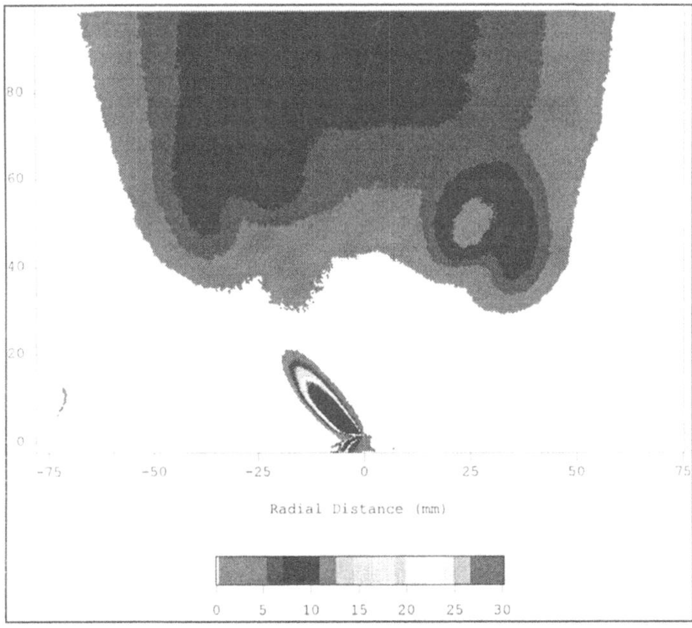

Figure 7 The mean flow field of the MPJ. Re = 20,000 $St_p = 0.0021$

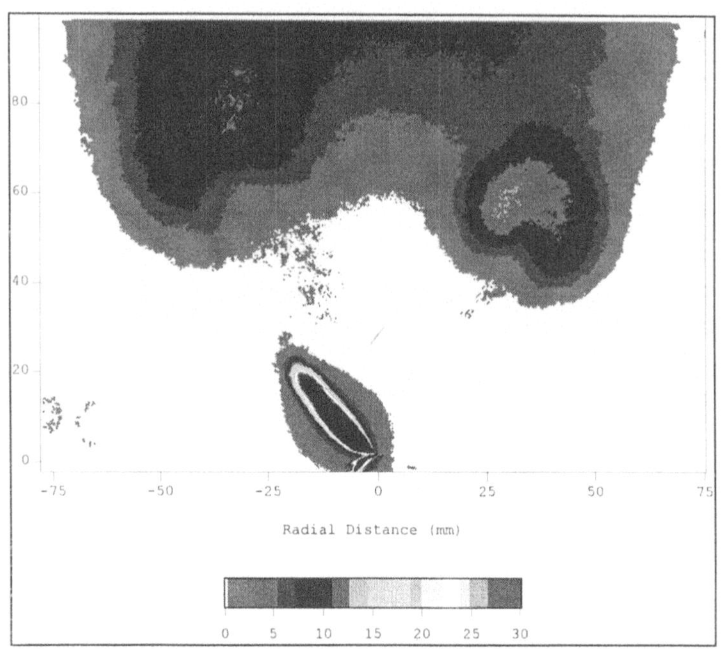

Figure 8 A low Strouhal/ Reynolds number case : Re = 10,000 St_p = 0.0021

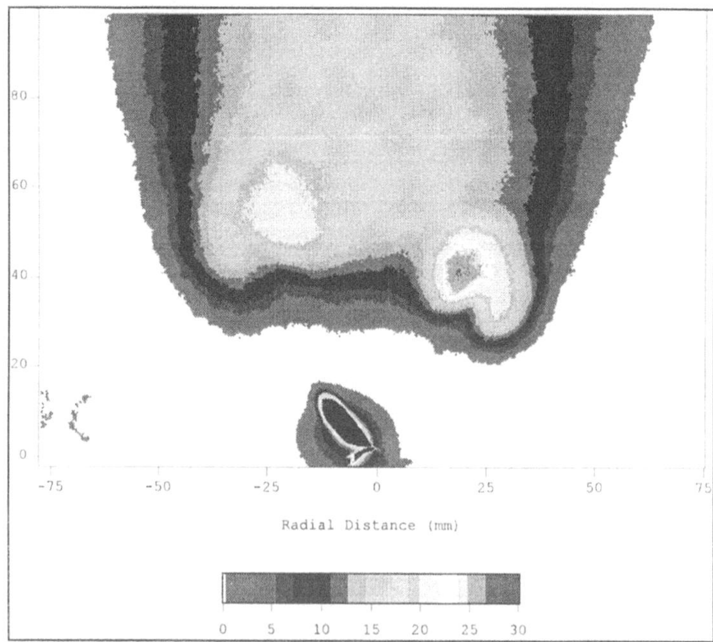

Figure 9 A low Reynolds, high Strouhal number case. Re = 10,000 St_p = 0.005

VIII. Vortical Structures

FORMATION OF VORTEX RINGS IN HELICOPTER ROTOR FLOW FIELDS

O. INOUE AND Y. HATTORI
Tohoku University, Sendai 980-77, Japan

Abstract. Unsteady flow fields of a model helicopter rotor are studied by direct numerical simulations of the three-dimensional, compressible Euler equations. "Vortex ring state" is captured numerically for the case of a descent velocity close to the average downward velocity in hover, and the existence of a vortex ring in this state is confirmed. The vorticity of the vortex ring is found to be supplied by tip vortices. The aerodynamic characteristics are in good qualitative agreements with experiments.

1. Introduction

In descending flight of a helicopter, a rotor descends into its own wake, and the blade tip vortices may interact with other blades (blade-vortex interaction, BVI). When the descent velocity of the rotor (v_d) is close to the average induced velocity of the rotor in hover (v_h), i.e., $v_d \simeq v_h$, the rotor may experience highly unsteady flow. This state, known as the vortex ring state, can lead to a significant loss of rotor control. A number of studies on these states have been performed and, for recent comprehensive reviews, readers are referred to McCroskey,[1] Leishman and Bagai,[2] Tung *et al.*[3] and Conlisk,[4] among others.

Experimental studies on low speed descending flight including the vortex ring state have been reported extensively, but relatively few experiments have documented the quantitative measurements of the unsteady aerodynamic characteristics.[5-7] One of the main reasons for this is that BVI is pronounced in descending flight and thus experimental measurements are difficult; the tip vortex dynamics in descending flight, especially in the vortex ring state, are not well understood, as noted by Leishman and Bagai.[2]

A number of computational works of model helicopter rotor flow fields have been reported. Most of the previous works adopted wake models[1] either for the entire flow field or for the far field. With improved performance of supercomputers as well as numerical algorithms in recent years, two-dimensional direct numerical simulations of the entire flow field without specifying any wake models have been practicable, and three-dimensional direct numerical simulations are beginning to emerge, using Euler and Navier-Stokes flow solvers.[8–10] Most of the computational works were focused on hover or forward flight, and computational works of the vortex ring state are very few.

In this paper unsteady flow fields of a model helicopter rotor are studied by direct numerical simulations using the three-dimensional, Euler equations. One of the main purposes of this study is to clarify the qualitative natures of the vortex ring state and to increase our understanding of the unsteady flow field of a helicopter rotor.

2. Numerical Methods

We consider a model helicopter rotor consisted of two blades. The origin of the coordinate system (x, y, z) employed in this study is located at the center of the two blades (see, Fig. 1). The coordinate system moves with the rotor (moving cell system), and we assume that for descending flight the rotor moves toward the negative z direction. The geometry of a model rotor adopted in this study is the same as that used in the experiment of hover by Caradonna and Tung.[11] The rotor employs two blades, each of which uses NACA 0012 profile and is untwisted and untapered. The aspect ratio is 6. The hub radius is equal to the chord length C. The hub assembly is omitted in this study, for simplicity.

The three-dimensional, compressible Euler equations are solved by a moving-cell finite volume method in an inertial frame.[10] An H-H type body-conforming cell system is adopted in this study (Fig. 1), in order to calculate the flow in the hub region. The computational domain is $-18 \le x/C \le 18, -18 \le y/C \le 18$, and $-15 \le z/C \le 15$. The number of cell is 152 in the chordwise direction (x) with 50 on each blade, 140 in the spanwise direction (y) with 30 on each blade, and 60 in the axial direction (z).

Throughout this study, the tip Mach number is fixed to be $M_{tip} = 0.44$, and the collective pitch angle is to be $\theta_c = 8,^\circ$ both of which are the same as those used in the experiment of hover by Caradonna and Tung.[11] The rate of descent of a rotor is prescribed to be $v_d/v_h = 0.0$ (hover),0.4,0.7,1.0,1.5 and 2.0. Only vertical descent is considered in this paper.

3. Results and Discussion

The chordwise surface pressure distribution and the spanwise sectional lift distribution for the case of hover ($v_d = 0$) are presented in Fig. 2 and Fig. 3, respectively, together with the experimental results (symbols) of Caradonna and Tung.[11] In Fig. 3, the sectional lift coefficient is defined as,

$$C_l = \frac{dl}{\frac{1}{2}\rho_\infty (r\Omega)^2 C dr} \tag{1}$$

where dl is the local sectional lift for the infinitesimal spanwise length dr and Ω is the angular velocity of the rotor. The symbol r denotes the distance from the rotor axis. Also presented in Fig. 2 are the thin-layer Navier-Stokes results (dashed lines) of Srinivasan *et al.*[8] We can see from Figs. 2 and 3 that the present results are in good overall agreements with the experiment as well as the Navier-Stokes results for all radial stations.

A typical example of a flow field in the vortex ring state is presented in Fig. 4 in which shown are instantaneous contour lines of vorticity in different planes of the blade rotation angle β around the rotor axis. The solid lines denote the anti-clockwise rotation while the dashed lines the clockwise rotation. In the figure, the numerals "1" and "2" denote each of the two blades, respectively. The $\beta = 0°$–plane coincides with the (y, z) plane, in which at this instant both the blade "1" and the blade "2" are located, as shown in Fig. 4(a). The location of the blade "1" is assumed to be $\beta = 0°$, while that of the blade "2" to be $\beta = 180°$. The direction of an increasing positive β is opposite to the sense of rotation of the rotor (Fig. 5). In the present rotor model, vortices are produced not only at the blade tips (hereafter referred to as tip vortices) but also at the edges of the root side ($r/C = \pm 1.0$) of the blades (referred to as root vortices). The sense of rotation of the root vortices is opposite to that of tip vortices; the root vortices induce the downward velocity on the blade side while the upward velocity in the hub region. The vorticity of the root vortices is weaker than that of the tip vortices. In addition to the tip and root vortices, the rotational motion of the blades at a finite collective pitch angle also produces the chordwise vorticity along the span (referred to as span vortices), because the rotational motion induces the downward velocity component $-w$ which varies along the span. Thus, the flow field around the rotor is characterized by three different kinds of chordwise vorticity (tip, root and span vortices), in addition to the spanwise vorticity emanated from the trailing edges of the blades. The sense of rotation of the span vortices is the same as the root vortices while opposite to the tip vortices. The so-called inboard sheets may consist of chordwise span vortices and spanwise vortices. In each β-plane of Fig. 4, the symbols TV_1 and TV_2

denote the tip vortices generated from the blade "1" and the blade "2", respectively, when the blades were located at the β-plane, and TV_1' and TV_2' are the older tip vortices generated at a half-period previous time. For example, in Fig. 4(a) TV_2' near the blade "1" was generated from the blade "2" at a previous time when the blade "2" was located at $\beta = 0°$, while TV_1' near the blade "2" was generated from the blade "1" at a previous time when the blade "1" was located at $\beta = 180°$. The root vortices (RV_1 and RV_2) as well as the span vortices (SV_1 and SV_2) are also denoted in the similar fashion. By examining the sequential variation of the flow field in a different β-plane according to the increasing β, we can follow approximately the time-development of the flow field. From Fig. 4, we can readily see the formation of a vortex ring (denoted by VR in Fig. 4) whose center is located slightly outside and also upper-side ($z > 0$) of the rotor blades. It is also seen from the figure, by tracking the locations of TV_1' and TV_2', that with an increasing β (thus with an increasing time) the tip vortices TV_1 and TV_2' (, and TV_2 and TV_1',) tend to merge and form a part of the vortex ring, indicating that the vorticity of the vortex ring is continuously supplied by the tip vortices. (For the case of hover, the tip vortices are swirling down toward the negative z-direction and thus do not form a vortex ring.)

Distributions of the mean vertical velocity component, w, against the distance r from the rotor axis are presented in Fig. 6 for various rates of descent. The velocity was measured at three different z-planes below the blades; $z/R = -0.12, -0.24$, and -0.47. Five measurement points in each z-plane are chosen, taking into account the experiment of Azuma and Obata,[6] such that the distance of each measurement point from the rotor axis is $r/C = 1.9, 2.7, 3.6, 4.4$ and 5.2, respectively. As seen from Fig. 6(a), for a low rate of descent at which a vortex ring does not appear, the downward velocity (i.e., $-w$) is very small near the tip region ($r/C > 5.0$ in Fig. 6(a)), increases with a decreasing r, takes its maximum in a midway region of the span ($3.0 < r/C < 5.0$), and then decreases toward the hub region. The velocity distribution shown in Fig. 6(a) is typical for a low rate of descent.[6] On the other hand, when the vortex ring exists, the induced velocity by the vortex ring modifies the flow field significantly; as shown in Fig. 6(b), the downward velocity becomes larger with a decreasing distance from the center of the vortex ring; that is, the velocity becomes larger with an increasing radial distance r from the rotor axis and also with a decreasing vertical distance z from the blade, because the center of the vortex ring in this case is located at a position of $r/C \simeq 8.5$ and $z/C \simeq 1.1$ which is slightly outside and upper side of the rotor blades (see Fig. 4). Figures 6(c) and 6(d) show that with a further increase in the rate of descent the radial variation of the downward velocity becomes smaller because the effect of the vortex ring becomes smaller.

The authors owe much to Dr. N. Uchiyama, Mitsubishi Heavy Industries, Co., Mr. K. Akiyama, Kobe Steel, LTD., and Mr. S. Onuma, Institute of Fluid Science, Tohoku University, for their kind advice and technical assistance. The first author expresses his sincere thanks to Asako Inoue for her continuous encouragement. Computations were performed on the CRAY C916 at the Institute of Fluid Science, Tohoku University.

References

[1] McCroskey, W. J., "Vortex Wakes of Rotorcraft," AIAA paper 95-0530, Jan., 1995.

[2] Leishman, J. G. and Bagai, A., "Challenges in Understanding the Vortex Dynamics of Helicopter Rotor Wakes," AIAA paper 96-1957, June, 1996.

[3] Tung, C., Yu, Y. H. and Low, S. L., "Aerodynamics Aspects of Blade-Vortex Interaction(BVI)," AIAA paper, 96-2010, 1996.

[4] Conlisk, A. T., "Modern Helicopter Aerodynamics," Annual Review of Fluid Mechanics, Vol. 29, 1997, pp.515-567.

[5] Washizu, K., Azuma, A., Kōo, J. and Oka, T., "Experiments on a Model Helicopter Rotor Operating in the Vortex Ring State," Journal of Aircraft, Vol. 3, 1966, pp.225-230.

[6] Azuma, A. and Obata A., "Induced Flow Variation of the Helicopter Rotor Operating in the Vortex Ring State," Journal of Aircraft, Vol. 5, 1968, pp.381-386.

[7] Xin, H. and Gao, Z., "An Experimental Investigation of Model Rotors Operating in Vertical Descent," Proceedings of the 19th European Rotorcraft Forum, Cernobbio, IT, 1993.

[8] Srinivasan, G. R., Baeder, J. D., Obayashi, S. and McCroskey, W. J., "Flowfield of a Lifting Rotor in Hover: A Navier-Stokes Simulation," AIAA Journal, Vol. 30, No. 10, 1992, pp.2371-2378.

[9] Strawn, R. C., "Wing-Tip Vortex Calculations with an Unstructured Adaptive-Grid Euler Solver," Proceedings of the 47th Annual Forum of the American Helicopter Society, American Helicopter Society, 1991, pp.65-76.

[10] Uchiyama, N. and Inoue, O., "Numerical Study of Helicopter Rotor Flowfield," Proceedings of the Third JSME-KSME Fluids Engineering Conference, Sendai, Japan, 1994, pp.737-741.

[11] Caradonna, F. X. and Tung, C., "Experimental and Analytical Studies of a Model Helicopter Rotor in Hover," NASA TM-81232, 1981.

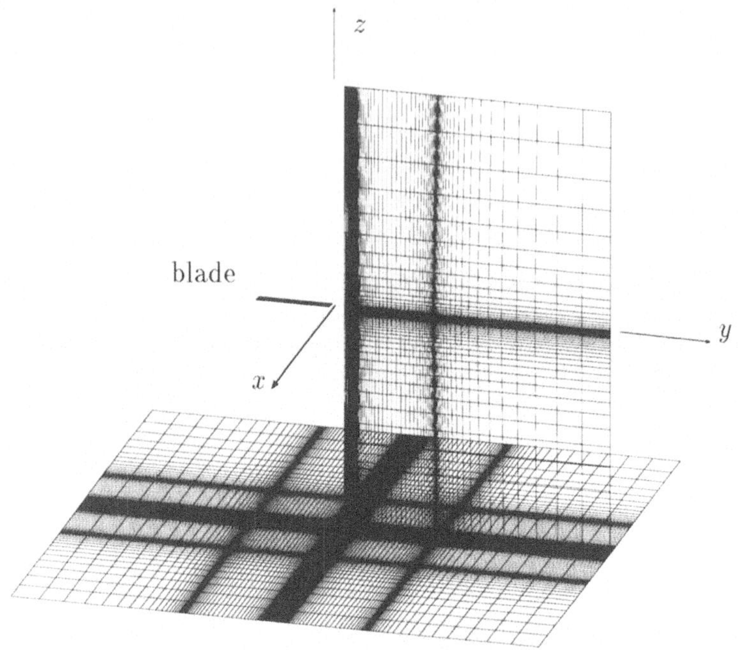

Figure 1: Coordinate system and H-H grid topology.

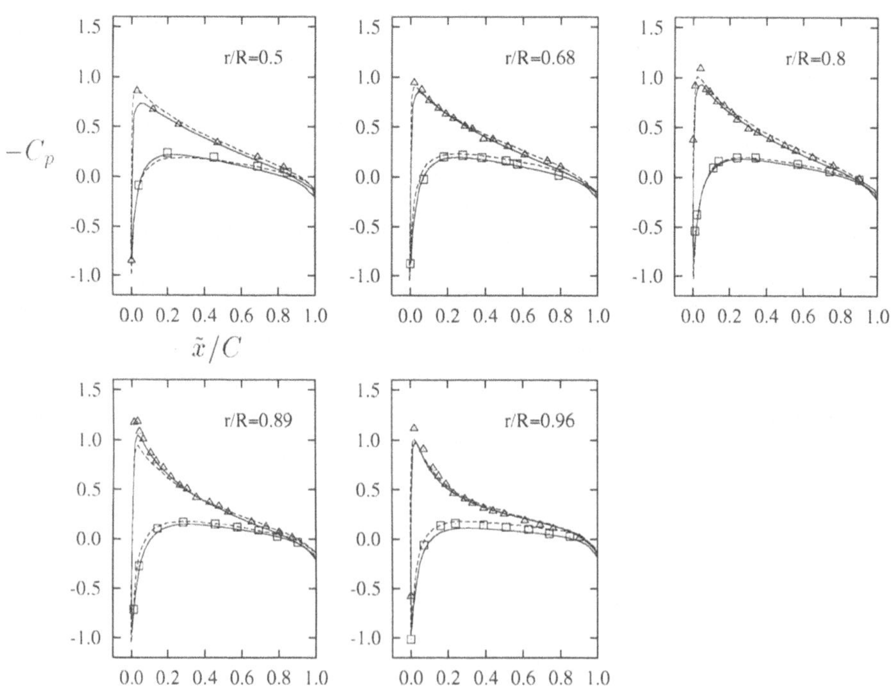

Figure 2: Chordwise surface pressure distributions for hover. $M_{tip} = 0.44, \theta_c = 8°$. ——— present, - - - - - thin-layer Navier-Stokes.[8] Symbols: experiment.[11]

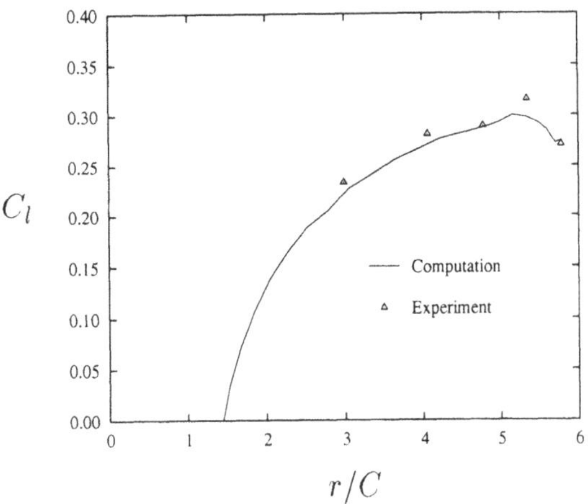

Figure 3: Spanwise sectional lift distribution for hover. $M_{tip} = 0.44, \theta_c = 8°$. ——— present, Symbols: experiment.[11]

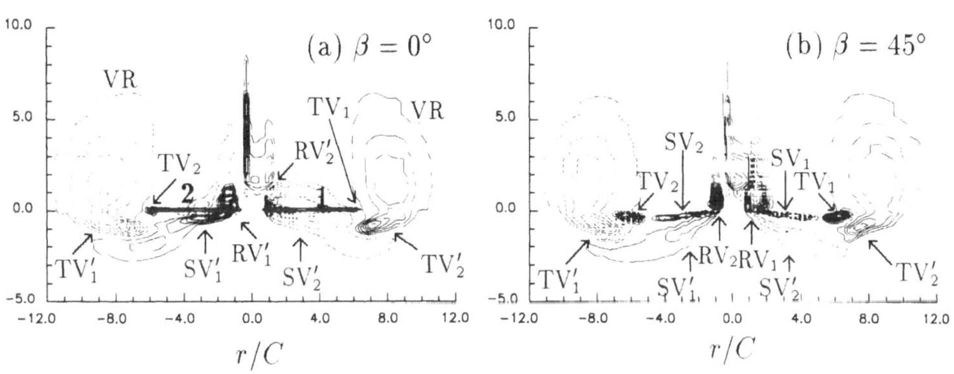

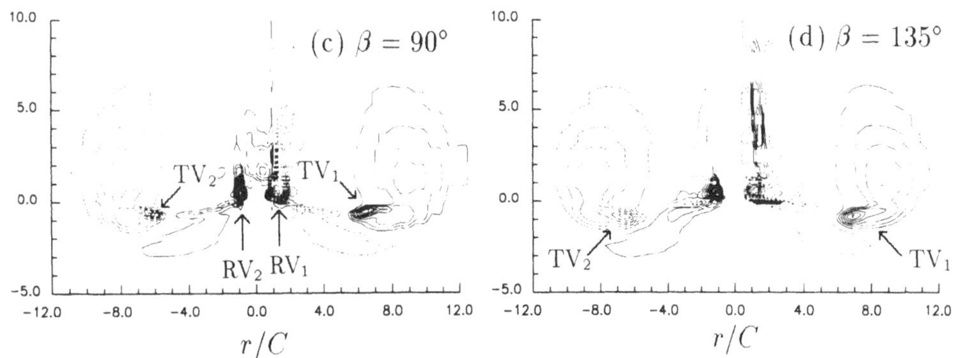

Figure 4: Flow field in the vortex ring state. Vorticity. $v_d/v_h = 1.0$. (a) $\beta = 0,°$ (b) $\beta = 45,°$ (c) $\beta = 90,°$ (d) $\beta = 135°$.

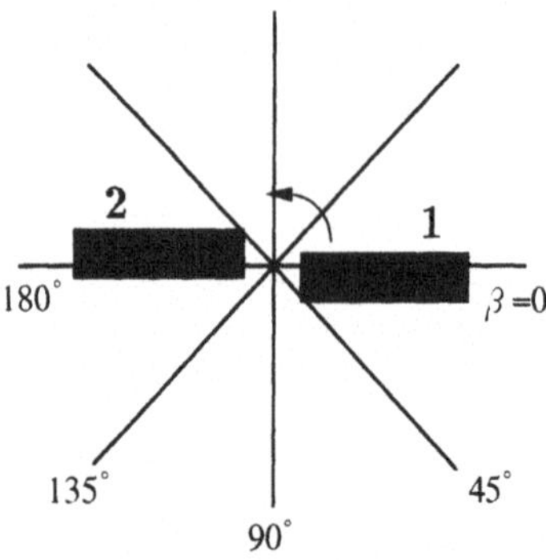

Figure 5: Schematic of blade rotation.

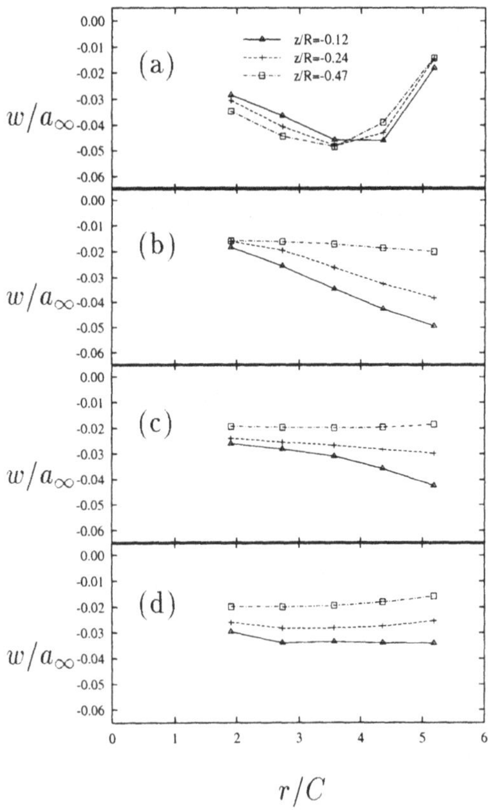

Figure 6: Spanwise distributions of the mean vertical velocity component w. (a)$v_d/v_h = 0.4$, (b) $v_d/v_h = 1.0$, (c) $v_d/v_h = 1.5$, (d) $v_d/v_h = 2.0$.

THREE-DIMENSIONAL VORTICAL STRUCTURE AND DIFFUSION MECHANISM OF AN EXCITED RECTANGULAR JET

K. TOYODA and R. HIRAMOTO
Hokkaido Institute of Technology
Maeda 7-15-4-1, Teine-ku, Sapporo 006, JAPAN

1. Introduction

The vortex rings evolving in noncircular jets deform three-dimensionally due to their self-induced velocity [1], and the three-dimensional deformations are very effective to enhance jet diffusion [2]. To clarify the details of the phenomena, we need a farther study of the relation between the vortical structures and the diffusion mechanism. In the present work, the three-dimensionally complicated vortical structure in a rectangular jet was detected by the measurements of the phase-average fluctuating static pressure [3], and the measured velocity field was discussed in relation to the vortical structure, focusing on the diffusion mechanism.

2. Experimental Apparatus and Procedures

The wind tunnel used in the experiment is shown in Fig.1. The air was issues from a sharp-edged rectangular orifice of aspect ratio 4. The equivalent diameter De of the rectangular orifice is 50mm, the velocity Ue at the jet exit center is 4 m/s and the Reynolds number Re (=UeDe/ν) is 1.3×10^4. The equivalent diameter is defined as a diameter of a circular orifice with the area of the rectangular orifice. The jet was excited by a loudspeaker at the side of settling chamber of the wind tunnel. The excitation frequency was a quarter of the most amplified frequency f_n (=564 [Hz]) of the shear layer generated from the orifice edge, and the excitation intensity u'/Ue (u': the rms value of fluctuating velocity) at the jet exit center was 3%. The intensity was decided so that the periodic and stable interaction of vortices was evolved. Under the excitation the large-scale vortices generating at $f_n/4$ were paired.

The pressure probe to measure fluctuating static pressure is shown in Fig.2. The static pressure tube with four small holes is connected at the end to the condenser microphone. The structure and the dimension of the probe were

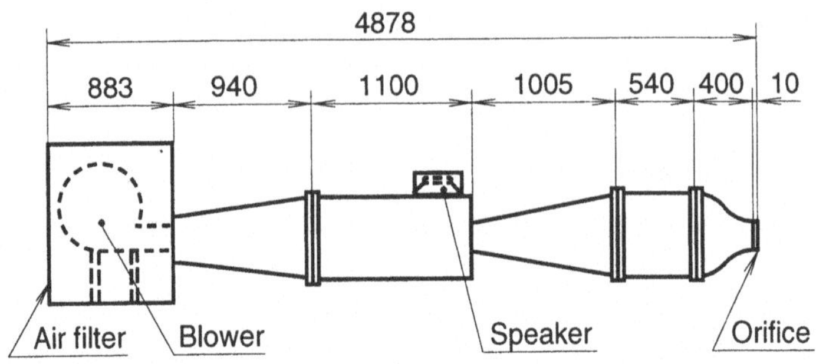

Fig.1 Wind tunnel

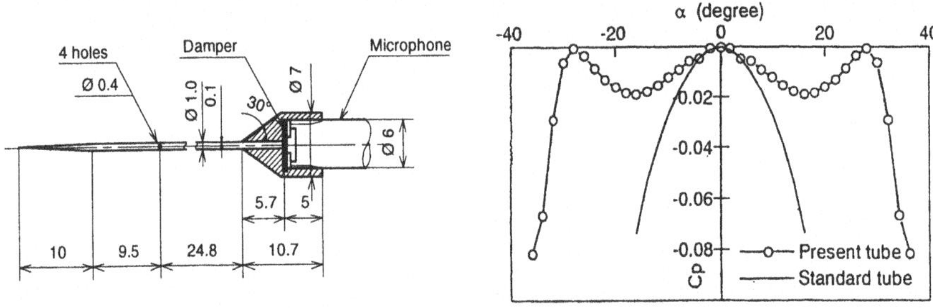

Fig.2 Pressure probe

Fig.3 Errors of static pressure in steady flow; $Cp = \Delta p/(\rho U_e^2/2)$

determined so as to minimize errors in the measurements of fluctuating static pressure. In order to damp the organ pipe resonance in the tube, a thin nylon gauze was inserted in front of the diaphragm of microphone, and the probe had a nearly flat frequency response to 2.0kHz. The cross-flow error of the static pressure tube was calibrated in the uniform flow by changing the flow attack angle to the tube. The result is shown in Fig.3 in comparison with that of a standard static pressure tube with hemispherical nose. The pressure drop Δp of the present tube is less than that of the standard one over the wide range of attack angle. The dynamic response of the probe is discussed by the authors [3].

The arrangement of the probes for the measurement of the phase-average fluctuating static pressure is shown in Fig.4. The reference velocity and the fluctuating static pressure were measured respectively by a single normal hot-wire probe fixed near the jet center at $x/De \doteqdot 1.0$ and by the pressure probe moved at 2160 points over the flow field. The phase-averaging of fluctuating pressure was carried out at 36 phase angles of the reference signal from $\theta=0$ to $\theta=2\pi$ with an increment of $\pi/18$. The sampling number for phase-averaging is about 210 for each phase.

The mean streamwise velocity and the turbulence intensity were measured by a single normal hot-wire probe. The measuring period at each measured

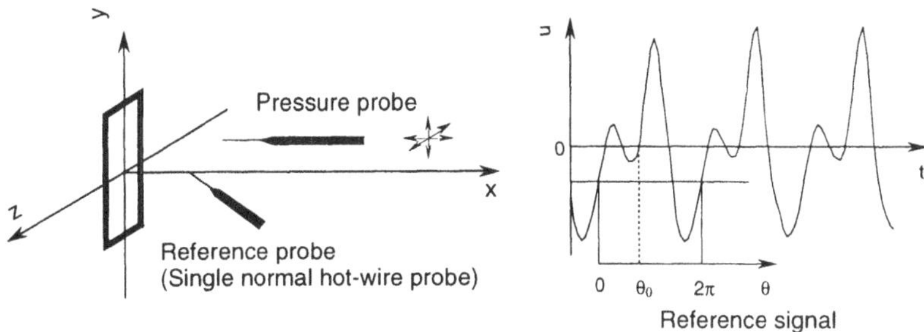

Fig.4 Arrangement of the probes for the phase-average measurement

point was 3 seconds, in which 423 large-scale vortical structures pass through the point. To discuss the excitation effect and to compare the results with those of a circular jet, the mean velocity and the turbulence intensity in unexcited rectangular and circular jets were also measured. The traversing of the pressure probe and the hot-wire probe, the data acquisition and the data processing were automatically controlled by the personal computers.

3. Results and Discussions

Figure 5 shows the contours of phase-average fluctuating pressure at six phases with an increment of $\pi/3$; the phase increases from the top to the bottom. The frames (a)-(f) indicate one cycle of the periodic vortex motion. The non-dimensional pressure $<p>$ is defined by

$$<p>=2p_c/\rho U_e^2 \qquad (1)$$

where p_c is the phase-average fluctuating pressure. The negative-pressure regions, shown by dark zones, correspond to the vorticity concentrated regions [4]. The figure shows how the leading (L) and trailing (T) vortices interact each other. In the major plane [Fig.5 (a)] the leading vortex (L) comes out from the left-hand side (see (a)), flows downstream, and engulfs the downstream trailing vortex (T') (see (c) and (d)). The parts of the trailing vortex (T) are observed in the inner and outer regions at x/De=1.0 (see (d)); the inner vortex rushes into the leading vortex.

Figure 5 (b) shows that the leading vortex stretches outward and splits into smaller vortices farther downstream (see L' in (f)), and the trailing vortex rushes into the leading vortex.

The contours in the yz plane at x/D_e=1.3 are shown in Fig.5 (c), which reveals the cross-sectional views of the interacting vortical structures. The leading vortex stretches in the minor-axis direction, and the trailing vortex passes through inside of the leading vortex. The contours at 36 phases enable

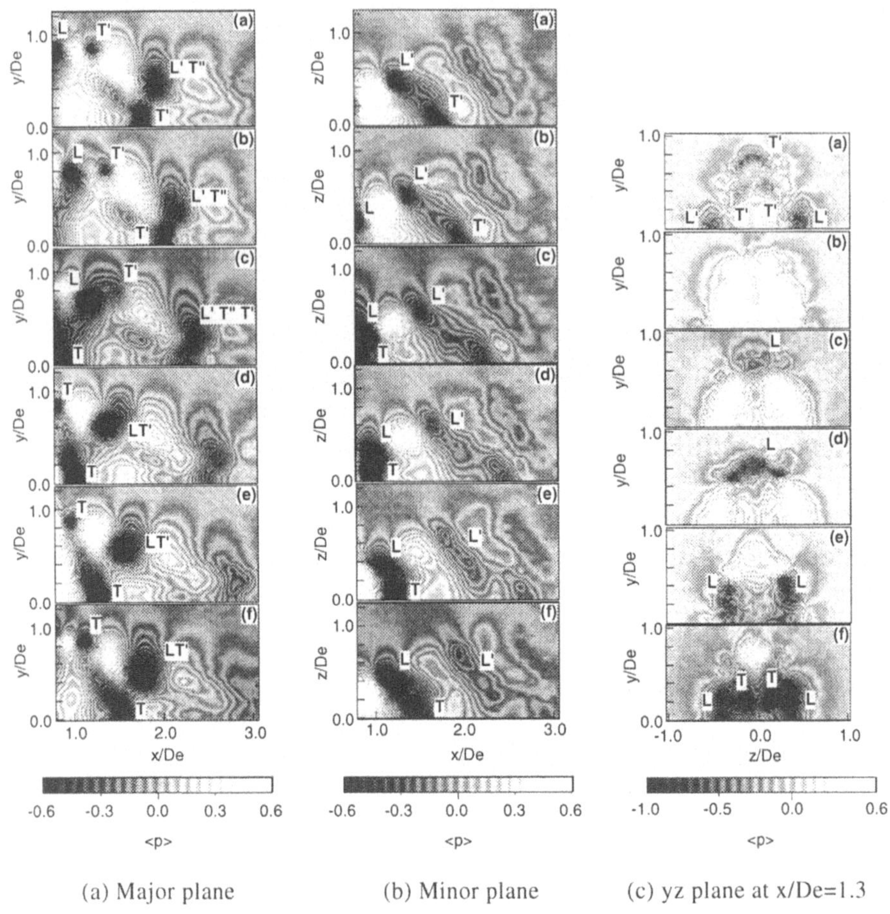

(a) Major plane (b) Minor plane (c) yz plane at x/De=1.3

Fig.5 Phase-average fluctuating static pressure contours

us to make the three-dimensional view of the vortical structure using the Taylor hypothesis. The results are shown in Figs. 6 and 7, where the convection velocity U_c of the vortical structure is assumed to be equal to $U_0/2$ (U_0 : jet-center velocity) . The interacted structure at x/D_e=1.0 in Fig.6 agrees well with that predicted by the visualization experiment [1]. The three-dimensional pressure field at x/D_e=1.5 in Fig.7 (a) indicates stretching of the leading vortex in the minor-axis direction and engulfing of the trailing vortex into the upstream leading vortex. The pressure field with the threshold level of lower pressure is shown in Fig.7 (b), which indicates the skeleton of the vortical structure. The figure suggests the partial merging of the leading and the trailing vortices. Such a complicated structure cannot be detected by other measurement techniques.

Figs.5-7 enable us to predict the three-dimensional vortical structure in the rectangular jet excited in the interaction mode. The predicted model is shown

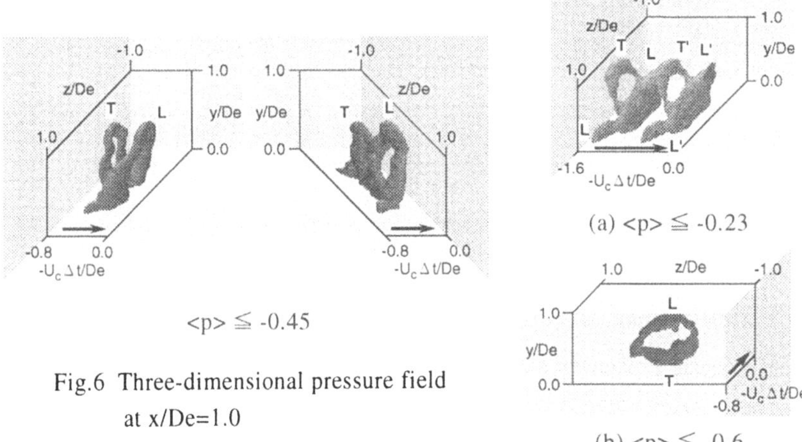

<p> ≦ -0.45

Fig.6 Three-dimensional pressure field at x/De=1.0

Fig.7 Three-dimensional pressure field at x/De=1.5

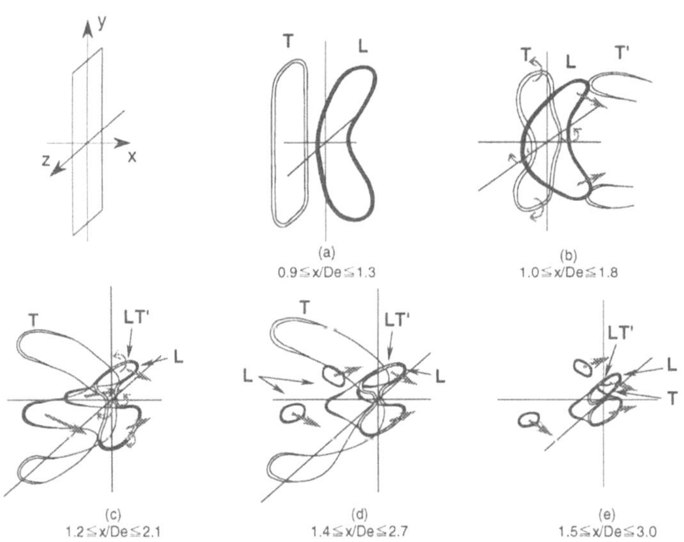

Fig.8 Three-dimensional vortical structure

in Fig.8. The leading vortex (L) bends downstream owing to the self-induced velocity and engulfs the downstream trailing vortex [T' in Fig.8 (b)]. Farther downstream the leading vortex stretches in the minor-axis direction [Fig.8 (c)] and splits into small vortex rings [Fig.8 (d)]. The outer part of the trailing vortex (T) stretches upstream [Fig.8 (b)], and the inner part rushes into the leading vortex near the jet center [Fig.8 (b) and (c)]. The vortex rings (L) spread outward in the minor-axis direction, and the compound vortical struc-

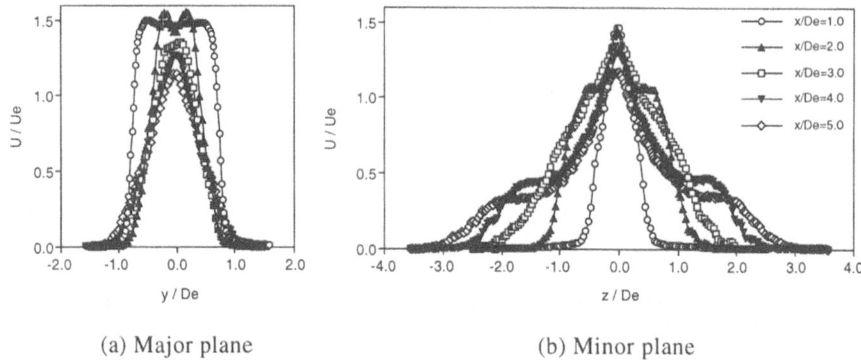

(a) Major plane (b) Minor plane

Fig.9 Mean streamwise velocity profiles of the excited rectangular jet

ture consisted of the parts of leading and trailing vortices is formed [Fig.8 (c)]. The compound vortex bifurcates in the major-axis direction via the cut-and-connect process [5] near the jet center [Fig.8 (e)].

The vortices evolved in the unexcited jet are not so stable as those in the excited jet, and various types of vortex interactions occur. The visualization experiments [1] showed that the interaction in Fig.8 was typical of the vortex interactions. Thus the present excitation enhances the vortex motion typical in the unexcited jet.

Figure 9 shows the mean streamwise velocity profiles of the excited rectangular jet. The jet width in the major plane decreases with increasing the distance from the jet exit, and the saddle-shaped velocity profile appears at x/De = 2.0 as shown in Fig.9 (a). The saddle-shaped velocity profile is well known in the unexcited rectangular jet [6], and the profile is more significant in the excited jet. The hairpin-shaped vortices generate the high velocity regions shown by the arrows in Fig.8 (c) and (d), and the regions cause the saddle-shaped velocity profile.

The jet width in the minor plane increases downstream and the profiles become very complicated at x/De $\geqq$ 2.0 as shown in Fig.9 (b). The velocity profile at x/De = 2.0 has the humps at z/De $\fallingdotseq \pm$ 0.5. The generating mechanism of the humps is interesting and can be explained as follows. As shown in Fig.8 the leading vortex stretches in the minor-axis direction, and splits into two vortex rings which move outward as shown in Fig.10. The induced flow by the vortex rings causes the high-speed humps. The humps of the profiles at x/De = 4.0 and 5.0 may be generated by the same mechanism.

The value of the induced velocity u_i in Fig.10 is estimated from the phase-average velocity obtained by the same technique as that mentioned in Section 3. The value at the vortex ring center is about 0.3Ue at x/De $\fallingdotseq$ 2.0.

In order to discuss the feature of the jet diffusion, the variations of the jet cross-sectional area, which is calculated with the mean streamwise velocity distributions in the yz planes, are shown in Fig.11, where $A_{0.1}$ is the area in-

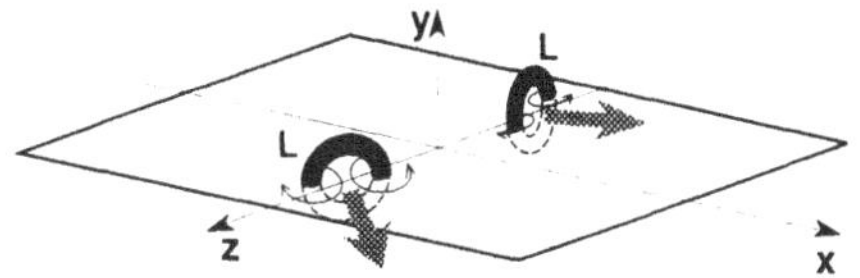

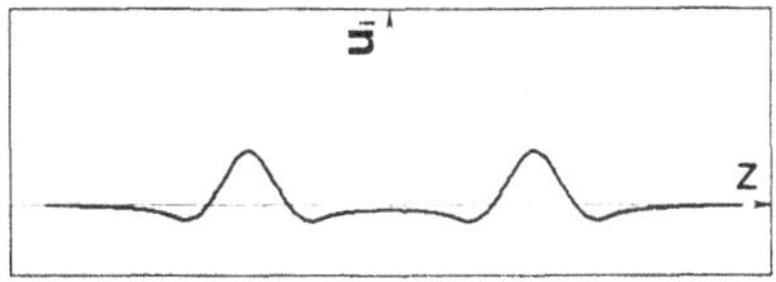

Fig.10 Induced velocity u_i by vortex rings

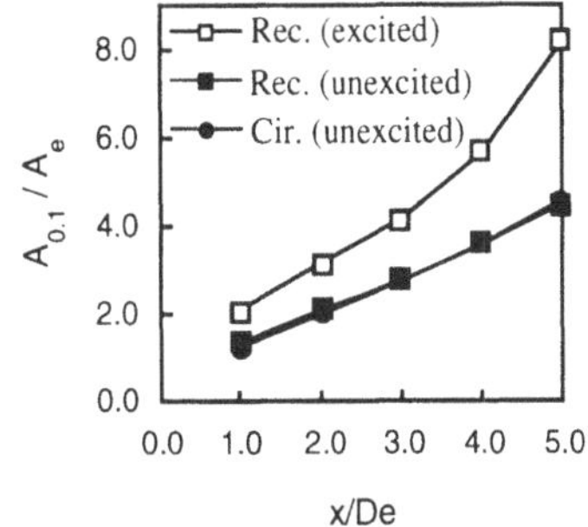

Fig.11 Variations of the jet cross-sectional area

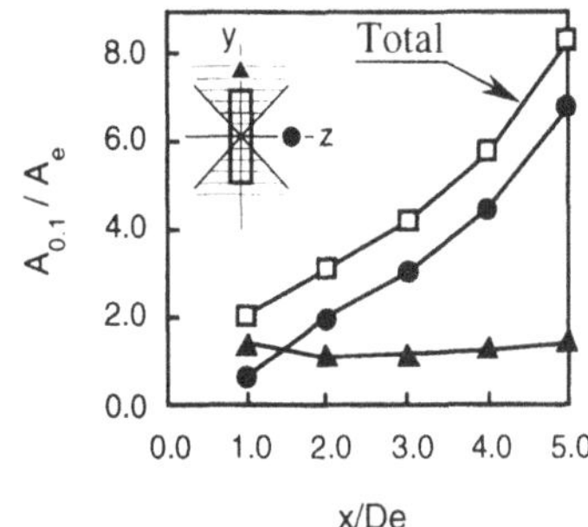

Fig.12 Variations of the jet cross-sectional area of the excited rectangular jet

side of U/Ue = 0.1 and A_e is the area of the jet exit. The variations for unexcited rectangular and circular jets are included in Fig.11. The areas of the unexcited rectangular jet are equal with those of the unexcited circular jet, while the areas of the excited rectangular jet are larger than those of the unexcited jets and the increasing rate becomes larger at x/De $\geqq$ 3.0.

The area at each streamwise location is divided into two sections which are the major and minor regions as shown in Fig.12. The areas in the minor region are larger than those in the major region at x/De$\geqq$2.0 due to the leading vortex spreading in the minor-axis direction [Fig.8 (c), (d)]. The areas in the major region decrease slightly at x/De=2.0 due to the leading vortex bending toward the jet center as shown in Fig.8 (b) ~ (d).

Figure 13 shows the turbulence intensity profiles of the excited rectangular jet. The turbulence intensity u' is defined as the rms value of fluctuating velocity. At x/De = 1.0 in the major plane [Fig.13 (a)], the turbulence intensity is increased at the jet center by the velocity fluctuation generated by the vortex pairs which are the parts of trailing vortices as shown in Fig.8 (c). The peaks on both sides of the jet center are caused by the leading vortices. In the profile at x/De = 2.0, small peaks appear inside the large outside peaks at y/De $\fallingdotseq \pm 0.4$. The inside peaks suggest small vortices passing near y/De = $\pm$0.2, because the increasing of turbulence is generally caused by vortex motion.

In the turbulence intensity profile at x/De = 1.0 in the minor plane [Fig.13 (b)], a peak at the jet center and two peaks on both sides of the jet center are

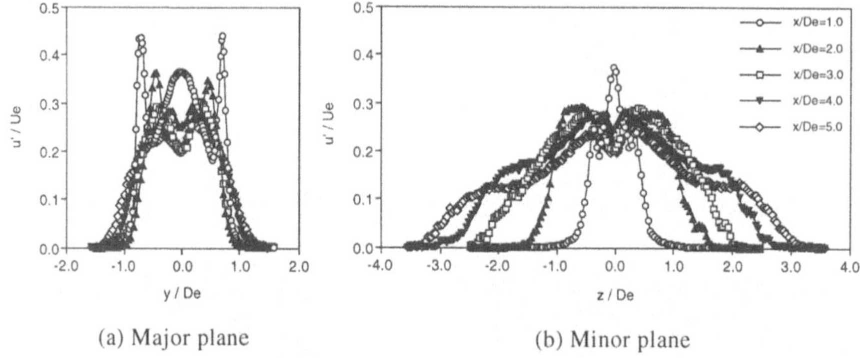

(a) Major plane (b) Minor plane

Fig.13 Turbulence intensity profiles of the excited rectangular jet

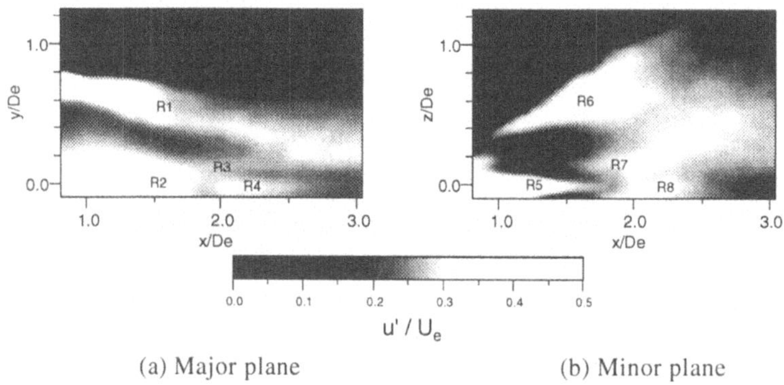

(a) Major plane (b) Minor plane

Fig.14 Turbulence intensity distributions

recognized. The peak at the jet center is generated by the same mechanism as that of the profile at x/De = 1.0 in the major plane. Two peaks on both sides of the jet center are caused by the leading vortices. The profile at x/De = 2.0 has four peaks which are generated by vortex motions. The outside peaks are generated by the leading vortices, and the inside peaks suggest vortex motions. The turbulence diffuses downstream significantly, similarly to the mean velocity in Fig.9 (b).

Figure 14 shows the turbulence intensity distributions in the major and minor planes. In the major plane [Fig.14 (a)], the leading vortices cause the high turbulence intensity region R1, and the trailing vortices increase the turbulence in the region R2 near the jet center. The peaks at y/De ≒ ± 0.2 in the turbulence intensity profile at x/De = 2.0 in Fig.13 (a) correspond to the region R3. The streak of R3 joins the upper high turbulence intensity region R1 at x/De ≒ 2.5. The high turbulence intensity region R4 appears near the jet center at 2.0 < x/De < 2.4.

In the minor plane [Fig.14 (b)], the high turbulence intensity region R5 is caused by the parts of trailing vortices passing near the jet center. The high

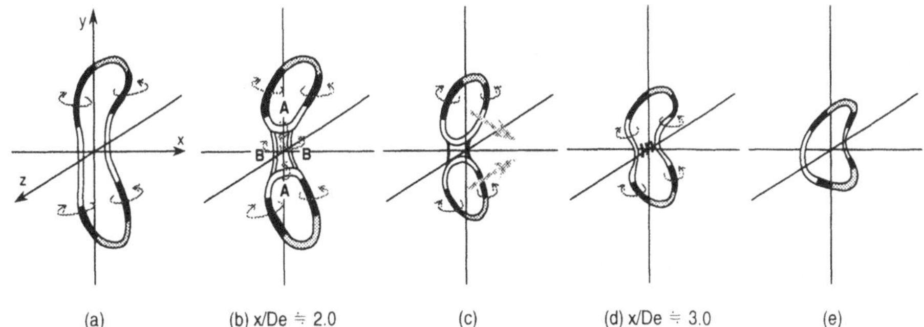

Fig.15 Bifurcation-and-reconnection model of the compound vortex ring

turbulence intensity region R6 is caused by stretching and splitting of the leading vortices, and the region R7 by the remnants of trailing vortex mentioned in Fig.15. The region R8 corresponds to R4 in the major plane.

Considering the results mentioned above, the authors propose the deformation model of the compound vortex ring formed by the interaction of leading and trailing vortices as shown in Fig.8. Figure 15 shows the bifurcation-and-reconnection model of the compound vortex ring. The deformation of the vortex ring generates the bridges (A) and the threads (B) shown in Fig.15 (b), which are caused by vortex annihilation and cross-linking [7] in the jet center region. This structure corresponds to the turbulence intensity distributions in Fig.14 as follows. The generation of the bridges causes the high turbulence intensity region R3 in the major plane, and decreases turbulence intensity near the jet center between R2 and R4. The threads, the remnants of trailing vortex, cause the region R7, corresponding to the inside peaks of the turbulence intensity profile at x/De = 2.0 in the minor plane in Fig.13 (b). At the next stage, two vortex rings are formed as shown in Fig.15 (c). The rings induce velocity fluctuation near the jet center, which causes the high turbulence intensity regions R4 and R8. The bifurcated vortex rings incline inside, move toward the jet center, collide each other, and form a vortex ring via the cut-and-connect process again as shown in Fig.15 (d) and (e).

4. Conclusions

The present study is concluded as follows.

(1) The velocity field is closely related to the vortical structure evolved in the rectangular jet.

(2) The diffusion enhancement of the excited rectangular jet is owing to the flow induced by the three-dimensional vortical structure.

(3) The deformation model of the compound vortex ring near the jet center is proposed in careful consideration of pressure and velocity fields.

References

1. Toyoda, K. and Hussain, F. (1989) Vortical Structures of Noncircular Jets, *Proc. of the Fourth Asia Congress of Fluid Mechanics*, A117-A127.
2. Toyoda, K., Shirahama, Y. and Kotani, K. (1991) Manipulation of Vortical Structures in Noncircular Jets, *ASME, FED* **112**, 135-140.
3. Toyoda, K., Hiramoto, R. and Shirahama, Y. (1995) Measurements of Fluctuating Static Pressure in a Rectangular Jet, *ASME, FED* **211**, 13-18.
4. Toyoda, K., Okamoto, T. and Shirahama, Y. (1994) Eduction of Vortical Structures by Pressure Measurements in Noncircular Jets, *Appl. Scientific Res.*, **53**, 237-248.
5. Kida, S., Takaoka, M. and Hussain, F. (1989) Reconnection of Two Vortex Rings, *Phys. Fluids A*, **1-4**, 630-632.
6. Tsuchiya, Y., Horikoshi, C. and Sato, T. (1986) On the Spread of Rectangular Jets, *Exp. Fluids*, **4**,197-204.
7. Hussain, F. and Husain, H. S. (1989) Elliptic jets. Part 1. Characteristics of Unexcited and Excited Jets, *J. Fluid Mech*, **208**, 257-320.

IDENTIFICATION OF STRONG, NEAR-WALL QUASI-STREAMWISE VORTICES AND THEIR BEHAVIOR

K. TSUJIMOTO AND Y. MIYAKE
Department of Mechanical Engineering, Osaka University
2-1, Yamada-oka, Suita, 565 Japan

1. Introduction

It is widely accepted that the near-wall region of wall turbulence is dominated by an organized structure the most important ingredients of which are quasi-streamwise vortices. In order to identify these vortices, several indices have been proposed. Among them, Q, which is the second invariant of velocity gradient tensor $\boldsymbol{A} = A_{ij} = \partial v_i/\partial x_j$ and which is defined by $Q = (1/2)\{[\mathrm{tr}(\boldsymbol{A})]^2 - \mathrm{tr}\boldsymbol{A}^2\} = -(1/2)(\Omega_{ij}\Omega_{ji} + S_{ij}S_{ji})$(Hunt *et al.*, 1988) where $\Omega_{ij} = (\partial v_i/\partial x_j - \partial v_j/\partial x_i)/2$, $S_{ij} = (\partial v_i/\partial x_j + \partial v_j/\partial x_i)/2$ and λ_2(Jeong *et al.*, 1995, Jeong *et al.*, 1997), which is the second largest eigenvalue of a symmetric part of second order tensor of Hessian $\partial^2 p/\partial x_i \partial x_j$ where p is pressure, are most widely used. In the near-wall layer of $y^+ = 10 \sim 50$, where $y^+(= yu_\tau/\nu)$ is the non-dimensional distance from the wall, Q and $-\lambda_2$ do not differ greatly in their ability to identify quasi-streamwise vortices, at least statistically. For example, the volume fractions of $Q > 0$ and of $\lambda_2 < 0$ are both about 40% of the whole volume in this layer, though that of $-\lambda_2$ is slightly smaller than that of Q, particularly in the buffer layer. Furthermore the wall normal distribution of both their mean values $Q_{mean} and \lambda_{2,mean}$, and the *r.m.s.* of their fluctuations $Q_{rms} and -\lambda_{2,rms}$ are also nearly identical.

In the dissipation *vs.* enstropy density map, or in the $S_{ij}S_{ij}$-$\Omega_{ij}\Omega_{ij}$ map(Soria *et al.*, 1994), points in the near-wall layer are concentrated around the straight line, $Q = S_{ij}S_{ij} + \Omega_{ij}\Omega_{ij} = 0$, which suggests that Q_{mean} is small and the number of points in the layer decreases quickly with Q, as shown in Fig.1(obtained from our own databases).

Not all of the vortices of positive Q control turbulence phenomena, such as strong ejection and/or downwash, but stronger vortices do control these

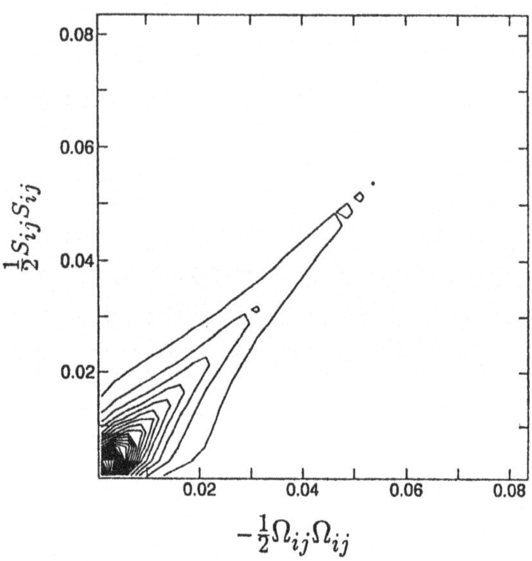

Figure 1. Iso-contours of $\frac{1}{2}S_{ij}S_{ij}$ vs. $\frac{1}{2}\Omega_{ij}\Omega_{ij}.y^+ = 10 \sim 50$

phenomena. Therefore, in order to identify 'strong' quasi-streamwise vortices, it is first necessary to specify some positive threshold value for Q. In the near-wall turbulence, the streamwise vorticity ω_x can also serve as an index (Brooke and Hanratty, 1993; Miyake *et al.*, 1997) to identify strong quasi-streamwise vortices, since Jeong *et al.* (1997) have shown that an adequately chosen threshold value of $-\lambda_2$ picks up correctly strong streamwise vortices which are identified by specifying a suitable threshold value for ω_x. In this work as well, we use $\omega_x^+(=\omega_x \nu / u_\tau^2)$, where u_τ is friction velocity on the wall. In the following, we discuss the property of strong quasi-streamwise vortices collected by $|\omega_x^+| = 0.2$.

2. Database

The materials used in this study were taken from the database of a channel flow generated by the authors using DNS. The code uses a spectral method based on Fourier modes in the streamwise and spanwise directions and on Chebyshev modes in the wall normal direction. The size of the computational box is $2\pi H$(streamwise, x)$\times H$(wall-normal, y)$\times \pi H/2$(spanwise, z), and the grid numbers are $256(x)\times 129(y)\times 64(z)$ for the Reynolds number $Re_\tau = u_\tau H/\nu = 300$, where H is channel width.

3. General properties of quasi-streamwise vortices with respect to streamwise vorticity ω_x

In this section, several kinds of properties representing quasi-streamwise vortices are examined with respect to combinations of streamwise vorticity ω_x and tilting angles $\alpha = \tan^{-1}(\omega_y/\sqrt{\omega_x^2+\omega_z^2})$ of vortex vector from the wall and $\beta = \tan^{-1}(\omega_x/\omega_z)$ from the spanwise direction. That is, a quantity f at each grid point in the layer $y^+ = 10 \sim 35$ weighted by streamwise vorticity ω_x^+ at that grid point is summed up for each combination of α and β, and then $\sum_n \omega_x^+ f dV$ (dV represents a weighted volume on a grid point) is presented as iso-contour maps in the α-ω_x^+ plane and β-ω_x^+ plane.

Indices Q and λ_2 and the production rate of streamwise vorticity are chosen, as the interested quantities f. Since the sustenance of vortices is one of our major concerns, the production rate P_x of the streamwise vorticity is of keen interest. In this regard, a new split of production rate term $\boldsymbol{A}\cdot\boldsymbol{\omega}$ is here proposed . That is, we split the term into two components, one in the direction of the vortex vector $\boldsymbol{f}_e$ and the other in a plane normal to the vortex vector $\boldsymbol{f}_t$.

The transport equation of vorticity $\boldsymbol{\omega}$ that is given by

$$\frac{D\boldsymbol{\omega}}{Dt} = \boldsymbol{A}\cdot\boldsymbol{\omega} + \frac{1}{Re_\tau}\nabla^2\boldsymbol{\omega}. \tag{1}$$

can be rewritten as

$$\begin{aligned}\frac{D\omega\boldsymbol{e}_\omega}{Dt} &= \frac{D\omega}{Dt}\boldsymbol{e}_\omega + \omega\frac{D\boldsymbol{e}_\omega}{Dt} \\ &= \boldsymbol{f}_e + \boldsymbol{f}_t + \frac{1}{Re_\tau}\nabla^2\boldsymbol{\omega} = f_e\boldsymbol{e}_\omega + f_t\boldsymbol{e}_t + \frac{1}{Re_\tau}\nabla^2\boldsymbol{\omega} \end{aligned} \tag{2}$$

where $\boldsymbol{e}_\omega$ and $\boldsymbol{e}_t$ are unit vectors in the direction of $\boldsymbol{\omega}$ and in some direction normal to $\boldsymbol{\omega}$, respectively, and $\omega = |\boldsymbol{\omega}|$. f_e, f_t are obtained by

$$f_e = (\boldsymbol{A}\cdot\boldsymbol{\omega})\frac{\boldsymbol{\omega}}{\omega} \tag{3}$$

$$f_t = |\boldsymbol{A}\cdot\boldsymbol{\omega} - f_e\boldsymbol{e}_\omega| \tag{4}$$

$\boldsymbol{f}_e$ is the production rate due to stretching and $\boldsymbol{f}_t$ that due to turning of a vortex line, or to tilting. $(\boldsymbol{f}_e)_i, (\boldsymbol{f}_t)_i$, $(i = x, y, z)$ are the components of each contribution in a direction of the co-ordinate axis. A clearer physical meaning is implemented in this split than in the conventional split(Brooke and Hanratty, 1993 ; Miyake and Tsujimoto, 1996), in which $\boldsymbol{A}\cdot\boldsymbol{\omega}$ is composed of the three components, $P_{x,str}$ (stretching), $P_{x,til}$ (tilting) and $P_{x,yaw}$ (yawing). These three terms are the first, second, and third, respectively of the right-hand side of the equation,

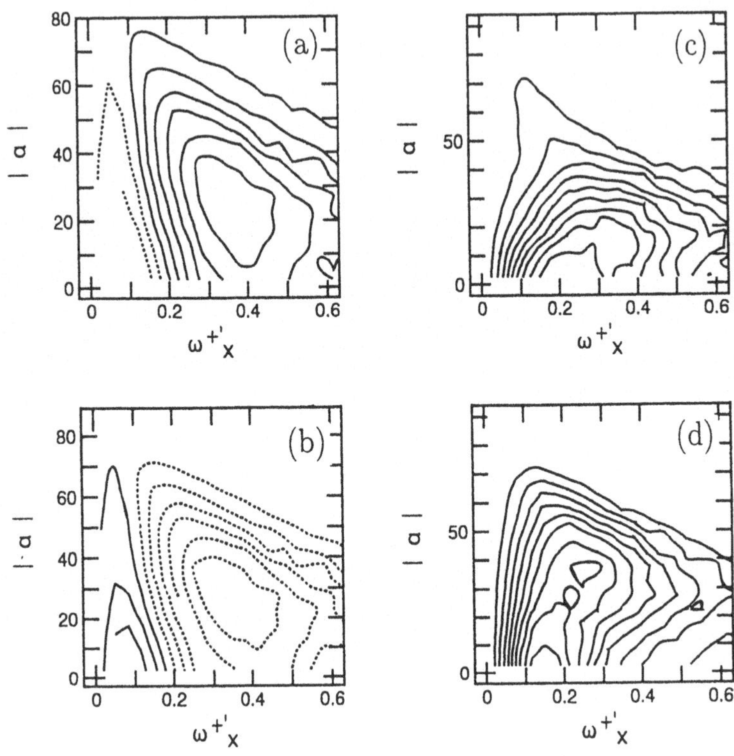

Figure 2. Iso-contours of (a) Q, (b) λ_2, (c) f_{ex}, (d) f_{tx} in α-ω_x^+ plane

$$\frac{D\omega_x}{Dt} = \omega_x\frac{\partial u}{\partial x} - \frac{\partial w}{\partial x}\frac{\partial u}{\partial y} + \frac{\partial v}{\partial x}\frac{\partial u}{\partial z} + \frac{1}{Re_\tau}\nabla^2\omega_x. \tag{5}$$

$P_{x,str}$ is fairly weakly correlated with f_{ex} and does not represent a pure contribution due to stretching ; $P_{x,til} and P_{x,yaw}$ also do not provide the pure turning but include some part of the stretching effect. However, $f_{ex} + f_{tx} = P_{x,str} + P_{x,til} + P_{x,yaw}$ holds.

Figure 2 shows a map for ω_x^+-α and Fig.3 shows a map for ω_x^+-β. In these figures, quantities related to vortices of negative rotation (counter clockwise rotation seen from the upstream) are reduced to those of the positive rotation. One finds that Q and λ_2 give a result similar to that for tilting angles of vortex lines and that $|\omega_x^+| \geq 0.2$ is a good measure for a identifying strong vortices. Vortex filaments of negative Q or positive λ_2, which are found at the top left corner of the map of ω_x^+-β(Fig.3(a),(b)), are mostly of weak

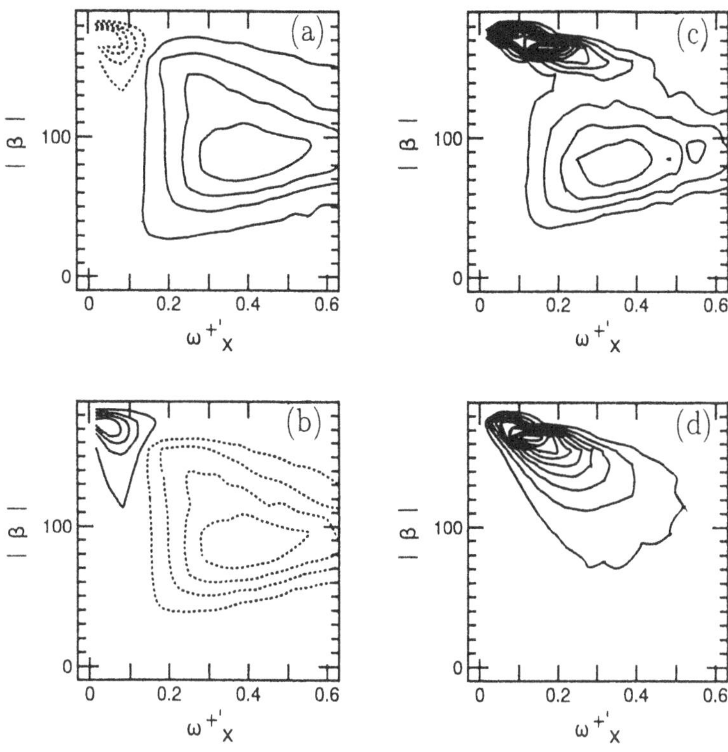

Figure 3. Iso-contours of (a) Q, (b) λ_2, (c) f_{ex}, (d) f_{tx} in β-ω_x^+ plane. $y^+ = 10 \sim 35$.

vorticity and oriented strongly to the spanwise direction($\beta = 180°$), and both f_{ex} and f_{tx} are quite strong. They are the vortex sheet contributing to sustain a vortex by converting the wall-normal and/or spanwise vorticity to a streamwise one. In Fig.3(c), it can be seen that the vortex filaments of $|\omega_x^+| \geq 0.2$ are of two types, one, strongly oriented to the spanwise direction, and the other, nearly in a streamwise direction. The latter type corresponds to the type at the central peak of Fig.3(a), which is of large Q and more vortex-tube-like than the former type, and which is more tilted to the wall as suggested in Fig.2 (a),(b).

4. Strong Quasi-Streamwise Vortices

We next examine the validity of identifying streamwise vortices in the near-wall layer $y^+ = 10 \sim 50$ by streamwise vorticity ω_x^+. For this purpose,

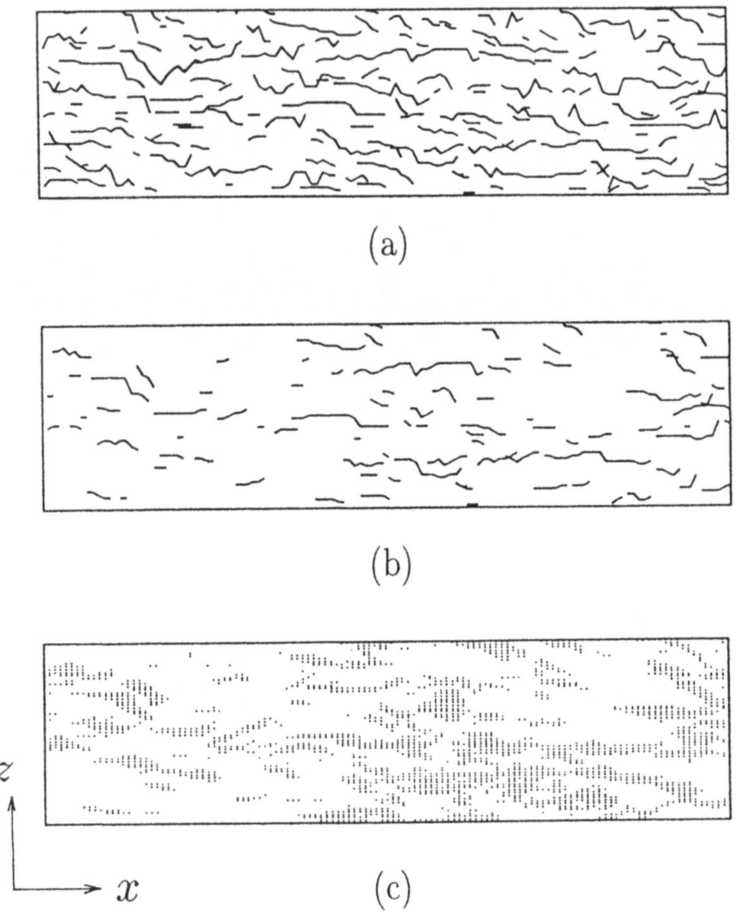

Figure 4. Plan view of streamwise vortices in a layer $y^+ = 10 \sim 50$. (a) centerlines of areas of $Q > 0$ and $L_x^+ > 50$, (b) centerlines of areas of $Q > 0$, $L_x^+ > 50$ and $|\omega_x^+| \geq 0.2$, (c) area of $|\omega_x^+| \geq 0.2$

correspondence of the vortices identified by $\omega_x^+ \geq 0.2$ with those found by the detector Q is examined as in Fig.4. Figure 4(a) is a plan view of the centerlines of long positive vortices ($\omega_x > 0$) identified by $Q \geq 0$ but with the additional condition that the streamwise length L_x of a vortex is long, i.e., $L_x^+ > 50$. Figure 4(b) shows the same view as Fig.4(a) but with the additional condition of $\omega_x^+ \geq 0.2$. That is, each line in Fig.4(b) represents a portion of a line in Fig.4(a) with trimming of the portion of $\omega_x^+ \leq 0.2$, which suggests that long vortices are not always strong ones. Figure 4(c) shows a plan view of the grid points where ω_x^+ is greater than 0.2. The dotted areas correspond to the centerlines of Fig.4(b).

Thus we can conclude that the condition $|\omega_x^+| \geq 0.2$ picks up almost every long and strong quasi-streamwise vortex that is expected to control major turbulence phenomena in the near-wall turbulence.

The strong vortices of $|\omega_x^+| \geq 0.2$ characterized in the manner shown above are our present concern. To have a sufficient number of samples to investigate the properties of these vortices, the following procedure was applied. The grid points having $|\omega_x^+| \geq 0.2$ in a cross section in the layer $y^+ = 10 \sim 50$ were identified in every cross section of the computational volume. A space composed of a group of consecutive points thus identified and which are neighboring both in a cross section and in a streamwise section is regarded as a body of a quasi-streamwise vortex. Among these vortices, those having a reasonably long streamwise length, i.e., $L_x^+ > 100$, were studied as follows so as to represent strong streamwise vortices.

In the database used in this study, approximately 310 samples were carefully chosen in order to avoid using one vortex twice. The properties of these quasi-streamwise vortices will be described in the following.

Figure 5 shows iso-contours of (a) vorticity ω_x^+, (b) production rate due to stretching f_{ex}, and (c) production rate of turning f_{tx} in a wall-normal plane, averaged in the spanwise direction of the ensemble mean of sampled vortices. Here, the ensemble average is determined in a manner that to superposes each flow field in a box of a size $\Delta x^+ \geq 100, y^+ = 0 \sim 50$, $\Delta z^+ = 50$ around an original point $x^+ = z^+ = 0$, which is the upstream end of the above-mentioned space of high vorticity.

The figure is similar to a side view of a typical single strong quasi-streamwise vortex, and the vorticity contours suggest that counter rotating vortices are arranged in a streamwise direction (Jeong and Hussain,1995 ; Miyake *et al.*, 1997), since the tail part of an upstream vortex is at the top. Also, it is found that, underneath a vortex, a vortex of opposite rotation exists which meets the no-slip condition of the wall. It is also found that the contribution of stretching f_{ex} to the production of streamwise vorticity is far smaller than that of turning f_{tx} and appears only in the core of a vortex. It is to be noted that only areas of low f_{ex} are found above, suggesting that the upstream-side vortex fluctuates substantially in both size and intensity. Meanwhile, the turning effect appears both above and beneath a vortex, particularly in the layer closer to the wall. The iso-contours of f_{tx} above a vortex strongly support the idea of a streamwise arrangement of the two vortices of opposite rotation.

We previously discussed a mechanism of regeneration of a quasi-streamwise vortex at its upstream end (Miyake *et al.* 1997), on the basis of a con-

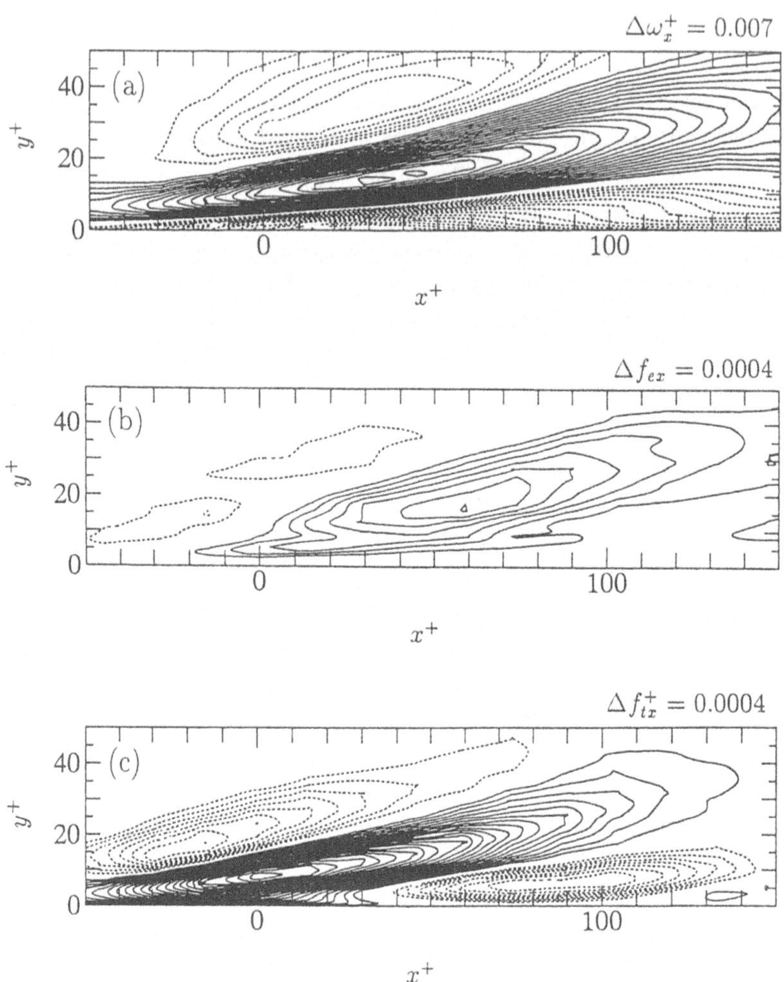

Figure 5. Iso-contours of (a) ω_x^+, (b) f_{ex} and (c) f_{tx} of ensemble averaged strong quasi-streamwise vortices (side view).

ventional split. In order to examine the usefulness of the conventional split, similar iso-contours are calculated using split $P_{x,str}$ and $P_{x,til}$ as shown in Fig.6, in which (a) is for $P_{x,str.}$ and (b) for $P_{x,yaw}$. The contours in Fig.6 are similar to those in Fig.5, despite the fact that $P_{x,str}$ and $P_{x,til}$ are only weakly correlated with f_{ex} and f_{tx}. However, $P_{x,str}$ is far greater than f_{ex} and, correspondingly, $P_{x,til}$ is fairly smaller than f_{tx}. So, the conventional split is misleading in that it overestimates the role of stretching in sustain-

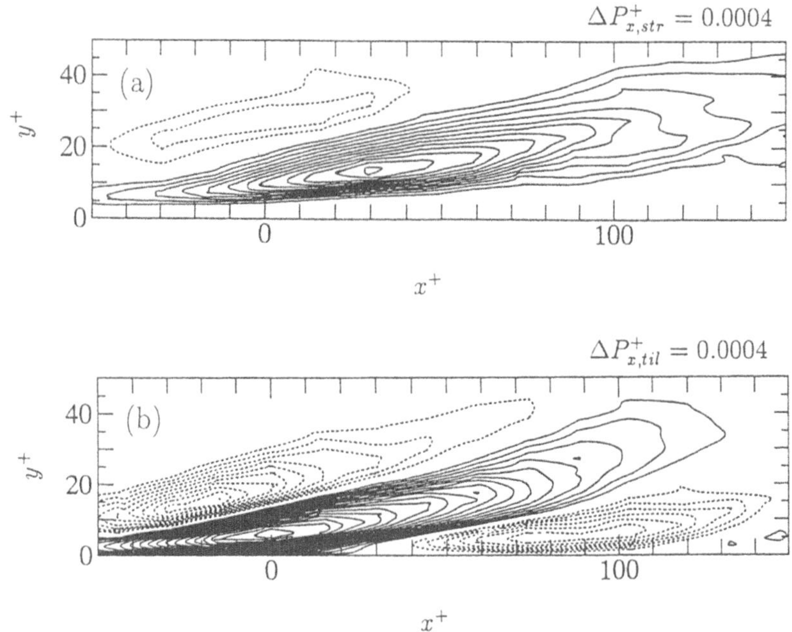

Figure 6. Iso-contours of (a) $P_{x,str}$ and (b) $P_{x,til}$ of ensemble averaged strong quasi-streamwise vortices (side view).

ing quasi-streamwise vortices. The turning effect which supplies the new streamwise vorticity is important in sustenance.

Then, it is interesting to note how the properties of the strong vortices defined above appear in the maps of Fig.2. The quantities at the points included in the long and strong vortices are processed in the same manner as in Figs.2 and 3, for which the resulting maps are very similar, except that a weak peak of $f_{ex} > f_{tx}$ appears in the β-ω_x^+ map at nearly the same place where the peak of Q-contours appears. This fact suggests the following properties of strong vortices. First, vortex lines generating streamwise vorticity by stretching f_{ex} are oriented nearly to the streamwise direction, at about 10° from the flow direction opposite to that of the axis of a strong vortex. This group of vortex lines represents those composing the downstream half of a strong vortex, as suggested in Fig.5(b). Secondly, vortex lines generating ω_x by turning f_{tx} are mostly those oriented strongly to the spanwise direction as $\beta \sim 145°$, and their tilting angle from the wall is large, i.e., $\alpha \sim 40°$. However, these vortex filaments are weaker than those of the previous group. This group of vortex filaments composes the upstream part of a strong vortex where in filaments of large wall-normal vorticity supply ω_x by turning.

Figures 2 and 3, which are maps including whole vortex filaments of $\omega_x \geq 0$, validate the threshold $\omega_x^+ > 0.2$ as a useful detector of strong quasi-streamwise vortices and suggest that layers of weak vorticity ω_x oriented strongly to the spanwise direction and having wall-normal vorticity component work jointly to sustain a strong streamwise vortex at the part closer to the wall. Therefore, the suppression of quasi-streamwise vortices is realized by managing the wall-normal vorticity found mainly at the upstream end portion of quasi-streamwise vortices.

5. Conclusion

1. In regard to the identification of strong quasi-streamwise vortices in the near-wall layer, the condition of streamwise vorticity $|\omega_x^+| \geq 0.2$ is as good a measure as the indices Q and λ_2, which give nearly the same identification.
2. The new split of production rate of ω_x here proposed works better than the conventional split to interpret the sustenance mechanism of a strong quasi-streamwise vortex.
3. The conversion of wall-normal and/or spanwise vorticity to a streamwise vorticity by turning is a dominant mechanism in sustaining of quasi-streamwise vortices, contrary to the conventional understanding that the stretching of vortex filaments by a wall-normal velocity gradient is dominant.

References

Brooke, J. W. and Hanratty, T. J. (1993), Origin of turbulence producing eddies in a channel flow, *Phys. fluids*, **A5**-4, pp.1011-1022.

Jeong, J. and Hussain, F. (1995), On the identification of a vortex, *J. Fluid Mech.*, **285**, pp.69-94.

Jeong, J., Hussain, F., Schoppa, W. and Kim, J. (1997), Coherent structures near the wall in a turbulent channel flow, *J. Fluid Mech.*, **332**, pp.185–214.

Hunt, J. C. R., Wray, A. A. and Moin, P. (1988), Eddies, stream and convergence zones in a turbulent channel flow, *Center for Turbulence Research Report CTR-S88.*

Miyake, Y. and Tsujimoto, K. (1996), Behavior of quasi-streamwise vortices in near-wall turbulence, *ASME FED, 1996 Fluids Engineering Division Conference*, **3**, pp.41-48.

Miyake, Y., Ushiro, R. and Morikawa, T. (1997), The regeneration of Quasi-Streamwise Vortices in the Near-Wall Region, *JSME Intern. J. ser.B*, **40**-2, pp.257-264.

Soria, J., Sondergaard, R., Cantwell, B. J., Chong, M. S. and Perry, A. E. (1994), A study of the fine-scale motions of incompressible time-developing mixing layers, *Phys. Fluids*, **6**-2, pp.871-884.

THREE DIMENSIONAL COHERENT STRUCTURES IN THE FLOW AROUND A CIRCULAR CYLINDER BY DIRECT NUMERICAL SIMULATION

H. PERSILLON*†, M. BRAZA* AND C. WILLIAMSON†

* *Institut de Mécanique des Fluides, UMR CNRS,*
Avenue du Professeur Camille Soula,
31400 Toulouse Cedex, France

† *Mechanical and Aerospace Engineering Department,*
Cornell University,
Ithaca, NY 14853, USA

1. Introduction

The simulation of three-dimensional organized coherent structures in the wake of a circular cylinder is a topic of major interest in the domain of fundamental research, concerning the transition to turbulence. In the last five years, a few numerical simulations of the tree-dimensional wake structure appeared in the literature. Rivet (1991) has simulated by a lattice-gaz method the oblique shedding pattern in the wake of a cylinder at $Re = 74$. This pattern has been found in accordance with the experimental studies of Gerrard (1966) and Williamson (1988). Karniadakis & Triantafyllou (1992) have simulated by a spectral-element Navier Stokes solver, Patera (1984) the three-dimensional flow around a cylinder at Reynolds numbers 200 – 550 range. However, this study has not reported distorsion of the main vortex rows in the third direction.

Experimental studies by Lasheras & Meiburg (1990) for the flow past a splitter plate and by Williamson (1992) for the flow past a circular cylinder show the appearance of an undulation in the alternating vortex rows simultaneously with the creation of streamwise vorticity. The wavy pattern called mode A in the cylinder's wake by Williamson selects preferred wavelength values, depending on Reynolds numbers. As Reynolds number increases above 220, the wavy pattern persists and interacts with smaller scale streamwise vortex filaments (mode B, Williamson (1992)).

Williamson (1992) suggests decreasing wavelengths as Reynolds increases and mentions a degree of scatter in the measured values. Mansy et al (1994) found values $s/D = 3$. For Reynolds number 200, the values are in the range 3 – 5. In the Reynolds number range 230 – 1000, the wavelength values are of order 1.

Thomson et al (1995-1996) have performed a direct Navier-Stokes simulation, by using a spectral method, for the Reynolds number range 180 – 300 and have shown a wavy pattern comparable to mode A and mode B. Zhang et al (1995) also report a wavy pattern of the vortex street comparable to mode A, in the viscinity of Reynolds number 180 – 190. Apart from the mentioned three-dimensional structures, the passage from Reynolds number 180 to 230 is characterized by a discontinuity region in the Strouhal-Reynolds relation, delimited by two steps, as observed by Williamson (1988), (1992). The majority of the mentioned studies do not predict the discontinuity region. Zhang et al (1995) predict a frequency drop in the viscinity of the first discontinuity. The DNS studies by Persillon et al (1995a,b) have predicted the frequency modulation in the overall discontinuity region, for the Reynolds number range 180 – 300. They have also provided the critical Reynolds number value ($Re = 187$) for the first discontinuity, by the fully non-linear DNS approach, Persillon et al (1996). The present study investigates the development of 3D coherent structures in the cylinder's wake, in respect to mode A and vortex dislocations, by the DNS approach.

2. Theoretical equations and numerical method

The governing equations for an incompressible viscous fluid past a circular cylinder are the continuity and the Navier-Stokes equations. The equations are written in a general curvilinear coordinates system normalized by the cylinder's diameter D and the uniform stream velocity U_∞, in velocity-pressure formulation. The numerical method is based on a predictor-corrector pressure scheme (Braza, 1986) and an Alternating Direction Implicit formulation is used. This method described by Braza (1991a,b) is done by extending the numerical scheme suggested by Douglas (1962) for the diffusion three-dimensional equation. According to this scheme, the u component equation is written:
First step, which gives the u-field at the time step $\bullet$,

$$
\begin{aligned}
&-2\frac{u^\bullet}{\Delta t}|J|\Delta z - \iint_{\Gamma_1} u^\bullet G_1^n \, d\eta dz + \nu \iint_{\Gamma_1} \frac{\alpha}{J} u_\xi^\bullet \, d\eta dz - \nu \iint_{\Gamma_2} \frac{\beta}{J} u_\xi^\bullet \, d\xi dz \\
&= -2\frac{u^n}{\Delta t}|J|\Delta z + \iint_{\Gamma_1} u^n G_1^n \, d\eta dz - \nu \iint_{\Gamma_1} \frac{\alpha}{J} u_\xi^n \, d\eta dz + 2\nu \iint_{\Gamma_1} \frac{\beta}{J} u_\eta^n \, d\eta dz \\
&\quad +2\iint_{\Gamma_2} G_2^n u^n \, d\xi dz - 2\nu \iint_{\Gamma_2} \frac{\gamma}{J} u_\eta^n \, d\xi dz + 2\nu \iint_{\Gamma_2} \frac{\beta}{J} u_\xi^n \, d\xi dz \\
&\quad +2\iint_{\Gamma_3} w^n u^n J \, d\xi d\eta - 2\nu \iint_{\Gamma_3} \frac{\partial u^n}{\partial z} d\xi d\eta - 2\iiint_{\Omega} S_u |J| \, d\xi d\eta dz \quad (1)
\end{aligned}
$$

Second step, which gives the u-field at the time step $\bullet\bullet$,

$$
\begin{aligned}
&-2\frac{u^{\bullet\bullet}}{\Delta t}|J|\Delta z - \nu \iint_{\Gamma_2} u^{\bullet\bullet} G_2^n \, d\xi dz - \nu \iint_{\Gamma_1} \frac{\beta}{J} u^{\bullet\bullet} \, d\eta dz \\
&+\nu \iint_{\Gamma_2} \frac{\gamma}{J} u_\eta^{\bullet\bullet} \, d\xi dz \,, = -2\frac{u^n}{\Delta t}|J|\Delta z - \nu \iint_{\Gamma_1} \frac{\beta}{J} u^n \, d\eta dz
\end{aligned}
$$

$$-\nu \iint_{\Gamma_2} u^n \, G_2^n \, d\xi dz + \nu \iint_{\Gamma_2} \frac{\gamma}{J} u_\eta^n \, d\xi dz \qquad (2)$$

Third step, which gives the u-field at the time step $(n+1)$,

$$2\,\frac{u^{n+1}}{\Delta t}\,|J|\,\Delta z + \iint_{\Gamma_3} u^{n+1}\, w^n \, J \, d\xi d\eta - \nu \iint_{\Gamma_3} \frac{\partial u^{n+1}}{\partial z} J \, d\xi d\eta \, ,$$

$$= 2\,\frac{u^n}{\Delta t}\,|J|\,\Delta z + \iint_{\Gamma_3} u^n \, w^n \, J \, d\xi d\eta - \nu \iint_{\Gamma_3} \frac{\partial u^n}{\partial z} J \, d\xi d\eta \qquad (3)$$

These equations are written for all the nodes of the unknown u.

3. Boundary conditions

On the surface of the cylinder, the boundary conditions are those of impermeability and non-slip $v = 0$. On the upstream section, a uniform velocity profile $u = 1$ is imposed. Absorption boundary conditions, derived from non-reflecting properties of the wave equation are used in the far field outlet boundary of the computational domain, Jin & Braza (1993).

These conditions are as follows:

$$\frac{\partial F}{\partial t} + u \frac{\partial F}{\partial x} - \nu \left(\frac{\partial^2 F}{\partial y^2} + \frac{\partial^2 F}{\partial z^2} \right) = 0$$

where F represents the velocity components u, v and w.

On the spanwise direction, in order to simulate an infinite cylinder, we have adopted periodical boundary conditions. Finally, on the y-direction, in order to not confine the flow, a Neumann type boundary condition has been adopted for the u-component and a Dirichlet type boundary condition has been taken for v ant w components.

4. Results

A H-type mesh of $150x80x33$ points is used. The size of the domain has been chosen after the tests carried out by Persillon (1995), (1996).

4.1. SPANWISE LENGTH S/D=3.72

Experimental results of Williamson (1988a,b) show the existence of a region of discontinuity in the St-Re relationship, where the Strouhal number diminishes, comparing to the 2D variation. Two discontinuities delimit this region for $170 < Re < 190$ and $240 < Re < 260$. The Reynolds number interval from the first discontinuity towards values 220−230 is associated with the 'Mode A' (Williamson, 1992). This phenomenon is accompanied by the inception of vortex loops caused by the deformation of the main vortices in the spanwise direction, during the process of vortex shedding.

In order to investigate the development of the three-dimensional instability, and the modifications of the coherent vortex structures in the spanwise direction, the calculation is performed for $Re = 220$, a Reynolds number value located between the two discontinuities.

The initial conditions for the present simulation are taken as the vortex shedding flow generated by a two-dimensional numerical simulation using the same code in its two-dimensional version.

The third component (w) is weakly perturbed by a random amount at each point (level of 10^{-4}). After a short computation time, w becomes well organized (figure 1) and starts to increase. When $w \sim 10^{-2}$, the main vortices (green and violet) show a slight undulation that becomes more and more pronounced as it is shown on figure 2. It can be seen that the wake is quite regular in spite of the appearance of three-dimensional structures. Figure 3 presents the first component of the vorticity and shows the inception of small streamwise structures just behind the cylinder. These structures are well organized and their regularity is maintained far from the wall. It is important to notice that their intensity is quite small compared to the main vortices one.

On figure 4, we have plotted the Strouhal number as a function of time. The Strouhal number is calculated on the lift coefficient. It appears that this number is constant for about 80 cycles (until $t = 800$) and equal to the two-dimensional one (0.201). During this phase, the main vortices undulation becomes more and more intense as it is shown on figure 5. Between $t = 800$ and $t = 1000$, the Strouhal number decreases suddenly. For this phase, the increase of the streamwise structures intensity (pink and yellow structures) can be observed.

Beyond $t = 1000$, the Strouhal number is about 0.19. Several visualizations of the flow are shown on figure 6. The main vortices present a strong oscillation that becomes stronger far from the cylinder. The streamwise structures are not as well organized as during the previous phase. Their intensity is much more strong. It can be observed that the yellow and pink streamwise structures form vortex loops comparable with the mentioned experimental visualization. The obtained configuration by the present simulation is comparable to the 'Mode A'.

By observing the near wake flow (figure 7), it can be noticed that these streamwise structures are born in the region between the two Kàrmàn vortices in the formation zone of the main eddies. Afterwards, they travel downstream and, leaving the region between the main vortices, they are located along the upper and lower shear layers.

4.2. SPANWISE LENGTH S/D=12

In the previous section, it has been discussed that a slight and regular oscillation of the vortex rows has been simulated. It is shown that this regularity is possible according to the use of a rather small spanwise cylinder. In this section, we present the results obtained for a spanwise length $s/D = 12$.

The Strouhal number as a function of time is plotted on figure 8. Its behavior is the same as on the previous case computation, but we can observe that the decrease of the Strouhal number starts earlier because of the higher degree of freedom provided for the present flow test. At the end of the computation the Strouhal number values appear to be more scattered than in the s/D=3.72 case.

The spanwise length scale of the mode A instability has been measured by Williamson among other (figure 9). Its value λ/D is between 3 and 4. Barkley & Henderson, by using a Floquet linear stability analysis, have predicted the most unstable wavelengths for the Reynolds number range $140-220$, being in the range 2 to 3.80. The present study provides the mode A instability by the DNS approach and gives a spanwise length of this instability λ/D=4, by the present simulation.

Figure 11 presents the instantaneous vorticity field ($\omega_x = \omega_z = 0.25$) for several time values. We can observe that the flow start to loose its regularity beyond t=800, when the Strouhal number values start to decrease. For t=820, a central part of the first main vortex delays its shedding as it is shown on figure 11c and this causes a discontinuity along this vortex filament. Furthermore, the existence of a cut in the pressure eddy can be observed (figure 10). This is qualitatively comparable to a vortex dislocation pattern, observed experimentally by Williamson (1992). In the present study, this pattern is associated with a decrease of the fundamental frequency. It is noticeable that the obtention of this pattern needs to perform the DNS computation over a very long time value. This phenomenon is the origin of a further destabilization of the wake pattern, where it can be seen that the flow is no longer as regular as previously. It seems that the Mode A persists but its wavelength is less regular. This phenomenon lasts over a significant duration. However, we have not observed any periodic occurrence of it in space or in time. The intensity of the streamwise vortex structures also is found to vary, during the present loss of irregularity phase.

According to these results, the spanwise undulation of the main, alternating vortex rows seem to be produced in a natural way by the present DNS. It is therefore reasonable to suggest that these transverse modes are intrinsic characteristics of the flow transition.

5. Conclusions

The present study offers the possibility to simulate by direct three-dimensional simulation (i.e by the fully non-linear approach), the establishment of three-dimensionality in the wake past a circular cylinder at low Reynolds number. Especially, it is found that the 3D alternating vortex rows are organized according to a wavy spanwise pattern under the action of the smaller-scale streamwise vorticity, compared with the experimentally reported configuration of mode A. It has been found that the smaller-scale streamwise structures are born in the formation region, between two alternating eddies. Furthermore, it has been found that the Strouhal number decreases, after a substantial duration, comparing to its 2D computations value. This drop is associated with an irregularity obtained on the waviness of mode A and in the streamwise vorticity and with a cut occurring along the transverse main vortex row. This break of continuity in the vortex rows may be associated with the vortex dislocation phenomenon, which is examined in detail in our studies in progress.

This study has been funded by DGA/DRET contract n0 93811-46/A000, concerning the post-doc stay of H. Persillon at Cornell University in the group of Professor Williamson. The CPU time is provided by CNUSC and IDRIS national centers of France and partly by Cornell Supercomputing Center.

References

BARKLEY, D. AND HENDERSON, R. 1996 "Three-dimensional Floquet stability analysis of the wake of a circular cylinder". J. Fluid Mech, Vol. 322, pp. 215-241.
BRAZA, M., CHASSAING, P. HA MINH, H. (1986) "Numerical study and physical analysis of the pressure and velocity fields in the near wake of a circular cylinder". J. Fluid Mechanics, 1986, Vol. 165, pp. 79-130.
BRAZA, M. (1986) Thèse de Doctorat d'Etat-ès-Sciences, I.N.P. T., Décembre 1986.
BRAZA, M. (1991a) Annexe II, rapport du contrat DRET: "Etude Expérimentale et numérique du décollement de la couche limite instationnaire sur modèle oscillant.
BRAZA, M. (1991b) "Code ICARE". Rapport Interne TELET-IMFT, N0 72, Juin 1991.
DOUGLAS, J. (1962) "Alternating Direction Methods for three-space variables". Numerische Mathematik, Vol. 4, PP. 41-63.
JIN, G. and BRAZA, M. (1993) "A non-reflecting outlet boundary condition for incompresible unsteady Navier-Stokes calculations". Journal of Computational Physics, vol. 107, N° 2, Aug. 1993.
KARNIADAKIS, G. E. and TRIANTAFYLLOU, G. S. (1992) J. Fluid Mech. Vol. 238, p.1.
LASHERAS, J.C. and MEIBURG, E. "Three-dimensional vorticity modes in the wake of a flat plate" Phys. Fluids A, Vol. 2 (3), March 1990, pp. 371-380.
MANSY, H., YANG, P., WILLIAMS, DR, 1994. "Quantitative measurements of spanwise-periodic three-dimensional structures in the wake of a circular cylinder. J. Fluid Mech. 270-277.
PATERA, A.T. (1984). J. Comp. Phys., Vol. 54, 468-488.
PERSILLON, H. and BRAZA, M.(1996a) "Physical analysis of the transition to turbulence in the wake of a circular cylinder by three-dimensional Navier-Stokes simulation". J. Fluid Mechanics, submitted June 1996, revised Feb. 1997.
PERSILLON, H (1996b) Rapport de stage post-doctoral n° 93811-46/A000, DGA/DRET, Cornell University & Institut de Mécanique des Fluides de Toulouse, CNRS/I.N.P.T.
PERSILLON, H., BRAZA, M. , JIN, G. (1995a) Proceedings 5th ISOPE, Int. Offshore Mechanics and Polar Engineering Conf. the Hague, June 11-16, 1995.
PERSILLON, H. , BRAZA, M. , HA MINH, H. , WILLIAMSON, C.H.K. (1995b) Proceedings, "Non-linear instability and transition in three-dimensional boundary layers", Ed. Kluwer, selected papers, IUTAM Symposium, Manchester, July 1995.
PERSILLON, H. (1995c) Thèse de Doctorat, I.N.P.T., 2 Novembre 1995.
RIVET, J.P. (1991) Comptes rendus de l'Académie des Sciences, vol. 313, p. 1991, pp.151-158.
THOMPSON, M. , HOURIGAN, K. , SHERIDAN, J. 1994. Int. Colloq; Jets, Wakes, Shear Layers, Melbourne, Austr. April 18-20, paper 10.
WILLIAMSON, C. H. K. (1988a) "Defining a universal and continuous S trouhal-Reynolds number relationship for the laminar vortex shedding of a circular cylinder" Phys. Fluids, Vol. 31, p.2742.
WILLIAMSON, C. H. K. (1988b) "The existence of two stages in the transition to three-dimensionality of a cylinder wake", Phys. Fluids, Vol. 31, pp. 3165-3168
WILLIAMSON, C. H. K. (1992) J. Fluid Mech. Vol. 243, p. 393.
ZHANG, H. , FEY, U. , NOACK, B.R. and ECKELMANN, H. (1995) Phys. Fluids, Vol. 7, pp. 779-794.

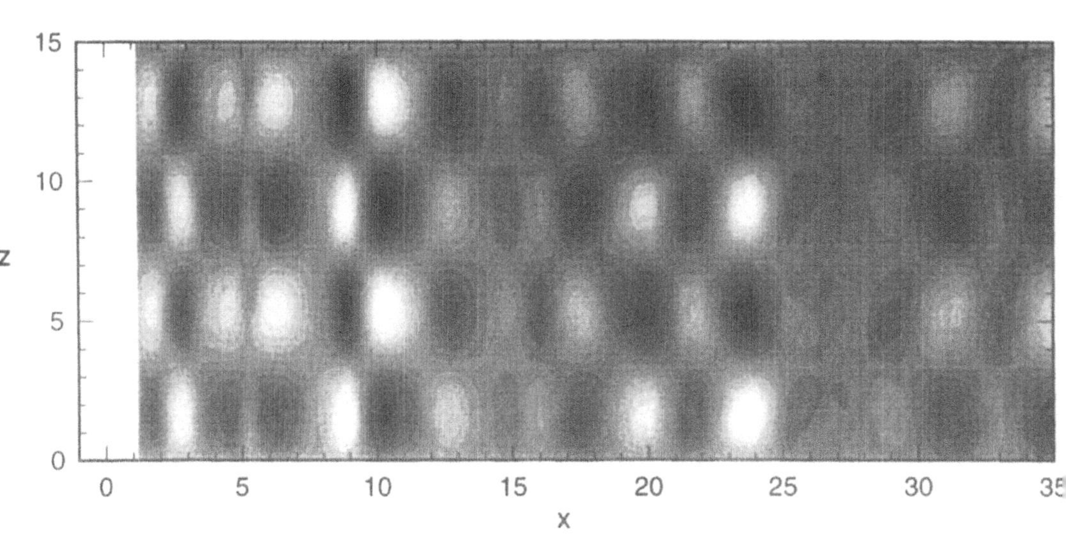

Figure 1. w-component, y/D=0

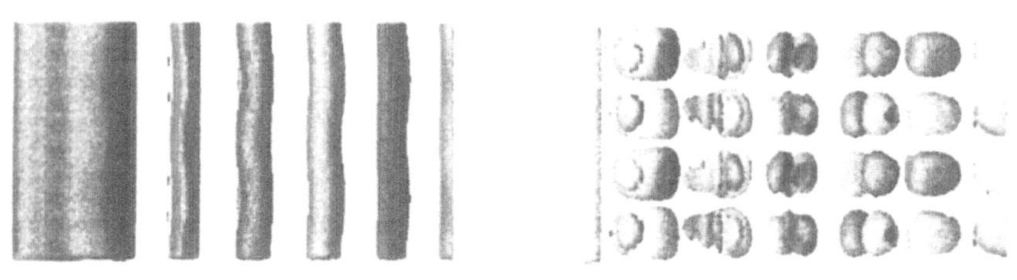

Figure 2. Isovorticity, t=780, $\omega_z = \pm 0.25$ (left), $\omega_x = \pm 0.025$ (right)

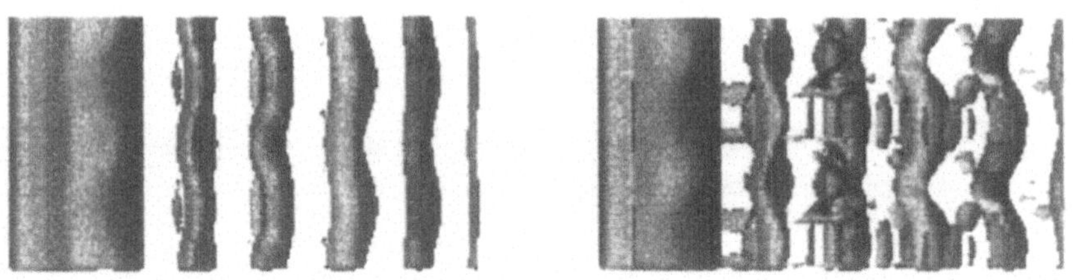

Figure 3. Isovorticity, $\omega_x = \omega_z = \pm 0.25$, t=890 (left), t=980 (right).

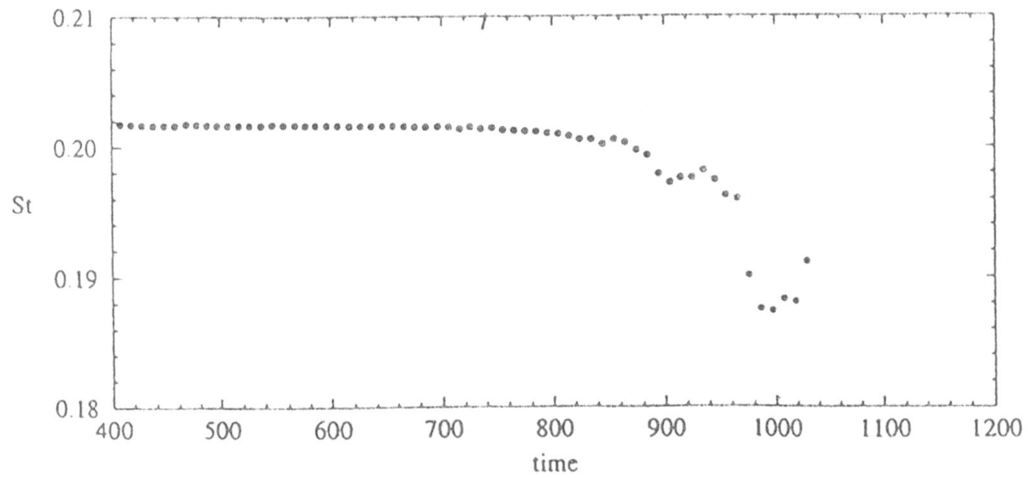

Figure 4. Strouhal number as a function of time - St=f(t).

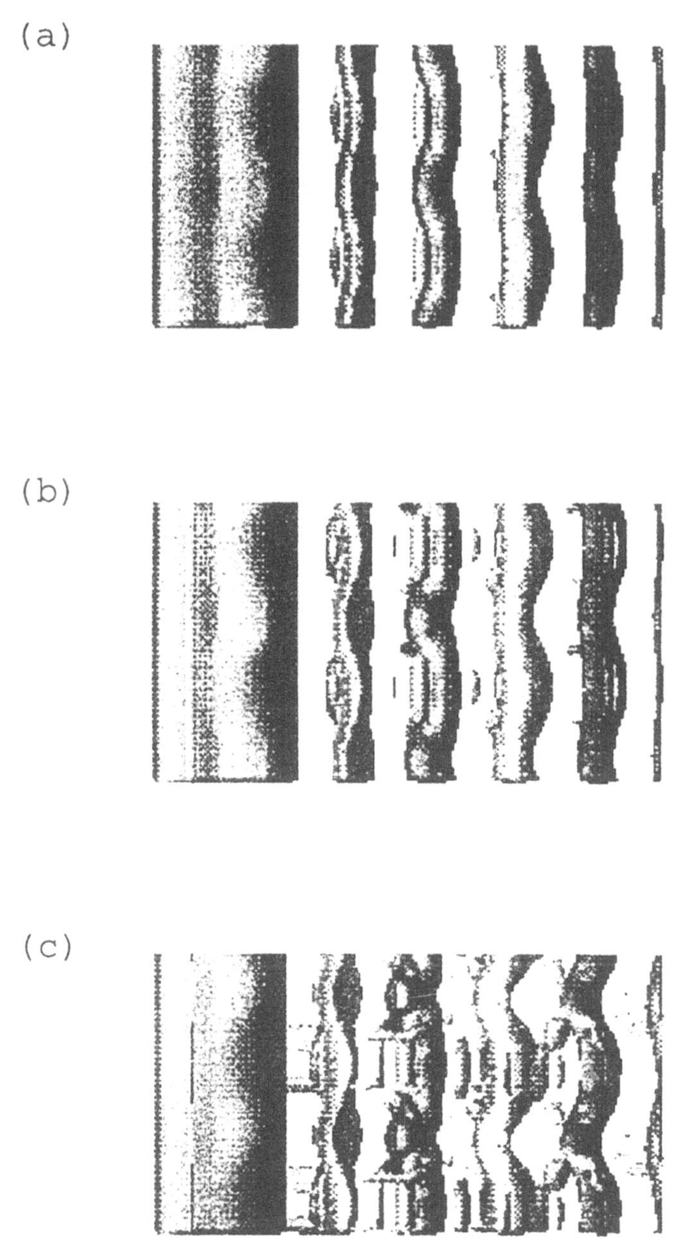

Figure 5. Isovorticity $\omega_x = \omega_z = \pm 0.25$, (a) t=890, (b) t=920, (c) t=980.

(a)

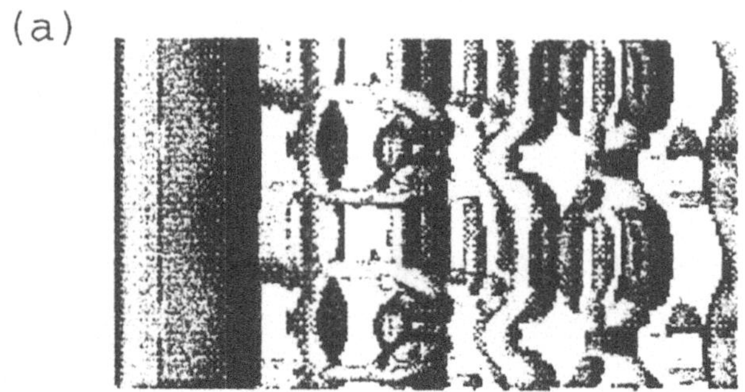

(b)

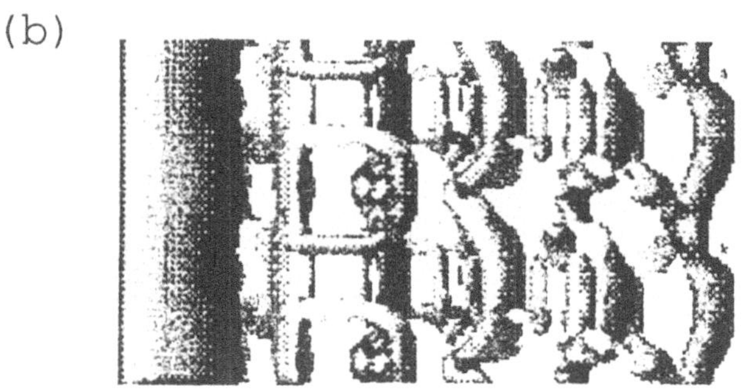

(c)

Figure 6. Isovorticity $\omega_x = \omega_z = \pm 0.25$, (a) t=1010, (b) t=1040, (c) t=1070.

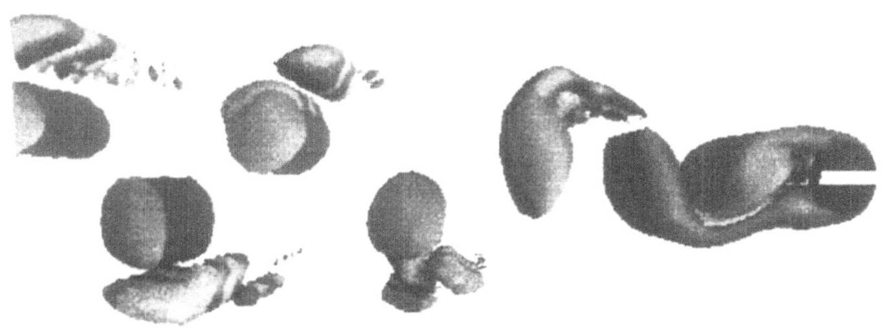

Figure 7. Isovorticity ω_x and ω_z, plane(x, y)

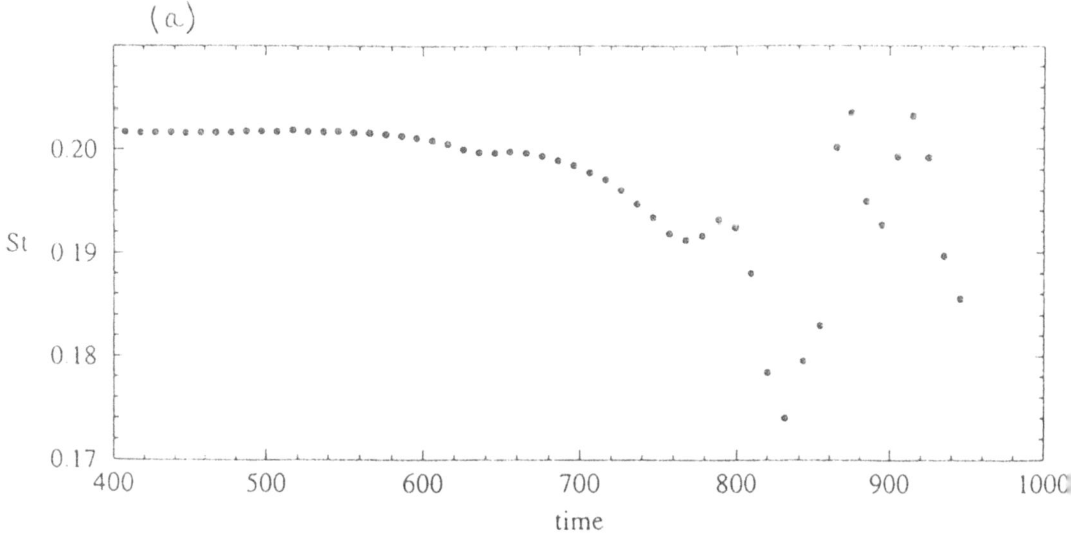

Figure 8. **Strouhal number as a function of time**

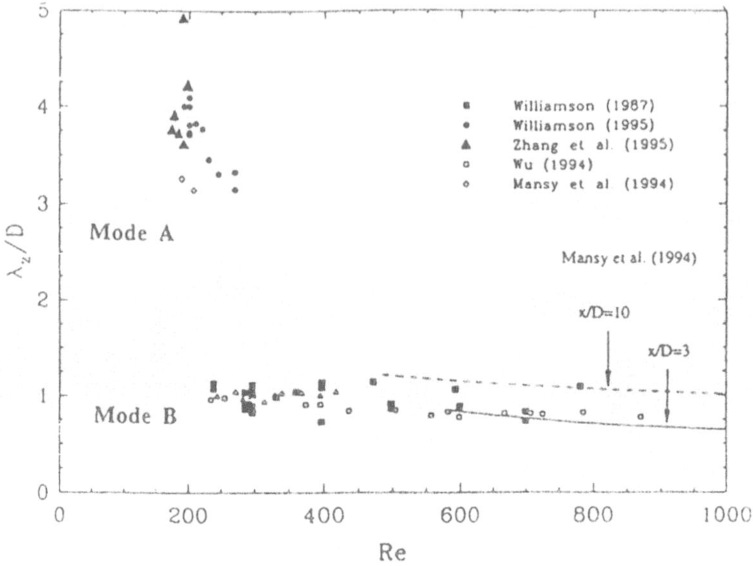

Figure 9. Spanwise instability wavelengths of the 3-D instabilities

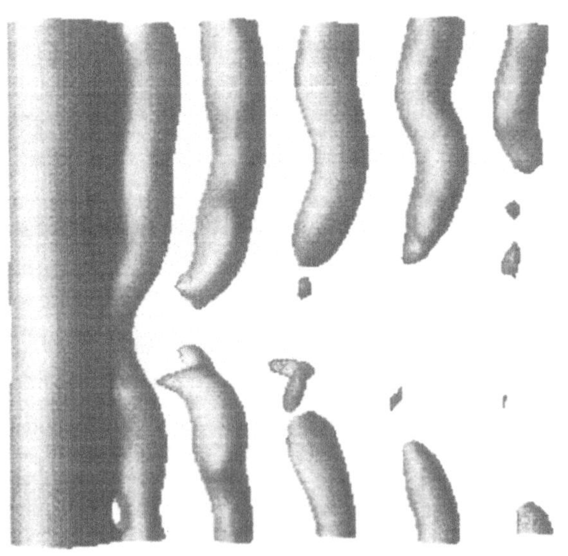

Figure 10. Isopressure $C_p = -0.25$, t=820

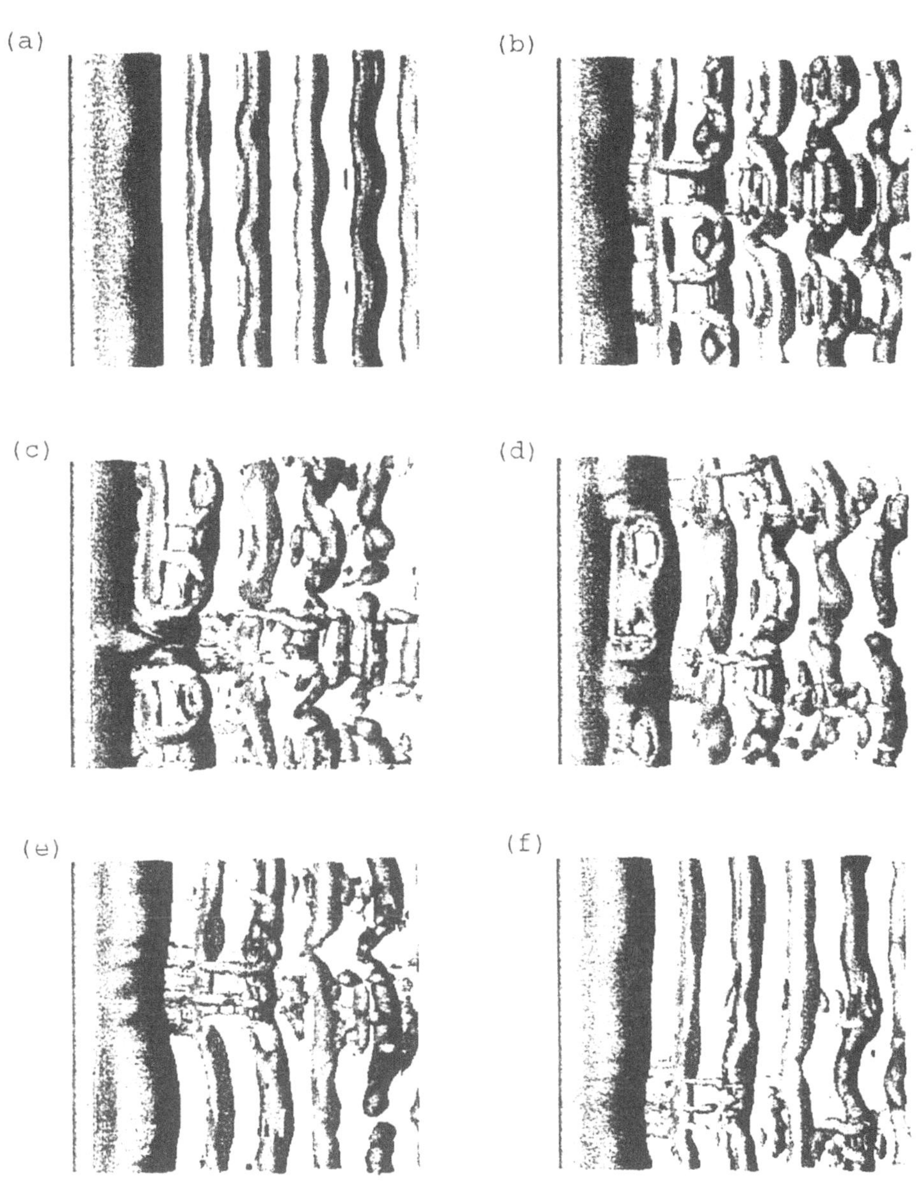

Figure 11. Isovorticity $\omega_x = \omega_z = \pm 0.25$, (a) t=680, (b) t=780, (c) t=820, (d) t=860, (e) t=900, (f) t=940.

IX. POD, LSE and Other Techniques

DETECTION and IDENTIFICATION OF NEAR-WALL COHERENT STRUCTURES THROUGH CONDITIONAL-SAMPLING

S. TARDU

Laboratoire des Ecoulements Géophysiques et Industriels
38041 Grenoble-Cédex France

1. INTRODUCTION

Local detection of coherent structures by simple means performed by Eulerian techniques through single or multiple probes, provide useful information for the near wall turbulence control, in an easy and feasible way. Such techniques are undoubtly uncomplete and suffer from the significant smearing produced by the three dimensionality of the random structure advecting past a sensor. Table 1 recapitulates the mostly used Eulerian schemes. Their commun feature is in their detection function. Indeed, almost all of them are based on the level crossings of fluctuating velocity signals, as for example the u-level , quadrant and to some extent VITA and WAG. One of the well known problems encountered in such techniques is the high sensitivity of the number of identified events to the threshold, even though the conditional averages are more or less "universal" in the inner layer, when they are properly normalized. We will first question this dependance and show that the level crossing activity may more or less be estimated by supposing that the signal is Gaussian. Based on this observation, we develop in 3.2 a simple formulation for the "ejection" frequency identified by Eulerian techniques and show the utility of this approach by considering WAG (Table 1). It is further argued that the most significant characteristic of the level crossings is the probability density function of their interarrival times, and that these characteristics differ a near wall signal from a normal one. The grouping methods dealing with the classification of the dynamicaly similar events (Tardu, 1995) are subsequently and briefly reviewed. Finaly some results concerning a simplified model linking the bursting mechanism to the shear stress distribution are presented and discussed.

2. RESULTS

2.1 - Normality of the level crossings

It is known for a while that the frequency of the zero-crossings (i.e the level crossings when the threshold is zero) of velocity signals in a boundary layer, behaves as if these signals were gaussian (Sreenivasan and al., 1983; Tardu et al., 1993; Kailasnath and Sreenivasan, 1993). For any continuous Gaussian signal one has:

$$\lambda = \sqrt{\overline{u^2} / \overline{(du/dt)^2}} = \Lambda = \frac{1}{2\pi f_0}$$

where λ is the Taylor time scale and Λ is the Liepmann scale connected to the zero crossing frequency f_0 (Rice, 1945; Ylvisaker, 1965). This equality is valid even in the buffer layer (Fig. 1) wherein the nonlinearity is strong. The zero-crossings of more intermittent signals such as the fluctuating wall normal velocity follow also quite satisfactorily this estimation as shown in Fig. 1 (the reader is referred to Tardu (1995) for the experimental set-up , data acquisition and data reduction used for the results presented here). This is quite surprising if one recalls that the flatness of v in this zone is as high as 5-6 . This good correspondance is due to a consequence of central limit theorem as also noted by Kailasnath and Sreenivasan (1993). Indeed high probability events in fluctuating velocity signals resulting in zero crossings follow approximately Gaussian results eventhough the overall pdf 's of u or v are not Gaussian.

We test here the hypothesis that the level crossings leading to the ejection and bursting frequencies are also governed by this principle. If this is the case the frequency f_L of the events identified when $u(t) < -L\sqrt{\overline{uu}}$ should be given by $f_L = e^{-L^2/2} f_0$ (Rice, 1945; p. 55). The "Gaussianity" in this expression is in the exponential term. Similarly if f_{L_1} and f_{L_2} are respectively the $L=L_1$ and $L=L_2$ level crossing frequencies, then $\ln\left(f_{L_2}/f_{L_1}\right) = \frac{L_1^2 - L_2^2}{2}$.

Fig. 2 shows the ejection frequency obtained through u-level, mu-level, Q2 and VITA techniques at $y^+= 10$ (hereafter, $^+$ denotes the variables nondimensionalized with the inner variables, i.e the viscosity ν and the shear velocity u_τ). The threshold L_2 is the convential value used in the literature, i.e $L_2= 1$ for the Q2 events and 1.3 for u-level. The modified u-level contains two thresholds, since this technique is a multiple level crossing scheme and the detection is turned on when $u(t) < -L\sqrt{\overline{uu}}$ and turned off once $u(t) > -L'\sqrt{\overline{uu}}$ with $L'=L/4$ (see Table 1). Finally the reference value for VITA in Fig. 2 is $L_1=0.4$ while the VITA integration time is $T_v^+=13$. VITA is based on the level crossings of V(t) defined in Table 1, but the threshold is proportional to $\sqrt{\overline{uu}}$ and not to $\overline{VV}$. Therefore, the level crossings of this scheme should satisfy $\ln\left(f_{L_2}/f_{L_1}\right) = \frac{L_1^2 - L_2^2}{2}\frac{\overline{uu}}{\overline{VV}}$ and this is what we plotted in Fig. 2.

The salient feature of Fig. 2 is the excellent agreement between the level crossing frequencies and the Gaussian model in a large range of thresholds values. This is quite striking for strongly non Gaussian signals as VITA and uv(t), but also for multiple level crossings such as those resulting from mu-level technique. At L=1.6 the difference between the estimated and measured Q2 frequencies is less than 10%. There are large departures from the Gaussianity begining with this critical value, and this is essentialy due to the high intermittency of v(t) in this flow region. The threshold values limiting the *normal* (Gaussian) behaviour of the level crossings are L= 2 for u-level and mu-level and k=1.5 for VITA. Similar results have been obtained further away in the inner layer.

This section may clarify some issues concerning the sensitivity of the ejection (and sweep) frequencies to the imposed thresholds.The level crossings are approximately normal as long as relatively low intermittent signals are concerned. The departure from Gaussianity is an indication of strong intermittency.

2.2 General formulation of the *linear* detection procedures

Fig.3a shows the block diagram representation of any detection procedure such as those quoted in the introduction. The input x(t) to the system is one or more flow quantities such as the fluctuating velocity components u'_i, the pressure p', the vorticity ω'_i etc.. The entries go through a detection scheme (DS) which is a linear system in most of the techniques used so far. Let $h(\tau)$ be the impulse response of the detection scheme and

$\rho(t) = \int_{-\infty}^{\infty} h(t+\tau)\, h^*(\tau)\, d\tau$ its correlation function, where * denotes the complex conjugate .

The output y(t) of the detection scheme is the input of the non linear detection function (DF) involving one or more thresholds (Fig. 3a). It is possible to obtain a general relationship for the frequency f_e of the detected events defined through $y'(t) < -L\sqrt{\overline{x'x'}}$ and, therefore to clarify the nature of the resulting conditional averages. We have shown, indeed that the level crossing characteristics of any signal follow more or less Gaussian results because of the Central Limit Theorem. One has therefore:

$$f_e = \frac{1}{\pi}\sqrt{\frac{-R''_{yy}(0)}{R_{yy}(0)}}\, e^{-L^2/2} = \frac{1}{\pi}\sqrt{\frac{R_{y'y'}(0)}{R_{yy}(0)}}\, e^{-L^2/2}$$

where ' denotes the time derivatives as usual and R stands for the autocorrelation function. Since $R_{yy}(\tau) = R_{xx}(\tau) \otimes \rho(\tau)$, with $\otimes$ beeing the convolution operator, one may easily show that:

$$f_e = \frac{1}{\pi}\sqrt{\frac{<\rho\,,\,R_{x'x'}>}{<\rho\,,\,R_{xx}>}}\;\exp\left[-\frac{L^2}{2}\frac{R_{xx}(0)}{<\rho\,,\,R_{xx}>}\right]$$

where the inner products < , > are :

$$<\rho\,,\,R_{x'x'}> = \int_{-\infty}^{\infty}\rho(\alpha)\,R_{x'x'}(\alpha)\,d\alpha \quad \text{and} \quad <\rho\,,\,R_{xx}> = \int_{-\infty}^{\infty}\rho(\alpha)\,R_{xx}(\alpha)\,d\alpha$$

The interesting physical feature of these results is straighforward. It is clearly seen that the time scale of the detected events is related to the *weighted integral scales* of the signal x(t) and its time derivative x'(t) by the detection correlation function $\rho(\alpha)$. If $h(\tau) = 1$ as in direct level-crossing techniques (u-level, quadrant etc..) then $\rho(\alpha) = \delta(0)$ and the interarrival period of the detected structures is "connected" to the corresponding Taylor scale. In other detection schemes, however, that is not as obvious. Consider for instance the window average gradient WAG developed by Antonia and Fulachier (1989). The aim of this technique is to detect the discontinuities in the streamwise u or wall normal v velocity components which characterize the large scale motion. The DS and DF of WAG applied to u(t) are given by:

$$W(t) = \frac{1}{2T_w}\left[\int_{t}^{t+T_w} u(t)dt - \int_{t-T_w}^{t} u(t)\,dt\right] \quad \text{and} \quad DF(t) = 1 \text{ if } W_t > L\sqrt{\overline{uu}}\ .$$

The WAG integration time is taken as the outer time scale $T_w \approx \frac{\delta}{U_\infty}$. We determined the impulse response function h_{Wg} of W(t) and its correlation function ρ_{Wg} (Fig. 3b). It is seen that the correlation function is negative for $\alpha > \frac{2}{3}\frac{\delta}{U_\infty}$ showing that there should be some transition from negative to positive u(t) in the detection procedure. Furthermore, the convolution through DS should accentuate the effect of the time derivatives and weaken somewhat that of the u autocorrelation, because the integral time scale of u'=du/dt is much smaller. It is expected therefore that WAG merely detects the strong accelerations. These observations are in agreement with Antonia and Fulachier (1989) and Krogstad and Antonia (1994).

There is a curious coincidence between the DS of WAG and the "mexican hat" used in wavelet analysis. Fig. 3b shows that the correlation function $\rho_{Wg}(\alpha)$ of WAG is well approximated by $\rho_{Wg}(\alpha) = \frac{3}{4k}(1-k^2\alpha^2)\exp\left[-k^2\alpha^2/2\right] = \frac{3}{4k}\,g(k\alpha)$ where g stands for the convolution operator occuring in mexican hat and $k = \frac{3}{2}T_w$. Consequently the wavelet transform w(k,t) of u is nearly equal to:

$$w(k,t) = \sqrt{k}\int_{-\infty}^{\infty} u(\tau)\,g(\,k(\tau-t))\,d\tau \;\approx \frac{4\,k^{3/2}}{3} W\left(t, T_w = \frac{2}{3}k\right) \otimes h_{Wg}(-t)$$

i.e WAG is in some way equivalent to the wavelet transform by mexican hat. The later detects local maxima, so does WAG. This example shows how the system approach may be useful in the interpretation of what we identify in the turbulent field.

It has to be emphasized here that this approach may be applied if and only if the DS is linear and the frequency of the detected events may be predicted within the hypothesis of the level crossing normality. It is meaningless for instance for VITA (Table 1) because of the square-law detectors occurring in :

$$V(t) = \frac{1}{T_v}\left[\int_{t-T_v/2}^{t+T_v/2} u^2(t)dt - \left(\int_{t-T_v/2}^{t+T_v/2} u(t)dt\right)^2\right]$$

This nonlinearity makes difficult if not impossible the investigation of the characteristics of VITA through linear system approach, eventhough it is known since a while that this scheme identifies strong shear layers. One may still make some comments. Consider the modified VITA method (called as the second modified VITA by Yuan and Dehghan, 1994) defined through $V_m(t) = \frac{1}{T_v}\int_{t-T_v/2}^{t+T_v/2} u^2(t)dt$. The entry of the system is now $x(t) = u^2(t)$ and the corresponding DS correlation $\rho_{V_m}(\alpha)$ is a triangle extending from $-2T_V$ to $2T_V$. The output of the system is related to $< \rho_{V_m},\ R_{u^2u^2} >$ and $< \rho_{V_m},\ R_{u^{2'}u^{2'}} > \infty < \rho_{V_m},\ R_{uu',uu'} >$ where uu' $= u(t)\frac{du}{dt}$. Assuming that the joint probability of u(t) and its time derivative is gaussian one may easily show that $R_{uu',uu'} = \overline{uu'}^2 + R_{uu}R_{u'u'} + R_{uu'}^2$. Therefore, the characteristics of the detected events depend not only upon the autocorrelations of u and its time derivative but also of the intercorrelation of u with $\frac{du}{dt}$. VITA looks for the large values of these correlations within a time interval of typically T_V. Note that, the autocorrelation of uu' has a specific meaning. Indeed, $\frac{\partial u}{\partial t} \approx -\overline{U}_C \frac{\partial u}{\partial x}$ by Taylor hypothesis and $\frac{\partial u}{\partial x} \approx -\frac{\partial v}{\partial y}$ by continuity (neglecting w). Now since near the wall $u \approx -\omega_z y$ and $v \approx -\frac{1}{2} y^2 \frac{\partial \omega_x}{\partial z}$ (Jiménez and Moin, 1991) VITA merely detects the spanwise vorticity zones correlated with the spanwise gradients of the streamwise vorticity layers within a given time interval, i.e $\overline{\frac{\partial \omega_x}{\partial z} \omega_z}$.

The analysis becomes quite complex when different schemes are combined (Morrison et al., 1992: VITA+ level crossing; Sullivan et al., 1994 : VITA+wavelets) This is also and unfortunately the case in pattern recognition techniques (Wallace et al., 1977: TPAV) .

2.3 Grouped and solitary events

We emphasized in 3.1 the approximate normality of the level crossing frequency of the near wall velocity signals. We question here the second important characteristic of this process, namely the probability distribution $p(\tau)$ of the crossing intervals.

One of the peculiarity of both zero and level crossings of near wall signals is that the cumulative probability distribution of their interrarival times consists of two different exponentials (Tiederman, 1988; Kailasnath and Sreenivasan, 1993). Fig. 4 shows $P(\tau)$ of the Q2 events at y^+=12. It is clearly seen that for small interarrival times $P(\tau) \infty \exp(-\alpha_1\tau)$ and that for large ones $P(\tau) \infty \exp(-\alpha_2\tau)$ with $\alpha_1 \neq \alpha_2$. The break point separating small and large interarrival times differentiate between cluster of events (also called bursts with multiple ejections, BME) and solitary occurences (bursts with single ejections, BSE).

It is know asked whether this feature of $P(\tau)$ is proper to the near wall turbulence events, or not. More clearly, does any normal process involve with similar pdf of level crossings? For a general Gaussian process, the exact axis-crossing interval distribution is unknown (Blake and Lindsey, 1973). First of all "the hypothesis that successive zero crossing intervals of Gaussian processes are independent" is not correct (Rainal, 1962). In other words $P(\tau)$ deviates from "Poissonianity" for some normal signals. An approximate form of the probability density $p(\tau)$ for small times has been obtained by McFadden (1956) :

$$p(\tau) \propto \frac{R(\tau)[R'(\tau)]^2 + [1 - R^2(\tau)]\, R''(\tau)}{[1 - R^2(\tau)]^{3/2}}$$

It is seen that the pdf of zero-crossings (and of the level crossings to some extend) depend on the autocorrelation fonction and its time derivatives. Several experiments conducted with large band and pass band noise do not reveal the existance of clearly identifiable multiple modes (local maxima) (Rainal, 1962, Mimaki and Munakata, 1978). Therefore, it would not be so mistaken (although with caution) to claim out that the double Poissonian trend observed in Fig. 4 is a trace of real physical phenomena proper to the near wall turbulence.

The local behaviour of $P(\tau)$ near the break point is a mixture of two different cpd which may false the grouping. The double structure of the pdf, on the other hand, is totaly absent in VITA interarrival times. Furthermore near the viscous layer the break point is never clearly seen. This is due to a stronger mixing of the two pdf's and not to the domination of one scale on the other as argued by Kailasnath and Sreenivasan (1993). The specific pattern recognition technique which split up the events in dynamicaly similar groups developed by Tardu (1995) shows clearly that the multiple and solitary events still exist in the viscous sublayer.

The present author has claimed for a while that the multiple and solitary events are not simply an artifact of the detection schemes, but that they result from real different mechanisms. Some of the results confirming this point of view have recently been published (Tardu, 1995, 1997).

2.4 Models for near wall turbulence

Haritonidis (1988) developed a simple near wall model which relates the mean velocity profile and the shear stress distributions to the bursting frequency in the buffer layer. Its model has similarities with that of Landahl (1990). The streamwise velocity is first related to the shear in the same way as Prandtl, i.e $u = -l \frac{\partial \overline{U}}{\partial y}$ where l is the instantaneous mixing length connected exactly to v through $v = \frac{Dl}{Dt}$. The Re shear stress is therefore $-\overline{uv} = \frac{1}{2} \overline{\frac{Dl^2}{Dt}} \frac{\partial \overline{U}}{\partial y}$. The second step is to suppose that $-\overline{uv}$ is produced intermittently and only during the bursts and to assume $\overline{\frac{Dl^2}{Dt}} = \frac{1}{\Delta t_b} \int_0^{\Delta t_b} \frac{Dl^2}{Dt}\, dt$ (Haritonidis, 1988; p. 303). Let us consider only the Q2 events, i.e the ejections and decompose $-\overline{uv}_{Q2} = -\overline{uv}_{MEB} - \overline{uv}_{SEB}$ into the contributions of multiple ejection bursts (MEB) with n ejections and solitary bursts (SEB). The direct consequence of the relationships given above is $-\overline{uv}_{MEB} = \frac{1}{2} n l_{MEB}^2 f_{MEB} \frac{\partial \overline{U}}{\partial y}$ and $-\overline{uv}_{SEB} = \frac{1}{2} l_{SEB}^2 f_{SEB} \frac{\partial \overline{U}}{\partial y}$ where $f_{(\,)}$ is the frequency and $l_{(\,)}$ is the corresponding mixing length. We measured the quantities involving in these relationships and deduced l_{MEB} and l_{SEB} . Fig. 5 shows the resulting profiles. Near the viscous sublayer, the mixing lengths behave like $l^+ = y^+$, i.e with a slop larger than the von Karman constant. This is clearly due to the strongly low shear zone created by the quasi-streamwise vortices at the ejection side (Fig. 6). One may show indeed that the spanwise averaged shear reads for:

$$\frac{D\omega'_z}{Dt} \approx -\left[\frac{\partial <u>}{\partial y} + \frac{\partial u'}{\partial y}\right]\frac{\partial w'}{\partial z} + \nu \nabla^2 \omega'_z \approx \frac{\partial <u>}{\partial y}\frac{\partial v'}{\partial y} + \nu \nabla^2 \omega'_z$$

near the quasi-streamwise structures with $\partial/\partial x \approx 0$. This relationship indicates that the positive and negative spanwise vorticity zones are reinforced near the stagnation points induced by the QSV's, respectively at the ejection and sweep sides, where $\frac{\partial v'}{\partial y}$ has the same sign as ω'_z. In the ejection side, therefore $u' \approx -\omega'_z y$ <<0 and that is detected by the Q2 events in this zone. The slop of the mixing length reaches the von Karman constant only in the high buffer layer $y^+> 15$. Two main results obtained here have to be emphasized. First the bursting frequencies obtained by Tardu (1995) lead nicely to the expected behaviour of the mixing length at $y^+>15$. Secondly, the steeping of the mixing lentgh related to the Q2 events in the low buffer layer reveals the increase of the near wall activity attached to the coherent structures.

3. Conclusion

* The level crossing frequency of the wall turbulent signals are Gaussian in a large range of imposed thresholds.

*The pdf of the interarrival times is a characteristic proper to the near wall phenomena.

*Clusters of ejections (MEB) and solitary events (SEB) result from different mechanisms.

* The instantaneous mixing length varies like y in the low buffer region and like κy at $y^+>15$ for Q2 events.

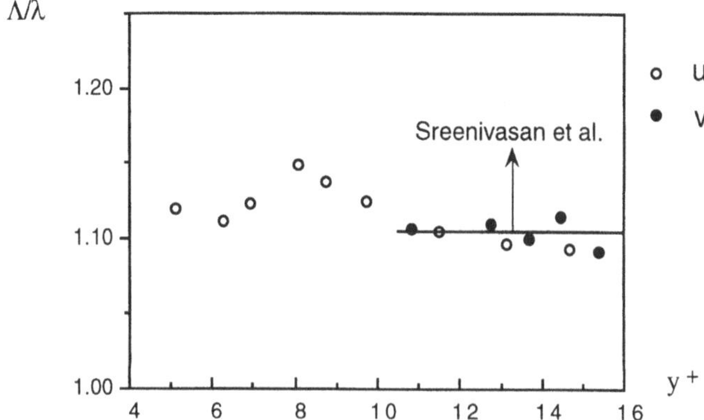

Figure 1 Ratio of the Liepman to Taylor time scales in the buffer layer. Data from Tardu et al., 1993.

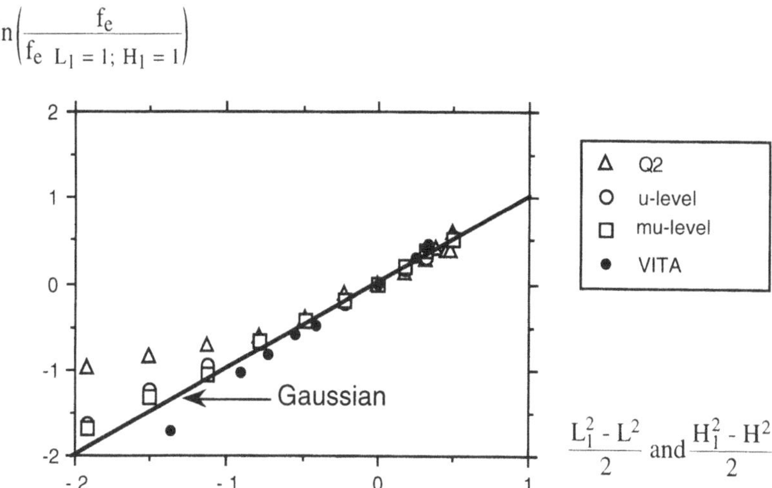

Figure 2 The ejection frequency obtained at $y^+=10$ by Q2, mu-level , u-level techniques and VITA ; $Re_\theta = 1000$;

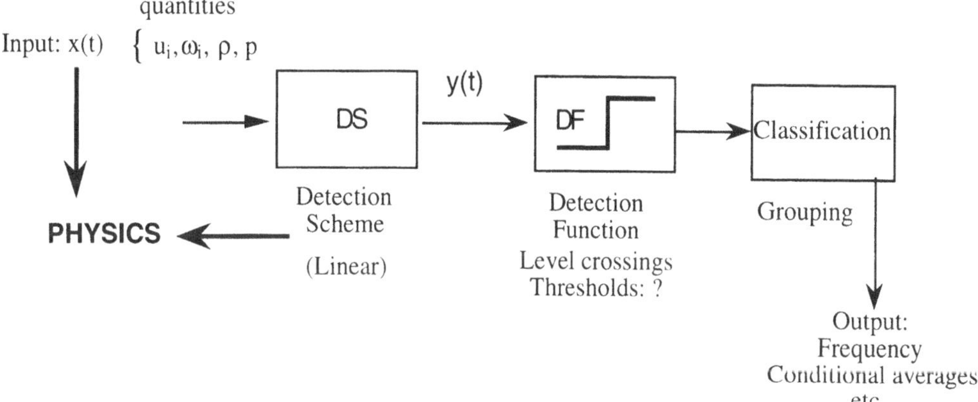

Figure 3a : Conditional averaging

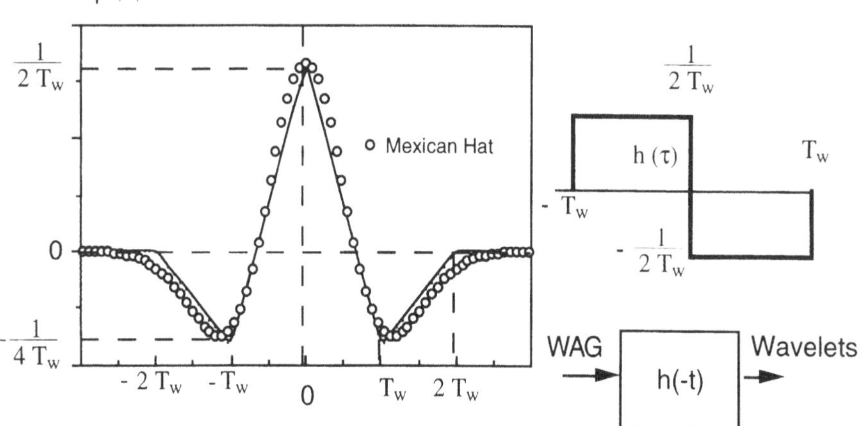

Figure 3b The impulse response and the autocorrelation of the WAG detection scheme. The circles show the corresponding Mexican Hat.

Mixture of two different probability densities:

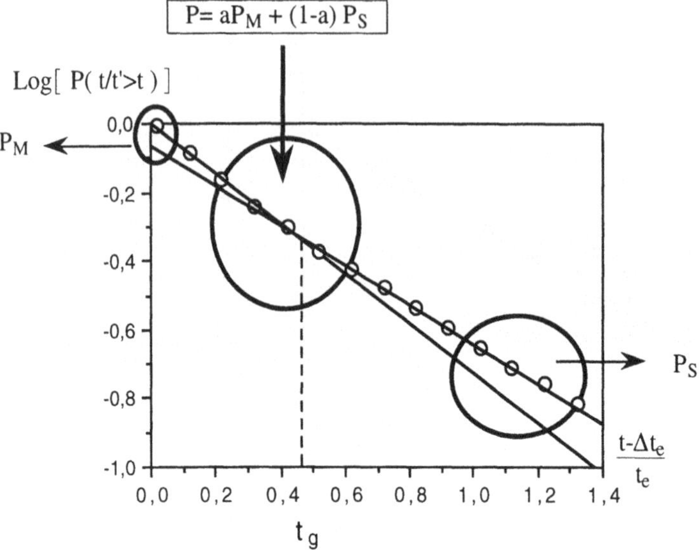

Figure 4 Cumulative probability density function of the interarrival times between Q2 events; $y^+=12$.

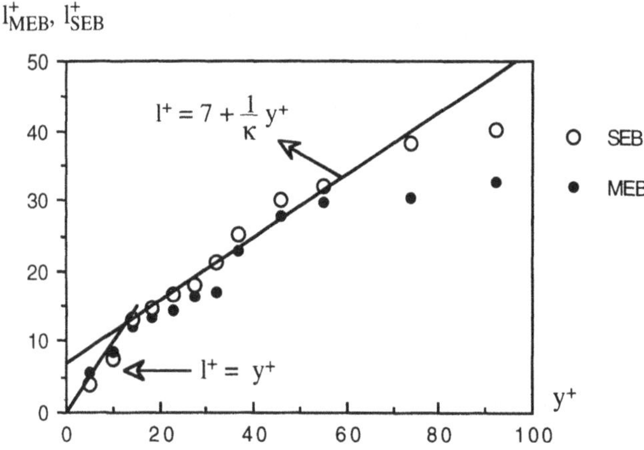

Figure 5 The "mean" mixing length resulting from Haritonidis' model.

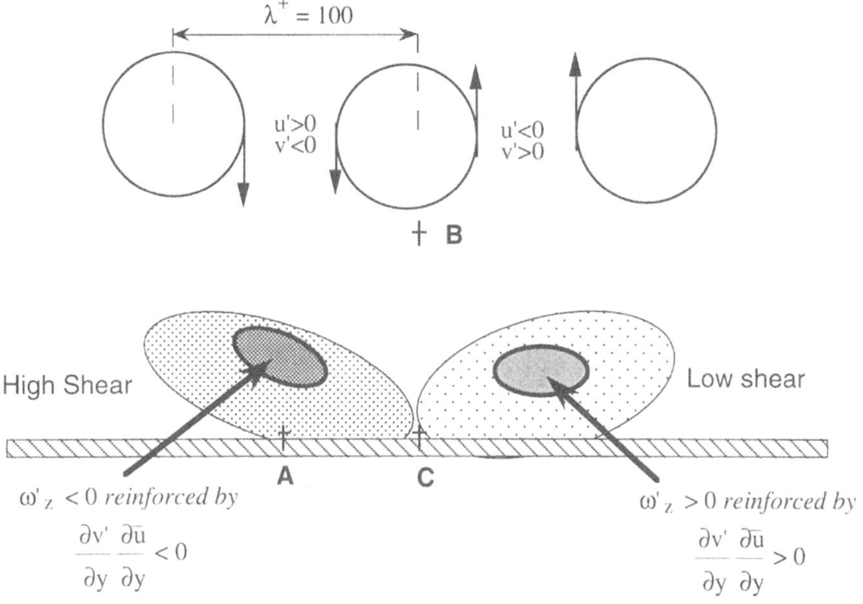

Figure 6 The quasi-streamwise structures and the shear they induce near the wall

References

Antonia R.,A., Fulachier, L. 1989 "Topology of a turbulent boundary layer with and without wall suction" J. Fluid Mech. 198, 429-451.

Blake I.F., Lindsey W.C., 1973 "Level crossing problems for random processes" IEEE Trans. Inform. Theory, 19, 295-315

Bogard, D.G, Tiederman, W.G."Burst detection by a single point velocity measurements" J.Fluid Mech., 162, 389.

Bogard, D.G, Tiederman, W.G."Characteristics of ejections in turbulent channel flow" J.Fluid Mech., 179, 1.

Jiménez J., Moin P., 1991 "The minimal flow unit in near-wall turbulence" J. Fluid Mech., 225, 213-240

Kailasnath P., Sreenivasan K.R. , 1993 "Zero crossings of velocity fluctuations in turbulent boundary layers" Phys. Fluids 5(11), 2879-2885

Krogstad, P.Ä , Antonia R.A. 1994 "Structure of turbulent boundary layers on smooth and rough walls" J. Fluid Mech., 277, 1-23

Luchik, T.S., Tiederman, W.G., 1987 "Time scale and structure of ejections and bursts in turbulent channel flows" J. Fluid Mech., 174, 529.

Morrison J.F., Subramanian C.S., Bradshaw P., 1992 "Bursts and the law of the wall in turbulent boundary layers" J. Fluid Mech.,241,75-108

Rice, S.O. , 1945 " Mathematical analysis of random noise" Bell Syst. Tech. J. 24, 46.

Sreenivasan K. R., Prabhu A., Narasimha R., 1983 "Zero-crossings in turbulent signals" J. Fluid Mech., 137, 251

Sullivan P., Day M., Pollard A.,1994 "Enhanced vita technique for turbulent structure identification" Exp. Fluids , 18, 10-16

Tardu S., Truong T.V., Tanguay B., 1993 "Bursting and structure of the turbulence in an internal flow manipulated by riblets" Appl. Sc. Research 50, 189-213

Tardu S. 1995 "Characteristics of single and clusters of bursting events in the inner layer; Part 1 : VITA events" Exp. Fluids, 20, 112-124

Tardu S., Binder G., 1997 "Reaction of bursting to an oscillating homogeneous pressure gradient" Eur. J. Mech. B/Fluids 16, 89-120
Tiederman, W.G. , 1989 " Eulerian detection of turbulent bursts. In Proceedings of Zaric International Symposium on Near Wall Turbulence" Ed. Kline and Afgan, Hemisphere.
Wallace J.M., Brodkey R., Eckelmann H., 1977 "Pattern-recognized structures in bounded turbulent shear flows" J. Fluid Mech.,83, 673-693
Yuan Y., Mokhtarzadeh-Dehghan M.R., 1994 "A comparison of conditionaly-sampling methods used to detect coherent structures in turbulent boundary layers" Phys. Fluids, 6, 2039-2057.

ORGANIZED STRUCTURE DYNAMICS IN A TURBULENT ROUND JET

JOSEPH H. CITRINITI
Thermo- and Fluid Dynamics
Chalmers University of Technology
S-412 96 Göteborg, Sweden

WILLIAM K. GEORGE
Dept. of Mechanical and Aerospace Engineering
State University of New York at Buffalo
Buffalo, NY 14260

Abstract

The dynamics of large scale azimuthally-coherent structures obtained in a high Reynolds-number axisymmetric mixing layer are presented. The structure dynamics are obtained by application of the Proper Orthogonal Decomposition (POD) to an ensemble of realizations of the streamwise velocity field obtained from 138 probes located 3 diameters downstream of the nozzle exit. The velocity field is measured at all 138 positions simultaneously, making it possible to obtain the instantaneous coefficients of the POD modes as well as to extract the large scale structure.

Pictures showing the dynamics of the structure interaction in the mixing layer demonstrate the importance of the mode-0, 3, 4, 5 and 6 azimuthal modes in both the entrainment and advection of fluid in the layer. Azimuthally coherent structures, which are remnants of circular vortex rings produced in the shear layer, appear near the potential core at regular intervals corresponding to the Strouhal frequency. The region between rings is dominated by a high azimuthal mode structures which appear near the outside of the layer. These structures do not appear to be singular but rather are made up of pairs of counter-rotating streamwise vortices that are very similar to the ribs seen in simulations of the axisymmetric layer and in plane mixing layers. The vortices advect high-momentum fluid toward the outside of the layer and low-momentum fluid toward the potential core.

1 Introduction

Understanding the dynamics of coherent structures in the axisymmetric mixing layer is of primary concern in many engineering applications such as understanding and controlling noise suppression, mixing enhancement, near-field entrainment and shear layer growth. Vital to the success of these endeavors is a clear understanding of the naturally occurring structures in the layer and, especially, the structure dynamics. The difficulty associated with eduction of large scale structures is the failing of most eduction techniques to provide an objective means to identify the structure while allowing a study of the structure dynamics. One technique which has been very successful in accomplishing both objectives is the Proper Orthogonal Decomposition (POD).

The POD has been applied to simultaneous measurements of the streamwise velocity field in the axisymmetric mixing layer at 138 positions with hot-wire anemometer probes. The POD technique is used to extract the large scale structure by using the first POD eigenmode to objectively define the large scale structure. The instantaneous velocity field is reconstructed using random coefficients obtained through Galerkin projection of the instantaneous velocity field on the POD eigenfunctions. The velocity field can then be partially reconstructed using only the first POD mode and, in this way, a view of the velocity field as produced by the large scale structure in the layer can be objectively analyzed. The temporal sequencing of events and a discussion of the naturally occurring structure in the layer are also presented.

2 Methods

2.1 Proper orthogonal decomposition

The POD seeks the most energetic fluctuations in a random vector field; its ability to extract large scale structure in a turbulent velocity field is well established, (Herzog 1986, Glauser *et al.* 1987, Moin and Moser 1989, Delville *et al.* 1990, Citriniti and George 1997) The structure is represented by an ordered set of orthogonal eigenfunctions, $\vartheta_i(\vec{x}, t)$, that are defined by the maximization of their normalized mean square projection on the velocity vector, $u_i(\vec{x}, t)$,(Lumley, 1967).

The maximization is performed via the calculus of variations and the result is an integral eigenvalue equation of the Fredholm type,

$$\int R_{i,j}(\vec{x}, \vec{x}', t, t')\vartheta_j(\vec{x}', t')d\vec{x}'dt' = \lambda\vartheta_i(\vec{x}, t) \tag{1}$$

where the symmetric kernel of this equation is the two point correlation tensor

$$R_{i,j}(\vec{x}, \vec{x}', t, t') = < u_i(\vec{x}, t)u_j(\vec{x}', t') > \tag{2}$$

and λ is the eigenvalue, (Lumley, 1967).

Solution of this equation produces the eigenfunctions and Galerkin projection of the instantaneous velocity on the basis determines the coefficients of the eigenfunctions. The velocity field can then be reconstructed via the POD; the form of this equation for the axisymmetric mixing layer is,

$$\hat{u}_i^{nmf}(r, m, f) = \sum_{n=1}^{N} \hat{a}_n(m, f)\phi_i^{(n)}(r, m, f) \tag{3}$$

where $n = 1, 2, 3\ldots$ represents the discrete nature of the solution set, $(x_1, x_2, x_3) = (x, r, \theta)$. The POD eigenfunctions, $\phi_i^{(n)}(r, m, f)$, and coefficients, $\hat{a}_n(m, f)$, are decomposed in frequency, f, and azimuthal mode number, m, (Citriniti, 1996). Performing partial sums in equation 3 (*i.e.* $N = 1, 2, 3, \ldots$) provides a way to visualize

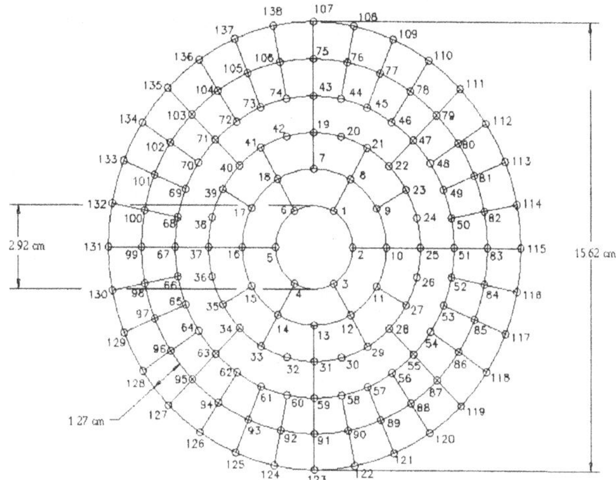

Figure 1: Spatial sampling grid for application of the POD to the axisymmetric mixing layer at $x/d = 3$ incorporating 138 simultaneous-sample, single-wire hot-wire probes.

different energy-weighted views of the flow. It will be shown that setting $N = 1$ effectively filters out the small scale structure and leaves an unobscured view of the large scale, or coherent, structure in the axisymmetric mixing layer.

2.2 Experiment

The flow field at 3 diameters downstream of the nozzle is representative of the fully developed mixing layer. At this position, the POD technique has been applied to realizations of the flow field made with 138 simultaneous operating single-wire hot-wire anemometer probes, *v.* figure 1. For the exit velocity of 12.5 m/s, the Reynolds number based on nozzle diameter, d, is 80,000. The free-stream turbulence intensity at the jet exit is 0.35% and the boundary layer at the jet exit was turbulent with an approximate thickness of 1.2 mm. The mean velocity profile was flat to within 0.1%.

The sampling rate of each of the simultaneously sampled 138 hot-wires was 2048 Hz to satisfy the Nyquist criterion which must be greater than twice the 800 Hz corner frequency on the low-pass anti-alias filters. There were 300 blocks of 1024 samples producing a bandwidth of 2 Hz and a block length of 0.5 s.

The statistics of the streamwise velocity field, as measured by the sampling grid, demonstrate that a fully-developed axisymmetric shear layer has been formed, *v.* figure 2(a). The spectral character of the velocity field is presented in figure 2(b). The energy distribution at the potential core, $r/d = 0.15$, is found to peak at about 100 Hz which is the natural passage frequency of the ring instability at

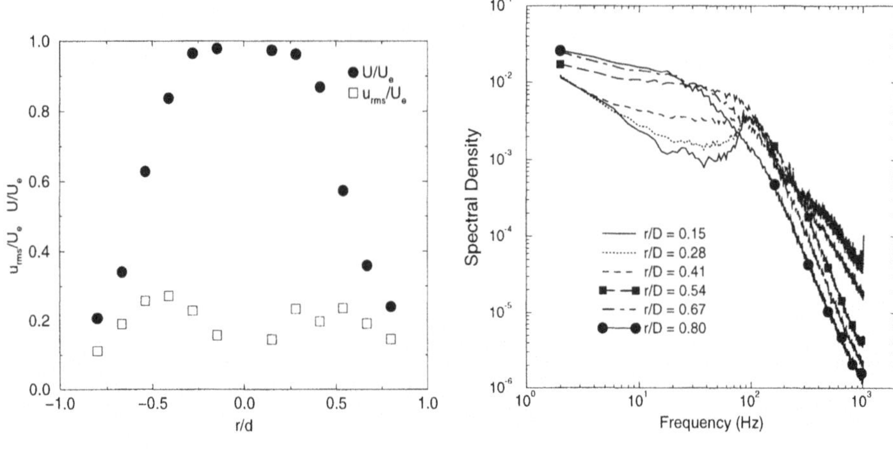

(a) Normalized mean and rms velocity statistics in the shear layer.

(b) Spectra at the 6 radial positions across the shear layer.

Figure 2: Statistical nature of the velocity field in the shear layer at $x/d = 3$.

the potential core, (Zaman and Hussain, 1984). This corresponds to a Strouhal frequency $(=fd/U_e) = 0.78$.

3 Results

3.1 Distribution of kinetic energy

The Hilbert-Schmidt theory which governs the solutions to equation 1 dictates that in this case the summation of all eigenvalues over n, m and f produces the total streamwise kinetic energy in the flow (Lumley, 1970). The distribution of the kinetic energy between radial POD and azimuthal Fourier modes can then be calculated by summing over various terms. A parameter, ξ, which is such a measure is defined by,

$$\xi = \frac{\sum_f \lambda^{(n)}(m,f)}{\sum_n \sum_m \sum_f \lambda^{(n)}(m,f)}. \tag{4}$$

The azimuthal mode energy distribution for the first 5 radial POD modes is plotted in figure 3 for the streamwise velocity field measured at $x/d = 3$, (Citriniti and George, 1997). The predominance of the first POD mode (as determined by the first eigenvalue) is indicated by the fact that 67% of the kinetic energy in the flow is contained in this eigenvalue. More importantly, the distribution of energy in azimuthal mode numbers within the first POD mode demonstrates the importance of the 0, 3, 4, 5 and 6 modes in the shear layer at $x/d = 3$.

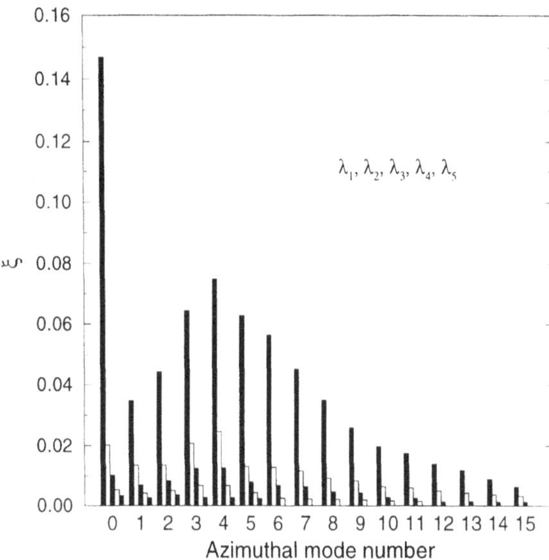

Figure 3: Azimuthal-mode kinetic energy distribution in the first 5 POD eigenvalues, from Citriniti and George (1997).

3.2 Extraction of large scale structure

The POD reconstructed velocity field can be compared to the original velocity to see if it is extracting structure from the seemingly random flow. Figure 4 shows surface mesh plots of the magnitude of the streamwise velocity at a single time step, $t_n = 413$, obtained directly form the hot-wire probes. Also shown are the corresponding velocity fields resulting from the POD reconstruction using only the first radial POD mode (*i.e.* $N = 1$ in equation 3). Part (a) of this figure shows the original velocity which contains many peaks. It is difficult to determine which peaks are due to spatially coherent structures and which are due to small scale perturbations. In part (b) of this figure, the velocity field resulting from the POD reconstruction is shown in which only the first radial POD mode and all azimuthal modes are included (*i.e.* $N = 1$ and $m = 0-15$ in equation 3). The POD has filtered out many of the small scale perturbations while retaining the features representative of the spatially coherent structure. Since the higher azimuthal modes ($m \geq 7$) will not contribute significantly to the spatially coherent structures, they can be removed as is done in figure 4(d).

Further refinement of the large scale structure is accomplished by removing other unimportant azimuthal modes from the POD reconstruction. The results of § 3.1 indicated that much of the energy in the flow was contained in azimuthal modes 0, 3, 4, 5 and 6 (*v.* figure 3). If only these modes are included in the representation, a further filtering of the small scale structure can be accomplished, *v.* (d) in figure 4. It should be noted that this decision is not made *ad hoc* since the POD has provided the

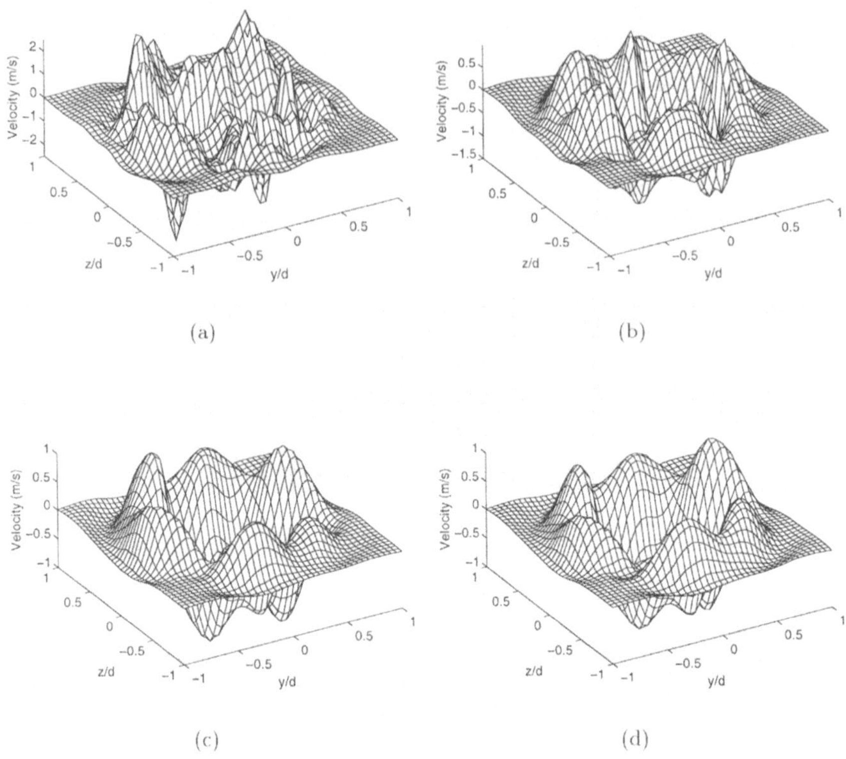

Figure 4: Surface mesh of streamwise velocity field at x/d = 3 with $t_n = 413$ from (a) the original hot-wire measurements and (b) the POD reconstruction (*v.* equation 3) using only the first POD mode and including; all azimuthal modes, *i.e.* $m = 0 - 15$ (c) $m = 0 - 6$ and (d) $m = 0, 3, 4, 5, 6$

information that led to the exclusion of these unimportant modes, namely through figure 3. Fig. 4(d) shows that the first POD mode including the azimuthal modes 0, 3, 4, 5 and 6 sufficiently recovers the large scale structure. Thus the POD has reduced an infinitely dimensional problem to one involving only 5 parameters.

3.3 Structure dynamics

Five sequential frames of the reconstructed streamwise velocity (with the mean subtracted off) using the first POD mode, *i.e.* $N = 1$ in equation 3, are shown in figure 5. The time index t_n is used to label the 1024 single images in the data blocks, the real time can be calculated by $t = t_n \Delta t$ where $\Delta t = 0.488$ ms. The portion of the mixing layer between $0.5 < r/d < 0.8$ is found to be dominated by a 5 and

6 mode structure, *v.* figure 5a. Citriniti and George (1997) have suggested that the structure causing this flow pattern are counter-rotating, streamwise vortex pairs similar to the Bernal-Roshko structures found in the plane mixing layer. Further, a model has been developed describing the dynamics of the large scale structure in the mixing layer which consists of two main events. The first is the passage of the azimuthally coherent ring, or remnant of a ring (*v.* figure 5c) which engulfs fluid from the irrotational outer flow (note that high streamwise-momentum fluid is represented by white in figure 5 while low momentum fluid is black). The second is the advection of fluid by the streamwise vortex pairs which "ride" on the induced velocity field of the rings and are stretched by the high extensional strain region between ring structures (Citriniti and George, 1997). The 6 mode structure of alternating high- and low-momentum fluid in the azimuthal direction (figure 5(a)) is produced by the counter rotating streamwise vortices (streamwise vortex centers are indicated by the plus and minus signs on the figure).

In figure 5(c) the remnants of a circular vortex ring pushes high-momentum fluid through its core. After the ring passes, see figure 5(d), the streamwise vortices are stretched and strengthened by the high extensional strain in the inter-ring regions, see figure 5(e). Finally, the shear layer returns to a state similar to that which started this life-cycle, compare figure 5(a) and figure 5(f), indicating the repetitive nature of the large scale structure life-cycle[1]. The total time from figure 5(a) to figure 5(f) is 9.76 ms which corresponds to a frequency of 100 Hz. This is the same as the peak frequency in the spectrum near the potential core which implies that the POD has indeed extracted the most energetic events in the layer.

4 Conclusions

The Proper Orthogonal Decomposition has been shown to be a useful tool in the analysis of large scale turbulent structure identification. The ability of the POD to extract structure objectively has been shown via visual and numerical means. It is possible to define a life cycle of the large scale structure dynamics in the mixing layer at $x/d = 3$ by using the POD. The first event in the life of the coherent structure is the bursting associated with the passage of an azimuthally coherent structure, or "ring", near the potential core. The induced velocity field between successive rings creates a high strain field that seems to be play an important part in the evolution of counter-rotating streamwise vortex pairs. These concepts are an extension of the model for the shear layer dynamics proposed by Glauser and George (1987).

The higher azimuthal Fourier mode structure, associated with azimuthal modes 3, 5 and 6, are created by streamwise vortex pairs which advect fluid into and out of the potential core of the jet, similar to the action of a side-jet. They are usually found near the outside of the layer ($r/d > 0.5$) and therefore are not convected very

[1]The reader is referred to the movies generated from this data set at the authors (JHC) internet web page accessible via the UB Turbulence Research Laboratory home page at http://www.eng.buffalo.edu/Research/trl/

quickly downstream. They tend to exist for many ring passages. The 0-mode and 4-mode structures appear most often near the potential core and thus are expected to represent the perturbed remnants of the Kelvin-Helmholtz "ring" generated in the shear layer. They are very fast moving and are highly energetic, advecting large amounts of fluid in the short time they are in the layer.

5 Acknowledgments

This work was completed with the support of the National Science Foundation grant number CTS-9102863.

References

Citriniti, J.H. and George, W.K. (1997). An application of the proper orthogonal decompostion to the axisymmetric mixing layer. part 2: reconstruction of the global velocity field. *J. Fluid Mech.* . Submitted for publication.

Citriniti, J.H. (1996), *Experimental Investigation Into the Dynamics of the Axisymmetric Mixing Layer Utilizing the Proper Orthogonal Decomposition.* PhD thesis, State University of New York at Buffalo.

Delville, J., Bellinin, S., and Bonnet, J.P. (1990). Use of the proper orthogonal decomposition in a plane turbulent mixing layer. In Metias, O. and Lesieur, M., editors, *Turbulence 89: organized structures and turbulence in fluid mechanics*, pages 75–90. Kluwer Academic Publishers.

Glauser, M.N. and George, W.K. (1987). An orthogonal decomposition of the axisymmetric mixing layer utilizing cross-wire measurements. In F. Durst *et al.* , editor, *Turbulent Shear Flows 6*, page 10.1.1. Springer Verlag.

Glauser, M.N., Leib, S.J., and George, W.K. (1987). Coherent structures in the axisymmetric mixing layer. In F. Durst *et al.* , editor, *Turbulent Shear Flows 5*, page 134. Springer Verlag.

Herzog, S. (1986), *The Large Scale Structure in the Near-Wall Region of Turbulent Pipe Flow.* PhD thesis, Cornell University.

Lumley, J. L. (1967). *Atmospheric Turbulence and Radio Wave Propagation*, chapter The structure of inhomogeneous turbulent flows. Nauka, Moscow.

Lumley, J. L. (1970). *Stochastic Tools in Turbulence.* Academic Press.

Moin, P. and Moser, R.D. (1989). Characteristic-eddy decomposition of turbulence in a channel. *J. Fluid Mech.* **200**, 471–509.

Zaman, K.B.M.Q. and Hussain, A.K.M.F. (1984). Natural large-scale structures in the axisymmetric mixing layer. *J. Fluid Mech.* **138**, 325–351.

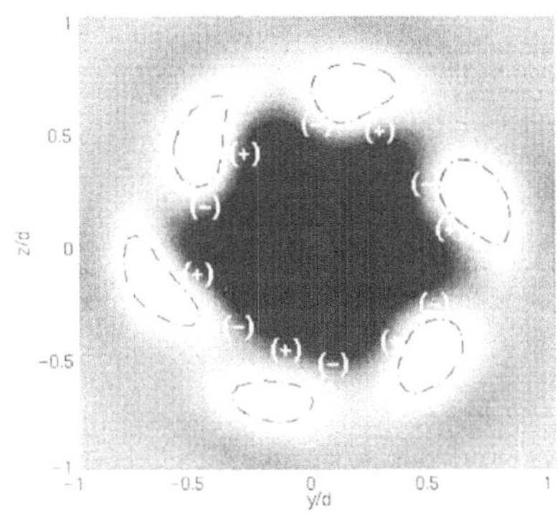

(a) $t_n = 413$

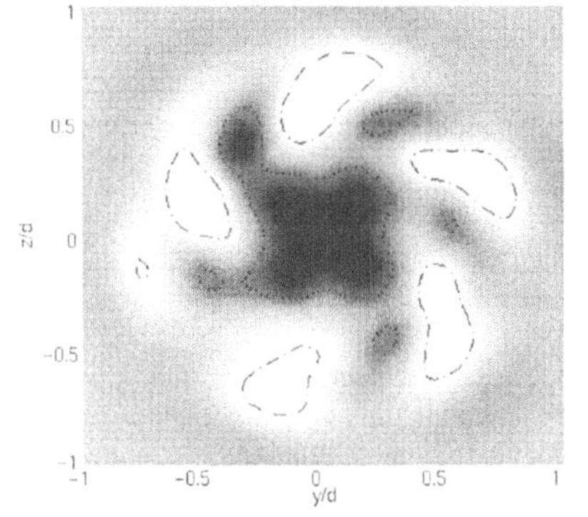

(b) $t_n = 419$

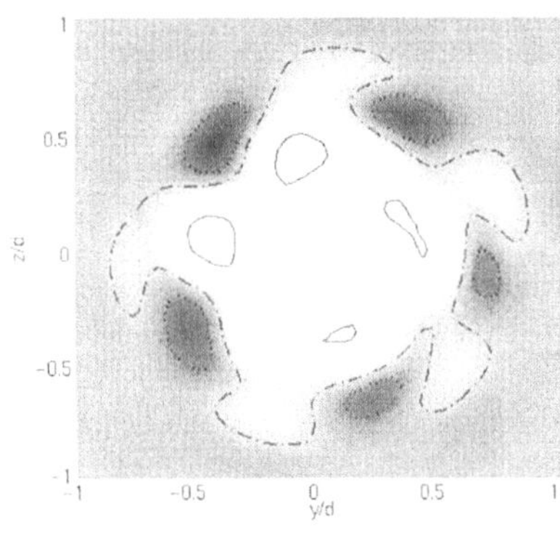

(c) $t_n = 422$

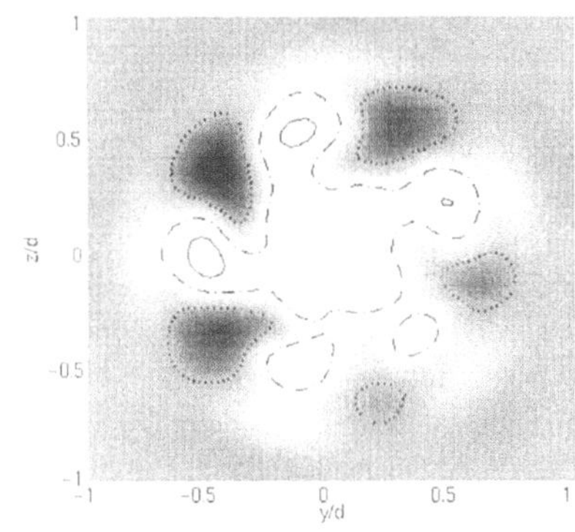

(d) $t_n = 427$

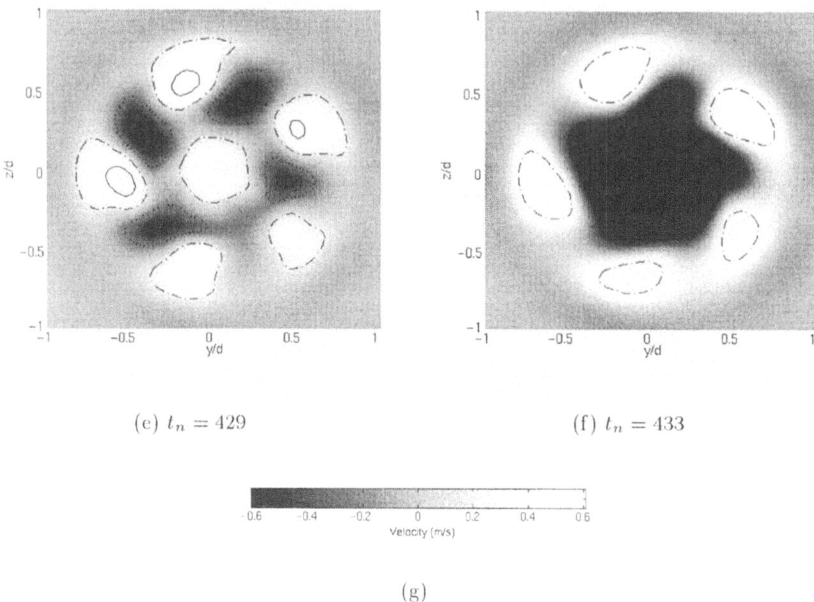

Figure 5: Two-dimensional projection of the POD reconstructed streamwise velocity field at $x/d = 3$. The temporal index is provided in the individual caption and the color bar denoting velocity above or below the local mean is given in 5(g).

Examination of a LSE/POD complementary technique using single and multi-time information in the axisymmetric shear layer

Dan Ewing
Department of Mechanical Engineering
Queen's University
Kingston, Ont., Canada.

Joseph H. Citriniti
Thermo- and Fluid Dynamics
Chalmers University of Technology
S-412 96 Goteborg, Sweden

Abstract

The data base measured by Citriniti (1996) is used to examine the accuracy of using Linear Stochastic Estimation to estimate the velocity field in the annular mixing layer. In the first case, the velocity information is estimated using information from a single instant in time on two radii. In the second approach, the signal is Fourier transformed in time and the frequency dependent coefficients are estimated using coefficients measured on two radii, thus effectively incorporating information from all points in time into the estimate. The estimated velocity fields are then used with the Proper Orthogonal Decomposition to yield low-dimensional reconstructions of the flow. It is found that both estimated fields are capable of accurately reproducing the gross features of both the ring structures and the inter-ring regions found in the shear layer, however, the single-time method does not accurately reproduce the amplitude of the fluctuating velocity.

1 Introduction

It is widely recognized that large energetic structures play an important role in the dynamics of the near field in many free-shear flows, such as the axisymmetric shear layer occurring near the exit of a round jet. Although there are many different techniques that can be used to identify the coherent structures (*cf* Bonnet and Glauser 1994), only a few of these can be used to study the dynamics of the structures. One technique that can be used for this purpose is the Proper Orthogonal Decomposition (POD) technique introduced by Lumley (1967).

One of the primary difficulties of studying the *instantaneous* dynamics of the large structures with this technique (and others) is that it requires the simultaneous measurement of the velocity over the region of interest. As a result, there have been few attempts to experimentally study the instantaneous dynamics of the structures on an entire plane in a free-shear flow. In one attempt to investigate these features Citriniti (1996) simultaneously measured the velocity field at 138 points in a high-Reynolds-number annular shear layer using hot-wire anemometers. Using the POD technique he found evidence that the shear layer was made up of both ring vortices and streamwise rib vortices that occur between the vortex rings.

Although this experiment yielded very useful physical information about the axisymmetric shear layer, the effort required to perform the experiment effectively precludes its use on a wide variety of flows. One alternative approach suggested by Ukeiley *et al.* (1993) is to measure the velocity at a small number of points and use Linear Stochastic Estimation

(LSE) to approximate velocity over the rest of the field. This estimated velocity field could then be used as an input into POD technique and an estimate for the dynamics of the large structures could be deduced. Ukeiley *et al.* (1993) and Bonnet *et al.* (1994) demonstrated that this technique was capable of accurately reproducing the dynamics of the large structures when applied to one non-homogeneous direction in a shear layer. They also suggested that the method could be extended to more spatial dimensions in a flow however, as yet, there has not been an attempt to confirm this.

The objective of this investigation is to examine if LSE can be used in conjunction with the POD technique to accurately estimate the dynamics of the large structures on a full $r-\theta$ plane in the axisymmetric shear layer. Here, two different approaches are used. In the first approach, only the information from a single point in time are used to estimate the velocity field on the plane at the same instant. In the second approach, the LSE is applied to the Fourier coefficients for the field (transformed in time) thus incorporating information from all points in time. In order to determine how accurately these estimated fields reproduce the dynamics of the large-scale motions, low-order reconstructions of the flows dynamics are computed by applying the POD to the estimated fields. These results are then compared to the analogous results computed from the actual velocity field (Citriniti 1996).

2 Background

2.1 Annular Shear Layer Data Base

In order to study the dynamics of the large structures in the axisymmetric shear layer Citriniti (1996)[2] constructed an array of 138 single-wire transducers that was placed at a distance of 3 diameters downstream of a round jet. Glauser (1987) had shown previously that the large structures in flow could be resolved using the POD if the velocity was measured at 6 radial positions across the layer at $r/d = 0.15, 0.28, 0.41, 0.54, 0.67$, and 0.80. He also found that the azimuthal dependence of the structures could be recovered if the velocity was measured at 6 azimuthal positions at the inner most radial position and 12, 24, 32, 32, and 32 positions at the other 5 radial positions, respectively. This information was used to locate the probes within the 138 wire array. The mean velocity normalized by the exit velocity ranged from 0.9 to 0.15 over these 6 radial points indicating that the array of hot wires measured the streamwise velocity component in a region that spanned most of the shear layer.

The exit velocity profile of the jet was a top hat to within 0.1% with a turbulence intensity of 0.35%. The outlet velocity of the jet was set such that the exit Reynolds number based on the nozzle diameter, d, was 80,000. For these conditions, Glauser (1987) demonstrated that the large structures could be well resolved in the first 800 Hz of the spectra so the signals from all of the wires in the array were filtered at 800 Hz and simultaneously sampled at 2,048 Hz. The data was gathered in 300 blocks of 1024 points, which allowed the accurate computation of spectral information.

2.2 Proper Orthogonal Decomposition

Lumley (1967) argued that the functions used to represent the large structures in a turbulent flow should be defined in an objective manner using information from the flow only. Lumley suggested these functions could be defined as those that make the largest contribution to the turbulent energy in the region of interest. Using this definition, it can be shown that

[2] See also the paper by Citriniti in this volume.

functions for homogeneous and stationary directions are simply Fourier modes so these directions can be transformed out of the problem in the standard manner.

For the axisymmetric shear layer, the functions that satisfy Lumley's definition for the radial direction, $\Phi^n(r,m,f)$, are solutions to the integral eigenvalue problem given by (Citriniti 1996)

$$\int_0^\infty F(r,r',m,f)\Phi^n(r',m,f)r'dr' = \lambda^n(m,f)\Phi^n(r,m,f), \tag{1}$$

where $F(r,r',m,f)$ is the Fourier transform of the two-point velocity correlation in both the azimuthal direction and time, m is the azimuthal mode number and f is the frequency. It is straightforward to show that these functions are orthogonal (*cf* Lumley 1967). Thus, the integral eigenvalue problem can be solved using only statistical measures of the flow to yield an orthogonal basis to optimally describe the flow in the radial direction.

The information about the dynamics of the motions in the annular shear layer can then be determined by computing the coefficients for the orthogonal functions, which are given by

$$a^n(m,f) = \int_0^\infty \hat{u}(r,m,f)\Phi^{n*}(r,m,f)rdr, \tag{2}$$

where $\hat{u}(r,m,f)$ is the Fourier transform of the *instantaneous* velocity field in the azimuthal direction and time. Thus, the coefficients can only be determined if the velocity field is simultaneously sampled over the region of interest.

The dynamics of the large-scale motions can then be studied by reconstructing the Fourier coefficients using a small number of modes; *i.e.*,

$$\hat{u}_{rec}(r,m,f) = \sum_{n=1}^{N} a^n(m,f)\Phi^n(r,m,f), \tag{3}$$

where N is a user specified parameter. If N is small then only the information from the most energetic modes is retained, thus reducing the small-scale information in the signal. The amount of small-scale information in the reconstructed velocity field can be further reduced by including the information from only a small number of azimuthal modes. The coefficients for the other modes are set to zero in order to remove them from the reconstructed velocity. This filtered signal can then be inverse Fourier transformed in time and the azimuthal direction to yield a low-order reconstruction of the instantaneous velocity field in the axisymmetric shear layer.

Citriniti (1996) demonstrated that the dynamics of the large-scale structures could be captured if only the first POD mode is used (*i.e.*, N=1) and only the coefficients for azimuthal mode numbers 0 and 3 – 6 are retained in the reconstructed field. Examples of the reconstructed instantaneous fluctuating velocity fields computed using these modes are shown in figure 1. In both of these figures the grey scale corresponds to the value of the instantaneous fluctuating velocity. The light colors correspond to positive fluctuating streamwise velocities while the dark colors correspond to negative fluctuating streamwise velocities.

It is clear that there is a band of faster-than-average moving fluid inside a band of slower-than-average moving fluid in figure 1(a), which is consistent with the pattern one would expect with the passage of a vortex ring like structure. On the other hand, slightly after the passage of the 'ring' there are neighbouring regions of fast and slow moving fluid (*v* figure 1(b)), suggesting that regions of fast flow moving out of the core neighbour regions of slow moving fluid being entrained into the core. This is consistent with the pattern one would expect in a braid region of counter-rotating streamwise vortices. Thus, the velocity

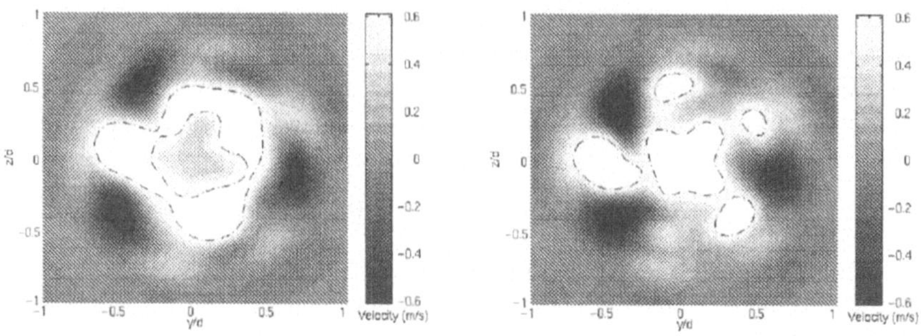

Figure 1: Partial reconstructions of the actual velocity field using the first POD mode and azimuthal modes $0, 3, 4, 5$, and 6. (a) ring-like structure (b) inter-ring region that appears to contain a series of counter-rotating streamwise vortices.

field reconstructed with only a few modes on the $r - \theta$ plane show evidence of both ring vortices and streamwise vortices in the inter-ring regions.

2.3 Linear Stochastic Estimation

Although these results shed insight into the dynamics of the structures in the axisymmetric shear layer, it is not trivial to simultaneously measure the velocity field at 138 points in the field. Thus, as Ukeiley *et al.* (1993) suggested, it would be useful if a similar picture of this flow (and others) could be deduced by applying the POD to an instantaneous field that was measured at a smaller number of points and estimated at the other points necessary to resolve the POD coefficients. Ukeiley *et al.* suggested using Linear Stochastic Estimation to extrapolate information from the measured points and fill out the field sufficiently to use the POD.

2.3.1 Single Time Estimation

For the current application, it is useful to estimate the Fourier coefficients of the velocity field, such as $\hat{u}(r_n, m, t)$, using Fourier coefficients measured at one or two other radial positions. For example, if the Fourier coefficients are measured at two radial positions, r' and r'', it follows that the coefficients at the other radial positions can be approximated as

$$\hat{u}_e(r_n, m, t) = A(r_n, m)\hat{u}(r', m, t) + B(r_n, m)\hat{u}(r'', m, t), \tag{4}$$

where $\hat{u}(r', m, t)$ and $\hat{u}(r'', m, t)$ are the measured Fourier coefficients at r' and r''. In this case, the mean square errors in the estimated coefficients, $E\{(\hat{u} - \hat{u}_e)(\hat{u} - \hat{u}_e)^*\}$, are minimized if

$$A(r_n, m) = \frac{\phi(r_n, r', m)\phi(r'', r'', m) - \phi(r_n, r'', m)\phi(r'', r', m)}{\phi(r'', r'', m)\phi(r', r', m) - |\phi(r'', r', m)|^2} \tag{5}$$

and

$$B(r_n, m) = \frac{\phi(r_n, r'', m)\phi(r', r', m) - \phi(r_n, r', m)\phi(r', r'', m)}{\phi(r'', r'', m)\phi(r', r', m) - |\phi(r'', r', m)|^2}, \tag{6}$$

where $\phi(r_n, r', m) = \overline{\hat{u}(r_n, m, t)\hat{u}^*(r', m, t)}$. The LSE coefficients, $A(r_n, m)$ and $B(r_n, m)$, are independent of time because the flow is stationary so that the same coefficients are used for both the ring and inter-ring regions of the flow.

It is also clear that the amplitude of the LSE coefficients is essentially determined by the correlation of the signals between the estimated point and the measured points. One difficulty of using the single time approach is that it combines contributions from both large-scale motions that are likely correlated across the layer and small-scale motions that are essentially uncorrelated since

$$\phi(r_n, r', m) = \int F(r_n, r', m, f)df. \tag{7}$$

Thus, the LSE coefficient may underpredict the large-scale motions because their contribution has been biased down by the contribution of the uncorrelated small-scale motions.

2.3.2 Frequency Mode Estimation

The problem can be reduced by incorporating information from different times thus shifting the weighting towards the large-scale motions that are correlated over longer times. This, however, has a tendency to increase the complexity of the expression for the estimated velocity and thus the equation set for the LSE coefficients. One technique to retain the time information efficiently is to Fourier transform the time signals and use LSE to approximate the individual Fourier coefficients; *i.e.*,

$$\hat{\hat{u}}_e(r_n, m, f) = C(r_n, m, f)\hat{\hat{u}}(r', m, f) + D(r_n, m, f)\hat{\hat{u}}(r'', m, f), \tag{8}$$

where $\hat{\hat{u}}(r', m, f)$ and $\hat{\hat{u}}(r'', m, f)$ are the measured Fourier coefficients at two radial positions, r' and r''. The mean square error in these coefficients are minimized if

$$C(r_n, m, f) = \frac{F(r_n, r', m, f)F(r'', r'', m, f) - F(r_n, r'', m, f)F(r'', r', m, f)}{F(r'', r'', m, f)F(r', r', m, f) - |F(r'', r', m, f)|^2} \tag{9}$$

and

$$D(r_n, m, f) = \frac{F(r_n, r'', m, f)F(r', r', m, f) - F(r_n, r', m, f)F(r', r'', m, f)}{F(r'', r'', m, f)F(r', r', m, f) - |F(r'', r', m, f)|^2}. \tag{10}$$

where $F(r_n, r', m, f) = \overline{\hat{\hat{u}}(r_n, m, f)\hat{\hat{u}}^*(r', m, f)}$. These coefficients are determined individually for each frequency and should be more capable of accurately extrapolating the information from the large scale motions.

2.4 Combining the LSE and POD

In the axisymmetric shear layer, the LSE technique could be used to interpolate or extrapolate information in two different spatial directions, r or θ. Since the signals are correlated over a greater distance in the radial direction than the azimuthal direction (relative to the characteristic length scales in those directions) it follows that the LSE should work more effectively in the radial direction than in the azimuthal direction. For this reason the application of the LSE to this direction is considered first. The application of this technique to the azimuthal direction will be reported elsewhere.

In all the cases considered here it is assumed that the Fourier coefficients $\hat{u}(r, m, t)$ or $\hat{\hat{u}}(r, m, f)$ are known at one or two radial positions in the annular shear layer. This information is then used to estimate the Fourier coefficients at the other radial positions. Physically, this corresponds to doing the experiment carried out by Citriniti (1996) using only one or two rings of hot-wires instead of the six used in that experiment.

The resulting estimated velocity fields can then be projected onto the orthogonal basis deduced by Citriniti (1996) in order to determine estimates of the POD coefficients; *i.e.*,

$$a_e^n(m,f) = \int_0^\infty \hat{\hat{u}}_e(r,m,f)\Phi^{n*}(r,m,f)rdr, \tag{11}$$

where $\hat{\hat{u}}_e(r,m,f)$ is the double Fourier transform of the velocity field. (Here, of course, it is necessary to solve a discrete version of this equation; *cf* Citriniti 1996.) This estimated Fourier coefficient is computed directly in the second method outlined above. In the first case, where the Fourier coefficient $\hat{u}_e(r_n,m,t)$ is estimated at each time, it is necessary to Fourier transform the resulting Fourier coefficient in time. These estimated POD coefficients can then be used to partially reconstruct the Fourier coefficients (eq. 3), which can then be inverse Fourier transformed to construct a low-dimensional model for the velocity field at all the spatial points in the field.

It is interesting to note that in this process it is actually more desirable to accurately estimate the POD coefficients (particularly the first one) than it is too accurately estimate the Fourier coefficient for the field. Thus, it might be more logical to argue that the coefficients for the LSE should be chosen to minimize the mean square error in the estimated POD coefficient; *i.e.*, $E\{(a^n - a_e^n)(a^n - a_e^n)^*\}$ where $*$ denotes a complex conjugate. It can be shown that this error is, in fact, minimized when the frequency Fourier coefficient $\hat{\hat{u}}(r_n,m,f)$ are estimated using the LSE (eq. 8-10).

3 Results

Initially the suitability of different configurations was examined by considering the normalized means square error in the estimated coefficients; *i.e.*,

$$\frac{\overline{|(\hat{u}_e(r_n,m,t) - \hat{u}(r_n,m,t))|^2}}{\phi(r_n,r_n,m)} \text{ or } \frac{\overline{|(\hat{\hat{u}}_e(r_n,m,f) - \hat{\hat{u}}(r_n,m,f))|^2}}{F(r_n,r_n,m,f)}.$$

It was found that no single radial positions could be used to accurately extrapolate the Fourier coefficients for both azimuthal mode 0 and modes $3-6$. On the other hand it was found that all the modes could be reasonable well approximated if the Fourier coefficients from two radii on opposite sides of the shear layer were used to estimate the coefficients at the other radii. The lowest error seemed to occur when the two radial positions were chosen near center of the shear layer, at $r/d = 0.41$ and $r/d = 0.67$ (analogous to the result outlined by Bonnet *et al.*). These two points (the third and fifth radial positions) were used for all the results presented here.

The normalized error in the estimated Fourier coefficients for the case where information from a single point in time are used to estimate $\hat{u}(r_\alpha,m,t)$ are shown in figure 2. The error is 0 at the two measured points but is quite large for all the radii. For example, even near the center, where the ring mode is dominant, the normalized error in the coefficients for mode 0 is approximately 40% while the error in the higher azimuthal modes is approximately 60% even at the outer radii where they play a dominant role. These levels of error, although high, are not inconsistent with the level of error that occur in other applications of the LSE to shear layers (*e.g.* Bonnet *et al.* 1994).

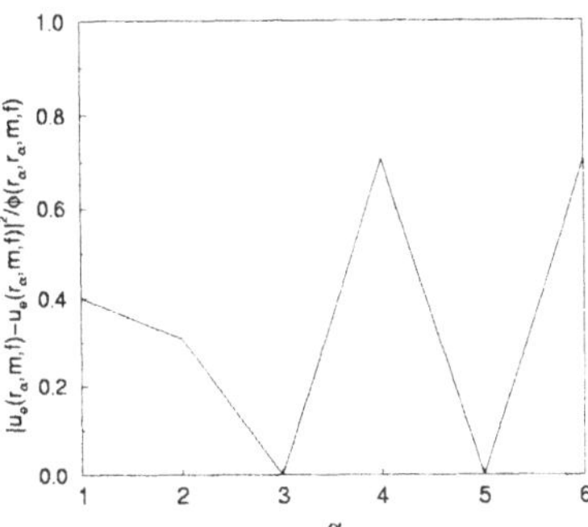

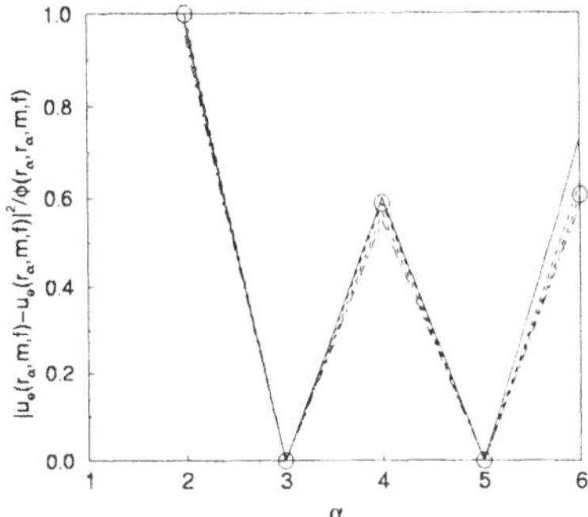

Figure 2: Normalized mean square error in the estimated Fourier coefficients at the points $r_\alpha/d = 0.15, 0.28, 0.41, 0.54, 0.67$, and 0.80 using information from a single point in time. (a) azimuthal mode 0 (b) azimuthal modes —— 3, ---- 4, −−− 5, and ---- ∘ 6.

On the other hand, examples of the normalized mean square error in the Fourier coefficient $\hat{u}_e(r_n, m, f)$ determined by directly estimating the Fourier coefficients are shown in figure 3. It is clear that the mean square errors for azimuthal mode 0 are small ($\sim 20\%$) for a range of frequencies around 100 Hz for all the radial positions except the fourth radial position. This corresponds to the peak in the mode 0 spectrum at the Strouhal frequency so the error in the coefficients are small over the energy containing range of the spectrum. It is not clear why the LSE performs more poorly at the central point but a similar result was noted by Bonnet *et al.* (1994).

Similarly the normalized error in the azimuthal mode 5 coefficients are shown in figure 3(b) (the errors for modes 3, 4 and 6 are very similar). It is important to note that the spectrum for this mode peaks at the lowest frequency (*cf* Citriniti 1996). Thus, the normalized mean square error is a minimum at the peak in the spectra. Although this error is larger than the error at the peak of mode 0 spectrum, it is smaller than the error for the higher modes for the single time case suggesting that the direct estimate of the frequency Fourier coefficients should provide a more accurate estimate of the dynamics in the inter-ring region.

Of course, the most important measure of these techniques is not the mean square error, but rather how accurately they reproduce the dynamics of the large-scale motions in the annular shear layer. In order to examine this question, a partial reconstruction of the field was computed using the estimated velocity fields from both techniques. Following the approach outlined by Citriniti (1996), only the first POD mode and only azimuthal modes 0, 3, 4, 5, and 6 were retained in the reconstruction. A realization from each of these reconstructions and the reconstruction from the original field are shown in figure 4. All three realizations are taken from the same point in time during the passage of a ring like structure. It is clear that the velocity field estimated from information at a single time yields a reconstructed velocity field that is much smaller in magnitude than the actual field ($\sim 50\%$). A similar level of error was found using realizations from the inter-ring region. On the other hand, the reconstructed velocity field computed using the field estimated by the frequency method more closely approximates the magnitude of the actual velocity field. Thus, as expected, the estimation in the frequency domain more accurately predicts the

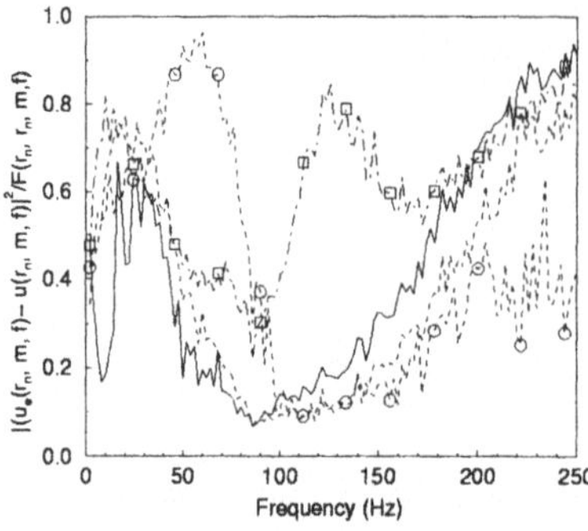

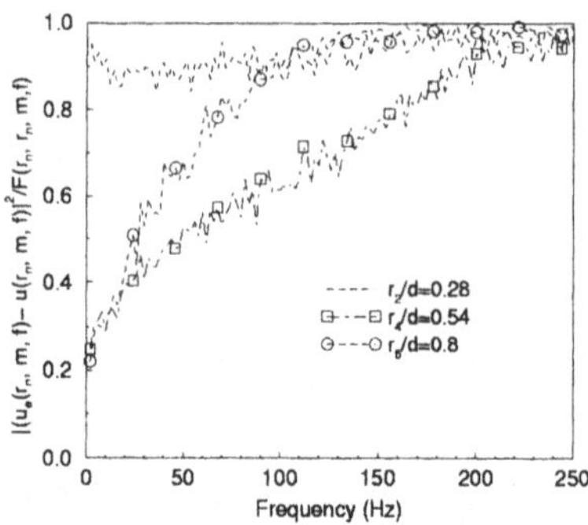

Figure 3: Normalized mean square error in the estimated Fourier in the estimated Fourier coefficients at the points —— $r_1/d = 0.15$, ---- $r_2/d = 0.28$, −−− $r_4/d = 0.54$, and ---- ∘ $r_5/d = 0.8$ (a) azimuthal mode 0 (b) azimuthal mode 5.

amplitude of the actual velocity signal.

In order to examine how accurately the two techniques reproduce the topology of the structures it is more useful to look at a two-dimensional contour plot of the instantaneous fluctuating velocity on the $r - \theta$ plane computed using both the estimated fields and the actual field. The estimated and actual reconstructed fields for realizations showing the ring region and the inter-ring region are shown in figure 5. The grey scale again corresponds to the fluctuating velocity but note that here the range of the grey scale for the single-time estimation technique has been modified in order to facilitate the comparison of the structures topology. Somewhat surprisingly the reconstructed fields from both the estimated fields have very similar topologies. It is also clear that the topology of the structure in the actual field is well represented by both the estimated fields. The estimated fields do not, however, capture all of the fine scale information of the structure as the estimation technique effectively acts as a filter.

4 Summary

The accuracy of using estimated velocity fields to study the dynamics of the large structures in the axisymmetric shear layer was examined using the data base measured by Citriniti (1996). It was found that the topology of the ring structures and inter-ring regions could be well reproduced if the Fourier coefficients were measured on 2 radii and estimated on 4 others. This was true whether the information from only a single time or all times was incorporated into the estimation process. It was found however that the amplitude of the fluctuating velocities in the structures were not well recovered when only information from a single time was used. The magnitude of the velocity field could be accurately predicted if the estimation was carried out on the Fourier transform of the signals in time thus effectively incorporating information from all times in the estimate of the velocity field. This process reduced the number of probes necessary to carry out the experiment from 138 to 56 thus effecting a 60 percent savings in the number of probes. Interpolation in the azimuthal direction will be considered in the future.

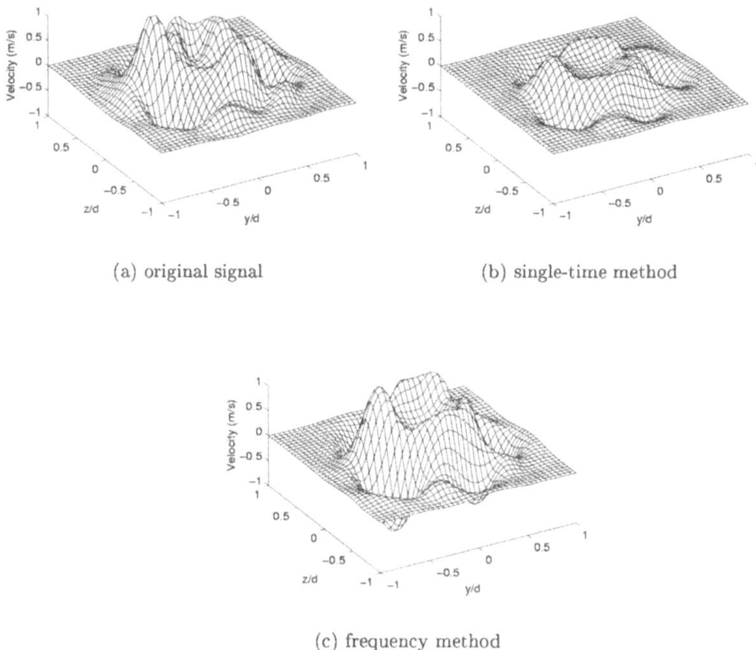

(a) original signal

(b) single-time method

(c) frequency method

Figure 4: Partial reconstruction of the velocity field using the first POD and azimuthal modes 0, 3, 4, 5, and 6. Surface height corresponds to the value of the fluctuating velocity.

References

Bonnet, J. P. and Glauser, M. N. (1994). *Eddy Structure Identification in Free Turbulent Shear Flows*. Kluwer Academic Press.

Bonnet, J. P., Cole, D. R., Deville, J., Glauser, M. N., and Ukeiley, L. S. (1994). Stochastic estimation and proper orthogonal decomposition: Complementary techniques for identifying structure. *Expts. in Fluids* **17**, 307–314.

Citriniti, J. (1996), *Experimental investigation into the dynamics of the axisymmetric mixing layer utilizing the Proper Orthogonal Decomposition*. PhD thesis, State University of New York at Buffalo.

Glauser, M. N. (1987), *Coherent Structures in the Axisymmetric Turbulent Jet Mixing Layer*. PhD thesis, State University of New York at Buffalo.

Lumley, J. L. (1967). The structure of inhomogeneous turbulence. In Yaglom, A. M. and Tatarski, V. I., editors, *Atmospheric Turbulence and Radio Wave Propogation*, pages 166–178. Nauka, Moscow.

Ukeiley, L. S., Cole, D. R., and Glauser, M. N. (1993). An examination of the asisymmetric jet mixing layer using coherent structure detection techniques. In *Eddy Structure Identification in Free Turbulent Shear Flows*. Kluwer Academic Publishers.

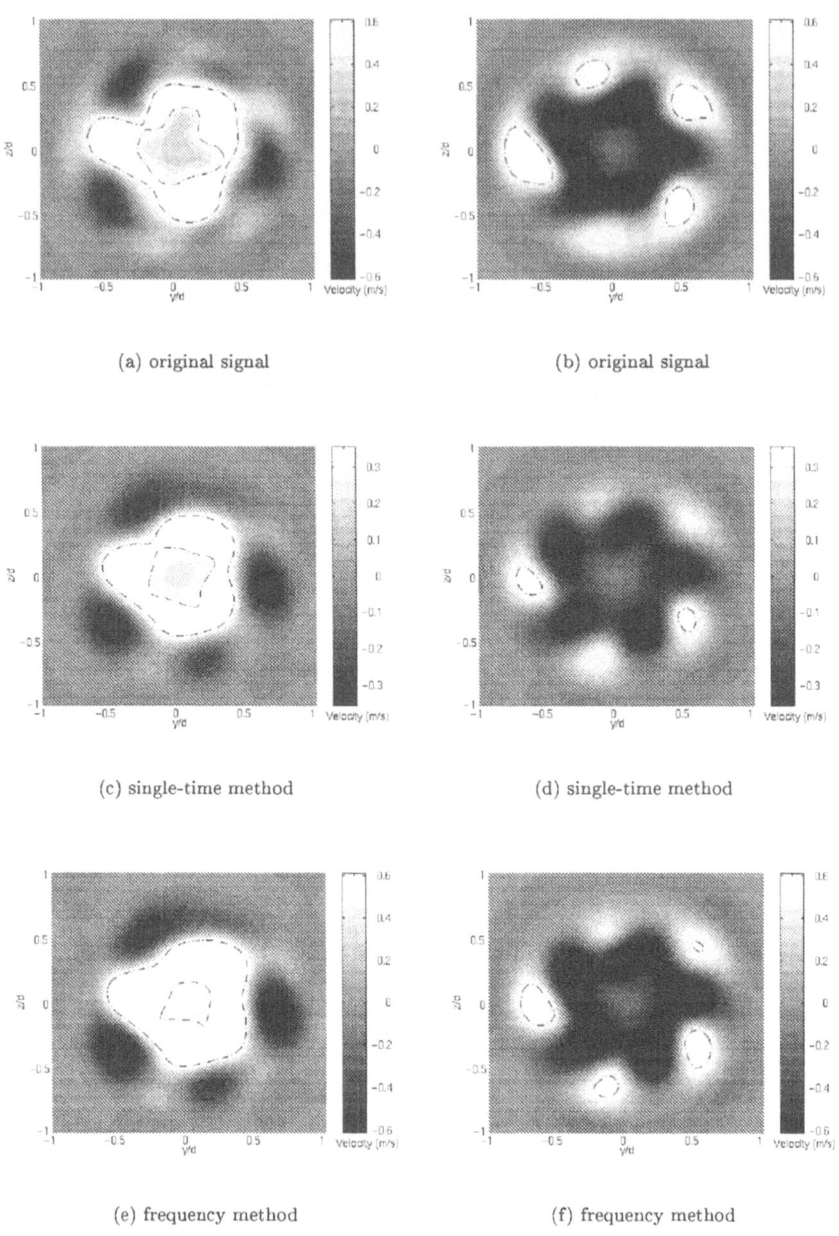

(a) original signal

(b) original signal

(c) single-time method

(d) single-time method

(e) frequency method

(f) frequency method

Figure 5: Partial reconstruction of the velocity field using the first POD and azimuthal modes 0, 3, 4, 5, and 6. Grey scale corresponds to the value of the instantaneous velocity in the reconstructed field. Figures (a), (c), and (e) correspond to $\tau = 422$ in Citriniti (1996) while (b), (d), and (e) correspond to $\tau = 431$

CONDITIONAL VORTICAL STRUCTURES OF A PLANE JET BASED ON THE COMPLEMENTARY LSE/POD TECHNIQUE

D. FAGHANI, A. SEVRAIN AND H.-C. BOISSON
Institut de Mécanique des Fluides de Toulouse
UMR CNRS/INP-UPS 5502
Av. du Prof. Camille Soula
31400 Toulouse - France

Abstract. The reliability of the complementary LSE/POD technique in an eddy detection scheme was investigated via X-probe hot wire measurements in the near field of a plane jet. Both natural and acoustically forced jets at Reynolds number 5600 were considered. LSE estimations proved to be satisfactory approximations of the velocity field as long as one reference probe was placed at the jet centre while the second was located near the maximum shear line.

An eddy eduction procedure, based on the projection of the estimated velocity on spatial POD eigenfunctions, was then applied and proved efficient even at a higher Reynolds number of 15000.

Symbols key

H	Exit slot width
U_0, Re	Jet exit velocity and Reynolds number U_0H/ν
δ_0, θ_0, δ_{ω_0}	Initial boundary layer, momentum and vorticity thicknesses
$U_i(x; y, t)$	Instantaneous velocity vector (at fixed X)
$\hat{U}_i(x; y, t)$	LSE-estimated instantaneous velocity vector (at fixed X)
$\overline{U}$, u, u'	Temporal mean, fluctuating and rms values
f_0, S_{θ_0}	Initial instab. frequency and Strouhal number $f_0\theta_0/U_0$
s^x, l^x_{ij}	Temporal and spatial BOD kernels (at fixed X)
R^x_{ij}	Temporal correlation tensor (at fixed X)
$\varphi_{i,n}$	n^{th} BOD topos or POD eigenvector
α^2_n, ψ_n	n^{th} BOD eigenvalue and chronos
a_n	n^{th} POD random process weight function

1. Introduction

During the last decade, as far as Coherent Structure (CS) study in turbulent flows is concerned, the Proper Orthogonal Decomposition (POD) has been of prime interest. Its use in turbulence was formalized by Lumley [10] while others introduced related extensions such as snapshot POD [14] or Bi-orthogonal Decomposition (BOD) [2]. In a typical eddy eduction scheme, instantaneous velocity field is projected on a few selected POD modes which are essentially the spatio-temporal correlation tensor eigenvectors. From an experimental point of view, the main difficulty in such a procedure is the measurement of instantaneous velocity over the whole spatial extent of interest in the flow. Hot wire rakes are commonly used though in most cases only limited low-dimensional POD can be achieved (see for instance [6]). To avoid hot wire rakes, Glauser, Bonnet and coworkers [9] suggested the use of an approximate velocity field obtained through Linear Stochastic Estimation (LSE) [1] which Cole et al. called pseudo-dynamic flow [5]. Bonnet et al. have shown that two-point estimations relying on only one pair of X-probe hot wires provided quite satisfactory results in a round jet and a shear layer [3].

The present study reports experimental results obtained by this complementary LSE/POD technique in a plane jet at several fixed near field X positions. First, reliability of LSE velocity fields was checked at moderate Reynolds number (5600) with and without acoustic forcing. Then, projecting the approximate velocity onto POD modes yielded temporal weight functions which contributed to prescribe eddy detection criteria and educe conditionally-averaged vortical structures. The same procedure was then applied at a higher Reynolds number (15000) and proved efficient in educing conditional eddies in a less organized situation.

2. Experimental apparatus and procedure

2.1. WIND TUNNEL FACILITY

Measurements were carried out in a low-speed open-return wind tunnel facility (fig.1) providing a clean plane jet at Reynolds numbers ranging from 5000 to 20000. The exit slot, 1-cm wide and 150-cm high, produced a flat-top initial velocity profile with quasi-laminar boundary layers presenting classic $1/\sqrt{Re}$ behaviour. Compared to Meyer et al.'s experimental set-up [11], the exit nozzle was stretched to produce thicker boundary layers. Meanwhile, θ_0/H was small enough to produce varicose (symmetrical) eddy roll-up at mean frequency S_{θ_0} in the non excited "natural" jet. This

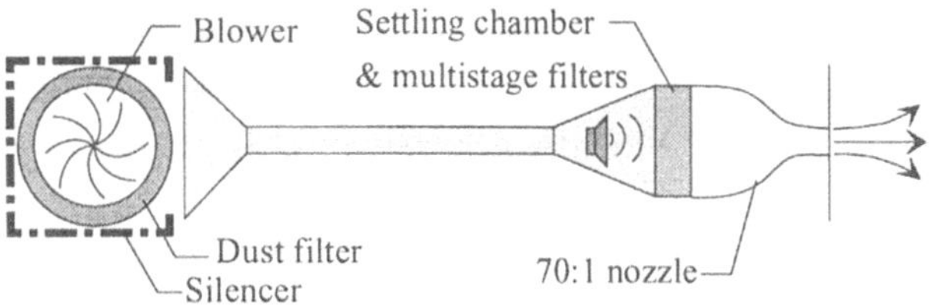

Figure 1. Schematic of air-jet facility.

instability frequency (half the theoretical value predicted by Michalke [12] for the hyperbolic tangent profile) was present in the initial mixing layers with a low rms magnitude. Measurements presented in this study mainly concern Reynolds number 5600 but some results at $Re = 15000$ are also reported. The initial flow characteristics at these two Reynolds numbers for the unforced jet are summarized in table 1.

TABLE 1. Initial flow characteristics.

Re	u'/U_0	θ_0	δ_0	δ_{ω_0}	$\delta_{\omega_0}/\theta_0$	S_{θ_0}
5600	0.8%	$0.028H$	$0.20H$	$0.12H$	4.3	0.0086
15000	2.4%	$0.018H$	$0.16H$	$0.08H$	4.4	0.0089

Acoustic forcing was applied at $Re = 5600$ with a loudspeaker placed before the settling chamber (fig.1). The frequency f_0 was chosen to meet S_{θ_0}. The rms magnitude of the induced fluctuations at this frequency was $u'/U_0 = 0.35\%$ in the initial shear layers. Mean profiles presented no alterations: all boundary layer thicknesses and global velocity fluctuations remained the same as in the "natural" case as listed in table 1. Meanwhile, instantaneous organization was severely altered exhibiting periodic eddy roll-ups and pairings taking place at fixed locations [7, 8].

2.2. DATA ACQUISITION

In order to perform orthogonal decomposition analysis, transverse velocity correlations were needed. Besides, linear stochastic estimation requires instantaneous two-point velocity measurements. Thus, two X-probe hot wires were used throughout this study to measure instantaneous velocities and spatial transverse correlations.

2.2.1. *Hot wire anemometry*

Velocity data were sampled at mid height by hot wire X-probes at a frequency far above Nyquist limit. A numerical scheme based on a modified cosine law was used to linearize anemometric signals. The over-all precision on velocity measurements was evaluated to be 2% for U and 8% for V. The hot wires were positioned with high accuracy ($\pm 0.05H$) at each X station while appropriate Y shifts allowed lateral probing of the jet. Special care was taken to avoid peripheral zones where hot wire measurements are unreliable due to recirculation and high turbulence intensity. Typically, transverse Y axis was covered by 10 to 15 equally spaced grid points around the jet axis.

2.2.2. *Phase-aligned measurements in acoustically forced jet*

For the forced jet at $Re = 5600$, data acquisition was triggered by the loudspeakers's driving signal enabling a posteriori phase alignment between signals. Special care was taken to minimize quarter and half period uncertainties which might arise from frequency halving entailed by subharmonic cascade. Sequentially sampled and phase-aligned velocity data $U_i(x; y, t)$ were comparable to hot wire rake measurements at each X position. The reliability of phase alignment was checked by considering the continuity equation for several transverse measurements performed at close X positions. The relative input-output flow difference was less than 10% at each time step and over the investigated spatial domain [7]. Appropriate correlation matrices could then be estimated from $U_i(x; y, t)$ for orthogonal decomposition and LSE analysis according to relations which will be presented later.

2.2.3. *Correlation measurements in natural jet*

In the case of the unforced jet at Reynolds numbers 5600 and 15000, correlation matrices at each X position were estimated via two simultaneous X-probes. Their lateral spacing was adjusted before each acquisition series. Then, they were shifted along Y axis at constant Δy so that correlation matrices were filled in along diagonals as illustrated in fig 2. Extensive use

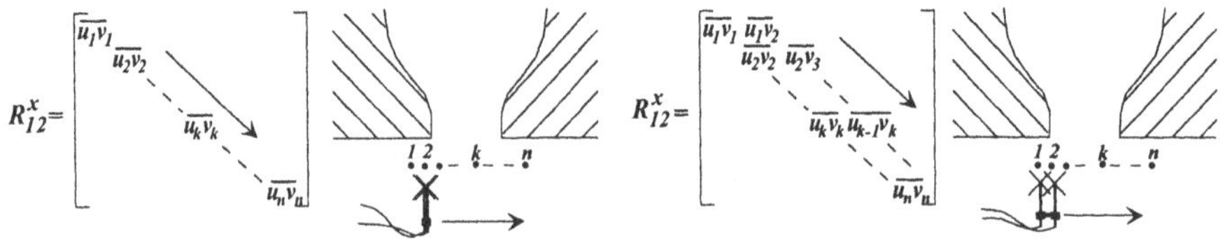

Figure 2. Correlations measurement procedure.

of symmetry relations was made to reduce the number of measurements. So, given n lateral grid positions yielding a $2n \times 2n$ correlation matrix, only $n(n+1)/2$ measurements were required. A more detailed description of this procedure can be found in reference [7]. Some simultaneous time series pairs were also saved at each X position to allow LSE computations.

2.3. LSE PSEUDO-DYNAMIC RECONSTRUCTION

LSE approximate or pseudo-dynamically reconstructed [5] velocity fields were computed via spatial correlations -measured according to the procedures described in the two previous sections- and two simultaneous reference velocities sampled at y_{ref_1} and y_{ref_2}:

$$\hat{U}_i(x;y,t) = \overline{U}_i(x;y) + A^x_{ij} u_j(x;y_{ref_1},t) + B^x_{ij} u_j(x;y_{ref_2},t) \tag{1}$$

where tensors A^x_{ij} and B^x_{ij} are defined for each y via spatial correlations $R^x_{ij}(y,y',0) = \overline{u_i(x;y,t)u_j(x;y',t)}$ and satisfy:

$$R^x_{ik}(y,y_{ref_1}) = A^x_{ij} R^x_{jk}(y_{ref_1},y_{ref_2}) + B^x_{ij} R^x_{jk}(y_{ref_1},y_{ref_2}) \tag{2}$$

$$R^x_{ik}(y,y_{ref_2}) = A^x_{ij} R^x_{jk}(y_{ref_1},y_{ref_2}) + B^x_{ij} R^x_{jk}(y_{ref_1},y_{ref_2}) \tag{3}$$

The choice of reference points y_{ref_1} and y_{ref_2} will be discussed later.

2.4. ORTHOGONAL DECOMPOSITION

Explicitly excluding complete space-time orthogonal decomposition, separable decompositions based on zero time lag spatial correlations were considered [7].

For the excited jet which exhibited rather deterministic T-periodic behaviour, we namely chose the bi-orthogonal decomposition. There, $U_i(x;y,t)$ was supposed to be a compactly supported deterministic function of Y and t defined over $[-Y_{max};Y_{max}] \times [0;T]$. Lateral boundaries $\pm Y_{max}$ were obviously imposed by hot wire limitations discussed earlier. Following Carrion [4], spatial and temporal X-dependant kernels were defined as:

$$l^x_{ij}(y,y') = \int_0^T u_i(x;y,t)u_j(x;y',t)dt \;,\; s^x(t,t') = \int_{-Y_{max}}^{Y_{max}} u_i(x;y,t)u_i(x;y,t')dy \tag{4}$$

The velocity field could then be reconstructed [2] thanks to spatial and temporal eigenfunctions namely the *topos* $\varphi^x_{i,n}(y)$ and the *chronos* $\psi^x_n(t)$:

$$U_i(x;y,t) = \sum_n \alpha^x_n \psi^x_n(t) \varphi^x_{i,n}(y) \tag{5}$$

It is important to note that topos, chronos and weight scalars α^x_n are all deterministic quantities.

The natural jet's velocity field $U_i(x;y,t)$ could reasonably be considered as a random stationary space-time function defined on $[-Y_{max};Y_{max}] \times I\!R$ and was modelled by:

$$U_i(x;y,t) = \sum_n a^x_n(t) \varphi^x_{i,n}(y) \tag{6}$$

where $\varphi_{i,n}^x$ are deterministic eigenvectors of $R_{ij}^x(y, y', 0)$ while a_n^x are stationary random processes.

3. Reliability of LSE approximation at moderate *Re*

Generally, LSE is used to extract organized features of a turbulent flow [1] and acts as a smoothing filter. Here, pseudo-dynamic reconstruction is rather used as an interpolator for the instantaneous velocity field. Estimated fields are indeed expected to follow real velocity time variations as closely as possible. The choice of reference velocities is therefore crucial for satisfactory results. As LSE is known to attenuate fluctuations with increasing distance between reference probe and estimation location [15] spatial u' and v' fields can indicate where the reference probes should be placed. Two rather obvious choices of reference points were considered. In case I, they were placed in the two symmetrical shear layers (maximum shear line). Since the forced flow was strongly symmetrical (varicose) [8], exact symmetrical positions had to be avoided to maintain a true 2-point estimation [7]. In case II, one of the references was placed on the jet axis. Results obtained in both natural and forced jet are presented and discussed in the next two subsections.

3.1. ACOUSTICALLY FORCED JET AT $Re = 5600$

Since quite comprehensive data were available from phase-aligned measurements, detailed comparisons could be performed with pseudo-dynamic flow through BOD. According to equation 1 the pseudo-dynamic field is a linear combination of two bi-component vectors $u_i(x; y_{ref_1}, t)$ and $u_i(x; y_{ref_2}, t)$. Therefore, its BOD spatio-temporal kernels have at most four non zero eigenvalues. Thus, comparison between the first four topos and chronos obtained from measured and estimated velocity fields could reveal the quality of LSE approximations and the best choice for reference points. Estimated instantaneous fields and their topos (not presented) were very similar to measured ones in both cases I and II. Projecting measured and estimated fields on their respective topos yielded their chronos the first two of which are shown on figure 3 at several X stations (case I ($\times$), case II (o), measured ($-$)). A very good agreement is obtained with a slightly better result when one reference probe is on the jet axis (case II, open circles). Here, more than gathering information about time and space scales, the central probe also reflects dramatic alterations induced by dynamics of two colliding shear layers at the centre of the jet.

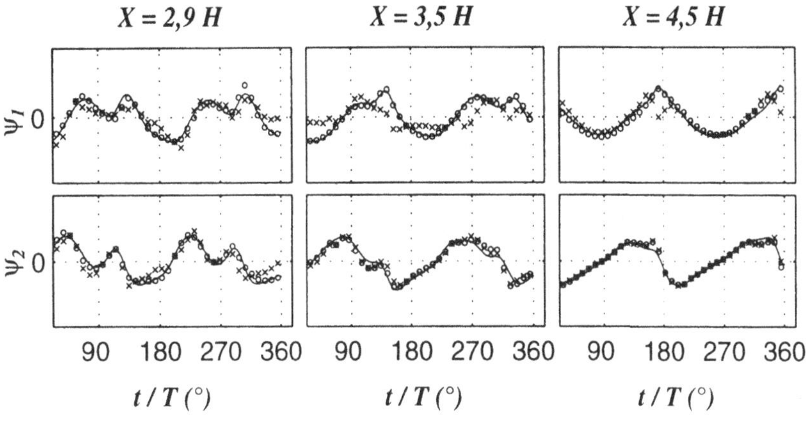

Figure 3. Measured and estimated chronos in forced jet.

3.2. NATURAL JET AT $Re = 5600$

For the natural jet, only time mean profiles could be compared because no multi-point instantaneous data were available. Figure 4 shows measure-

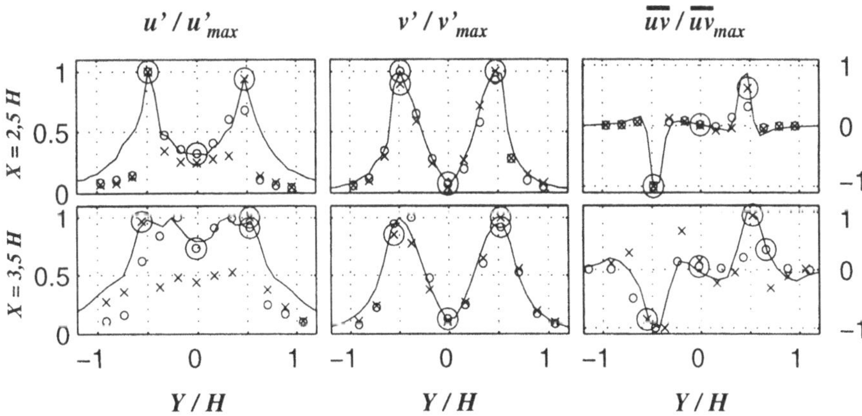

Figure 4. Measured and estimated mean profiles in natural jet.

ments and estimates at two X positions. Large open circles spot reference points. The general quality of LSE approximation is striking. Meanwhile, even though v' and $\overline{uv}$ are well estimated by both LSE schemes, significant discrepancy appears for u'. It is clearly underestimated at the centre of the jet when reference probes are located in the shear layers (case I ($\times$)) and none detects centre jet velocity fluctuations. Measurements made further downstream confirmed this trend since eddy interaction entails high u' levels at the jet centre [7].

4. Conditional eddies in natural jet

In the previous section it was shown that pseudo-dynamic reconstruction allows an accurate mimic of the plane jet's velocity field. Though instantaneous comparisons were not possible for the natural case, second order moments were well approximated. Moreover, the remarkable temporal behaviour agreement in the forced jet was a good indication of LSE approximation quality. Therefore, pseudo-dynamic velocity field was substituted to non accessible real velocity field. Besides, given the previous results, reference probes were placed at the jet centre and in the shear layer (case II). It should be noticed that this reference probe positioning does not prevent non symmetrical instantaneous flow configurations (e.g. sinuous eddies) from being well approximated [7].

4.1. DETECTION PROCEDURE

At moderate and transitional Reynolds numbers, highly organized eddies dominate plane jet kinematics. Their temporal signature at a fixed X station appears to be randomly (y, t)-dependant because of eddy trajectory, size and shape jitter. Eddy interactions such as pairings and multiple coalescences could introduce further variability. However, in the present study Y jitter was not significant and randomness could mainly be attributed to time intermittency. Recalling equation 6, eigenvectors $\varphi^x_{i,n}$ could be identified with (single, paired ...) eddy convection signature while random scalars a^x_n could reflect intermittency. Assuming that the first few eigenmodes (i.e. the most energetic ones) were related to eddies or pairs of coalesced eddies, a^x_n were good candidates for a detection scheme. In fact, forced jet

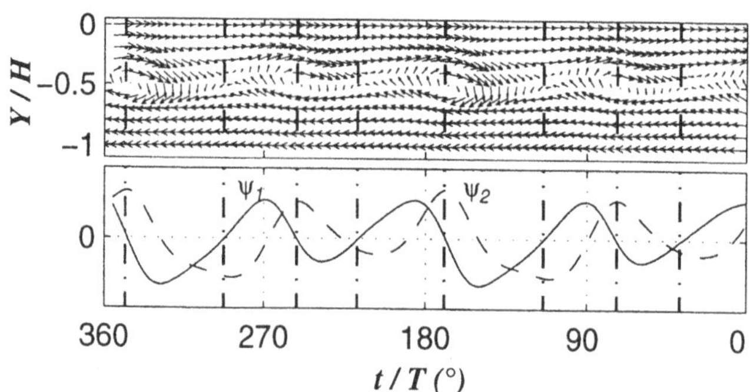

Figure 5. Eddy detection criteria in forced half-jet at $X = 1.9H$.

results were here to justify such assumptions and, moreover, provided detection criteria. It can be shown that separable orthogonal decompositions produce pairs of eigenmodes with $\pi/2$ time lag which mimic eddy convection [8, 13]. Figure 5 shows how eddy centres could be detected via local maxima and zero crossings (with negative slope) of the first two chronos in the forced jet at $X = 1.9H$. In the natural jet, one expects that despite random features, the same criteria remain valid. However, several characteristic vortical structures might be present (e.g. eddies, paired eddies etc.) and should be discriminated before averaging.

4.2. CONDITIONAL AVERAGES

Conditional averaging was performed to educe eddy signature at different X stations for Reynolds numbers 5600 and 15000 according to the aforementioned procedure. Two types of events were detected namely single pairs of symmetrical eddies and double pairs of eddies at the beginning of a symmetrical pairing. The first event was simply detected on the same basis as in figure 5. However, detection criteria were somewhat relaxed: a_1^x still had to cross zero with a negative slope but a_2^x only had to remain positive. Indeed, due to the stochastic nature of the velocity field, a_2^x

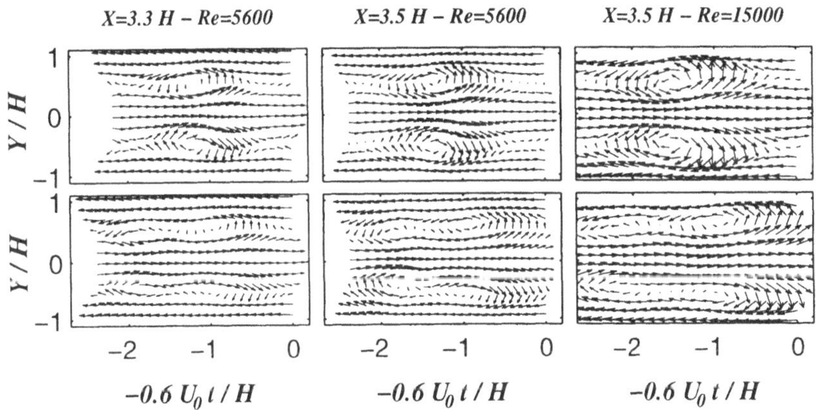

Figure 6. Conditional eddies in natural jet.

maxima were not always synchronous with a_1^x zero crossings. The second type of events was in fact detected by spotting two consecutive close pairs of eddies. After several attempts, it appeared that thresholding a_1^x above $0.4\sigma_{a_1^x}$ was a sufficient condition. Some typical results are shown on figure 6 with time reference set at the structure's centre. Taylor hypothesis was used ($U_c = 0.6U_0$) to render a spatial image of vortical eddies crossing the considered X sections. Without performing any rescaling, it appears that conditional vortical structures exhibit very similar features at moderate and high Reynolds number. This similarity was also present in spatial eigenvectors (not presented). In fact, examination of instantaneous pseudo-dynamic velocity showed that with increasing Re, more and more non symmetrical eddies were convected past a fixed X section while their shape did not undergo significant changes. Unfortunately, non symmetrical eddy detection could not be performed since they were related to high order eigenmodes which were not resolved by the pseudo-flow velocity (see section 3.1).

5. Discussion and concluding remarks

The reliability of 2-point pseudo-dynamic approximation was investigated against the choice of reference probes. It appears that they should be located near high fluctuating regions to minimize smoothing effects. Comparison between measured and estimated spatial eigenvectors is a simple indicator of their good positioning. Concerning pseudo-dynamic flow symmetries, reference positions do not play a direct role. The real flow statistical symmetries are incorporated in the spatial correlations and affect LSE estimation via spatial eigenvectors.

Regarding organized structure identification, it is common use to define coherent motion as the smoothed velocity field obtained through its projection onto the first few POD eigenmodes. It should be noted that the resulting field is still a random process. This is why Lumley [10] and others suggested a priori statistical treatments such as shot noise modelling which require higher order (multipoint) statistics. Here, we chose a posteriori statistics achieved by conditional averaging which obviously suffers from the need to prescribe detection criteria. However, POD's optimal convergence properties are expected to provide very few such criteria. Besides, for the most frequently occurring events who deeply affect the eigenvectors, simple criteria such as local maxima or zero crossing are sufficient. Of course, the more turbulent and less organized the flow, the lesser the chances of such schemes to succeed in eddy eduction.

In the present study, the conditional averaging of POD filtered pseudo-dynamic velocity fields proved quite encouraging. Besides, some eigenfunction invariance could be inferred from the similarity of conditional averages at Reynolds numbers 5600 and 15000 since they were indeed based on spatial eigenvectors. However, more investigation is necessary since even at $Re = 15000$, the near field of the plane jet remains well organized and dominated by vortical structures. One simple way to check this issue is to examine the correlation tensor's Reynolds number dependance.

Acknowledgements
We gratefully acknowledge Professor J.-P. Bonnet, Dr. J. Delville and their coworkers for the interest they showed in our work during its progress. We also wish to thank J.-F. Alquier and the electronics laboratory staff at

IMFT for their technical support in the realization of our data acquisition equipment.

References

1. ADRIAN R. J. Conditional eddies in isotropic turbulence. *Physics of Fluids*, 22 (11):2065–2070, 1979.
2. AUBRY N., GUYONNET R., and LIMA R. Spatio-temporal analysis of complex signals: Theory and applications. *Journal of Statistical Physics*, 64(3/4):683–739, 1991.
3. BONNET J.-P., COLE D. R., DELVILLE J., GLAUSER M. N., and UKEILEY L. S. Stochastic estimation and proper orthogonal decomposition: complementary techniques for identifying structure. *Experiments in Fluids*, 17:307–314, 1994.
4. CARRION S. Reliability of bi-orthogonal decomposition applied to a rotating disk boundary layer. *Experiments in Fluids*, 14:59–64, 1993.
5. COLE D. R., GLAUSER M. N., and GUEZENNEC Y. G. An application of the stochastic estimation to the mixing layer. *Physics of Fluids*, A 4 (1):192–194, 1991.
6. DELVILLE J. *La Décomposition Orthogonale aux Valeurs Propres et l'Analyse de l'Organisation Tridimensionnelle des Ecoulements Turbulents Cisaillés Libres.* Thèse de l'université de Poitiers, janvier 1995. Poitiers, France.
7. FAGHANI D. *Etude des Structures Tourbillonnaires de la Zone Proche d'un Jet Plan: Approche non Stationnaire Multidimensionnelle.* Thèse de l'Institut National Polytechnique de Toulouse, novembre 1996. Toulouse, France.
8. FAGHANI D., SEVRAIN A., and BOISSON H.-C. Vortical structure of an acoustically forced plane jet: bi-orthogonal eddies vs. physical eddies. In *11th Symposium on Turbulent Shear Flow*, Grenoble, France, 1997.
9. GLAUSER M. N., COLE D. R., UKEILEY L. S., BONNET J.-P., and DELVILLE J. Stochastic estimation and proper orthogonal decomposition: complementary technics for identifying structure. In *9th Symposium on Turbulent Shear Flow*, Kyoto, Japan, 1993.
10. LUMLEY J. L. *Stochastic tools in turbulence.* Academic Press, 1970.
11. MEYER J. *Structures Organisées et Transition dans la Zone Proche du Jet Plan: Synthèse d'Analyse Expérimentale, Visuelle et Numérique.* Thèse de l'Institut National Polytechnique de Toulouse, février 1989. Toulouse, France.
12. MICHALKE A. On spatially growing disturbances in an inviscid shear layer. *Journal of Fluid Mechanics*, 23:521–544, 1965.
13. RAJAEE M., KARLSSON S. K. F., and SIROVICH L. Low-dimensional description of free-shear-flow coherent structures and their dynamical behaviour. *Journal of Fluid Mechanics*, 258:1–29, 1994.
14. SIROVICH L. Turbulence and the dynamics of coherent structures, Parts I,II & III. *Quarterly of Applied Mathematics*, XLV(3):561–582, 1990.
15. VINCENDEAU E. *Analyse Conditionnelle et Estimation Stochastique Appliquées à l'Etude des Structures Cohérentes dans une Couche de Mélange.* Thèse de l'Université de Poitiers, décembre 1995. Poitiers, France.

APPLICATION OF POD TO PIV IMAGES OF FLOW OVER A WALL MOUNTED FENCE

BO H. JØRGENSEN
Dept. of Energy Engr., Fluid Mech.,
DTU, DK-2800 Lyngby, Denmark

Abstract. Large sets of digital PIV measurements (each containing 500 images) of the flow over a wall mounted fence have been analyzed by POD. The results show coalescing of naturally occuring vortices. Furthermore, the growth and decay of the vortical structures originating from the perturbations generated by an upstream oscillating plane jet is indicated by the energy and vortex locations in different POD modes. For the perturbed flow, it is possible to extract information by conditional sampling of PIV data, which significantly affects the modes obtained by POD analysis.

1. Introduction

Proper Orthogonal Decomposition (POD) is by now a well established tool. See for instance the review article by Berkooz, Holmes and Lumley (1993). An important application of POD consists in constructing optimal orthogonal vectors used to derive reduced dynamical systems via Galerkin projection of the Navier-Stokes equations as presented in the pioneering work of Aubry *et al.* (1988). See also Christensen *et al.* (1993). Recently, POD has been applied to data from Large Eddy Simulations (LES), for example by Manhart and Wengle (1993) and to Direct Numerical Simulations (DNS) (Christensen *et al.*, 1993). Yet another application of POD is analyzing hot wire measurements (Delville, 1993).

With the advent of real time digital PIV recording, it is now possible to acquire large numbers of vector maps, say 500 or more, of 2D sections of a flow. Such data is now being analyzed by POD. The system currently investigated is the flow over a solid fence mounted on a wall in a wind tunnel. Coherent structures in the flow are evident when the upstream flow is perturbed by a flat pulsed jet. Schmidt (1997) has used this setup for LDA measurements. He finds that the length of the separation zone is clearly affected by changes in the perturbation frequency and energy. The details of the PIV experiment will be presented by Ullum, Schmidt, Larsen and McCluskey (1997), and by Ullum, Schmidt and Larsen at this IUTAM symposium. The results described by the authors are based on their PIV measurements. It is of great interest that Orellano and Wengle (1995) have already done large eddy simulations of a similar system. Further results by Orellano and Wengle are reported at the present IUTAM symposium.

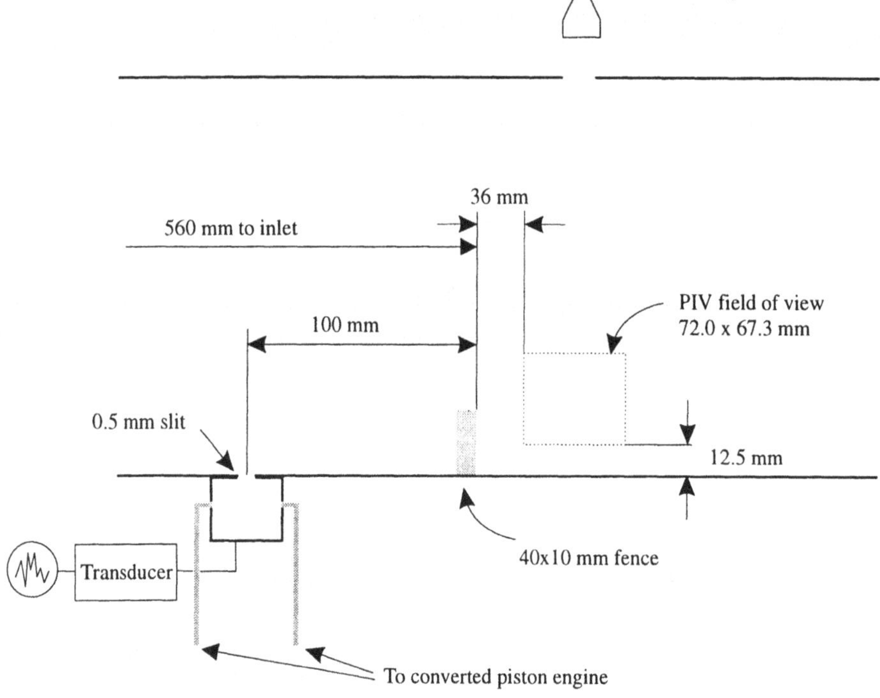

Figure 1. Test section with 40×10 *mm* wall-mounted fence and 0.5 *mm* wide wall-slit for periodic perturbation flow from converted piston engine. PIV field-of-view located downstream of fence, $0.90 < x/h < 2.70$ by $0.31 < y/h < 2.00$.

2. The Test Rig

The test rig consists of a fence (10 wide by 40mm high) mounted at the wall 560mm downstream of the inlet to a wind tunnel test section (600mm wide by 300mm high), see Fig. 1. The channel bulk velocity $U = 0.5\frac{m}{s}$ gave a Reynolds number $Re_h = 1300$ based on the fence height. The flow could be perturbed by oscillating blowing/suction through a 0.5mm wide slit parallel to the fence, 100mm (2.5h) upstream of the trailing edge of the fence. The PIV field of view (72.0mm wide by 63.3mm high) was positioned 12.5mm above the wall and 36mm downstream from the trailing edge of the fence. The PIV images were captured by a Kodak Megaplus ES1.0 CCD-camera connected to a DANTEC FlowMap PIV 2000 processor with a vector map acquisition rate of 5Hz. More details are given in Ullum *et al.* (1997).

3. Measured Data

Both unperturbed and perturbed flows were investigated. The perturbed flow was given an oscillation of 1.87Hz, which corresponds to a Strouhal number of $St_h = 0.15$ based on fence height. The squared RMS bulk velocity of the jet was $5.94\frac{m^2}{s^s}$. In Schmidt (1997) more details are given about choices of perturbation frequency and amplitude. For the perturbed flow, both unconditional and conditional sampling was applied. The conditional sampling was triggered by an electrical signal from the blowing and suction mechanism. In this way, four distinct time phases separated by 130ms were utilized for capturing PIV images, each phase yielding one data set. As the sampling rate of the PIV equipment was set to approx. 100ms, sampling was skipped whenever the PIV equipent was not ready for an incoming triggering signal.

In total, six different data sets were sampled. Each contains 500 vector maps of 62x58 (u,w)-components of velocity, i.e. 500x7000 data points.

4. The POD Method

Proper Orthogonal Decomposition is a method that has been well established for a long time. It is also known as the Karhunen-Loeve transformation, and in pattern recognition and statistical image processing it is known as Principal Component analysis (PC). A description of POD can be found in Berkooz *et al.* and in Aubry *et al.*. The method is outlined below.

Given a data set $\vec{x}^1, ..., \vec{x}^k$, the ensemble average is subtracted

$$\vec{u}^k = \vec{x}^k - < \vec{x}^k >$$

Estimating the auto-covariance matrix

$$R_{ik} = \vec{u}^i \cdot \vec{u}^k$$

and determining the eigenvectors $\vec{g}^k$ of R allows calculating the normalized orthogonal basis

$$\vec{\phi}^k = \frac{\sum_{i=1}^{m} g_i^k \vec{u}^i}{\| \sum_{i=1}^{m} g_i^k \vec{u}^i \|}$$

which solves the problem

$$\lambda_k = \max_{|\vec{\phi}^k|=1} \sum_{i=1}^{m} (\vec{\phi}^k \cdot \vec{u}^i)^2$$

$$\vec{\phi}^k \cdot \vec{\phi}^l = 0, \qquad l = 1, 2, ..., k-1$$

with eigenvalues $\lambda_1 \geq \lambda_2 \geq \ ... \ \geq \lambda_m > 0$. The eigenvalues are sometimes called energies, and the basis vectors are refered to as modes.

5. POD Analysis of PIV Images

All velocity components were placed in one vector for each PIV image. The ensemble mean image was calculated and subtracted from all PIV images. The auto-covariance matrix was estimated by simple multiplications and additions, and the eigenvalue problem was solved by an eigenvalue routine for symmetric matrices based on QR-factorization. The normalized orthogonal basis vectors were calculated by matrix multiplication, and the relative energies of each basis vector were calculated from the eigenvalues. Also, the coefficients for reconstructing each PIV image were calculated by projection of the PIV images on the basis modes.

The output of the above analysis were for each data set: One mean image of 7000 data points, 20 modes of 7000 data points, and 500 sets of 20 coefficients.

6. Results and Discussion

6.1. UNPERTURBED FLOW

In Fig. 2, the first four POD-modes for the unperturbed flow is shown. Together, they represent 51% of the total energy of the data set. One of the first things that captures the eye is the seemingly isolated islands of large vectors, which are represented by long arrows. Thus, the POD modes are not divergence free in 2D. It is simple to prove that if the velocity fields corresponding to the original PIV images are divergence free, then also the POD-modes will be divergence free. Since the 2D continuity equation is not satisfied, 3D-effects are present in the flow. The first picture depicts a

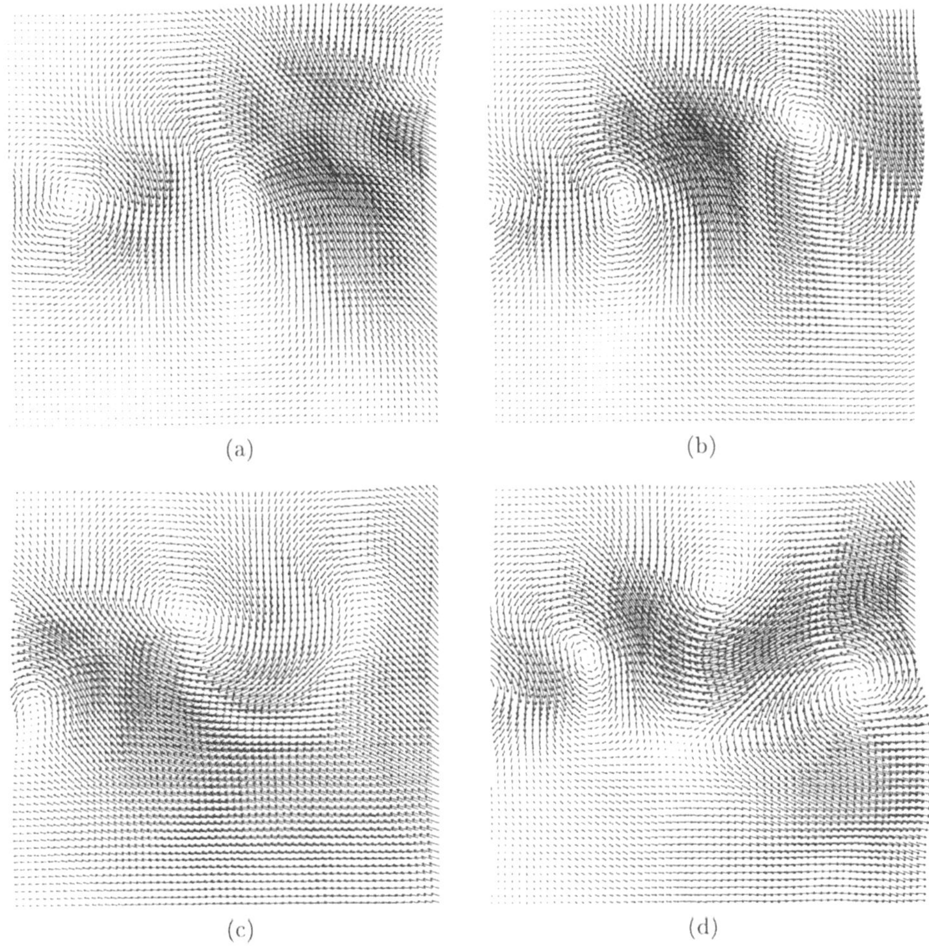

Figure 2. The first four POD modes of unperturbed flow. (a) Mode 1 contains 19% energy, (b) mode 2 contains 15% energy, (c) mode 3 contains 11% energy, and (d) mode 4 contains 6% energy.

vortex in the center, a vortex to the left and part of a vortex to the upper right. Notice that the distance between the vortices in the left half of the picture is smaller than the distance between the vortices in the right half of the picture. The same property can be observed on the other three pictures. A plausible interpretation is vortex coalescing. Note that the vortices do not travel at the channel flow bulk speed, and that they need not travel at the same speed. One observes that there is a stagnation point a little down and to the left from the center of the fourth picture.

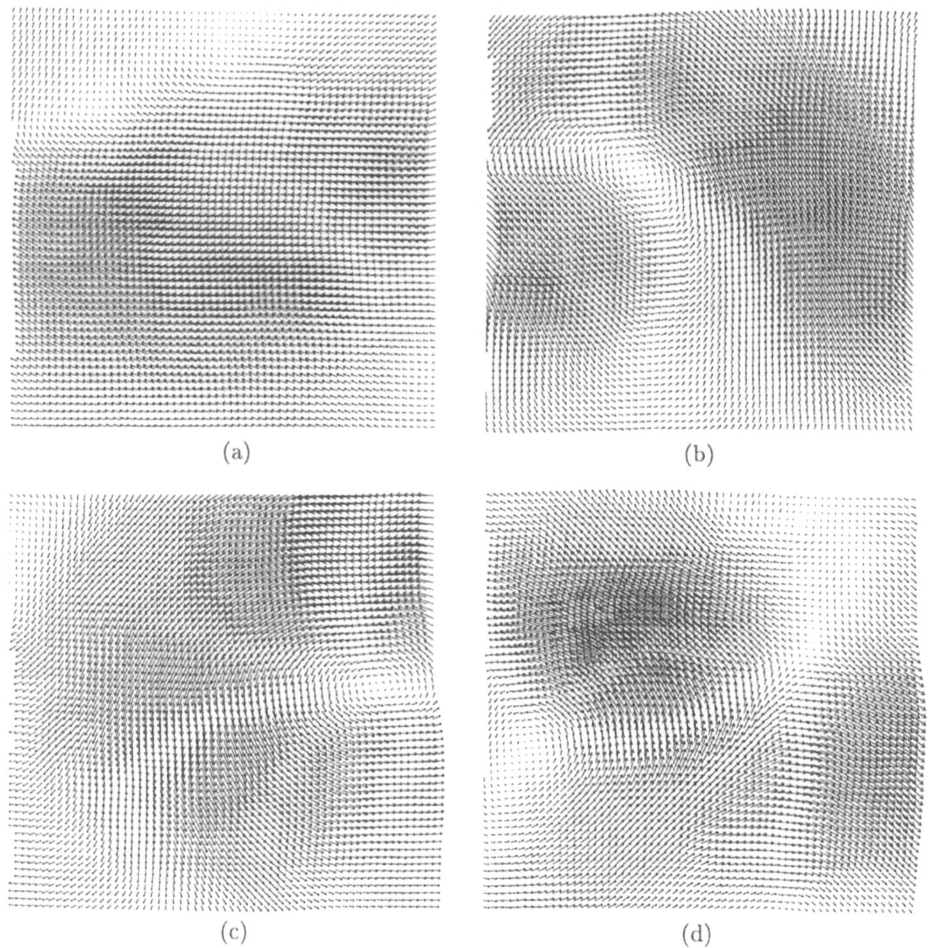

Figure 3. The first four POD modes of perturbed flow (without conditional sampling). (a) Mode 1 contains 14% energy, (b) mode 2 contains 13% energy, (c) mode 3 contains 10% energy, and (d) mode 4 contains 6% energy.

6.2. PERTURBED FLOW

In Fig. 3, the first four modes of the perturbed flow are shown. Together, they represent 43% of the total energy of the data set. The most striking difference between these modes and the modes shown in Fig. 2 is that no vortices appear in the first mode, which seems to represent a bulk flow from right to left. In addition, only a single vortex appears in both the second and the third modes. Mode 4 seems to be more complicated. It contains several vortices and a stagnation point. A close observation of mode 2, 3, and 4 reveals that the vortex closest to the center is found in mode

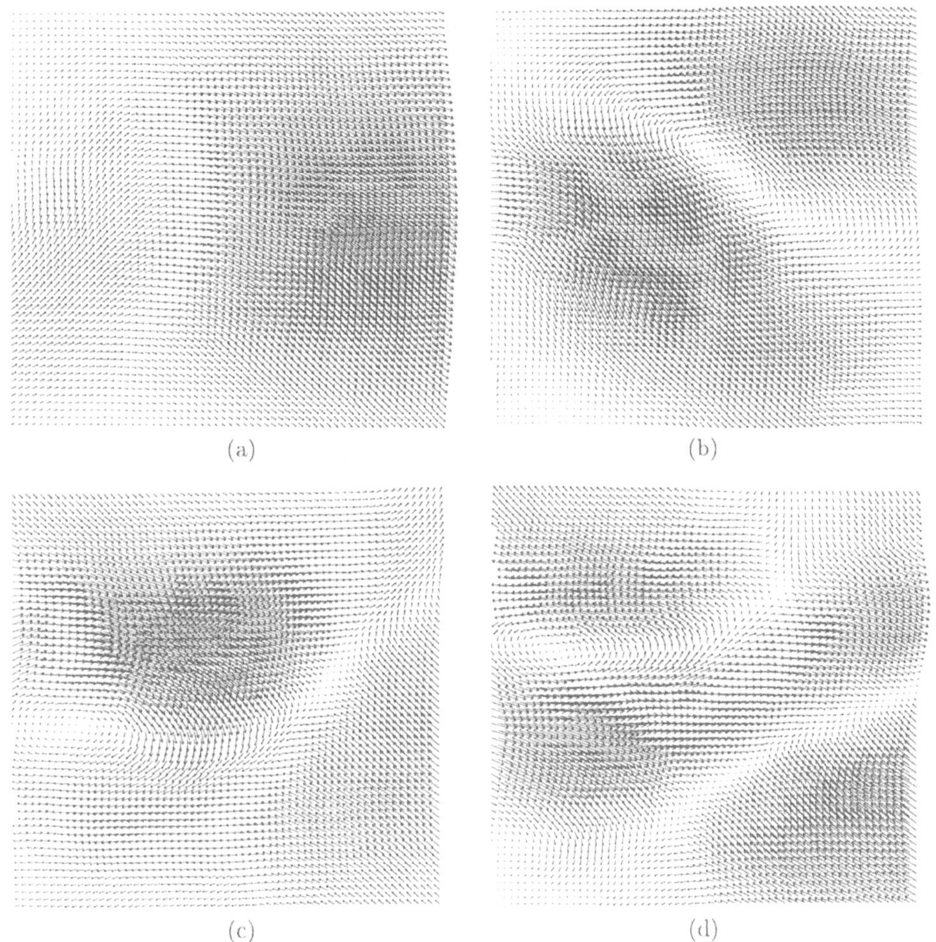

Figure 4. The first four POD modes for phase 1 of conditionally sampled perturbed flow. (a) Mode 1 contains 17% energy, (b) mode 2 contains 9% energy, (c) mode 3 contains 9% energy, and (d) mode 4 contains 5% energy. Here, 100% means the total energy of phase 1.

2, which contains 13% of the energy. The vortex located to the most far right is found in mode 3, which contains 10% of the energy. Finally, the vortex located to most far left and vortices located to the right are found in mode 4, which contains 6% of the energy. If one defines a POD mode as a structure, this implies that vortex structures grow as they move from the left towards the center, and then start decaying when they reach the right half of the picture. The horizontal position of the left edge of the field of view corresponds to approximately one fence height from the trailing edge of the fence, and the right edge of the field corresponds to a position of three fence heights. The growth and decay of the vortex structures is confirmed by the time correlations measured by LDA at 1h, 2h, and 4h (Schmidt, 1997) for the flow perturbed at 1.87Hz with the same perturbation energy as before ($< v^2 >= 5.94\frac{m^2}{s^2}$).

6.3. CONDITIONALLY SAMPLED PERTURBED FLOW

In Fig. 4, the first four modes of phase 1 of conditionally sampled PIV images of the perturbed flow is shown. Together, they represent 40% of the energy of phase 1. The bulk movement in mode 1 is to the right. On mode 2, 3, and 4 there are more stagnation points than on the corresponding pictures in Fig. 3. This is a feature which is seen mostly on higher modes with lower energy. Important information has already been extracted in the ensemble mean image, which contains a distinctly located mean vortex, which does not appear on the unconditionally sampled mean image (not shown).

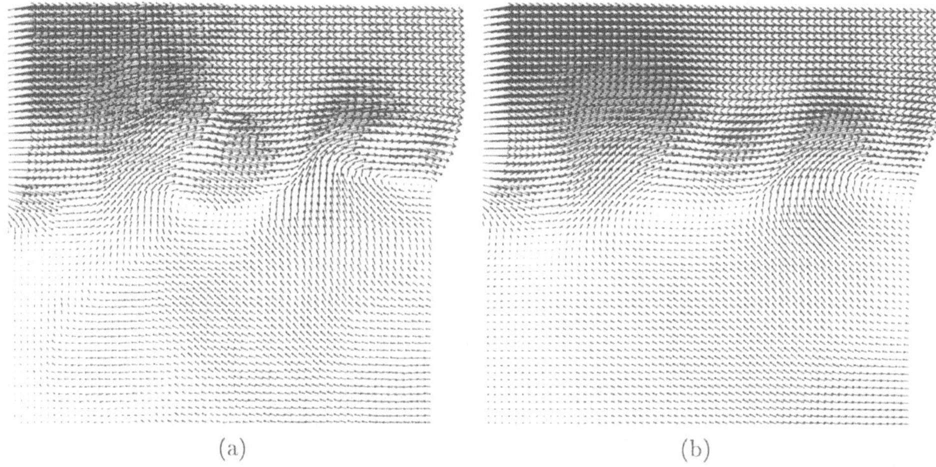

Figure 5. Unperturbed flow. (a) PIV image, (b) Image recontructed from 20 POD modes containing 81% of the energy.

6.4. RECONSTRUCTED IMAGES

In Fig. 5, (a) the PIV image from the unperturbed flow is reconstructed in (b) from 20 POD-modes containing 81% of the energy. This reconstruction seems fairly good. The PIV image (a) from the perturbed flow in Fig. 6 is reconstructed in (b) from 20 modes containing 73% of the energy. Many details and 3D effects seem hard to represent from these modes. In Fig. 7 a PIV image from phase 1 of the perturbed flow is shown (a) with a

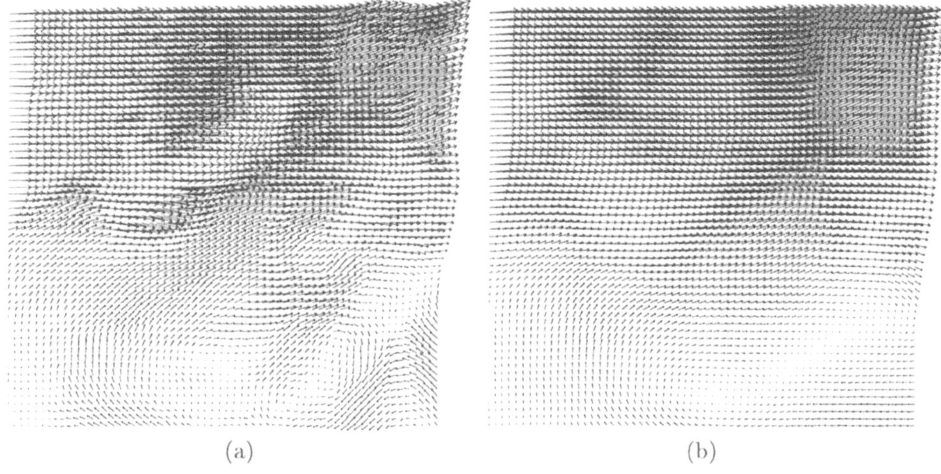

Figure 6. Perturbed flow (without conditional sampling). (a) PIV image, (b) Image recontructed from 20 POD modes containing 73% of the energy.

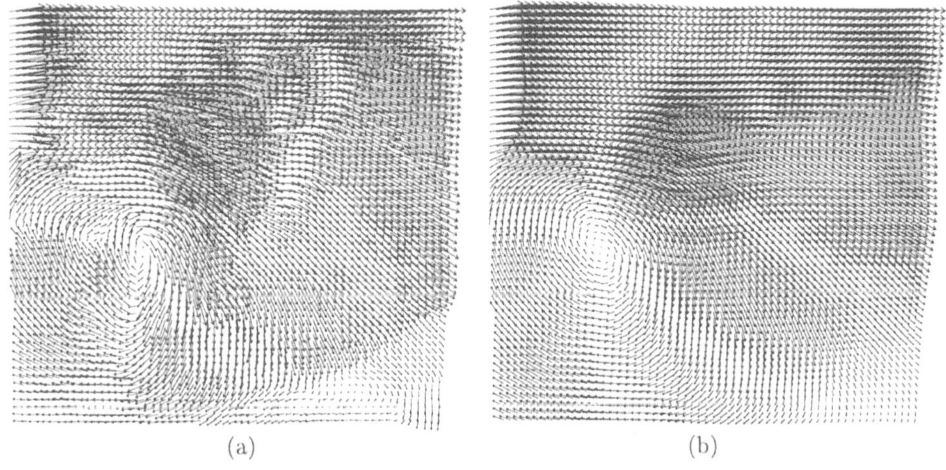

Figure 7. Phase 1 of conditionally sampled perturbed flow. (a) PIV image, (b) Image recontructed from 20 POD modes containing 68% of the energy of phase 1.

corresponding image (b) reconstructed from 20 modes containing only 68% of the energy of phase 1.

6.5. ENERGY SPECTRA

It is interesting to note the shapes of the energy spectra for unperturbed flow (a), perturbed flow (b) and phase 1 of the perturbed flow (c) shown

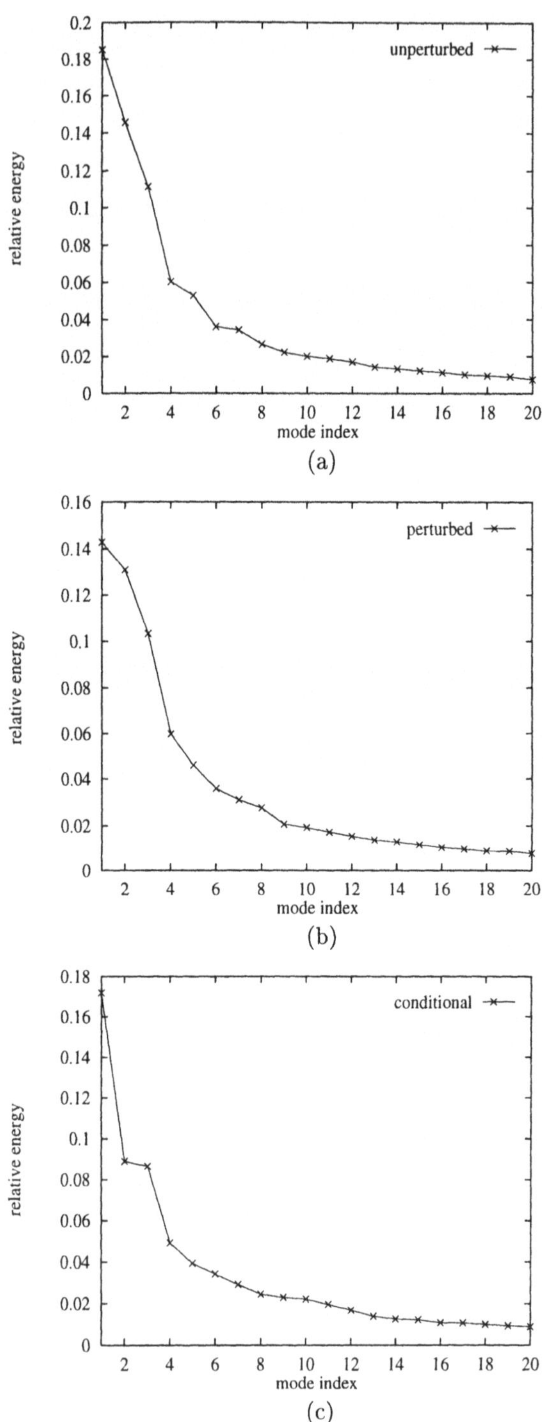

Figure 8. Energy vs. POD mode number for three different data sets. (a) The first 20 modes of unperturbed flow contains 81% of the energy, (b) the first 20 modes of perturbed flow (without conditional sampling) contains 73% of the energy, and (c) the first 20 modes of phase 1 of conditionally sampled perturbed flow contains 68% of the energy of that phase.

in Fig. 8. The perturbed flow has a smaller drop in energy from mode 1 to mode 2 compared to the unperturbed flow. Phase 1 of the perturbed flow has a large drop in energy from mode 1 to mode 2, and the energy level of mode 1 is improved. This tendency is common for all four phases (not shown).

7. Summary

For the unperturbed flow, vortex coalescing has been found, and good representation is achieved by 20 modes containing 81% of the energy. The perturbed flow suggests a growth and decay of vortical structures, and 3D effects are evident. This type of flow seems hard to represent by 20 modes containing only 73% of the energy. The conditionally sampled perturbed flow images show high energy content of mode 1 compared to mode 1 of the unconditionally sampled images. In conclusion, the application of the POD technique to PIV data is a promising tool for extraction of coherent structures from complex flows.

References

Aubry, N., Holmes., P., Lumley, J. L., and Stone, The dynamics of coherent structures in the wall region of a turbulent boundary layer, *J. Fluid Mech.* (1988), Vol. 192, pp. 115-173.

Berkooz, G., Holmes, P., and Lumley, J. L., The Proper Orthogonal decomposition in the Analysis of Turbulent Flows, *Annu. Rev. Fluid Mech.* 1993, 25: 539-575.

Christensen, E. A., Sørensen, J. N., Brøns, M., and Christiansen, P. L., Low-Dimensional Representation of Early Transition in Rotating Fluid Flow, *Theoret. Comput. Fluid Dynamics* (1993) 5: 259-267.

Delville, J., Characterization of the Organization in Shear Layers via Proper Orthogonal Decomposition, from Bonnet, J. P. and Glauser, N. (eds.), *Eddy Structure Identification in Free Turbulent Shear Flows*, 225-237, 1993, Kluwer Academic Publishers.

Manhart, M. and Wengle, H., A Spatiotemporal decomposition of a Fully Inhomogeneous Turbulent Flow Field, *Theoret. Comput. Fluid Dynamics* (1993) 5: 223-242.

Orellano, A. and Wengle, H., Numerical Simulation of Manipulated Flow over a Fence, *IMACS-COST Conference on Three-dimensional Complex Flows*, Sep. 13-15, 1995.

Schmidt, J. J., Experimental and Numerical Investigation of Separated Flows, Ph.D. Thesis, 1997, DTU, Dept. of Energy Engr., Denmark.

Ullum, U., Schmidt, J. J., Larsen, P. S., McCluskey, D. R., Temporal Evolution of the Perturbed and Unperturbed Flow Field Behind a Fence: PIV Analysis and Comparison with LDA data, to be presented at the *7th International Conference: Laser Anemometry - Advances and Applications*, Sep 8-11, 1997, University of Karlsruhe.

STRUCTURE DETECTION IN DRIVEN DRIFT WAVE TURBULENCE

V. NAULIN
Association EURATOM - Risø National Laboratory
Optics and Fluid Dynamics Department
Risø National Laboratory
PO Box 49
DK-4000 Roskilde

1. Introduction

The appearance of long lived coherent structures in turbulent drift-wave dynamics is a long known feature. The existence of these structures is believed to be important for the anomalous plasma diffusion across magnetic field lines as observed in experiments. Stationary localized solutions are known for adiabatic electron density response and their self-organization was shown numerically (Kono and Miyashita, 1988). The statistical properties of the turbulence were examined in detail by e.g. Horton (Horton, 1986). Nevertheless the dynamics of the structures in the flow is not yet well examined nor understood. Here a model incorporating an instability leading to quasistationary self-sustained turbulence will be considered. It can be understood as a paradigm for 2D-turbulence excited by an internal instability at low k and energy dissipation at large k.
The paper is organized as follows: After a short description of the model and the numerics in section 2, we will use tracking techniques to gain insight into the statistical properties of the transient structures in the flow in section 3. We finally use conditional sampling in section 4 to get knowledge on the average shape and the dynamics in some situations.

2. Model and Numerical Results

As the derivation of the model has been presented in detail before (Crotinger and Dupree, 1992; Naulin *et al.*, 1995) we here just state the equation used in the numerical simulations. We consider a plasma with a density gradient in the x-direction in a strong, homogeneous magnetic field pointing in the

z-direction. The high parallel mobility of the electrons makes the system two-dimensional. Using the drift-scaling we get from the two-fluid equations

$$\begin{aligned}&\partial_t(1-\nabla^2)\varphi+\partial_y\varphi+\mu\nabla^2\nabla^2\varphi+\{\nabla^2\varphi,\varphi\}\\&=\delta\partial_t\left(\partial_t+\partial_y\right)\varphi+\delta\left\{\varphi,\left(\partial_t+\partial_y\right)\varphi\right\}\ ,\end{aligned}\tag{1}$$

where the density fluctuations n can be computed from the electrostatic field φ through

$$n=\varphi-\delta\left(\partial_t+\partial_y\right)\varphi\ .\tag{2}$$

Quantities were normalized using

$$\begin{aligned}&\frac{e\varphi}{T_e}\frac{L_n}{\rho_s}\rightarrow\varphi\ ,\quad\frac{\delta n_e}{n_{00}}\frac{L_n}{\rho_s}\rightarrow n\ ,\quad\frac{tc_s}{L_n}\rightarrow t\ ,\\&\rho_s\nabla_\perp\rightarrow\nabla\ ,\quad\frac{\mu_iL_n}{\rho_s^2c_s}\rightarrow\mu\ ,\quad\delta=\frac{1}{k_{||}^2L_{||}^2}\ ,\end{aligned}\tag{3}$$

where $\rho_s=c_s/\Omega_i$ is the ion gyroradius at electron temperature T_e and the sound speed is given by $c_s=(T_e/m_i)^{1/2}$. An average density n_{00} is used for normalization, while within the so-called local approximation the background density n_0 is assumed to have an exponential decay in the x-direction with a decay length L_n. Furthermore the Poisson bracket $\{\varphi,\psi\}\equiv\vec{z}\times\nabla\varphi\cdot\nabla\psi$ was used. An effective parallel length

$$L_{||}=\left(\frac{L_nT_e}{m_e\,c_s\,\nu_{ei}}\right)^{1/2}\ ,\quad L_{||}\nabla_{||}\rightarrow\nabla_{||}\ .\tag{4}$$

is introduced to normalize $k_{||}$. Based on experiments showing a very long correlation length along the magnetic field, it is assumed that there is only one parallel mode exited, thus reducing the model to two dimensions. The finite $k_{||}$ makes the energy in the background density gradient accessible to the system by driving some modes at $k\approx1$ unstable. The term with μ is the ion-viscosity which dissipates energy in the large k region. With $\delta=\mu=0$ we get the Hasegawa-Mima Equation

$$\partial_t(1-\nabla^2)\varphi+\partial_y\varphi+\{\nabla^2\varphi,\varphi\}=0\ .\tag{5}$$

This equation can be translated to the Charney-Obukhov-Equation for Rossby-Waves using the following dictionary: The electrostatic potential corresponds to the stream function, while the background density gradient takes the same role as the gradient of the coriolis force.
Equation (1) is solved numerically using a pseudo-spectral code with periodic boundary conditions. The time derivatives on the rhs of Equation (1)

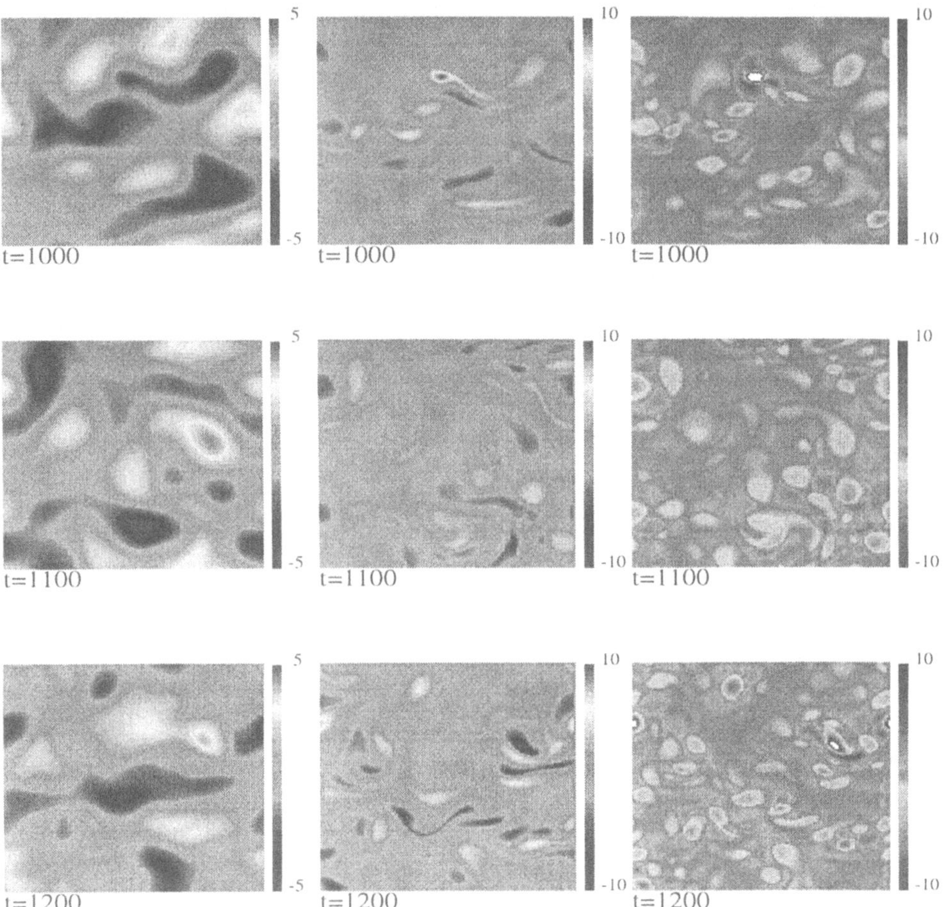

Figure 1. Potential (left), vorticity (center), and Weiss field (right) for times $t = 1000$ to $t = 1200$. (For reasons of clarity only values between 10 and -10 were color-coded in the Weiss field, although it has valucs down to -25)

are evaluated by iterating for $\delta = 0$, e.g. the Hasegawa-Mima Equation (5). We present a run with a low spatial resolution of 128 times 128 points as we are interested in a long timeseries. Parameters for the run are $\delta = 0.5$ and $\mu = 0.03$ while the domain of integration is 21×21 length units in size and initialized with low amplitude ($|\phi_k| \sim 10^{-7}$) random fluctuations, having a Gaussian shape spectrum centered at the zero mode in k-space. Approximately 25 modes around $k_x = 0$ and $k_y = \pm 1$ are linearly unstable with growth rates $\gamma_{max} \approx 0.05$.

The overall picture for the dynamic is as follows: After the linear growth of the most unstable modes leading to amplitudes of order unity the nonlinearities become important and the system gets into a saturated turbulent

state as depicted in Fig. 1. This turbulent state is characterized by fluctuations up to 50% around the average values. For details see (Naulin, 1995). Visual inspection of the time development of the turbulence – looking at a movie – shows transient blobs of potential moving through the domain.
After the turbulence has reached a saturated state 5000 time-samples are taken each $\Delta t = 0.5$, thus covering 2500 time-units forming the data-set to be considered in the following.

3. Tracking Coherent Structures

We define coherent structures as entities confining plasma-volume for a time much longer than the typical nonlinear time - corresponding to an internal eddy turnover time - given by $t_{nl} \approx \frac{1}{|\phi|_{max} k_{av}^2} \approx 1$, with the spectrum being peaked at an average wavenumber $k_{av} \approx 1$. To have a confined area a bunch of nested closed isopotential-lines – which are the streamlines of the $E \times B$-velocity – is necessary, therefore a coherent structure is localized by definition.
The Weiss field

$$Q_W = \frac{1}{4}\left(\sigma^2 - \omega^2\right) \tag{6}$$

gives an additional criterion to identify a structure in the flow. It balances the square of the strain $\sigma^2 := (\partial_x u - \partial_y v)^2 + (\partial_y u + \partial_x v)^2$ (u and v are the velocities in x and y respectively) and the square of the vorticity $\omega := \nabla^2 \varphi$. This balance was introduced by Weiss (Weiss, 1991) and separates the flow into elliptic regions defined by $Q_W < 0$ and hyperbolic regions defined by $Q_W > 0$. In the latter fluid elements are separated from each other by the dominating strain ($Q_w > 0$), while in the elliptic regions ($Q_w < 0$) the rotation dominates the dynamics and keeps fluid elements together, thus distinguishing between areas with and without particle confinement.
However, as the Weiss field only analyses the frozen in flow, it is still necessary to take the time evolution of the system into account. This can be done by identifying and tracking events and setting a minimum lifetime before they qualify for the notation coherent structure.
The strategy is as reported previously (Naulin and Spatschek, 1997). In every time-slice produced by the numerics candidates for coherent structures are identified using the Weiss criterion. For every event in time-slice t_n it is then tested if there is an predecessing event in time-slice t_{n-1} localized in some ball around the position of the event at t_n. Only events which can be tracked are identified as belonging to coherent structures. Using the Weiss-field we assign a size L to each structure over which $Q_W < 0$, and by integration of the density – assuming a Boltzmann distribution $n \approx \phi$

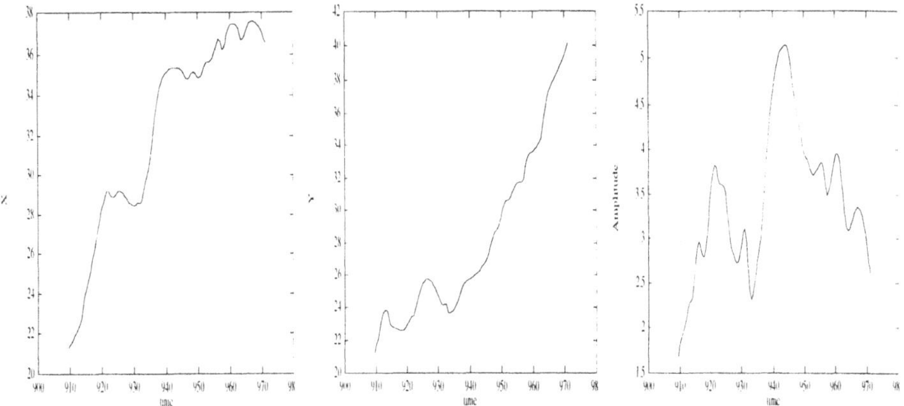

Figure 2. x-position (left), y-position (center) and amplitude (right) of a single structure over time.

to lowest order – a positive mass M is defined by

$$M = \text{sign}(n(\vec{r}_0)) \int_A n(\vec{r})\, H\left(-Q_w(\vec{r})\right) d^2 r \tag{7}$$

where $H(\zeta)$ is the Heavyside-function. The integration is carried out over an area A including only one event and the exact position $\vec{R}$ of a structure is given by its center of mass

$$\vec{R} = \frac{1}{M} \int_A \vec{r}\, |n(\vec{r})|\, H\left(-Q_w(\vec{r})\right) d^2 r \,. \tag{8}$$

assuming, that the sign of the density fluctuation does not change over the extend of a structure. The track of a single longer lived event is shown in Figure 2. Average quantities taken from all events found can be taken from Table 3. We refer to events with a positive potential (respectively density) as "humps" and to events with a negative potential as "dips". One notes, that dips and humps appear symmetrical in the flow. The distribution of velocities is widespread (Figure 3), so that an individual structure path can be very different from the average picture we get. Further it is interesting to notice, that structures of all masses (Figure 3) and lifetimes (Figure 4) appear, and that there is only a weak correlation between the maximum mass a structure achieves during its existence M_{max} and its lifetime. As the structures not only have a finite velocity component in the direction perpendicular to the background density gradient, but also parallel to the latter they account for particle transport. Regarding humps and dips separately and averaging over all events the mass transport is evaluated:

$$\vec{\Gamma}_{D,H} = \frac{1}{T} \int \sum_{D,H} M_{D,H} \vec{v}_{D,H} dt \,. \tag{9}$$

TABLE 1. Average lifetime t_l, velocity v, size L, Mass M and transport Γ of potential humps and dips with standard deviation σ and number of observations N.

	Humps			Dips		
	Value	σ	N	Value	σ	N
Lifetime t_l	14.5	12	1515	14.7	13	1531
v_x	0.15	0.47	43898	-0.16	0.47	45074
v_y	0.41	0.40	43898	0.41	0.39	45074
L_x	2.67	0.88	43898	2.66	0.88	45074
L_y	2.52	0.89	43898	2.48	0.86	45074
M	10.2	8.8	43898	9.9	8.6	45074
Γ	0.0051	4.0	43898	0.0047	3.8	45074

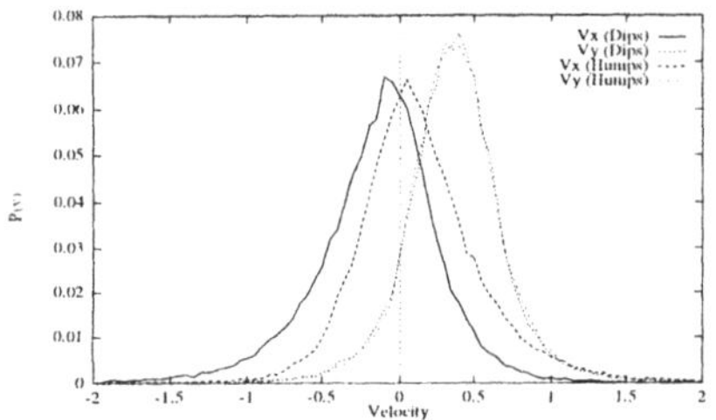

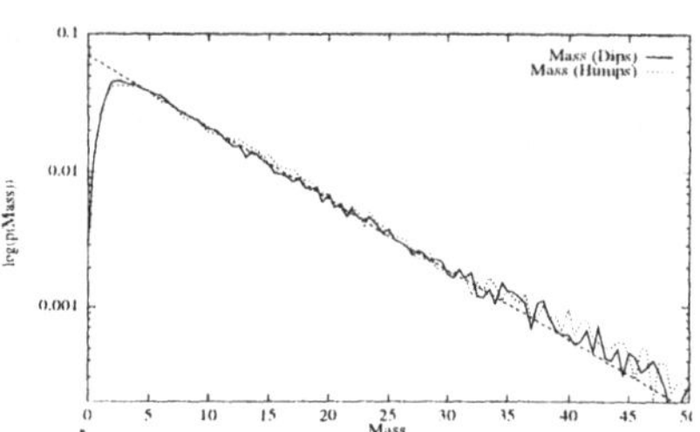

Figure 3. Probability densities of velocities (left) and mass (right) of structures.

Some implications on the total transport were discussed in detail in (Naulin and Spatschek, 1997). Here we note that the structures change the statistical properties of the transport although they account only for a small fraction of the total transport. (Figure 4).

4. Conditional Sampling

Until now the tracks and the movement of the structures were evaluated. In this section we will focus on the spatial configuration of the structures. To that extend an additional tool to analyze the flow is introduced, the so-called conditional sampling technique (CST). The CST has been often and successfully applied in plasma-experiments and fluid dynamics (Nielsen *et al.*, 1996; Pécseli and Trulsen, 1989), the basic ideas are discussed in

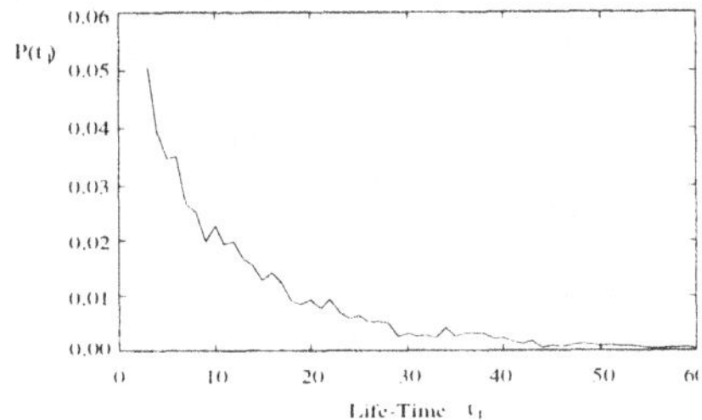

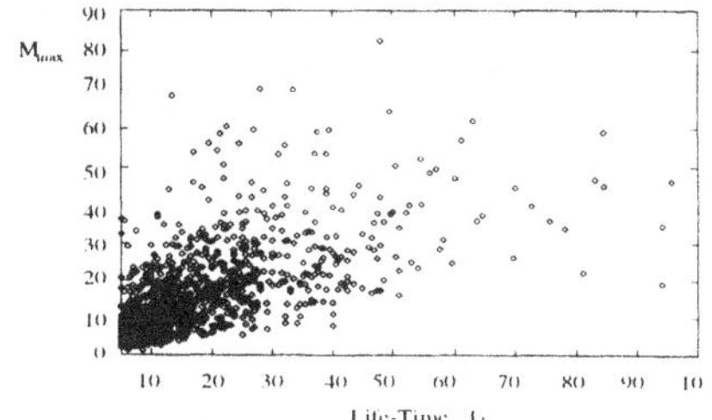

Figure 4. Probability densities of life-time t_l (left) and maximum mass M_{max} (right) over life-time.

(Johnsen *et al.*, 1987) and (Adrian, 1979). Originally the CST is applied to derive from independently taken time-series a picture of the spatial organization of a flow.
Here we will use it in a different way: In each time-slice we again search for events E located at positions $\vec{r}_E$ and time t_E. Then a subsequence of time-slices is sampled around each event and normalized by the number of events N_E

$$\Phi_C(\vec{R},T) := \sum_E \phi(\vec{r}_E + \vec{R}, t_E + T)/N_E = < \phi(\vec{r},t)|E > . \tag{10}$$

If the dynamics in the vicinity of an event is reproducible, Φ_C should reflect this average dynamics. When used for the analysis of time-series large scale structures which are reproduced in the experiment can be found by this kind of analysis, while small scale erratic turbulence averages out. Here we know what kind of structures we are looking for, concerning sizes and velocities. As the spread in the velocities is high, we expect that the coherent sample smears out over times T being of the order $L/\sigma_v \approx 5$. While individual paths of the structures vary a lot, we can use sharper criteria for the sampling, thus trying to sample more similar situations. The cost for this is that with a smaller number of events to be sampled the statistics gets worse.
First we choose local maxima or minima as trigger for the sampling. The result is depicted – for the sampled maxima only, as the sampled minima look symmetric – in Figure 5 and compared for $T = 0$ to the two point correlation function

$$C(\vec{r}) = \frac{1}{< \phi^2 >} \int \exp(i\vec{k}\vec{r}) \left(\frac{1}{T} \int_0^T |\phi_k(t)|^2 dt \right) d\vec{k} \, . \tag{11}$$

The correlation function is – as can easily be seen – dominated by the structures in the flow as they account for most of the correlations and occupy a significant amount of the total area. The evolution of Φ_C with T reflects the movement of the maxima in the y and x direction as determined

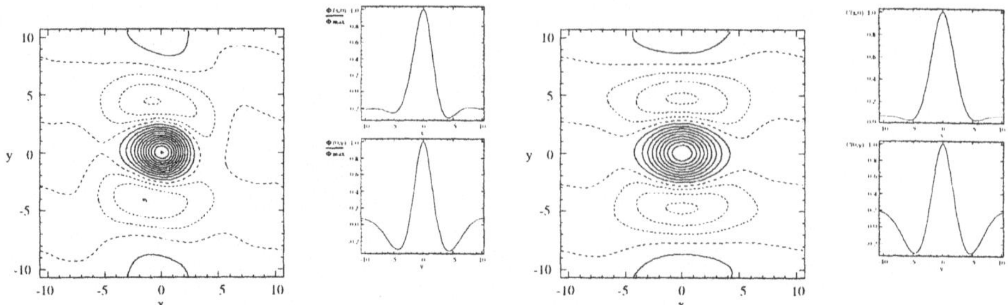

Figure 5. Normalized Conditional Sample (left) at $T = 0$ sampled using the maximum criterion as a trigger and Two Point Correlation (right). On each side: Contourplot (left) with 0.1 spacing between levels and sections in x (top, right) and y (bottom, right).

before. Due to the spread in velocity the area of positive correlation smears out as expected. Additionally one can see, that there is a large probability for other structures of the same size to be at the same y-position, whereas structures of the opposite sign appear with large probability below or over the reference point. This indicates the existence of a global zonal flow, with the localized structures riding on top of this background.
To investigate the behaviour of two monopolar structures which are aligned as as dipole, the maximum value of the electric field $\vec{E} = \nabla\varphi$ being anti-parallel to the x-axis was used to trigger the sampling, selecting dipolar structures where the positive ($v_x > 0$) and negative part ($v_x < 0$) are colliding. This criterion alone leads to insufficient results, as the dipole may be turning right or left, giving a very weak conditional sample for $T \neq 0$. Thus only events were considered for the sampling process, where the absolute value of the positive potential a length L_x away from the trigger point was more than 5% larger than that of the negative potential the same distance away in the other direction. The result of that sampling process is shown in Figure 6. It is seen, that in this combination the weaker monopol is torn around by the stronger one until the both components are able to continue their way in opposite x-directions. For $T = 0$ the sample reminds strongly to the well known dipole solution of the Hasegawa-Mima equation. In this context it is interesting to note, that the conditional sampling at $T = 0$ shows a well organized scatter plot (Figure 7) with a nonlinear relation between ϕ and ω, demonstrating that the CST revealed a highly organized state.

5. Summary

Using different techniques for the detection of structures we were able to clarify their behaviour in driven, quasistationary drift-wave turbulence. The following picture for the overall dynamics of structures in the considered

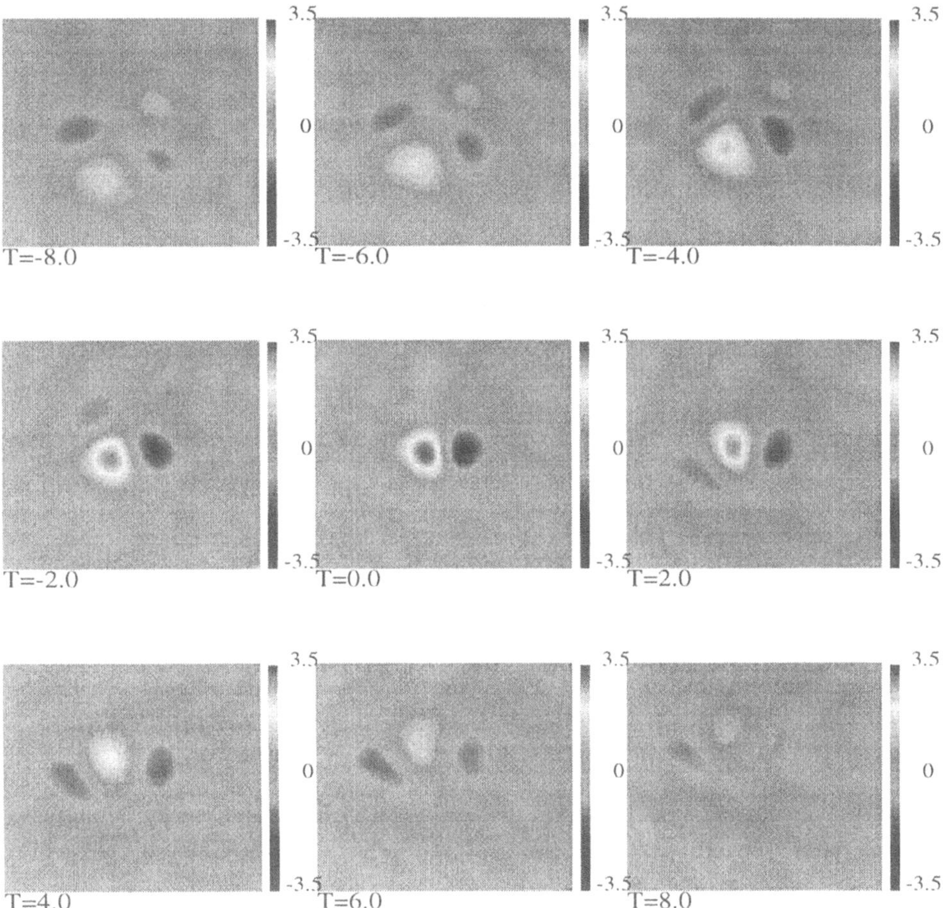

Figure 6. Conditional Sample sampled using maxima of the electric field and a stronger positive monopol for different time lags T.

flow emerges: Transient monopolar structures move through the flow. Although the individual paths of dips and humps are complicated, they move on the average in opposite directions parallel to the background density gradient. In collisions they may form temporary dipoles, which after half a rotation separate again. On their way through the plasma they transport confined density, thus changing transport properties. We showed that the conditional sampling technique can be a useful tool in the analysis of numerical data, when additional information about the flow is available.

Acknowledgement

Discussions with J.J. Rasmussen, K. Rypdal, A.H. Nielsen, and K.H. Spatschek are greatfully acknowledged.

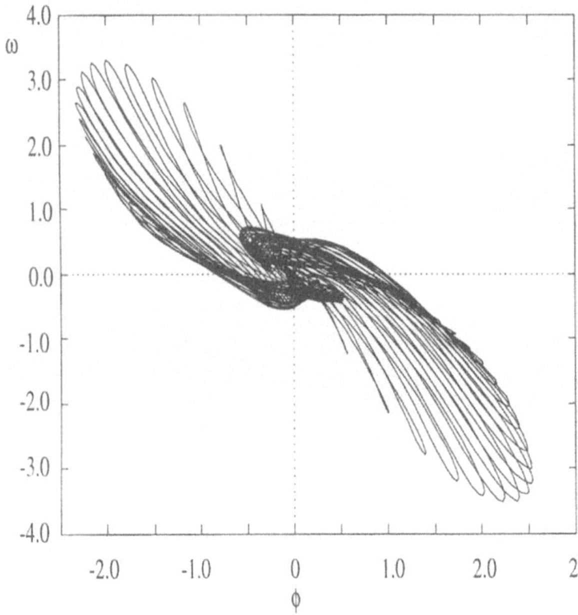

Figure 7. Scatter plot showing ω over ϕ for the conditional sample of Figure 6 at $T = 0$, values of same location x are connected.

References

R. J. Adrian, Phys. Fluids **22** (1979) 2065.

J. Crotinger and T. H. Dupree, *Trapped structures in drift wave turbulence*, Phys. Fluids B **4** (1992) 2854–2870.

W. Horton, *Statistical properties and correlation functions for drift waves*, Phys. Fluids **29** (1986) 1491–1503.

H. Johnsen, H. L. Pécseli, and J. Trulsen, , Phys. Fluids **30** (1987) 1609.

M. Kono and E. Miyashita, *Modon formation in the nonlinear development of the collisional drift wave instability*, Phys. Fluids **31** (1988) 326–331.

V. Naulin, *Nichtlinearer Transport in ebenen Driftwellenmodellen*, PhD thesis, Heinrich-Heine Universität, Düsseldorf, 1995.

V. Naulin and K. H. Spatschek, *Modon Saturation in Collisional Driftwave Turbulence*, Contrib. Plasma Phys. **35** (1995) 33–43.

V. Naulin and K. H. Spatschek, *Structures and Transport in Two-Dimensional Drift-Wave Turbulence*, accepted Phys. Rev. E (1997).

V. Naulin, K. H. Spatschek, S. Musher, and L. I. Piterbarg, *Properties of a two-nonlinearity model for drift-wave turbulence*, Phys. Plasmas **2** (1995) 2640–2652.

A. H. Nielsen, H. L. Pécseli, and J. J. Rasmussen, *Turbulent transport in low-β Plasmas*, Phys. Plasmas **3** (1996) 1530–1544.

H. L. Pécseli and J. Trulsen, *A statistical analysis of numerically simulated plasma turbulence*, Phys. Fluids B **1** (1989) 1616–1636.

J. Weiss, Physica D **48** (1991) 273.

A COHERENT STRUCTURE DETECTION METHOD USING THE WAVELET TRANSFORM

P. MUSCAT, P. DUSSOUILLEZ, P. DUPONT AND J. LIANDRAT
Institut de Recherche sur les Phénomènes Hors équilibre
Unité Mixte de Recherche C.N.R.S. No 6594
Universités d'Aix-Marseille I & II
12, Avenue Général Leclerc, 13003 Marseille, France

Abstract.

We propose a detection method of coherent structures based on wavelet transform which has been implemented on multipoint experimental data. Given 2 signals, we first define a notion of time and scale localized inter-correlation. We then define, in the context of multipoint measurements, the notion of time and scale localized convection velocity. We validate the method on simulated signals and we use it to detect and to analyse organized energetic structures in a turbulent compressible mixing layer with a convective Mach number of 0.62 . In particular the defined convection velocities are used to split the signal in different parts. The spectral analysis of each part provides informations on the scale distribution of the signal.

Key words: Coherent structure, wavelet transform, mixing layer, compressible flow, convection velocity

1. Introduction

The presence of large organized structures in turbulent shear flows has been reported by many authors. The possibility to influence their development opens the possibility to control macroscopic properties of the flow such as the rate of mixing processes or the amplitude of noise generation.

In subsonic mixing layer, for example, the spatial development of layer has been related to the amalgamation of spanwise coherent large scale vortices (Winant 74).

Such coherent structures have been extensively studied both using optical methods (Brown 74; Winant 74; Bernal 86) and multipoints hot wire measurements (Delville 89). Two dimension visualizations of subsonic mixing layers have confirmed the quasi two dimensional aspect of these structures. That explains why optical method such as shadowgraph can be used to visualize the structure of supersonic mixing layer in spite of their space integration (Brown 74).

On the opposite, what is known on coherent eddies in high speed flows is much limited. Their existence, even for high convective Mach numbers ($M_c > 1$) has been etablished both by visualizations (Bonnet 93; Clemens 95; Papamoschou 88) and by spectral analysis of hot wire measurements (Dupont 95a). However, it has been found that they are not as well defined, at least from visualizations, as in subsonic cases and that they are essentialy three dimensional. An important effort has been made to relate the dynamic of these vortices to the drastic reduction of the spreading rate of the compressible mixing layer compared to equivalent subsonic case with same velocity and density ratios. An important parameter to quantify the compressibility effects (Papamoschou 88; Bogdanoff 83) is the convective Mach number defined by $M_{c_i} = |U_i - U_c|/a_i$, where the index i identifies the two external flows, U_i is the corresponding velocity, a_i is the corresponding sound speed and U_c is the convection velocity of the Kelvin-Helmholtz type structures. Isentropic models gives a symmetric behavior ($M_{c_1} = M_{c_2} = M_c$), but they are in contradiction with early determinations of U_c which shows a strong dissymmetric behaviour ($M_{c_1} \neq M_{c_2}$). Dimotakis (Dimotakis 91) and Barre (Barre 94) tried to explain the departure from the isentropic relation for $M_c > 0.5$ by the creation of shocklets due to transonic or supersonic velocity differences in the layer. They obtained a good collapse with the results of various authors, at least for confined layers. Unfortunately there are only few experimental results on the kinematics of the large coherent structures in compressible mixing layer, mainly due to technical limitations. As already mentioned, their eduction in high velocity flows appears more difficult than in the low velocity case, even when using recent optical visualization methods. Multipoints hot wire methods (rake of probes) are quite difficult to use for two main reasons, the multi probes interactions and the requirement of a sufficient bandwidth on the anemometers (typically up to 100 kHz for the energetic scales). Moreover, it is important, for compressible cases, to obtain an information on the kinematics through U_c, but this longitudinal convection cannot be inferred from transversely distributed measurements.

These remarks have been the starting point of the present work. The goal is indeed to develop a detection method for coherent structures, efficient in supersonic turbulent mixing layer, giving their convection velocity as well as their characteristic length and time scales. As a convective information was required, two points measurements along the mean direction of the flow have been performed. The problem is then to derive a definition for the coherent structures which can be used whith only two probes signals.

A general concept is to define coherent structures as a set of fluid particles ($\mathcal{D}$) with a phase correlated vorticity in the domain $\mathcal{D}$ (Hussain 83). In some cases a simpler definition based on the kinematics of the coherent structures can be considered. For example, the Kelvin Helmholtz type structures developing in mixing layer or Karman's vortices in a wake are known to move with a convection velocity U_{c_1}. It is characteristic of the coherent structures and, in general, it is locally (in space and/or in time) different from the convection velocity of the non coherent part of the flow U_{c_2}. This has been verified in previous works (Dupont 95a; Dupont 95b), where experimental spectra, space-time correlation functions ($R(\xi, \tau)$) and dispersion relationship $\omega(k)$ ($U_c(\omega) = \omega/k$ whith k the wave number), were measured in a compressible mixing layer ($M_c = 0.62$). The results suggested the existence of two media: one with a characteristic convection velocity of the order of the isentropic value, and a second one with a convection velocity following the mean local velocity. Fourier analysis gave indications on the temporal properties of the flow (Dupont 95b) and suggested that a local convection velocity U_c could be used as a tracer of the coherent and uncoherent part, at least in such flows. But the classical spatio temporal tools are not suitable to give such an information, and it seems more appropriate to decompose turbulent signals on elementary components well localized both in Fourier and physical spaces. That is why we use the wavelet transform of the two signals to generalize the classical space-time tools and to define a convection velocity localized in time and scale. Thus we may study the kinematics of the different parts of the wavelet plane and define coherent parts of the signal embedded in incoherent motions with a different convection velocity.

2. Method description

In turbulent flows, convection velocities can be deduced with two points measurements from spatio-temporal intercorrelation function $R(\xi, \tau)$ ($U_c = \xi/\tau_{opt}$, where ξ is the probe separation and τ_{opt} the optimal lag time) or from its Fourier transform through the dispersion relationship $\omega(k)$ ($U_c(\omega) = \omega/k$) where k is the wave number extrapolated from the phase shift between the two signals. But these classical tools give only a global information on the signal and are unable to isolate the coherent structures wich have locally a convection velocity U_{c_1} different from the convection velovity U_{c_2} of the remaining part of the signal.

The time-scale analysis deduced from the wavelet transform allows to achieve a characteristic information of the different parts of the flow localized in time and scale.

The following definitions will be used to derive a concept of convection velocity localized in scale and time.

BASIC DEFINITIONS

Let $\{W_s(t,a) = \int s(t')1/\sqrt{a}\psi^*(t'-t/a)dt', t \in I\!R, a \in I\!R^{+*}\}$ be the wavelet transform of a function $s(t)$ where ψ is the complex wavelet, t is the scale parameter and b is the position parameter. Since the wavelet transform is isometric we may define a localized variancy as:

$$E_l(t,a) = 1/C_\psi |W_s(t,a)|^2/a^2$$

where C_ψ depends only on ψ.

For stability requirements, it is more suitable (Liandrat 96) to consider $E_m(t,a)$ which is obtained by averaging $E_l(t,a)$ according to $\chi_a(t)$ a rescaled bump function $(\chi_a(t) = \chi(t/a))$ taking into account the support of $\psi(x/a)$. We get:

$$E_m(t,a) = \int El(s,a)\chi_a(t-s)ds.$$

Considering two functions $s(t)$ and $s'(t)$ and the corresponding wavelet coefficients W_s and $W_{s'}$, we define:

$$R_l^{ss'}(\tau,t,a) = 1/C_\psi R_e[\frac{W_s(t,a).W_{s'}^*(t+\tau,a)}{a^2}]$$

where $*$ denotes complex conjugate.

$R_l^{ss'}(\tau,t,a)$ is the normalized intercorrelation of $W_s(t,a)$ and $W_{s'}(t,a)$. Moreover, we get that :

$$R^{ss'}(\tau) = \int_{-\infty}^{+\infty} s(t)s'(t+\tau)dt = \int_0^{+\infty}\int_{-\infty}^{+\infty} R_l^{ss'}(\tau,t,a)dtda$$

where $R(\tau)$ is the intercorrelation of s and s'.

As previously, we prefer: $R_m^{ss'}(\tau,t,a) = R_l(\tau,t,a) * \chi_a(t)$ that preserves:

$$R^{ss'}(\tau) = \int_0^{+\infty}\int_{-\infty}^{+\infty} R_m^{ss'}(\tau,t,a)dtda.$$

If A is a subdomain of $I\!R^{+*}$ (usually an interval of the form $[a_{min}, a_{max}]$), we introduce:

$$R^{ss'}_{m_A}(\tau,t)=\int_A R^{ss'}_m(\tau,t,a)da.$$

As shown on figure 1, $R^{ss'}_{m_A}(\tau,t)$ is the intercorrelation between the subdomain S_A of the wavelet plane related to s (defined by A, χ and t) and the subdomain S'_A related to s'(defined by A, χ and $t+\tau$).

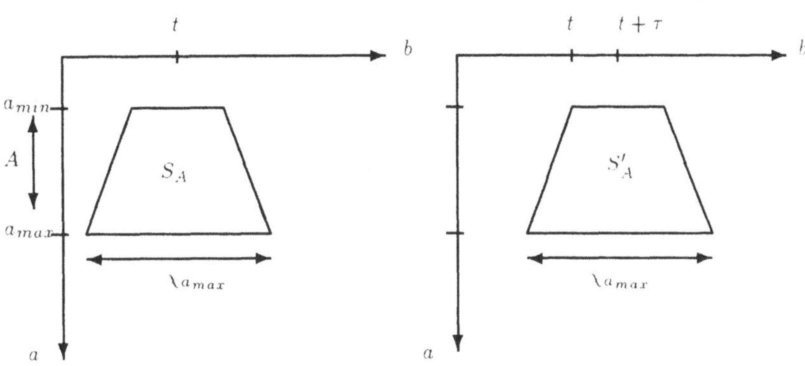

Figure 1. Wavelet planes and subdomains S_A and S'_A for $A=[a_{min},a_{max}]$

The localized correlation coefficient $\rho_A(\tau,t)$ is then defined by:

$$\rho_A(\tau,t)=R^{ss'}_{m_A}(\tau,t)/\sqrt{R^{ss}_{m_A}(0,t)R^{s's'}_{m_A}(\tau,t)}.$$

We define τ_{opt_A} as:

$$\forall\tau\neq\tau_{opt_A}, |\rho_A(\tau,t)|\leq|\rho_A(\tau_{opt_A},t)|.$$

Since $R^{ss'}_{m_A}$ can be seen as a scalar product in a suitable space, we get $|\rho_A(\tau,t)|\leq 1$.

INTERPRETATION OF THE PROPOSED DEFINITIONS AND DEFINITION OF THE LOCAL CONVECTIVE VELOCITY

The proposed definitions can be interpreted in term of dynamics when s and s' are velocity signals at two different positions x and $x+\xi$. Indeed, in such a case, τ_{opt_A} can be used to define a localized convection velocity as:

$$U_c(\xi,t,A)=\xi/\tau_{opt_A}(t).$$

U_c can then be interpreted as the convection velocity related to the subdomain of the wavelet plane S_A(see figure 1). Moreover, $\rho_A(\tau_{opt}, t)$ can be considered as a local version of the classical coherence function.

The analysis of the function $U_c(t)$ provides the segmentation of s.

3. Validation

The validation of the method has been performed on the two following generated signals.

First we construct a signal S composed of alternative parts of two white noises of same energy(figure 2a) related to distinct velocities: to simulate the convection velocities, we generate a second signal s', in which each part is shifted alternatively by $t1 = 20$ points and $t2 = 30$ points. We can see on figure 2b that the evolution of $\tau_{opt}(t)$ reproduces nicely the segmentation of the signal except in the interface zones.

The second case corresponds to filtered white noises with different energy levels (figure 3a). As previously, the detection of each medium is very good (figure 3b) and allows us to build the conditional spectrum of each medium from its wavelet coefficients. The results are presented on figure (4) with the spectrum of the signal. Each conditionnal spectrum collapses with the energy content of each medium.

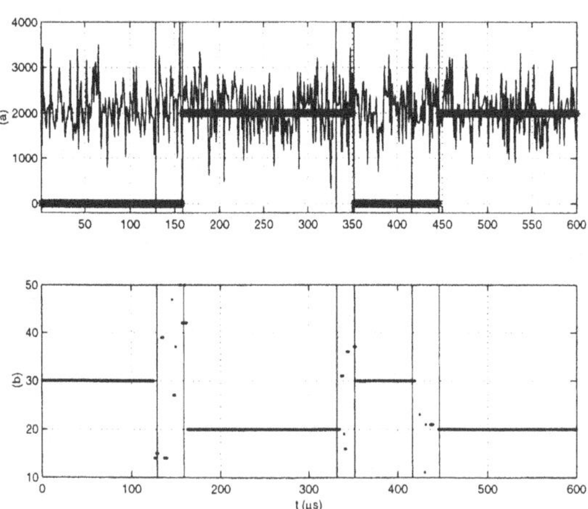

Figure 2. (a) simulated signal composed of random noise of same energy each part being shifted of a different value (b) $\tau_{opt}(t)$. Vertical lines: limit of interface zone

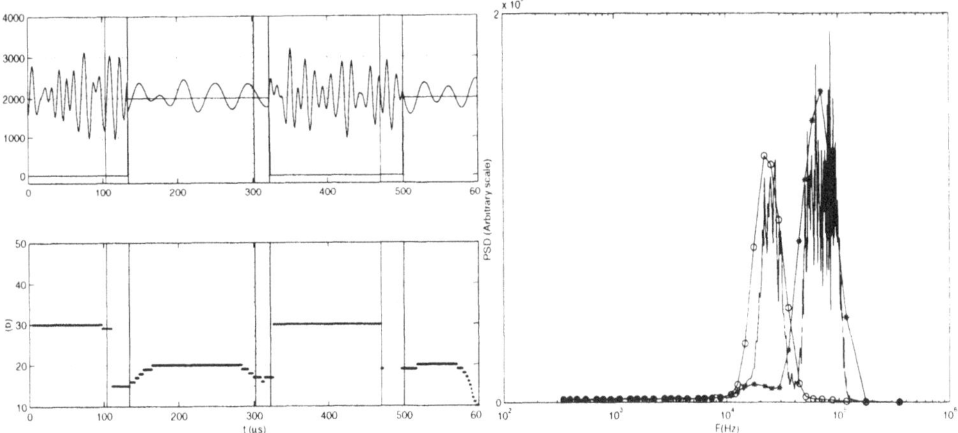

Figure 3. simulated signal: Narrow band random noise

Figure 4. simulated signal: Narrow-band random noise(spectra), PSD of the signal '-', conditional PSD of media 1 '*' and 2 'o'

4. Application to a turbulent compressible mixing layer

The method of detection has been tested in a mixing layer with a supersonic-subsonic combination, in the continuous supersonic wind tunnel at IRPHE. The convective Mach number deduced from the isentropic relation was 0.62. Two points measurements were performed using Dantec "Streamline" constant temperature hot wire anemometers. The longitudinal probe separation ξ was 9mm for a local layer thickness of 23.8mm. The probe separation acts as a spatial low-pass filter. Its equivalent cut off frequency determined using coherence function is 70kHz. Therefore, only scales ranging between $20\mu s$ and $1ms$ are physically meaningfull. 80% of the total energy of the original signal s (see figure 5) was contained in the corresponding filtered one.

The convection velocity $U_c(t)$ is ploted on figure 6. Since $Uc(t)$ is considered as a characteristic function of the different media, its temporal properties can be compared to the spectrum of the hot wire signal $s(t)$. The spectra S_{U_c} of Uc and S_s of s are presented figure 7. The characteristic frequencies identified on both spectra confirm the presence of structures with different convection velocity and with quasi periodic distribution in time.

As the probes stand on the the high velocity side of the layer, the eddy signature should be related to local minima of $U_c(t)$ (see figure 6). Time scales associated with the structures are deduced from the evolution of $U_c(t)$ between two consecutive gradients of $U_c(t)$ negative then positive.

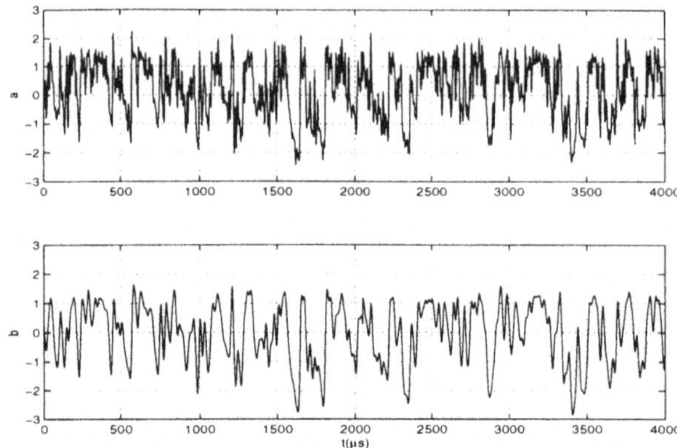

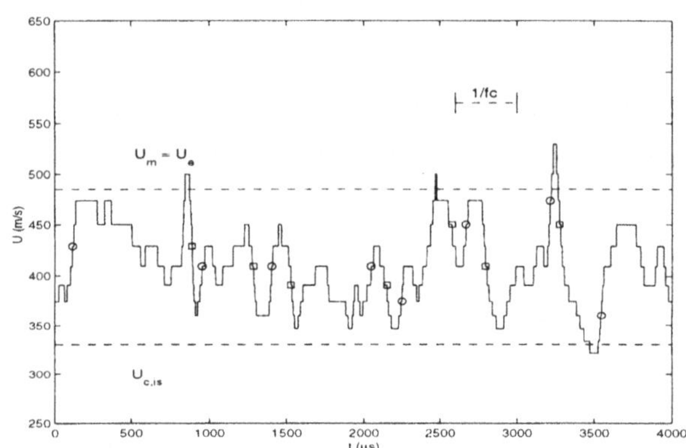

Figure 5. Hot wire longitudinal velocity signal in a supersonic turbulent mixing layer $M_c = 0.62$, $y^* = 0.5$ a) original signal b) filtered signals(only scales between $20\mu s$ and $1ms$ are considered)

Figure 6. Localized convection velocity $U_c(t)$. Same location as in figure 5. U_e is the external velocity and U_m is the mean velocity(pitot measurements), $U_{c,is}$ is the isentropic estimate of the convection velocity. o: location of the maximum of the derivative $\partial U_c/\partial t$; □: location of the minimum of the derivative $\partial U_c/\partial t$

respectively localized at the instant t_{G-} et t_{G+}. The quantity $p_i = (U_{max} - U_{c_i})/\sum(p_j)$, where U_{max} is the maximum convection velocity in the interval $[t_{G-}, t_{G+}]$, is considered as a weighting function. The time t_i at which the middle of the structure passes on the probe is defined as : $t_i = p_j t_j$ for $t_{G-} < t_j < t_{G+}$ and the convection velocity of the structure is defined as the average value of $U_c(t)$ in the interval $[ti - \sigma(t), t_i + \sigma(t)]$, where $\sigma(t)$ is the standard deviation of $t : \sigma(t) = [p_j(t_j - t_i)^2]^{1/2}$.

In a similar way, the incoherent part, is associated to local maxima of $U_c(t)$, and the weighting function is $p_j = (U_{min} - U_{c_j})/\sum(p_i)$, where U_{min} is the minimum convection velocity in the interval $[t_{G-}, t_{G+}]$. Note that no conditions on the absolute value of U_c is used to estimate the convection velocity U_{c_i} associated with the media.

A conditional analysis of the coherent part of the signal (medium 1) and the incoherent part (medium 2) has been performed. The mean value of U_{c1} is $353ms^{-1}$, close to the isentropic estimation ($330ms^{-1}$). On the opposite, medium 2 has a mean convection velocity of the order of the local mean velocity ($\pm 4\%$). This confirms previous results obtained from the dispersion relationship. The level of energy of both media are about the same. It is clear from the condtional spectra (figure 7) that the same temporal scales are involved in both media. This shows that a criterion based on the local level of energy or on the local distribution of energy cannot be efficient to detect the periodic scales of a mixing layer in the case of a fully turbulent flow. This was expected as , in our case, no specific partition of energy has been observed in the wavelet energy plane.

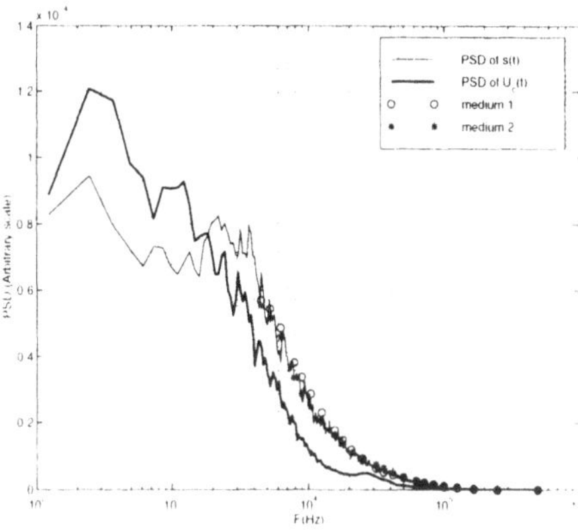

Figure 7. Power spectra of signal s, $U_c(t)$ and conditional spectra of the fluctuations in the media 1 (o) and 2 (*)

The validity of the conditional analysis has been tested by comparison with the Fourier analysis. It may be shown that when two non dispersive media (same phase velocity for each frequency) are considered with finite non-zero convection velocities respectively denoted U_{c_1} and U_{c_2}, the resulting signal is not dispersive and has a convection velocity such as : $U_{c_{1+2}} = 2U_{c_1}U_{c_2}/(U_{c_1} + U_{c_2})$. This relation was applied to the present measurements ; the results are given in figure 8. There is very good agreement with the group velocity deduced from the experimental dispersion relationship for frequencies equivalent to the considered time scales.

Finally, the time average value of $\rho_A(t, \tau_{opt}, \xi)$ is nearly the same in each medium and has practically the value of the coherence function considered at an equivalent frequency. This suggests that the spatial decay of coher-

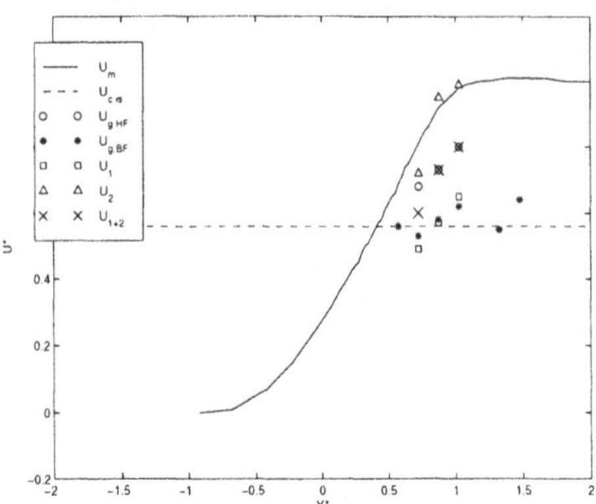

Figure 8. Dimensionless velocity profiles $U_{g,BF}$ and $U_{g,HF}$ are group velocities of the signal in the low frequency and the high frequency ranges

ence due to the cascade of turbulence for the considered energetic scales is roughly the same in the different parts of the flow.

5. Conclusions

We proposed a method to localize and analyse coherent structures in turbulent signal from multiple point wavelet transform. It is based on the determination of a local convection velocity and it supposes that the coherent structures have locally a convection velocity different than the rest of the flow. The results in a compressible mixing layer at convective Mach number of 0.62 showed that a physically meaningfull segmentation was produced. Moreover, conditional analysis of the coherent an incoherent part of the flow has shown that each medium contains the same energy with similar spectra and the same spatial decay of coherence. Finally, the conditional results have been used to reinterpret results obtained from the Fourier spectra and dispersion relationship.

References

DAUBECHIES, I., 1992, Ten lectures on wavelets, SIAM Philadelphia, Pennsylvania, U.S.A.

HUSSAIN, A.K.M.F., 1983 Coherent structures - reality and myth , Phys. Fluids 26,pp .2816-2850.

LIANDRAT, J., 1996 , Some Wavelet Algorithms for Turbulence Analysis and Modeling in Wavelets, Theory and Applications, Oxford Univ. Press.

BARRE S., 1994 "Estimate of convective velocity in a supersonic turbulent mixing layer ", A.I.A.A. J.,Vol. 32, N 1, pp 211-213 .

BOGDANOFF D.W., 1983 "Compressibility effects in turbulent shear layers ", A.I.A.A. J . 21, 926-927.

BONNET J. P., DEBISSCHOP J.R. ,CHAMBRES O., 1993 "Experimental studies of the turbulent structures of supersonic mixing layers", A.I.A.A. Paper 93-0217.

BROWAND F.K. and TROUTT T.R. 1985 "The turbulent mixing layer : geometry of large vortices", J. Fluid Mech. 158, pp. 489-509.

BROWN G. L., ROSHKO A. 1974, "On density effects and large structure in turbulent mixing layers". J. Fluid Mech. 64, part 4,pp. 775-816.

CLEMENS N. T., MUNGAL M. G., 1995 "Large-scale structure and entrainment in the supersonic mixing layer", J. Fluid Mech. 284, pp. 171-216.

DELVILLE J., BELLIN S., BONNET J.P.,1989 , in ,Advances in Turbulence II, Fernholdz, Fiedler ed., springer, 1989

DIMOTAKIS P. E. 1991 " Turbulent free shear layer mixing and combustion, Vol. 137 of Progress in Astronautics and Aeronautics", A.I.A.A. Washington D. C.. Murthy and Curran Editors.

DUPONT P , MUSCAT P and DUSSAUGE J.P., 1995a " Time and space-time statistics in a supersonic mixing layer". Second symposium on transitional and turbulent compressible flows, Joint ASME/JSME Fluids Engineering Conference, Hilton Heads, S.C. , August 1995.

DUPONT P , MUSCAT P and DUSSAUGE J.P., 1995b "Properties of energetic scales in a supersonic mixing layer". IUTAM Symposium on Combustion in Supersonic Flows, Poitiers, October 1995.

PAPAMOSCHOU D. ROSHKO A., 1988 "The compressible turbulent shear layer : an experimental study ". J. Fluid Mech. 197, pp. 453-477.

WINANT C.D., BROWAND F.K., 1974 "Vortex pairing : the mechanism of turbulent mixing-layer growth at moderate Reynolds number ", J. Fluid Mech. 63, part 2, pp. 237-255

BERNAL L.P., ROSHKO A., 1986,"Streamwise vortex structure in plane mixing layers",J. Fluid Mech. 170, pp. 499-525

X. Low-Dimensional Modelling

MULTI-POINT MEASUREMENTS AND LOW-DIMENSIONAL MODELS: TOOLS FOR THE CHARACTERIZATION AND CONTROL OF TURBULENT FLOWS

L.S. UKEILEY
National Research Council Fellow
NASA Langley, Hampton, VA

AND

M.N. GLAUSER
Department of Mechanical and Aeronautical Engineering
Clarkson University, Potsdam, NY 13699-5725

1. Introduction

The existence and importance of large scale structures in turbulent flows has long been established. One of the most important tools of the fluid dynamicist for structure identification has been the ability to visualize the fluid motions in the problem of interest. Flow visualization by tagging fluid material has a long and rich history, and has stimulated scientist, engineer and artist alike. It is an unfortunate fact of nature that the same fluids whose mysterious motions we seek to unravel, conspire through molecular diffusion and other processes to invalidate these simple tools. Thus most have had to recognize the futility of such efforts in a large number of flow environments, and have had to settle for far less information about what the flow is really doing.

For the experimentalist, this has usually meant settling for measurements of the average properties of the flow at a relatively small number of points in it. These limited experimental capabilities have undoubtedly had a negative effect on the efforts of theoreticians, both because of the lack of possibility for comparison with experimental data and the failure to overturn theories which were incorrect. One needs only witness the role being played by supercomputer simulations of flows (in combustion, for example) to realize what might have been if the experimentalist was not forced to settle for so little. Since it will be some time before full Navier-Stokes simulations of most engineering problems will become possible further need for development of experimental techniques is necessary.

One of the opportunities presented by the advances in electronics and computers over the past two decades is the possibility of making many measurements at many points in the flow simultaneously. Experiments utilizing tens of probes together have become routine, and now experiments using hundreds of probes have been performed. The transducers vary from thin film gauges to hot-wires, from optical scanners to PIV and holographic interferometry; and the flow environments vary from gas turbines to high speed heated jets and boundary layers as well as in combustion chambers. All have the common objective: to obtain by computer imaging and statistical means, a picture of what is really happening.

Many of the techniques for studying coherent structures are dependent on utilization of multi-point measurement techniques. These include conditional sampling, pseudo flow visualization, stochastic estimation, and the Proper Orthogonal Decomposition (POD). To appreciate the need for these one must understand the character of high Reynolds number, often turbulent, motions. As modern full Navier-Stokes computer simulations have made clear, knowing the data at many points in the flow does little by itself to make clear what is happening because of the chaotic nature of the flow. The key to understanding usually lies in what is done to the data to bring the underlying order to the foreground. For a review of how measurements at many points can be used to infer the structure of the flow and what constraints must be placed on the measurements to ensure that proper interpretation is possible the reader is referred to Glauser and George (1992).

The seminal paper of Aubry et al. (1988) provided the first glimpse of the potential of POD based low-dimensional models. With a 10 equation model for the near wall region of the turbulent boundary layer, they were able to extract many of the features of the flow such as bursting times etc. Since the early Cornell work several additional efforts by different research groups have applied similar ideas to various flows as reviewed by Berkooz et al. (1993). A recent Cambridge Monograph by Holmes et al. (1996) provides a nice overview on how to develop such dynamical systems. This monograph, combined with the review of Glauser and George (1992), provides the engineer and scientist with most of the necessary background to develop such systems; from the design of the experiments to obtain the experimental basis set to their use in a Galerkin projection for the development of POD based low-dimensional models.

The intent of this paper is to present results of such a POD based low-dimensional model for the 2D mixing layer. This collaborative effort between CEAT/Poitiers and Clarkson University utilizes the Multi-Point hot wire measurements obtained in the Poitiers mixing layer facility to provide the POD basis set on which the low-dimensional model is based.

2. Proper Orthogonal Decomposition

One of the most popular techniques for characterizing flow structure in turbulence is the Proper Orthogonal Decomposition. The POD, at least in its conventional formulation, requires knowledge of the two point correlation tensor. Applications of the POD have grown extensively over the past decade as the ability to acquire multi-point measurements have become more practical.

2.1. GENERAL THEORY

The POD was introduced to the turbulence field by Lumley (1967) and is based on the Karhunen-Loeve expansion. In the POD the coherent structure is chosen to be the structure with the largest mean square projection on the velocity field. The inner product that defines this maximization can be manipulated to form the following integral eigenvalue problem,

$$\int_D R_{ij}(\vec{x},\vec{x}')\psi_j(\vec{x}')d\vec{x}' = \lambda\psi_i(\vec{x}) \tag{1}$$

where the kernel of equation 1 is the velocity cross-correlation tensor, $R_{ij}(\vec{x},\vec{x}') = \overline{u_i(\vec{x})u_j(\vec{x}')}$ and the eigenvalue, λ, is representative of the integrated turbulent kinetic energy in a given POD mode. Since R_{ij} is a symmetric function the solutions to equation 1 can be discussed using the Hilbert-Schmidt theory (Lumley, 1967).

One of the more interesting artifacts of the Hilbert-Schmidt theory is that an infinite number of orthonormal solutions can be used to express the original random velocity field $u_i(\vec{x},t)$ as

$$u_i(\vec{x},t) = \sum_{n=1}^{\infty} a^n(t)\psi_i^{(n)}(\vec{x}) \tag{2}$$

where

$$a^n(t) = \int_D u_i(\vec{x},t)\psi_i^{(n)*}(\vec{x})d\vec{x}. \tag{3}$$

The solution for the integral in equation 3 implies that the velocity field is known at all locations simultaneously. Using multi-point measurements such as rakes of hot wires makes it possible to directly calculate the expansion coefficients for the POD at least along a line. However, this is generally not the case if more spatial dimensions are included and one must utilize techniques such as the shot noise decomposition or dynamical models to examine the results of the POD in physical space.

2.2. APPLICATION OF THE POD FOR THE PLANE MIXING LAYER

In this paper we are using a plane mixing layer for the demonstration of the discussed techniques. The data was acquired at the CEAT in Poitiers, France and descriptions of the experiment can be found in Delville and Ukeiley (1993) or Delville (1995). In the mixing layer, time can be considered stationary and the spanwise direction can be assumed homogeneous. If a direction is stationary or homogeneous the POD reduces to a harmonic decomposition in that direction (George, 1988) and is typically handled through Fourier analysis. This is done by Fourier transforming the cross-correlation tensor, yielding the cross-spectral tensor which is used in the following integral eigenvalue equation to solve for the POD modes.

$$\int_D \Phi_{ij}(y, y', k_1, k_3)\phi_j^{(n)}(y', k_1, k_3)dy' = \lambda^{(n)}(k_1, k_3)\phi_i^{(n)}(y, k_1, k_3) \quad (4)$$

Note that the eigenvalues and eigenfunctions are streamwise and spanwise wavenumber dependent.

The eigenvalues and eigenfunctions from the above equation will be used for developing the low-dimensional model discussed in the next section. In the rest of this section we will briefly discuss the results from this application of the POD. See Delville (1995), Ukeiley (1995) or Delville and Ukeiley (1993) for a more complete analysis.

As expected convergence of the POD modes is extremely rapid. The first mode contained 49% of the turbulent kinetic energy with 99% being represented by the summation of the first 7 modes. Therefore their use in a low-dimensional model is justified. Figure 1 shows a surface plot of the eigenvalues as a function of wavenumber for both the first mode and summation of all modes. The shape of the two plots is quite similar showing that the first mode has the same energy distribution in wavenumber space as the entire flow field.

3. Low-Dimensional Model For the Plane Mixing Layer

The low-dimensional model developed in this section will use POD eigenfunctions in a Galerkin projection with the Navier-Stokes equations. This results in a set of ODE's for the expansion coefficients of equation 2.

3.1. MOMENTUM EQUATIONS

After a Reynold's decomposition the Navier-Stokes equations for an incompressible fluid without body forces can be written as;

$$\frac{\partial}{\partial t}u_i' + u_j'\frac{\partial u_i'}{\partial x_j} + u_2'\frac{\partial U_1}{\partial x_2} + U_1\frac{\partial u_i'}{\partial x_1} - \overline{u_j'\frac{\partial u_i'}{\partial x_j}} = -\frac{1}{\rho}\frac{\partial p'}{\partial x_i}\nu\frac{\partial^2 u_i'}{\partial x_j \partial x_j} \quad (5)$$

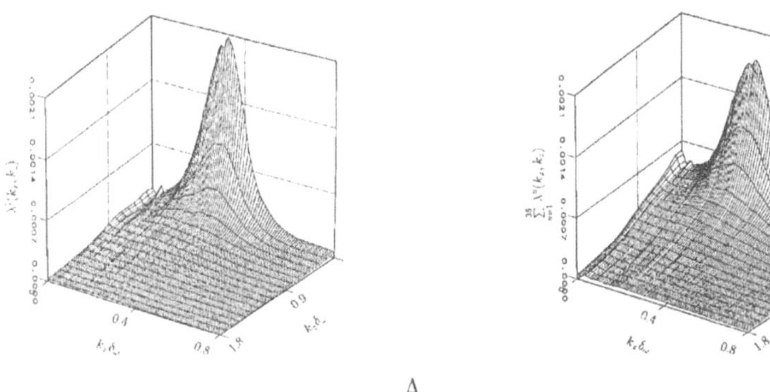

A. B.

Figure 1. Eigenspectra A) λ^1 B)$\sum_N \lambda^N$

To arrive at the previous equation four assumptions specific to the mixing layer were made. The first two assumptions, $U_3 \approx 0$ and x_3 homogeneous, are fairly standard for the mixing layer and forced by some of the symmetries which were applied to the experimental data (Ukeiley, 1995). The third assumption, $U_2 \approx 0$, has been examined and it was shown that the terms involving U_2 were small in comparison to the other terms kept in the model. The fourth assumption, $\partial\overline{(\text{mean quantity})}\partial x_1 = 0$, implies little or no growth of the mixing layer and is forced by the application of Taylor's Hypothesis to map time to streamwise information. This is a reasonable assumption given that the mixing layer grows like 0.04 x and the fairly small spatial window in the streamwise direction that will be used for the calculations.

Before proceeding, a relationship must be developed for the mean streamwise velocity present in equation 5. This relation must be developed because, in the severely truncated system which will be modeled, the measured mean velocity profile will be incorrect and there is no direct way to represent the mean velocity in terms of the POD modes. In this study an eddy viscosity relationship was used to balance the mean streamwise velocity with the Reynolds stress as shown below;

$$U_1 = \frac{1}{\nu_e} \int_{-L_2}^{x_2} -\overline{u_1' u_2'} dx_2 + U_c - U(y=0), \tag{6}$$

In the above equation U_c is the convection velocity and $U(y=0)$ is the value of the calculated mean velocity at $y=0$. Using this type of closure for the mean velocity was studied further by Cordier (1996).

Three different ways of introducing this term into equation 5 were examined. The first approach was to calculate a value for the mean streamwise velocity for the truncated system by writing equation 6 in terms of the

POD modes. This method forces the mean contribution to be a constant and appears in the linear term discussed in the next section. The downside of this method is that there is no mechanism for feedback from the turbulence to the mean flow. This mechanism is important because the presence or absence of a structure will alter the mean velocity gradient, thus altering the production mechanism (Aubry et al. 1988 and Wick et al. 1994).

The second approach, termed Feedback, was to substitute equation 6 into equation 5 and then proceed with the Fourier Transform and Galerkin projection. In this equation the term representative of the mean streamwise gradient will evolve in time and allow for the feedback between the mean streamwise velocity and the turbulence. A mathematical consequence of this is that this term will be cubic in nature, instead of linear. This will result in a set of equations that does not have a term to contribute to linear growth (Cordier, 1996). This method has been shown not to yield dynamically interesting results (Ukeiley, 1995) and will not be discussed further.

The final approach, termed the filter method, is a combination of the first two methods discussed. In this method a cutoff wave number is determined and the contribution of the wavenumbers lower than this cutoff are used in a static sense (no-feedback) while those greater than the cutoff are dynamically simulated in time (feedback). By splitting the contribution in this manner we account for the need to have both production through the linear term and feedback between the turbulent and mean velocity fields via the cubic term.

3.2. GALERKIN PROJECTION

The Galerkin projection is essentially used to expand the dependent variables in terms a finite series of independent basis functions, the POD modes. The basis functions must form a complete basis for the relevant class of functions and satisfy the relevant boundary conditions which these functions are known to do (Lumley, 1967). Using this projection with the POD modes yields a set of ordinary differential equations for the expansion coefficients.

The Galerkin projection is performed on the Fourier transform of the Navier-Stokes equations and is defined as the following inner product;

$$(N, \phi^l) = \int_D N_i(t, x_2)\phi_i^{l*}(x_2)dx_2 = 0. \qquad (7)$$

After performing the above operation and using the orthogonality of the eigenfunctions a set of ordinary differential equations arises whose full form can be seen in Ukeiley (1995). This equation can be written in a simplified form as follows,

$$\frac{da}{dt} = Bl + B2l + Cq + Dc + \text{pressure term} \tag{8}$$

where B, $B2$, C and D are matrices calculated from the POD eigenfunctions. The first two terms on the right hand side of equation 8 are linear. The first is a direct result of the viscous diffusion term in the Navier-Stokes equations. The second linear term consists of two parts, a production term and a convection term. This term involves the contribution to the mean velocity from all the modes in the truncation for the no-feedback case, only those smaller than the cutoff for the filter case and will be zero for the feedback case. The next term is quadratic. This term is representative of the fluctuation interactions and exhibits the transfer of energy in between the Fourier and POD modes in the dynamical system.The fourth term is cubic and is a result of the modeling of the mean velocity and will vary depending on the closure. It will reduce to zero for the no feedback case, and has slight differences between the feedback and filter case (Ukeiley, 1995). The final term is a pressure term which will be eliminated in the rest of the analysis because the integrations are over the whole mixing layer (Rajaee et al., 1994).

3.3. ENERGY TRANSFER MODEL

In order to limit the degrees of freedom in the dynamical system, the number of streamwise/spanwise wavenumber modes kept in the model will be truncated. When the truncation is performed, the system is no longer complete because it will not account for the energy transfer between the modes kept and those lost to the truncation. As a first attempt to account for this, a Heisenberg spectral model similar to one utilized by Aubry et al. (1988) and Glauser et al. (1989) will be used. In this approach, it is assumed that the small scales (neglected modes) remove energy from the larger ones (the modes kept) by a global viscous action.

The relationship gets introduced to equation 8 in the following manner;

$$\frac{da}{dt} = B(1 + \frac{\alpha \nu_T}{\nu})l + B2l + Cq + Dc. \tag{9}$$

where ν_T is the turbulent viscosity and its full form can be seen in Ukeiley (1995). The coefficient α controls the portion of energy which the small scales, neglected by the truncation, acquire. In terms of dynamical systems, α is considered like a bifurcation parameter with large values of α corresponding to a stable system.

4. Results and Discussion

The results from many different truncated versions of equation 9 have been examined. All of the truncated systems examined in this study used only the first POD mode. This is justified by its dominance, as discussed earlier. The selection of the streamwise and spanwise wavenumbers to keep is dominated by the need to keep as few nodes as possible that represent a significant amount of turbulent kinetic energy. In the process of picking these nodes the eigenspectral plot presented in figure 1 is used to determine the optimal nodes. Once the truncation was determined a Runge-Kutta technique with a time step of $8.0e^{-5}$ was used to carry out the integrations. Several symmetries and invariant subspaces are used to simplify the actual equations which need to be solved, for a detailed description see Ukeiley (1995). The initial conditions for the system of equations will be based on the magnitude of the eigenvalues. Essentially since a goes like $\lambda^{0.5}$ the initial values for the coefficients were taken to be the square root of the experimentally determined eigenvalues. The effects of varying the initial conditions have been studied and shown to have little effect on the results. A change in amplitude of the initial conditions, by as much as an order of magnitude, only affected the amount of time before the results settled into the same behavior. Below a brief summary of some of the results found for this type of approach are reported. A more detailed account can be found in Ukeiley (1995).

4.1. TRUNCATION FOR $K_3 = 0$

An 8 node system consisting of the first spanwise wavenumber and 8 streamwise wavenumbers spaced evenly to represent the peak in figure 1 was evaluated. This was done by varying the value of α and examining the results of equation 9.

4.1.1. *No Feedback*

The dynamical behavior as a function of α varied from trivial to unstable solutions. All of the stable non-trivial solutions exhibited an underlying periodic behavior. This periodic behavior is always buried in the time traces of the random coefficients, to some scale, regardless of larger time scale events. The frequency increases with larger values of k_1; since the streamwise wavenumbers are mapped from frequency this seems intuitively correct. Other complex behavior, which will be discussed in further detail, sits on top of the underlying periodicities. This is consistent with the results of Rajaee et al. (1994) where a forced mixing layer was studied.

The other behavior that is observed regardless of the value of α is that the node associated with the third spanwise wavenumber appears to lead

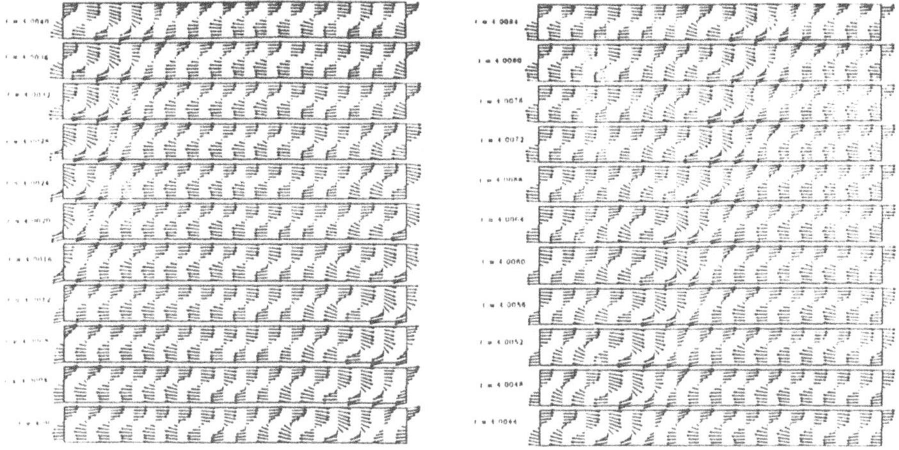

Figure 2. Velocity Vector Plot for $\alpha = 1.85$

to the temporal evolution. This is consistent with the results of Cordier (1996) who performed a linear stability analysis and found this node to be the most linearly unstable. It is also interesting to note that this streamwise wavenumber is associated with energy in the center of the mixing layer. This is also consistent with the results of Metcalfe et al. (1987) where it was postulated that disturbances from the center of the mixing layer manifest into the instabilities that cause the flow to exhibit three dimensionality.

Figure 2 shows velocity vector plots for $\alpha = 1.85$ In these plots the flow goes from left to right and the time increment between plots is $4.0e^{-4}$ seconds which is five times the resolved time step. The dimensions of the window are 66 mm in the x_2 direction and 345 mm in the x_1 direction. The figures are plotted in a frame of reference moving at the convection velocity. In the pseudo real space plotted here there is a periodic event which appears to be spanwise aligned vortices passing through the window. Another event can be observed in the first window ($t = 4$). This structure is much longer in the streamwise direction and appears to have two cores. This is how a pairing event would appear plotted with velocity vectors. Although this organization does not seem to have the energy wrap around as would be truly indicative of a pairing event. The scenario of these two events passing through the window has a regular period.

4.1.2. *Partial Feedback*

In this set of equations the mean velocity is split into a steady part that is held constant for all times and a time dependent part that manifests itself as a cubic term in the dynamical equations. The reason for the introduction of this filtering technique was discussed previously, and allows time dependent interaction between the coherent structures and mean velocities. This technique will also allow for the scaling of the production term.

Since the simulations of the No-Feedback equations yielded reasonable results it seemed like a natural first step to start by using only the largest streamwise wavenumber to represent the feedback. This allows for some contribution to the cubic term without much loss in mean streamwise velocity gradient. Upon examination of the calculated mean velocities it was apparent that there was little change in the production term if the two largest streamwise wavenumbers were used to allow the feedback, and in fact these two values of the cutoff wavenumber yielded similar dynamics.

Initial results from the numerical simulations showed that the introduction of the cubic term had a stabilizing effect on the system, as was expected. For large values of α the behavior is the same as for the No-Feedback. However, now for small values of α the system no longer exhibits the blow up mentioned in the discussion of the No-Feedback case. For values of α in this range the system now exhibits a complex behavior in full space on top of the underlying periodicities.

As one would expect, the smaller the value of cutoff wavenumber the more closely the system mimics the Feedback relationship discussed previously. As long as there was some contribution of the production term the systems never reached a blowup state, although they would not exhibit interesting dynamics even for $\alpha = 0$. The reason for this is again the lack of linear growth. It is also noticed that as the contribution to the mean velocity from nodes smaller than the cutoff, is decreased, the production term is reduced in magnitude. This allows for the complex secondary behavior to occur with the amplitude of the coefficients being smaller.

4.2. TRUNCATIONS FOR $K_3 \neq 0$

In an attempt to better mimic the full dynamics of the plane mixing layer non-zero spanwise wavenumbers were included in the truncations. Specifically in the truncation studied here had 6 streamwise wavenumbers spaced similar to the $k_3 = 0$ model for each of the first four spanwise wavenumbers. Integrations of this model with the no-feedback closure showed interesting dynamics for different values of α. However when the velocity field was reconstructed the rms profiles from the truncated system overestimated the original measurements. This behavior is believed to be due to the fact that in the no-feedback closure there is too much production and no mechanism for the turbulence to feedback to the mean. This prompted the use of the filter technique for the mean velocity closure.

With this larger system more nodes could be used for the feedback without largely reducing the mean velocity gradient. The best case allowed

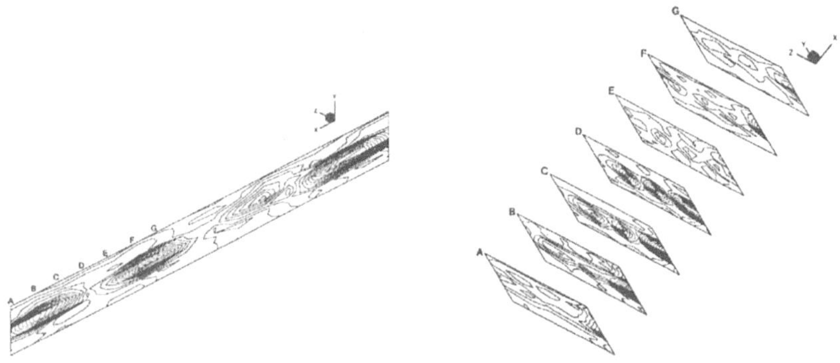

Figure 3. Vorticity for a filter setting of (40, 2) (39 node system)

for all the nodes from the largest streamwise and spanwise wavenumbers to be used for the feedback term. As with the smaller model the behavior between the no-feedback and filter closure schemes were similar. In this case all nodes exhibited the underlying periodicities, with a more complex second periodic behavior sitting on top. The mean square amplitude of the coefficients is also much closer to the experimentally observed values when compared to the No-Feedback case. Therefore, this is an appropriate filter setting for examining the reconstructed flow field.

Figure 3 shows plots of the vorticity for the x_1, x_2 and x_2, x_3 planes. In figure 3 a slice of the flow at $x_3 = 0$ is displayed. In this figure there is strong evidence of several spanwise vortices, e.g., the two contour peaks at locations B and F. In the x_2, x_3 there are streamwise aligned vortices at the locations, in x_1 between the two spanwise vortices. These are represented by the contour lines at locations C and D which lie between the spanwise vortex centers at locations B and F. This result is in good agreement with previous studies of the mixing layer, where there is strong evidence that there are spanwise vortex tubes connected by streamwise aligned vorticity (Metcalfe et al., 1987).

5. Conclusions and Perspective

A low dimensional dynamical systems model based on the POD modes from a plane turbulent mixing layer was developed. Several methods of closing the equations in terms of the mean streamwise velocity were examined. In the no feedback closure a Boussinesq approximation was used to calculate the contribution from the truncation to the mean streamwise velocity. This quantity was then held constant for all times causing a fixed level of production. In the other closure scheme a cutoff wavenumber was chosen and wavenumbers greater than this value are allowed to vary with time while the contribution from wavenumbers less contribute as a constant similar to the no feedback scheme. Initial application of this method was encouraging. It served two purposes; firstly it allows feedback between the turbulence and the mean flow, secondly it scales the mean velocity gradient which will, in turn, scale the contribution from the production term.

Several truncations were evaluated with this model. The first truncation involved $k_3 = 0$ and 8 nodes evenly spaced in k_1. With this truncation the system exhibited a rich range of dynamics. Along with this rich behavior an underlying periodicity was always present. This underlying periodicity occurred at realistic frequencies for the truncation studied here and is representative of the spanwise vortex tubes which are known to be a periodic event. While the complex behavior on top of the underlying periodicity appears to control the structure interaction. The wavenumbers associated with the energy at the center of the mixing layer tended to lead to the temporal dynamics, which is consistent with the results of Metcalfe et al. (1987).

A truncation which involved spanwise wavenumbers not equal to zero was also examined. Applying the Filter technique to the truncation involving non zero spanwise wavenumbers also had a pronounced effect. The scaling of the mean streamwise velocity gradient reduced the production term thus resulting in magnitudes for the coefficients that were in good agreement with the experimentally determined ones. With the reduced energy in the system the amplitudes of the reconstructed mean square velocity profiles were consistent with what one would expect from the severely truncated system. Plots of the vorticity showed strong evidence of the known flow organizations. These plots showed spanwise vortex tubes being connected by streamwise aligned vorticity. This type of structure is consistent with what has been reported in many previous studies (Metcalfe et al., 1987).

References

Aubry, N., Holmes, P., Lumley, J.L. and Stone, E.: 1988, 'The Dynamics of Coherent Structures in the Wall Region of a Turbulent Boundary Layer'. *J. Fluids Mech.* **Vol. no. 192**, pp. 115-173.

Berkooz, G., Holmes, P. and Lumley, J.L.: 1993, 'The Proper Orthogonal Decomposition in the Analysis of Turbulent Flows'. *Annual Review of Fluid Mech.* **Vol. no. 25**, pp. 539-575.

Laurent, C.: 1996, "Etude De Systemes Dyanmic Bases Sur La Decomposition Orthogonale Aux Valeurs Propres. Application A LA Couche De Melange Turbulents Et A L'Ecoulement Entre Deuxe Disques Contra-Rotatifs", Thesis, University of Poitiers.

Delville, J. and Ukeiley, L.: 1993, 'Vectorial Proper Orthogonal Decomposition, Including Spanwise Dependency, in a Plane Fully Turbulent Mixing Layer'. *Ninth Symposium on Turbulent Shear Flows.* Kyoto, Japan.

Delville, J.: 1995, ' La Décomposition Orthogonale Aux Valeurs Propres et L'Analyse

De L'Organisation Tridimensionelle Des Écoulements Turbulents Cisaillés Libres', Thesis, University of Poitiers.

George, W.K.: 1988, 'Insight into the Dynamics of Coherent Structures from a Proper Orthogonal Decomposition'. *Proceedings, Symposium From Near Wall Turbulence*, Dubrovnik, Yugoslavia, May 16-20.

Glauser, M.N., Zheng, X. and Doering C.R.: 1989, 'The Dynamics of Organized Structures in the Axisymmetric Jet Mixing Layer'. in *Turbulence and Coherent Structures*,(eds. O. Metais and M. Lesieur), Kluwer Academic Publishers.

Glauser, M.N. and George, W.K.: 1992, 'Application of Multipoint Measurements for Flow Characterization', *Experimental Thermal and Fluid Sciences* **Vol. no. 5**, pp. 617-632.

Holmes, P.J., Lumley, J.L. and Berkooz, G. *Turbulence, Coherent Structures, Dynamical Systems and Symmetry*, Cambridge Monographs on Mechanics, Cambridge University Press, 1996.

Lumley, J.L.: 1967, 'The Structure of Inhomogeneous Turbulent Flows'. *Atm. Turb. and Radio Wave Prop.*. Yaglom and Tatarsky eds. Nauka, Moscow, pp. 166-178.

Metcalfe, R. W., Orszag, S. A., Brachet, M. E., Menon, M. and Riley, J. J.: 1987, 'Secondary Instability of a Temporally Growing Mixing Layer'. *J. Fluid Mech.*, **Vol. no. 184**, pp. 207-243.

Rajaee, M., Karlson, S. and Sirovich, L.: 1994, 'Low-Dimensional Description of Free-Shear-Flow Coherent Structures and Their Dynamical Behavior'. *J. Fluid Mech.*, **Vol. no. 258**, pp. 1-29.

Ukeiley, L.S.: 1995, 'Dynamics of Large Scale Structures in a Plane Mixing Layer'. Ph.D Dissertation, Clarkson University.

Wick. D. P., Glauser, M. N. and Ukeiley, L. S.: 1994, 'Investigation of Turbulent Flows via Pseudo Flow Visualization Part 1: Axisymmetric Jet Mixing Layer'. *Experimental Thermal and Fluid Science* **Vol. no. 9**, pp. 391-404.

LARGE EDDY SIMULATION OF A SPATIALLY DEVELOPING 3D SHEAR LAYER IN INCOMPRESSIBLE FLOW: COMPARISONS WITH DETAILED EXPERIMENTS

R. LARDAT, A. DULIEU, W.Z. SHEN, L. TA PHUOC AND C. TENAUD
L.I.M.S.I. - UPR CNRS 3251
B.P. 133, 91403 ORSAY CEDEX, FRANCE

AND

L. CORDIER AND J. DELVILLE
L.E.A. / C.E.A.T. - UMR CNRS 6609
43, Route de l'Aérodrome,
86036 POITIERS CEDEX, FRANCE

1. Introduction

The active control of fully developed turbulent flows is of particular interest for many industrial applications. In such flow fields, the large-scale coherent structures contain most of the turbulent kinetic energy and are mainly responsible for vibrations, noise generation, etc... Therefore, in term of control, it seems important to describe correctly the characteristics of these structures and to predict precisely their time evolution using models as simple as possible. One of the methods proposed to mimic the dynamics of the flow is to develop low-order dynamical systems (Aubry *et al.*, 1988) (Glauser *et al.*, 1989) derived from the Proper Orthogonal Decomposition (POD) (Lumley, 1967; Sirovich, 1987).

The aim of the present study is to develop a L.E.S. of a 3D, plane mixing layer spatially developing downstream of a flat plate in order to apply such procedures. The L.E.S. has been performed on the flow configuration studied experimentally in details in earlier works (Delville, 1994; Ukeiley, 1995; Cordier, 1996; Cordier *et al.*, 1997). Several gains can be derived from such a L.E.S. computation: mainly the whole flow field is knowm in a deterministic way, thus the efficiency of the POD can be analyzed and the time evolution of the POD modes studied in details. The data base generated by this simulation can be used to check assumptions that are required when experimental procedures are applied: mainly the pressure, experimentally difficult to obtain, can be taken into account and the feedback relations (Aubry *et al.*, 1988) between the modeled structures and the mean field can be sorted out.

We present here preliminary results concerning the L.E.S. that has been performed. A comparison with the experimental results is provided. It is shown that the main statistical values, expansion factor, Reynolds stress are well predicted. Instantaneous snapshots of the flow show that this simulation of the 3D flow exhibit spanwise lengthscales in agreement with experiments. Endly a snapshot POD is derived, and a comparison is performed with experimental results.

2. Numerical Approach

Basic Equations: The L.E.S. of the mixing layer is performed using the filtered Navier-Stokes equations by means of the Reynolds decomposition. Using the velocity-vorticity formulation, these equations are written as follows:

$$\begin{aligned} \frac{\partial \overline{\omega}}{\partial t} - \nabla \times (\overline{v} \times \overline{\omega}) &= -Re^{-1} \nabla \times \nabla \times \overline{\omega} + \nabla \times \tau \\ \overline{\omega} &= \nabla \times \overline{v} \\ \nabla \cdot \overline{v} &= 0 \end{aligned} \tag{1}$$

where Re is the Reynolds number ($Re = \frac{U_1 \theta_1}{\nu}$), based on the external velocity (U_1) of the high speed boundary layer and its momentum thickness (θ_1). In these equations, $\overline{v}$ is the resolved-part (filtered part) of the velocity field, $\overline{\omega}$ is its rotational field and τ stands for the subgrid scale contribution: $\tau = \overline{v \times \omega} - \overline{v} \times \overline{\omega}$

Subgrid Scale Models: In order to take into account the participation of the small-scale structures in the fluid motion, the subgrid scale vector (τ) is modeled using the vorticity transfer theory of Taylor (Taylor, 1932), by means of an eddy viscosity (ν_t) : $\tau = -\nu_t \nabla \times \overline{\omega}$. In many subgrid scale models, the eddy viscosity is often related to the local vorticity. Hence, these models give subgrid contribution even in laminar flows which does not correspond to the right behaviour. The idea sought after in the *mixed-scale model* (Ta Phuoc, 1994), (Sagaut, 1995), is to dump smoothly the eddy viscosity (ν_t) in the regions where all the eddy structures are well captured. Using the *mixed-scale model*, ν_t is then calculated using two different velocity scales (Ta Phuoc, 1994), (Sagaut, 1995).

Computational Domain and Grid: The computational domain starts at the trailing edge of the flat plate and spreads over 20 δ_{ω_0} far downstream (δ_{ω_0} refers to the experimental vorticity thickness at a prescribed location, $\delta_{\omega_0} = 30 \times 10^{-3}$ m). The size of the mesh is 401×71×55. The grid uses 401 points equally spacing in the streamwise direction (x). In the inhomogeneous (vertical) direction (y), the domain lays over 6 δ_{ω_0}. The mesh is tightened around the centerline of the mixing layer, following a cosine distribution using 71 grid-points. The flow is supposed to be periodic in the spanwise direction (z) ; the domain lays over 5 δ_{ω_0} using 55 points equally spaced. Using this mesh, the grid filter width ($\overline{\Delta} = (\Delta x \times \Delta y \times \Delta z)^{1/3}$) is close to the Taylor micro-scale estimated following the experimental results (Delville, 1995).

Boundary Conditions: At the inflow surface, (ie. the trailing edge of the splitting plate ($x = 0$)), as a staggered grid is used we need to prescribe the value of the normal component of the mean velocity ($\overline{v_x}$) and the mean vorticity tangential components ($\overline{\omega_y}$, $\overline{\omega_z}$). $\overline{v_x}$ is initialized using two turbulent Whitfield profiles, for the boundary layers from each side of the flat plate. The profiles of $\overline{\omega_y}$ and $\overline{\omega_z}$ are then deduced from $\overline{v_x}$ profiles. A white noise is superimposed on $\overline{\omega_y}$, $\overline{\omega_z}$. The perturbation magnitude is equivalent to an amplitude of $10^{-3}\, U_1$ on the streamwise velocity component ($\overline{v_x}$). At the outlet boundary ($x = 20\ \delta_{\omega_0}$), the tangential components of the vorticity are calculated by extrapolation along the characteristic directions. The normal component of the velocity ($\overline{v_x}$) is then deduced from the vorticity profiles, prescribing that the mass conservation is satisfied. At the upper and lower surfaces of the domain ($y = \pm 3\ \delta_{\omega_0}$), slip conditions are imposed: $\overline{v_y} = 0$ and $\dfrac{\partial \overline{v_x}}{\partial y} = 0$, $\dfrac{\partial \overline{v_z}}{\partial y} = 0$.

Resolution: The spatial discretization uses a M.A.C. staggered grid where the velocity components are defined at the center of the cell faces, the vorticity components are prescribed at the middle of the vertices and the pressure as well as the eddy viscosity are defined at the center of the cell. The system of equations (1) is solved in two steps (Lardat *et al.*, 1997). First we solve the transport equation of the vorticity. The time discretization uses a Crank-Nicholson scheme. The convective terms are estimated by means of an Adams-Bashforth extrapolation which is 2nd order accurate in time. The space discretization uses a 2nd order finite difference method by means of a QUICK scheme for the convective terms and a centered scheme for the diffusive ones. Secondly, we solve the Cauchy-Riemann problem. Knowing the new vorticity field ($\overline{\omega}^{n+1}$), the velocity field is solved using a projection (or fractional step) method, following the work made by F. Bertagnolio and O. Daube (Bertagnolio *et al.*, 1996). In the first step, an intermediate velocity field is obtained which satisfies the curl equation ($\nabla \times v = \omega$). This intermediate velocity is then projected onto the space of divergence free vector fields (Bertagnolio *et al.*, 1996), (Lardat *et al.*, 1997).

3. The Snapshot-POD Method

In order to analyze the structure organization within the mixing layer, the Snapshot-POD method (Sirovich, 1987) is applied on the L.E.S. numerical results. This method based on spatial averaged quantities, has been preferred to the "standard" approach (Lumley, 1967), since simulations use an important number of grid points but rather limited temporal samples compared to experiments. Every spatio-temporal event $v_i(x,t)$ can be written using a discrete basis of the flow:

$$v_i(x,t) = \sum_{n=1}^{N_{POD}} a^{(n)}(t)\, \Phi_i^{(n)}(x) \tag{2}$$

where N_{POD} is the number of modes solved in the POD. In the snapshot-POD approach, the purely temporal $a^{(n)}(t)$ eigenfunctions are calculated by means of a Fredholm integral equation:

$$\int_T C(t,t')\, a^{(n)}(t')\, dt' = \lambda^{(n)}\, a^{(n)}(t) \tag{3}$$

where $C(t,t')$ is the tensor of the temporal velocity-correlations:

$$C(t,t') = \frac{1}{T}\int_{\mathcal{D}} v_i(x,t)\, v_i(x,t')\, dx = \frac{1}{T}\sum_{n=1}^{N_{POD}} a^{(n)}(t)\, a^{(n)}(t') \tag{4}$$

and $\lambda^{(n)}$ are the (real, positive) eigenvalues of this tensor ; each eigenvalue is associated to an energy density contained in each mode and the sum of $\lambda^{(n)}$ are equal to the turbulent kinetic energy included in the integral domain ($\mathcal{D}$). The spatial eigenfunctions are then deduced from:

$$\Phi_i^{(n)}(x) = \frac{1}{T}\left(\lambda^{(n)}\right)^{-1}\int_T v_i(x,t)\, a^{(n)}(t)\, dt$$

4. Results and discussions

The L.E.S. is performed on the same flow configuration as the experimental one (Delville, 1995). The velocity ratio is $r = U_2/U_1 = 0.59$, where $U_1 = 42.8$ m/s and $U_2 = 25.2$ m/s are the magnitudes of the external velocities

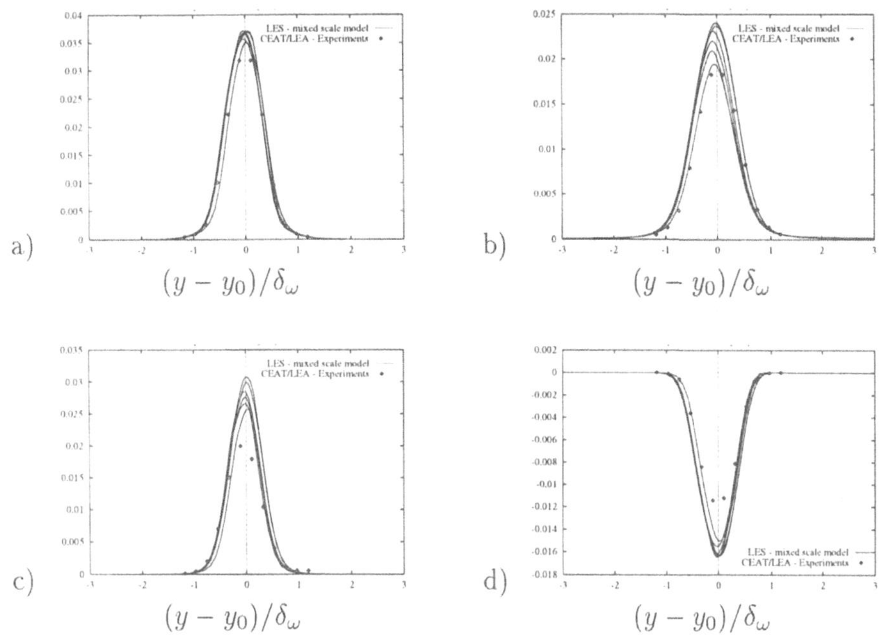

Figure 1. Profiles of the Reynolds stress components (a) $(\overline{u'^2}/\Delta U^2)$, (b) $(\overline{v'^2}/\Delta U^2)$, (c) $(\overline{w'^2}/\Delta U^2)$, (d) $(\overline{u'v'}/\Delta U^2)$: —— L.E.S. ; $\diamond$ Experimental data.

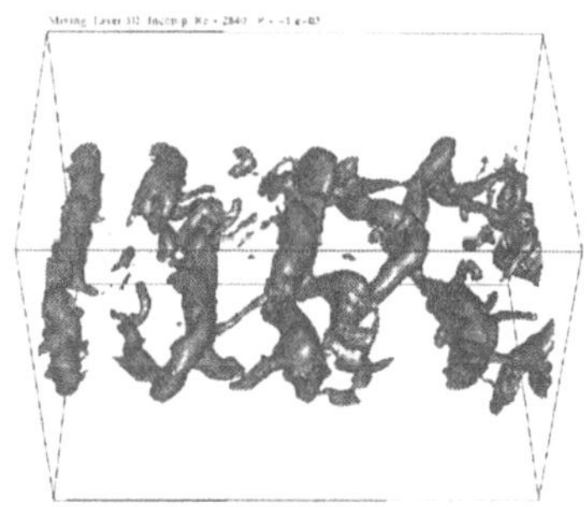

Figure 2. Isobaric surface of the instantaneous calculated pressure field at a dimensionless time t = 250. (region clipped: $10\ \delta_{\omega_0} < x < 20\ \delta_{\omega_0}$).

of the boundary layers at the trailing edge of the flat plate. The Reynolds number, based on δ_{ω_0} and on $\Delta U = U_1 - U2$, is 35000. A portion of the computational domain has been selected to compare the numerical results with the experimental data. This region starts from 10 δ_{ω_0} downstream of the trailing edge of the flat plate and lays over 5 δ_{ω_0} to minimize the influence of the exit boundary condition.

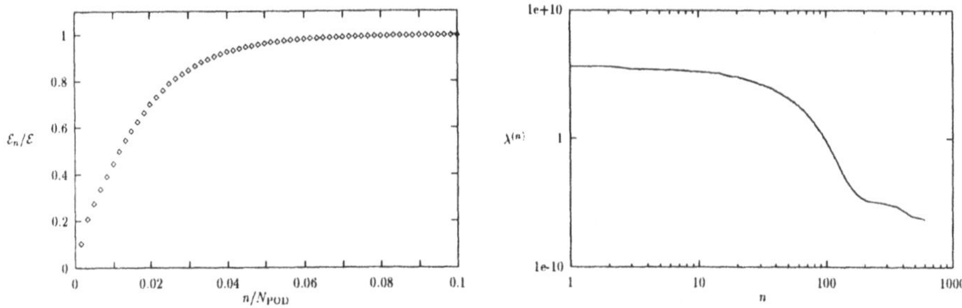

Figure 3. Energy contained in the first n-POD modes (on the left) and distribution of the energy density of the eigenvalue of each POD modes (on the right).

Validations of the Simulation: A comparison of the numerical results has been performed with the experimental data on the mean and Reynolds stress profiles. The average has been calculated on a dimensionless time $T = 206$ (T is based on U_1 and δ_{ω_0}) and by integrating in the spanwise direction: $< \bullet >= \frac{1}{L_z T} \int_{L_z} \int_T \bullet \, dt \, dz$. The self-similarity behavior is recovered by the L.E.S. ; the vorticity thickness (δ_ω) and its longitudinal evolution ($d\delta_\omega/dx$) are correctly predicted by the computation (L.E.S.: $d\delta_\omega/dx = 4.22 \ 10^{-2}$; Experiments: $d\delta_\omega/dx = 4.1 \ 10^{-2}$). Very good agreements are achieved on the profiles of the Reynolds stress components. The corresponding profiles calculated at 6 streamwise locations in the selected region are plotted in Fig. 1. The maximum value on the center line of the layer is very well recovered by the L.E.S. on the streamwise components of the stress tensor, while it seems that the simulation overestimates the other component maxima. One explanation might be that the similarity region, according to the Reynolds stress components, is not totally reached. As regards the instantaneous organization of the mixing layer, we plot the isobaric surface of the pressure field at a threshold $\overline{p} - P_0 = -10^{-3}$ for a dimensionless time $t = 250$. (Fig. 2). The large scale arrangement, strongly tridimensionnal, is clearly visible on this figure. A streamwise length scale has been recorded close to $\Lambda_x \sim 3.25 \ \delta_\omega$. In the spanwise direction (z), two large scale patterns have been recorded leading to an estimation of the spanwise length-scale close to $\Lambda_z \sim 2/3\Lambda_x$, which is in rather good agreement with the one generally admitted (Bernal *et al.*, 1986).

Application of the Snapshot-POD Following the good agreements between simulations and experiments, the snapshot-POD method is then applied on the temporal fluctuating velocity field provided by the numerical simulation. The POD analysis is performed on $N_{POD} = 600$ temporal events representing a dimensionless time $T = 10$. We apply the POD on a box with a streamwise extent of $5\delta_\omega$, corresponding to about two streamwise

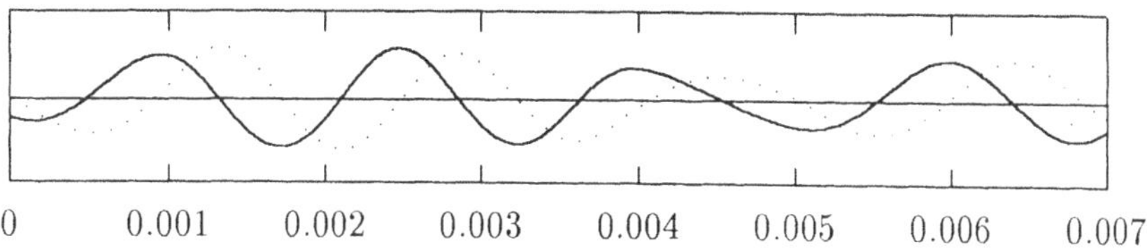

Figure 4. Temporal evolution of the eigenfunctions of the modes 1:___ and 2: . . . . .

length scale. The convergence of the snapshot-POD is presented on the ratio $(\mathcal{E}^{(n)}/\mathcal{E})$ between the energy contained in the first n-POD modes and the turbulent kinetic energy in the flow (Fig. 3). Note that on this figure, only 10 % of the N_{POD} modes used are plotted. The convergence is rapid since 10 % of the turbulent kinetic energy is contained within the first mode, 63 % is within the first 10 modes and the first 40 modes contain 99 % of the turbulent kinetic energy. This convergence is confirmed by the distribution of the energy in each mode since an important decrease in the energy density, with an order of magnitude close to 10 decades, occurs for modes greater than 40 (Fig. 3).

The first two temporal eigenfunctions are plotted in Fig. 4. A time shift is visible between $a^{(1)}(t)$ and $a^{(2)}(t)$, which is due to the convective nature of the flow (Rempfer, 1993). In fact the first two modes of the snapshot-POD have to be considered simultaneously if one wants to represent the mean convection in the streamwise direction. Due to the fact that a relatively small number of events have been retained for applying the POD, the obtained spatial eigenfunctions are somewhat noisy. Iso-surfaces of $\Phi_i^{(n)}(x)$ are presented in Fig 5 for the first two modes. In these figures, the dark grey surfaces correspond to positive values while the light grey ones corresponds to negative ones. Several general features can be noticed when considering these eigenfunctions. One can notice a streamwise shift when comparing modes 1 and 2. Whatever the velocity components, the mode 2 is shifted downstream when compared to mode 1. This behavior has to be related to the temporal shift already noticed for the $a^{(1)}$ ans $a^{(2)}$ coefficients in Fig. 4. These eigenfunctions exhibit a preferred organization in the spanwise direction. Particularly, $\Phi_v^{(n)}$ is clearly aligned in the spanwise direction. $\Phi_w^{(n)}(x)$ exhibits lambda-shape like structures (Fig. 5). These behaviors are similar to those observed in the analysis of the experimental data by (Delville, 1995).

The good agreement between the computations and experiments can be shown when the two-point space correlations $R_{ii}(x_0, y_0, x', y')$ are consid-

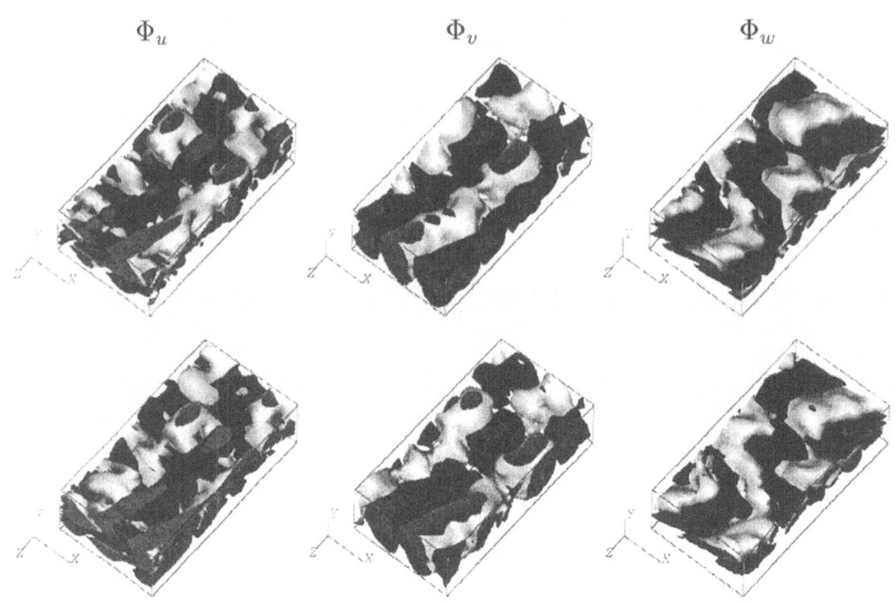

Figure 5. Iso-surface of the spatial eigenfunctions of the three components of the velocity for modes 1 (top) and 2 (bottom).

ered. From the snapshot POD, this correlation can be written:

$$R_{ii}(x_0, y_0, x', y') \propto \sum \lambda^{(n)} \Phi^{(n)}(x_0, y_0) \Phi^{(n)}(x', y')$$

In order to validate the eigenfunctions, they can be compared to the space-time correlations experimentally determined by using Taylor hypothesis (Delville, 1995). On figure 6, the space-time correlations experimentally measured are compared to a slice in a plane z=cst of the spatial eigenfunctions of the first POD mode. Very good agreements are achieved and the overall shape of the measured correlations is well described by considering only the first POD mode. This comparison cross-validates computations as well as experiments. Note that the u component can be described by fluctuations of opposite sign from part to part of the mixing layer, while the v component can be described by fluctuations in phase all over the y direction, alternated in sign in the x direction. These features are in agreement with the typical organization that large scale vortices with spanwise axis, could produce. On the other hand, the shape of the w component indicates the presence of a strong organization for the streamwise component vorticity, this organization being recovered from both experiments and L.E.S.

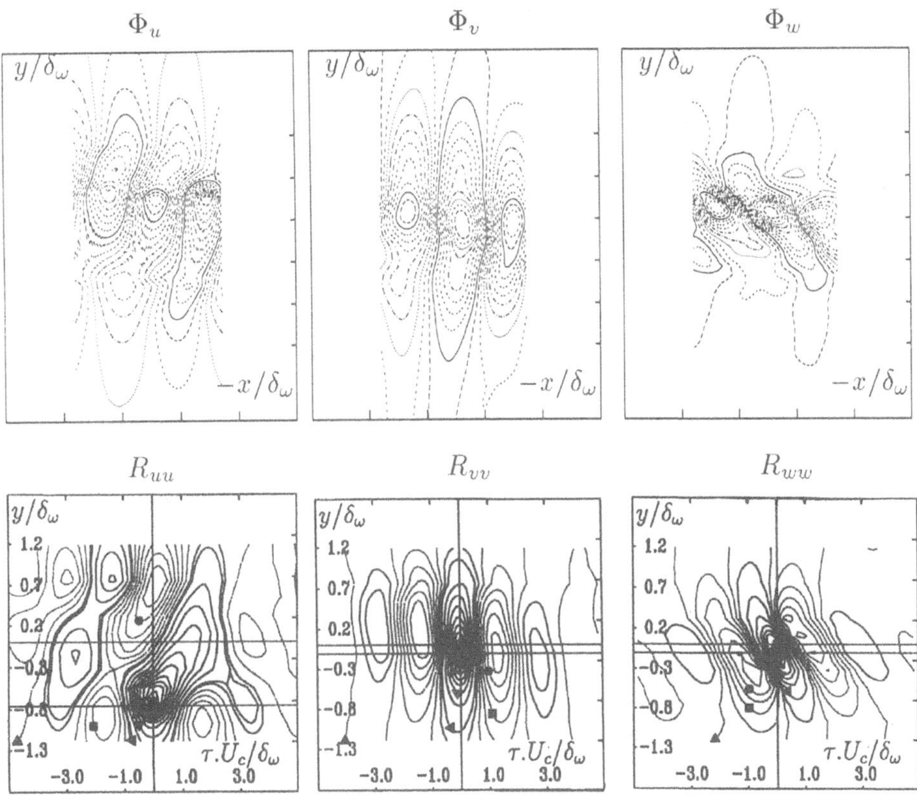

Figure 6. Comparison in a plane (z =cst) of the experimentally measured space time correlations $R_{ii}(y, y'; \tau)$ (bottom) with a vertical slice of the eigenfunctions $\Phi_i^{(1)}(x, y)$ of the first POD mode from L.E.S. (top)

5. Conclusions

The validation of the L.E.S. of the spatially developing 3D mixing layer has been performed successfully. Very good agreements between experiments and calculations have been recorded either on the mean velocity profiles, expansion factor, or on the the Reynolds stress components. Examination of individual snaphots from the L.E.S. indicates that the simulated flow possesses a highly 3D feature. Following these preliminary results, a first application of the snapshot-POD has been performed, whose results can be favorably compared to the experimental results. The first spatial eigenfunctions mimic very well the coherent structures organization since a good agreement is achieved between these eigenfunctions, calculated on results from the L.E.S., and the space-time correlation tensor given from the experiments.

In a following stage, a dynamical system will be derived from this L.E.S. based on the POD.

Acknowledgements

The calculations have been performed on the Cray C90 of the Institut du Développement et des Ressources en Informatique Scientifique (I.D.R.I.S / C.N.R.S) Orsay, France. The authors greatly acknowledge the support of these institutions.

References

Aubry N., Holmes P., Lumley J.L., & Stone E. (1988) *J. Fluid Mech.*, **192**, 115–173.

Bernal L.P.& Roshko A. (1986) *J. Fluid Mech.*, Vol. **170**, 499–525.

Bertagnolio F. & Daube O. (1996) "Velocity-Vorticity Formulation of the Incompressible Navier-Stokes Equations on Non Orthogonal Grids." *Proceedings of the Third ECCOMAS Computational Fluid Dynamics Conference*, pp. 644.

Cordier L. (1996) "Etude de Systèmes Dynamiques Basés sur la Décomposition Orthogonale aux Valeurs Propres (POD) : Application à la Couche de Mélange Turbulente et à l'Ecoulement Entre Deux Disques Contra-rotatifs." *Thèse de Doctorat de l'Université de Poitiers.*

Cordier L., Manceau R., Delville J. & Bonnet J.-P. (1997) *C. R. Acad. Sci. Paris*, **t. 324**, Série II *b*. 551-557

Corke T.C., Glauser M.N., & Berkooz G. (1994) *Applied Mechanics Review*, **47**, N° 6, part 2, S132–S138.

Delville J. (1994) *Applied Scientific Research* **53**, 263–281.

Delville J. (1995) "La Décomposition Orthogonale aux Valeurs Propres et l'Analyse de l'Organisation Tridimensionnelle des Ecoulements Turbulents Cisaillés Libres." *Thèse de Doctorat de l'Université de Poitiers.*

Glauser M.N., Zheng X. & Doering C.R. (1989) "The Dynamics of Organized Structures in the Axisymmetric jet mixing layer". *Turbulence and Coherent Structures*, (eds. O. Metais and M. Lesieur), Kluwer Academic Publishers, p. 253.

Lardat R., Bertagnolio F. & Daube O. (1997) *C. R. Acad. Sci. Paris*, **t. 324**, Série II *b*.

Lumley J.L. (1967) "The structure of inhomogeneous turbulent flows." *Atmospheric Turbulence and Radio Wave Propagation*, A.M. Yagom and V.I. Tatarski, 166–178, Moskow:Nauka.

Rempfer D. (1993) *Phys. Fluids*, **6**, (3), 1402–1404

Sagaut P. (1995) "Simulations Numériques d'Ecoulements Décollés avec des Modèles de Sous-maille." *Thèse de Doctorat de l'Université PARIS VI.*

Sirovich L. (1987) *Quarterly of Applied Mathematics*, **XLV**, N⁰ 3, 561–590.

Ta Phuoc L. (1994) "Modèles de sous maille appliqués aux écoulements instationnaires décollés." eds. DGA/DRET, *Journée thématique DRET: Aérodynamique instationnaire turbulente - Aspects numériques et expérimentaux.*

Taylor G.I. (1932) "The Transport of Vorticity and Heat through Fluids in Turbulent Motion." *Proc. of the London Math. Soc. Series* **A 151**, pp 421

Ukeiley L. (1995) "Dynamics of large scale structures in a plane turbulent mixing layer" *PhD Dissertation, Clarkson Univ. NY.*

LOW-DIMENSIONAL STUDY OF THE FLOW BETWEEN TWO COUNTER-ROTATING DISKS

CORDIER L., DELVILLE J. AND PÉCHEUX J.
L.E.A. UMR CNRS 6609
Université de Poitiers
43 Route de l'Aérodrome
F-86036 POITIERS, FRANCE

1. Introduction

Some recent developments in Fluid Mechanics are dealing with the use of active control for flows. It is now well known that, for flows possessing coherent structures, the main features can be captured by a low-order dynamical system. An efficient method for obtaining this system is the use of the POD technique (Proper Orthogonal Decomposition) in classical form (Lumley, 1967) or by mean of snapshots (Sirovich, 1987). This last approach is generally used with direct numerical simulations. The POD provides a basis (set of eigenfunctions) from which a dynamical system can be derived through a Galerkin projection procedure. In terms of active control of turbulence, the low order models has to be robust when the control parameter of the flow is changed (e.g. boundary conditions, Reynolds number,...). At least the effects of the control parameter onto the eigenfunctions need to be analyzed.

2. Flow under Study

In this study, a specific flow configuration is selected, exhibiting rich and various dynamics as the control flow parameter varies (Pécheux and Vigor, 1995). We consider the flow inside a close container of height e extending between $-e/2$ and $e/2$ in the z direction and from r_0 to r_1 in the r direction (Fig. 1). The top and bottom walls rotate with angular velocity ω and $\lambda\omega$ ($-1 \leq \lambda < 0$), while the peripheral one is fixed. The characteristic flow parameters are λ, the radius ratio $\eta = r_0/r_1$, the aspect ratio $A=r_1/e$ and the Reynolds number (flow control parameter) $Re = \omega e^2/\nu$, where ν is the

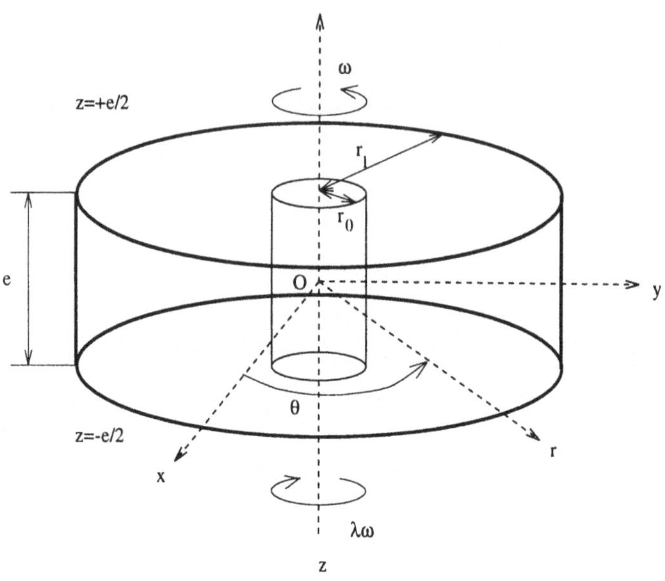

Figure 1. Flow configuration

kinematic viscosity. In the following, λ=-1, η=0, A=15 and we restrict the study to the axisymmetric case.

3. Numerical Simulation

The pseudo-spectral numerical algorithm used (Le Quéré and Alziary, 1985) integrates the time dependent Navier-Stokes equations in primitive variables (u, v, w, p), where u, v and w (or alternatively u_1,u_2 and u_3) correspond respectively to the radial, tangential and axial velocity components. The space discretization relies on spectral spatial expansions along r, z in a double truncated series of Chebyshev polynomials. The time stepping scheme is of finite type of second order combining a semi implicit treatment of the diffusive terms and an explicit extrapolation of the non linear terms (Vanel *et al.*, 1986). This leads to Helmholtz problems solved in the spectral space using the tau-method and an algorithm derived from a partial diagonalization algorithm (Heidvogel and Zang, 1979). The incompressibility constraint is maintained through the use of the influence matrix technique. The numerical simulations were obtained with a 2D (129×33) spatial mesh.

4. Flow Dynamics

The diagram shown in Fig. 2 summarizes the extensive computations that have been performed. Two zones appear: one for low and moderate Re val-

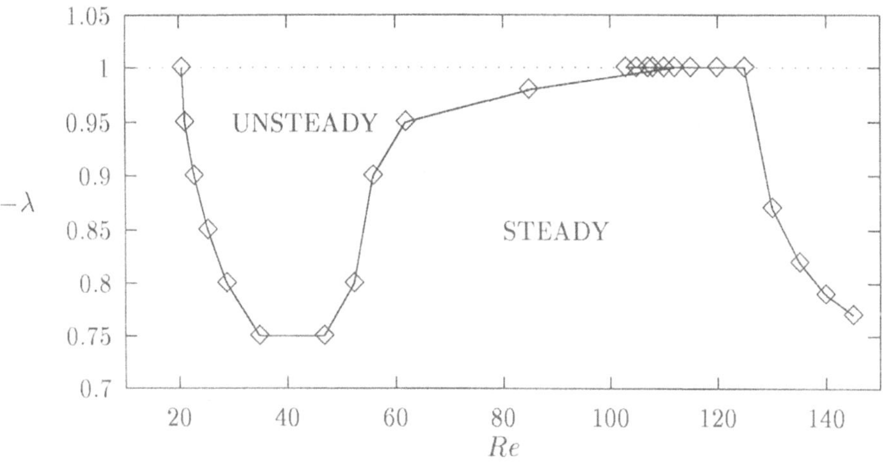

Figure 2. Stability diagram for $A = 15$

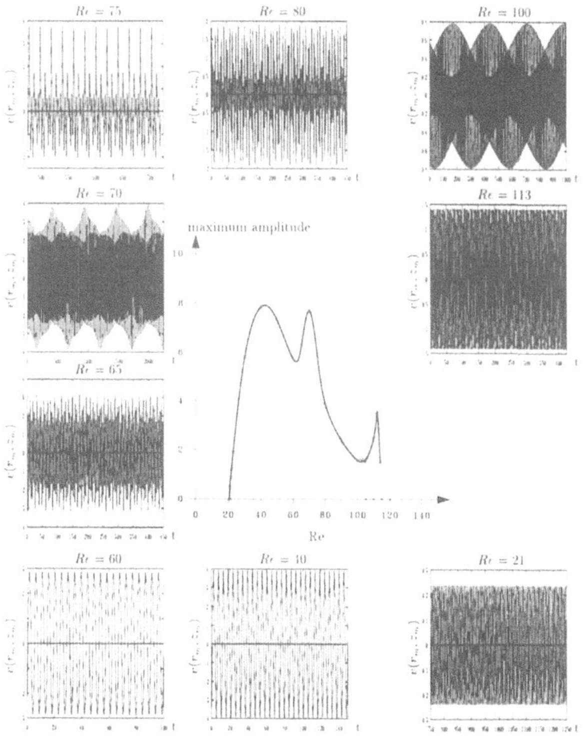

Figure 3. Bifurcation diagram for $\lambda = -1$, based on amplitude of $v(r_m, z_m)$.

ues and the other one for Re >125 which will be excluded of the present study. For low Re values (depending on λ) the basic flow is stable and stationary. The instability of this regime appears first for λ=-1 and Re=20.8 through supercritical Hopf bifurcation (traveling waves). Figure 2 also shows that this first instability does not take place for values of λ greater than -0.73 approximately and it can be seen that the solutions always revert to a steady state for large enough Re.
The bifurcation diagram (Fig. 3) shows for λ =-1 the amplitude of the fluctuating tangential velocity, at a fixed monitoring point located at (r_m=0.7r_1, z_m=0), versus Re. The corresponding temporal evolutions at some values of Re also show the wide variety of the features of the dynamics.
- For $0 < Re < 20.8$: an axisymmetric and stationary solution is found.
- For $20.8 < Re < 50$: traveling waves appear, with a large amplification rate, located mainly near the edges of the container.
- For $50 < Re < 60$: a perturbed motion appears ($r \geq 0.4r_1$)
  ▷ $0.4r_1 < r < 0.75r_1$: traveling waves without deformation,
  ▷ $0.75r_1 < r < 0.95r_1$ propagating waves with dislocation,
  ▷ $r > 0.95r_1$ stationary waves (at monitored point, the energy of the 2nd harmonic increases).
- For $60 < Re < 70$: a second frequency appears (f_2), modulated by a low frequency (f_0) (quasi-periodic + harmonics). The dislocation appears for smaller r.
- For $70 < Re < 100$: when Re increases, f_0 and f_2 have comparable energy and $f_2 \simeq f_0$ (period doubling).
- For $100 < Re < 105$: the same feature is found, however the largest period (T_l) becomes very large (Re=103: T_l=1305)
- For $105 < Re < 111$: f_0 becomes less energetic and disappears at Re=111.
- For $111 < Re < 113.8$: a mono-periodic regime is reached for $Re \simeq 113$. The amplitude of $v(r_m, z_m)$ decreases quickly.
- For $114 < Re < 125.2$: a steady stationary configuration is obtained after a sub-critical bifurcation: two solutions coexist symmetric and antisymmetric relatively to the median plane (z=0).

In the present study we will focus on the first Hopf bifurcation ($Re = 20.8$).

5. POD Approach

The results discussed in the following have been obtained from fluctuating flow fields: $u_i'(r,z,t) = u_i(r,z,t) - \overline{u_i}(r,z)$, where $\overline{\bullet}$ is a temporal average: $\overline{\bullet} = \frac{1}{T}\int_T \bullet \, dt$. A snapshot POD is used and the integral problem to be solved is:

$$\int_T C(t,t')a^{(n)}(t')dt' = \lambda^{(n)}a^{(n)}(t)$$

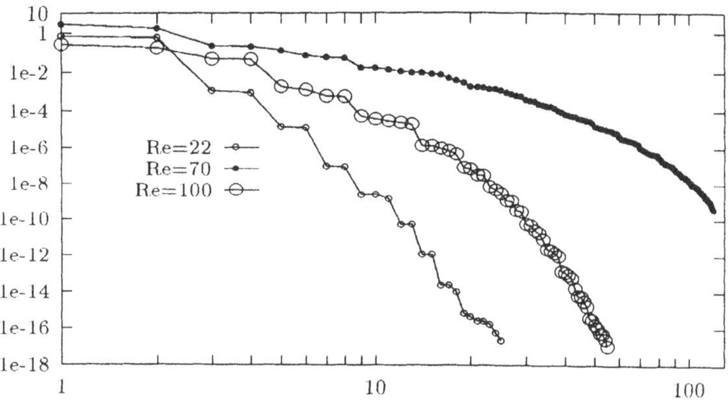

Figure 4. $\lambda^{(n)}$ vs. (n), for 3 Reynolds numbers: $Re = 22, 70, 100$

where $C(t, t')$ is the temporal two point correlation tensor:

$$C(t, t') = \frac{1}{T} \int_{\mathcal{D}} u_i'(r, z, t) u_i'(r, z, t') \, dr dz$$

The spatial eigenfunctions are calculated by:

$$\Phi_i^{(n)}(r, z) = \frac{1}{T} (\lambda^{(n)})^{-1} \int_T u_i(r, z, t) a^{(n)}(t) \, dt$$

The number of snapshots (N_S) used in this study was comprised between 30 and 120, according to the complexity of the dynamics. Figure 4 presents the distribution of eigenvalues found for selected Reynolds numbers. It is noticeable that the number of POD modes required to describe the flow increases with the complexity of the dynamics. The good convergence of POD is illustrated by the Karhunen-Loeve dimension D_{KL}, defined as the number of modes required to describe more than 90% of the energy. Whatever Re, it has been found that this dimension is at most 4.

Re	22	30	40	50	60	70	80	90	100
D_{KL}	2	2	2	2	2	4	4	4	3

When the dynamics is close to a limit cycle (Re=22 or Re=100), the eigenvalue spectrum exhibits pairs of modes of same amplitudes (Fig. 4). But when the dynamics becomes more complex, this feature disappears. This feature is related to the radially convective character of the dynamics of the flow when traveling waves are present (Rempfer, 1993). The POD allows to analyze the spatial feature relative to the dynamics of the fluctuating field.

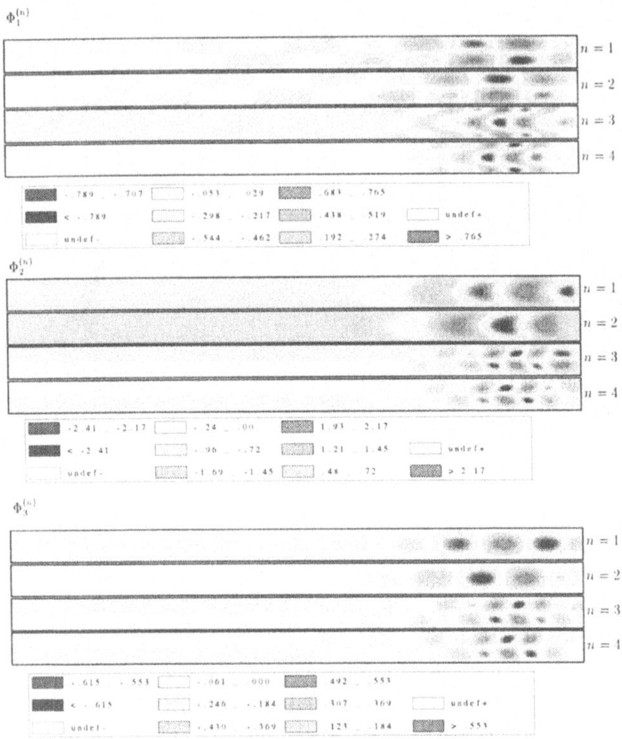

Figure 5. Isovalues of $\Phi_i^{(n)}(r, z)$ for Re=22

The spatial eigenfunctions obtained for $Re = 22$ are plotted in Fig. 5. The convection in the radial direction can then be retrieved from the translation between modes 1 and 2 or between modes 3 and 4 that can be noticed. The analysis of the temporal evolution of these modes has confirmed the propagating character of the dynamics, for this Reynolds number. It is interesting to note that different flow organizations can be reached by the same mode: for the first POD mode, the eigenfunction of the radial component of velocity is antisymmetric relatively to the plane z=0, while for the other two components, the eigenfunctions are symmetric. However when higher modes are considered, smallest spatial scales are described, for example the radial wave number is approximately multiplied by a factor 2 when passing from the first pair of POD modes to the second one.

6. Dynamical System

For each value of Re, a dynamical system is obtained by projecting the Navier Stokes equations onto the POD basis previously determined. The governing equations retained:

$$\frac{\partial \underline{u}}{\partial t} = \underline{N}(\underline{u}) \text{ with } \underline{u} = \underline{u}(\underline{x}, t) \quad \underline{x} \in \mathcal{D}, t \geq 0 \tag{1}$$

are expressed in pressure-velocity formulation ($u_i = \overline{u_i} + u_i' \quad i = 1, 2, 3,$ and $P = \overline{P} + P'$). However, in the present case, there is no explicit contribution of the pressure term to the dynamical system, due to the imposed boundary conditions (Rempfer, 1993). The Galerkin projection procedure consists on writing (1) in terms of the POD eigenfunctions

$$\begin{aligned} \sum_{m=1}^{N_{gal}} \frac{da^{(m)}(t)}{dt} \Phi_i^{(m)}(\underline{x}) &= \underline{N}(\sum_{m=1}^{N_{gal}} a^{(m)}(t) \Phi_i^{(m)}(\underline{x})) \\ &= \sum_n \underline{\mathcal{N}}^{(n)}(a^{(1)}, \cdots, a^{(n)}(t)) \Phi_i^{(n)}(\underline{x}), \end{aligned} \tag{2}$$

and to consider the orthogonality relation: $\int_{\mathcal{D}} \Phi_i^{(n)}(r, z) \Phi_i^{(m)}(r, z)\, dr\, dz = \delta_{nm}$. Here, the POD is applied on fluctuating fields, then a constant term appears which can be modeled by using the eigenvalues (Rajaee *et al.*, 1994). The set of equations of the dynamical system finally takes the quadratic form:

$$\dot{a}^{(n)} = \sum_{m=1}^{N_{gal}} B_{nm} a^{(m)} + \sum_{m=1}^{N_{gal}} \sum_{k=1}^{N_{gal}} C_{nmk} (a^{(m)} a^{(k)} - \lambda^{(m)} \delta_{mk})$$

where N_{gal} is the number of Galerkin modes retained in deriving the low order dynamical system.

7. Results and Discussion

7.1. DYNAMICAL SYSTEM AT FIXED *RE*

For each given Re_0, a dynamical system DS(Re_0) has been derived. These dynamical systems were individually able to describe properly the evolution of the projection coefficients $a^{(n)}$, that can be derived by directly projecting the instantaneous flow field arising from the direct simulation onto the POD eigenfunctions. Examples of such results are shown in Fig. 6. In these cases, corresponding to relatively simple dynamics, keeping only 4 POD modes in the Galerkin projection is sufficient to correctly represent the main features of the flow.

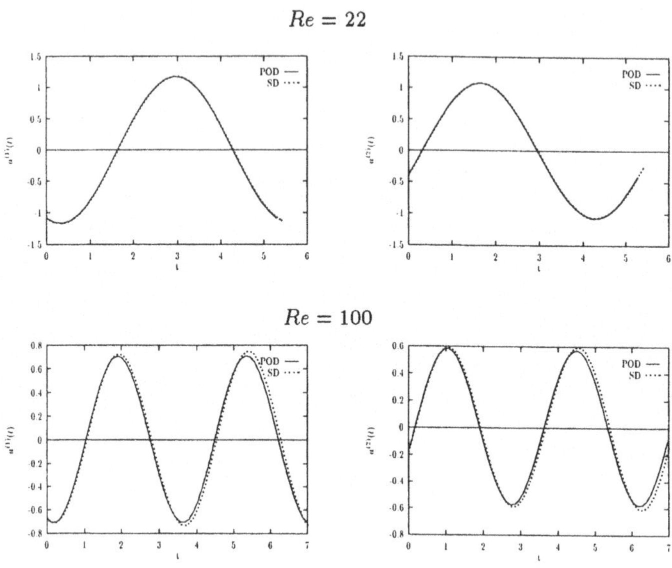

Figure 6. Temporal evolution of the first two $a^{(n)}(t)$ for the fluctuating field.

7.2. PREDICTION OF THE FIRST HOPF BIFURCATION

In a second stage, a stability analysis of the individual dynamical systems has been performed by means of AUTO94 (Doedel, 1981). In this analysis, the control parameter is the Reynolds number. For DS(Re_0=22) a critical value of Re is found, in good accordance with that found numerically (Re = 21.79); but for higher Re_0, the stability analysis gives unrealistic critical values for the first Hopf bifurcation. In fact for each DS(Re_0), the critical value of Re was found to be a few percent smaller than Re_0. This result confirms the fact that, due to the variation of the eigenfunctions with the Reynolds number, the DS becomes unable to describe properly the system.

In an attempt to improve these results, we tried to apply classical methods used in some earlier studies to take into account the variations of the eigenfunctions. We used either modeling of the mean flow fields (Deane *et al.*, 1991) or introduced realizations arising from various Re_0 in the snapshots (Christensen *et al.*, 1993). Contrarily to what was obtained by these authors, our results remain very unsatisfactory (Cordier, 1996) in the sense that only a slight improvement has been found, the critical Re being always related to Re_0.

In order to analyze more precisely this feature, we consider the projection of flow realizations (from direct simulation) at $Re_0 = 22$ onto the eigenfunctions obtained for various Reynolds numbers Re_p. To get an esti-

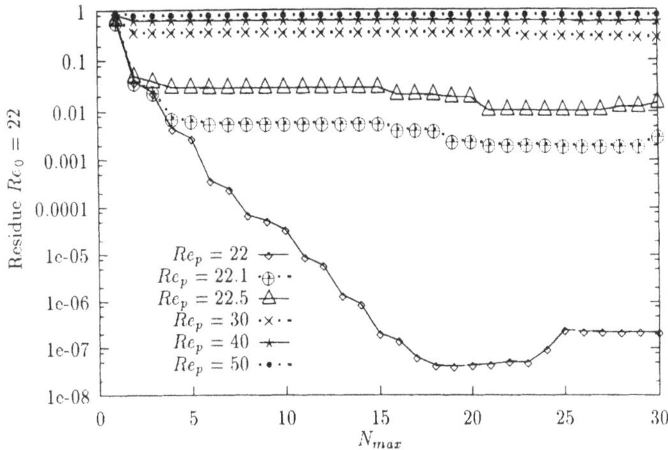

Figure 7. Evolution of **Res**, function of N_{max}. Re_0=22, Re_p in [22–50]

mate of the accuracy of these projections, the following residue is defined:

$$\mathbf{Res}(N_{max}) = \frac{\langle |u_i^{Re_0}(r,z,t) - \sum_{j=1}^{N_{max}} (u_i^{Re_0}(r,z,t), \Phi_i^{(j)Re_p}(r,z))\Phi_i^{(j)Re_p}(z,r)|\rangle}{\langle |u_i^{Re_0}(z,r,t)|\rangle}$$

where N_{max}, is the number of POD modes kept for reconstructing the case Re_0 by using different set of eigenfunctions: $\Phi_i^{(n)Re_p}$. These set being a basis of the flow, **Res** should tend towards zero when N_{max} increases. As partly shown in Fig. 7, comparing to a reference case $Re_p = Re_0$, the eigenfunctions at $Re_p = Re_0 + \delta Re$ cannot accurately describe the dynamics at $Re_0 = 22$, even if $|\delta Re|$ is small. For example, for Re_p=22.1 a residue of 1% is asymptotically reached. Examination of the spatial eigenfunctions (Fig. 8) shows that their spatial localization can be quite different, depending on Reynolds number. The existence of non-overlapping regions implies that small changes occurring with small Re variations suffice to dramatically increase the value of **Res**.

8. Concluding Remarks

As shown, POD is a very efficient method for extracting coherent structures of weakly or fully turbulent flows and for deriving a low-order dynamical system able to describe the flow dynamical evolution. However, many theoretical as well as experimental work is still required before to conclude definitively to the practical efficiency of POD based dynamical system to achieve active control.

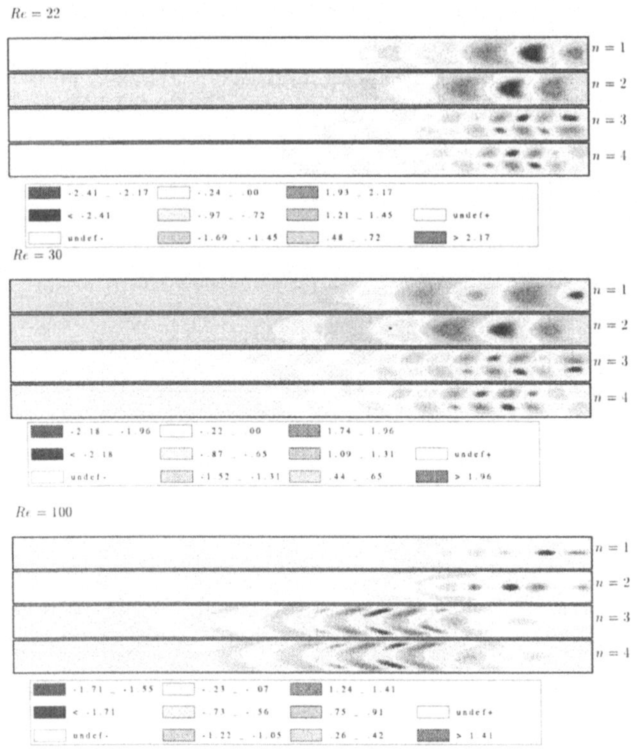

Figure 8. Isovalues of $\Phi_2^{(n)}(r, z)$ for Re=22, 30, 100

References

Christensen E.A., Sørensen J.N., Brøns M. and Christiansen P.L. (1993), *Theoret. Comput. Fluid. Dynamics* **5**, pp. 259–267.

Cordier L. (1996) "Etude de Systèmes Dynamiques Basés sur la Décomposition Orthogonale aux Valeurs Propres (POD) : Application à la Couche de Mélange Turbulente et à l'Ecoulement Entre Deux Disques Contra-rotatifs", *Thèse de Doctorat de l'Université de Poitiers.*

Deane A.E., Kevrekidis I.G., Karniadakis G.E. and Orszag S.A. (1991), *Phys. Fluids A* **3** (10), pp. 2337–2354.

Doedel E.J. (1981), *Congressus Numerantium* **30**, pp. 265-284.

Heidvogel D.B. and Zang T.A. (1979), *J. Comput. Phys.* **30**, p. 167.

Le Quéré P. and Alziary de Roquefort Th. (1985), *J. Comput. Phys.* **57**, p. 210.

Lumley J.L. (1967), *Atmospheric Turbulence and Radio Wave Propagation*, A.M. Yaglom et V.I. Tatarski, Moskow:Nauka, pp. 166–178.

Pécheux J. and Vigor H. (1995) "Instabilité de l'écoulement dans une cavité cylindrique à parois contra-rotatives", 2ème *Congrès de Mécanique, Casablanca.*

Rajaee M., Karlsson S. and Sirovich L. (1994), *J. Fluid Mech.* **258**, pp. 1–29.

Rempfer D. (1993), *Phys. Fluids* **6** (3), pp. 1402–1404.

Sirovich L. (1987), *Quarterly of Applied Mathematics* **XLV** N^o3, pp. 561–590.

Vanel J.M., Peyret R. and Bontoux P. (1986), *Num. Meth. for Fl. Dyn. II* edited by W. Morton and M.J. Baines (Clarendon Oxford 1986), pp. 463-475.

SINUOUS AND VARICOSE MODES IN PHASE-LOCKED INTERACTION

G. SCIORTINO AND M. MORGANTI

Dipartimento di Scienze dell'Ingegneria Civile
III Università di Roma,
Via Segre 60,
00146 Roma, Italy

AND

M.A. BONIFORTI

Dipartimento di Idraulica, Trasporti e Strade
Università di Roma La Sapienza,
Via Eudossiana 18,
00184 Roma,Italy

Abstract. The aim of this paper is to propose an identification technique of dynamical systems in order to describe the local evolution of modes which dominate the dynamics of transitional shear flows. By using the linear eigenmodes of viscous flows, a numerical investigation was performed to model the dynamical evolution of disturbances which arise in a transitional symmetric shear flow when a planar mode interacts in a phase-locked mechanism with one or more oblique three-dimensional modes. Numerical results highlighted a good qualitative agreement with the experimental ones and showed furthermore some interesting correspondences with the phenomenological conclusion of recent theoretical investigations obtained by spatial stability of inviscid flows. Our investigation confirmed that only some nonlinear triadic interactions can be active, depending on the sinuous or varicose nature of the selected modes. In particular, mechanisms of coupling among more triadic systems turned out to be of remarkable interest, in that they can induce a preferential amplification of oblique modes in spite of their dumped-varicose nature.

1. INTRODUCTION

Symmetric shear flows are characterised both by symmetric and antisymmetric instability modes. The sinuous modes have a symmetric vertical velocity distribution, while the varicose modes are antisymmetric in this velocity distribution. As shown by many experimental investigations, natural transition of symmetric shear flows is activated by two-dimensional sinuous disturbances, in agreement with linear instability theory. On the contrary, three-dimensional disturbances become dominant in the later stage of transition, sufficiently far downstream.

In recent years, in order to explain the rapid development of three-dimensional disturbances two important mechanisms have been proposed. The first is the secondary instability theory proposed by Herbert (1988), in which three-dimensional waves superimposed on a mean flow plus a two-dimensional disturbance, can become instable. As observed by Wu (1996), this mechanism does not have a solid mathematical justification except in some particular cases, so that it does not furnish a sufficiently general theory to explain the arising of three-dimensional disturbances. Another important mechanism is the subharmonic resonant triad interaction originally proposed by Raetz (1959) and Craik (1971). According to this scenario, nonlinear quadratic interactions among three modes, namely a planar mode and a pair of symmetric oblique modes, produce an explosive evolution of the disturbance in a finite time. While such a *blow-up* evolution could be connected in this approach with the rapid development of oblique modes, one should remember that the drastic truncation of the governing equations makes sense only for small growth rates. Furthermore, it can be observed that subharmonic resonance can take place, in symmetric shear flows, among appropriate varicose modes. Given that varicose modes are not the most unstable, it is difficult to understand the reason why this theory could be a viable mechanism of natural transition.

Recently, a phase locked modal interaction was proposed by Wu and Stewart (1996) as a new mechanism for promoting the growth of three-dimensional disturbances in shear flows. This mechanism can operate under much less restrictive conditions than those required by subharmonic resonance and secondary instability theory. The phase-locked interaction takes place between a planar mode and an oblique mode with the same phase speed. In some particular cases the forced difference mode can coincide with an eigenmode, so that the phase-locked interaction becomes a resonant triad interaction (Wu, 1996). This provides a possible explanation of the preferential amplification of oblique sinuous and varicose modes in the later stages of symmetric shear flow transition.

All these considerations show that the dynamics of transitional shear flows is characterised by the existence of some three-dimensional *leading modes* whose evolution can dominate the physical process. In this paper, we analyse the problem of selecting these leading modes and defining the related nonlinear evolution equations. Keeping this aim in maind, we consider flows which evolve at moderate Reynolds numbers, for which eigenfunctions of the linear problem constitute a suitable basis to decompose the disturbance. Next, an operative criterion able to characterise univocally leading modes with the minimal number of experimental information was formulated. A criterion to select eigenmodes which has found some experimental confirmations, is the triadic temporal resonance proposed by Craik (1985). Nevertheless, there are some restrictions affecting the hypotheses under which the latter was deduced. First, it is a *linear* resonance criterion, in that it involves only parameters of the linear problem (eigenvalues), so that it is effective only in a weakly nonlinear theory. Furthermore, it does not allow a univocal selection of modes verifying the resonance condition. In fact, this can be, a priori, verified by any combination of sinous or varicose modes without the nature of modes affecting directly or indirectly the criterion. Here, a *nonlinear* resonance criterion allows us to eliminate the above restrictions, which, together with the phase-locked condition, permit to characterize univocally a low-dimensional dynamical system whose solution describes the local temporal evolution of the selected modes. In order to verify the validity of this criterion in selecting the leading modes, we used the experimental results of Williamson and Prasad (1993) as test data together with some theoretical consideration on the phase-locked modal interaction.

2. FORMULATION AND IDENTIFICATION OF THE MODEL

We consider a solenoidal disturbance $\mathbf{V}(\mathbf{x}, t)$, superimposed on a basic shear flow $\mathbf{V}_0(\mathbf{x}) \equiv (U(y), 0, 0)$, $\mathbf{x} \equiv (x, y, z)$ where x, y and z are streamwise, transverse and spanwise coordinates respectively. By using the eigenfunctions of the linear problem, we decompose $\mathbf{V}(\mathbf{x}, t)$ into a sum of eigenfunctions :

$$\mathbf{V} = \nabla \wedge \sum_{\mathbf{k}} \sum_{n} \epsilon_{\mathbf{k}}^{(n)}(t) \mathbf{w}_{\mathbf{k}}^{(n)}(y) e^{i\mathbf{k}\cdot\mathbf{x}}$$

where $\mathbf{w}_{\mathbf{k}}^{(n)} = \tilde{\mathbf{w}}_{-\mathbf{k}}^{(n)}$, $\tilde{}$ indicating complex conjugation, $\mathbf{w}_{\mathbf{k}}^{(n)}(y)$ is the n-th eigenvector and $\epsilon_{\mathbf{k}}^{(n)}(t)$ is the n-th unknown eigenmode amplitude related to the wave vector $\mathbf{k} \equiv (\alpha, \beta, 0)$. With a suitable Galerkin projection of the Navier-Stokes equations onto the basis of adjoint eigenfunctions, the following nonlinear dynamical system holds (Boniforti *et al.* 1996):

$$(\frac{d}{dt} + i\omega_{\mathbf{k}}^{(m)})\,\epsilon_{\mathbf{k}}^{(m)}(t) = -\sum_{\mathbf{l}}\sum_{p,q=1}^{\infty} \sigma_{\mathbf{kl}}^{(m,p,q)} \epsilon_{\mathbf{k-l}}^{(p)}(t)\epsilon_{\mathbf{l}}^{(q)}(t) \tag{1}$$

where $\sigma_{\mathbf{kl}}^{(m,p,q)}$ are complex coefficients which account for nonlinear interactions and $\omega_{\mathbf{k}}^{(m)}$ are the eigenvalues associated with the eigenvectors $\mathbf{w}_{\mathbf{k}}^{(m)}(y)$.

In order to investigate the evolution of the leading modes which capture most of kinetic energy of the disturbance, an appropriate truncated form of System (1) with at least one unstable mode (im$[\omega] > 0$) and $\sum_{\mathbf{k},m}$ im$[\omega_{\mathbf{k}}^{(m)}] < 0$, namely a dissipative system, is used (im[#] is the imaginary part of [#]). Owing to the structure of the system itself, a generic truncated form of System (1) is characterised by one or more triadic coupled systems. If we consider a generic system of M coupled triadic equations, the spatial resonance condition among the modes of each triad

$$\mathbf{k}_{1,j} + \mathbf{k}_{2,j} = \mathbf{k}_{3,j} \qquad j = 1, 2, \ldots, M \tag{2}$$

must be fulfilled. Not all $\mathbf{k}_{p,j}$ are distinct because we are considering coupled triads, namely triads with some modes in common.

The following coupled equations can be obtained from System (1):

$$\begin{aligned} \frac{d\epsilon_{1,j}}{dt} &= -i\omega_{1,j}\epsilon_{1,j} + \sum_{s_1} \lambda_{1,s_1}\tilde{\epsilon}_{2,s_1}\epsilon_{3,s_1} \\ \frac{d\epsilon_{2,j}}{dt} &= -i\omega_{2,j}\epsilon_{2,j} + \sum_{s_2} \lambda_{2,s_2}\tilde{\epsilon}_{1,s_2}\epsilon_{3,s_2} \\ \frac{d\epsilon_{3,j}}{dt} &= -i\omega_{3,j}\epsilon_{3,j} + \sum_{s_3} \lambda_{3,s_3}\epsilon_{1,s_3}\epsilon_{2,s_3} \\ j &= 1, 2, \ldots, M \end{aligned} \tag{3}$$

where the index $s_p\,(p = 1, 2, 3)$ denotes those spatial resonance equations (2) which have the $\mathbf{k}_{p,j}$ wave vector in common (for the sake of simplicity we have written λ instead of the corresponding σ of System (1)). In order to characterize System (3) we link the mathematical properties of these equations to physical observations by using a *nonlinear* resonance criterion which takes into account effects of coupled nonlinear interactions. This, together with the phase-locked condition allows us to univocally select eigenmodes activated by the physical process. At this aim, we consider under which conditions System (3) admits solutions of the form :

$$\epsilon_{p,j}(t) = \rho_{p,j} e^{i(-\Omega_{p,j}t + \phi_{p,j})} \tag{4}$$

where $\rho_{p,j}, \Omega_{p,j}$ and $\phi_{p,j}$ are real constants with $\rho_{p,j} > 0$. Substituting (4) into equations (3), it can be shown that the conditions

$$\Omega_{1,j} + \Omega_{2,j} = \Omega_{3,j} \qquad j = 1, 2, \ldots, M \tag{5}$$

together with the following set of nonlinear algebraic equations which involves the unknowns $\rho_{p,j}$, $\Omega_{p,j}$, and $\phi_{p,j}$

$$\begin{aligned}
\Omega_{1,j} &= \omega_{1,j} + i \sum_{s_1} \frac{\lambda_{1,s_1} \rho_{2,s_1} \rho_{3,s_1} e^{i(\phi_{3,s_1} - \phi_{1,j} - \phi_{2,s_1})}}{\rho_{1,j}} \\
\Omega_{2,j} &= \omega_{2,j} + i \sum_{s_2} \frac{\lambda_{2,s_2} \rho_{1,s_2} \rho_{3,s_2} e^{i(\phi_{3,s_2} - \phi_{2,j} - \phi_{1,s_2})}}{\rho_{2,j}} \\
\Omega_{3,j} &= \omega_{3,j} + i \sum_{s_3} \frac{\lambda_{3,s_3} \rho_{1,s_3} \rho_{2,s_3} e^{i(\phi_{1,s_3} - \phi_{3,j} + \phi_{2,s_3})}}{\rho_{3,j}} \\
j &= 1, 2, \ldots, M
\end{aligned} \tag{6}$$

must be satisfied.

The nonlinear algebraic equations (6) together with the conditions (5) define the resonance equations. If these admit solutions $(\rho_{p,j}, \Omega_{p,j}, \phi_{p,j})$ in the set of real numbers with $\rho_{p,j} > 0$, System (3) is taken to verify the *nonlinear* resonance condition. In this case, System (3) admits the above solutions (4). Nevertheless, they arise only when the initial conditions are chosen on a suitable set of zero measure in the state space of the system or can occur asymptotically if they are attractive solutions (Boniforti *et al.* 1996). As shown from the above set of algebraic equations, frequencies $\Omega_{p,j}$ differ from the linear ones (re$[\omega_{p,j}]$) by a correction term linked to nonlinear interactions. Numerical tests performed on a number of dynamical systems have revealed that if there are dominant peaks in the power spectra linked to some frequencies, they are closer to $\Omega_{p,j}$ than re$[\omega_{p,j}]$. Therefore, in order to select eigenmodes, we require that the frequencies $\Omega_{p,j}$ (and not re$[\omega_{p,j}]$) be close to the experimental values. The difference between $\Omega_{p,j}$ and re$[\omega_{p,j}]$ are negligible in weakly nonlinear dynamics where the nonlinear coupling coefficients λ are small. Nevertheless, it is possible to show that in shear flows with a symmetric profile, when the λ coefficient models the interaction among three sinuous modes or that among two varicose modes and one sinuous one, the value of λ is zero. In this case resonance equations cannot be satisfied owing to the fact that the eigenvalues $\omega_{p,j}$ are in general complex while the nonlinear frequencies $\Omega_{p,j}$ are real. This means that a necessary condition to satisfy this *nonlinear* resonance criterion is to consider only triads constructed by two sinuous modes and a varicose one or three varicose modes. When a single triadic system is taken into account only the case of two sinuous modes and a varicose one remains, owing to the fact that at least one sinuous mode must be unstable. The dynamical evolution of such triads, as demonstrated by Wu *et al.* (1996), seems to be a key to understand the preferential development of three-dimensional modes in the later stage of transition.

The fulfillment of the phase-locked condition together with the nonlinear resonance criterion, allows us to select univocally eigenmodes which are good candidates capable of describing the physical process. In fact, among modes which verify (5) and (6), if they exist, we choose those which minimize the following function:

$$\sum_{p,q}\sum_{j,k}\left(\frac{\Omega_{p,j}}{\alpha_{p,j}}-\frac{\Omega_{q,k}}{\alpha_{q,k}}\right)^2 \qquad p,q=1,2,3 \quad j,k=1,2,\ldots,M \tag{7}$$

This function is zero for eigenmodes which evolve in phase-locked modal interactions.

3. NUMERICAL RESULTS AND DISCUSSION

The subject of this investigation is the identification of suitable dynamical systems able to simulate the temporal evolution of disturbances which arise in symmetric shear flows. Modes whose evolution seems to dominate the local temporal dynamics were selected by using the nonlinear resonance criterion together with the phase-locked condition. The experimental results recorded in a cylinder far wake by Williamson and Prasad (1993) are used here as test data to validate the theoretical approach. A non-dimensional mean wake profile $U = U_0/U_* + \tanh^2(2y/\delta)$ - where U_0 is the centreline velocity, $U_* = U_\infty - U_0$, U_∞ is the free-stream velocity and δ is the width of the wake - was used to obtain eigenmodes from the linear problem. This analysis was performed for a Reynolds number $R = U_\infty D/\nu = 150$ at a distance $x/D = 150$, where D is the diameter of the cylinder.

In particular, we examined two different scenarios, one corresponding to the dynamical evolution of three phase-locked interacting modes and the other one describing five phase-locked coupled modes. These cases are particularly significant since they show how a nonlinear interaction among oblique and planar modes can produce one or more oblique difference mode(s), which dominate the spectral energy distibution in spite of the varicose nature of the oblique modes.

In the first case we analyse the local temporal evolution which arises from the interaction among three modes linked to a free-stream frequency f_T, a Kármán shedding frequency f_K and a difference frequency $f_{K-T} = f_K - f_T$, with $f_T/f_K = .55$. The iteraction between the planar mode linked to the f_T-frequency and the oblique Kármán mode was modelled employing a dynamical system of type (3) with $M = 1$, corresponding to a single triadic system. Among the wave vectors of these modes, the spatial resonance condition $\mathbf{k}_{K-T} + \mathbf{k}_T = \mathbf{k}_K$ must be fulfilled. The selection of eigenmodes was carried out in the following way. For an observed shedding angle $\theta = 14^\circ$, we selected among the three-dimensional modes with a wave vector of the form $\mathbf{k}_K = (\alpha_K, \alpha_K \tan\theta, 0)$ (α_K varying), those whose eigenvalue has a real part near the Kármán frequency occurring the particular Reynolds number. From the resonance spatial condition it follows that $\mathbf{k}_{K-T} = (\alpha_K - \alpha_T, \alpha_K \tan\theta, 0)$ and $\mathbf{k}_T = (\alpha_T, 0, 0)$. After this, we locate possible triadic interactions of eigenmodes related to these wave vectors requiring the fulfilment of the nonlinear resonance condition, which, in this case can, be analytically solved. After this first screening, a set of triadic systems was found. Among triads of this set we choose that which minimizes Function (7) in order to take into account the phase-locked condition.

As a result of this analysis, we locate three modes $K-T, K, T$ corresponding to the nonlinear frequencies $\Omega_{K-T}, \Omega_K, \Omega_T$ close to the experimentally observed f_{K-T}, f_K, f_T and the corresponding dynamical system. The oblique $K-T$ mode was found to be sinuous with a positive temporal growth rate in the linear regime. The planar T mode was found to be sinuous and dumped. Finally, the K mode resulted in being varicose and strongly dumped owing to the fact that the Kármán frequency is much higher than those of locally unstable modes in this wake region. The numerical integration of the dynamical system reveals an energy spectral distribution characterised by a dominant peak of the oblique difference mode and two secondary peaks at frequencies f_K and f_T(fig. 1).

This result, even if deduced by a local analysis, supports the fact that in the later stage of transition three-dimensional sinuous modes can undergo a preferential amplification as some theoretical and experimental considerations suggest (Wu 1996, Corke *et al.* 1992).

In the second analysis, corresponding to the case $f_T/f_K = .31$, the experimental spectrum highlights five dominant peaks at frequencies f_K, f_T, $f_K - f_T, f_K - 2f_T, f_K - 3f_T$. In order to model the temporal evolution of this physical process, we make use of three coupled triadic systems corresponding to a dynamical system of type (3) with M=3. Following a methodological approach analogous to the above case, we selected a set of eigenmodes linked to the previous frequencies with corresponding wave vectors $\mathbf{k}_K, \mathbf{k}_T, \mathbf{k}_{K-T}, \mathbf{k}_{K-2T}, \mathbf{k}_{K-3T}$. These must verify three spatial resonance conditions $\mathbf{k}_{K-T} + \mathbf{k}_T = \mathbf{k}_K$, $\mathbf{k}_{K-2T} + \mathbf{k}_T = \mathbf{k}_{K-T}$ and $\mathbf{k}_{K-3T} + \mathbf{k}_T = \mathbf{k}_{K-2T}$. Requiring the fulfilment of the resonance condition using numerical methods, together with the phase locked condition, we locate five modes $K, T, K-T, K-2T, K-3T$ with nonlinear frequencies $\Omega_K, \Omega_T, \Omega_{K-T}$, $\Omega_{K-2T}, \Omega_{K-3T}$ close to the experimental ones. In this case we find two forcing modes, namely the planar T mode and the oblique $K-3T$ mode, which are sinuous modes. The oblique sinuous $K-T$ mode and varicose $K-2T$ mode are linearly dumped. Finally, the oblique K mode is varicose and strongly dumped owing to its high frequency. The power spectrum of the numerical solution of the dynamical System (3) is shown in fig. 2.

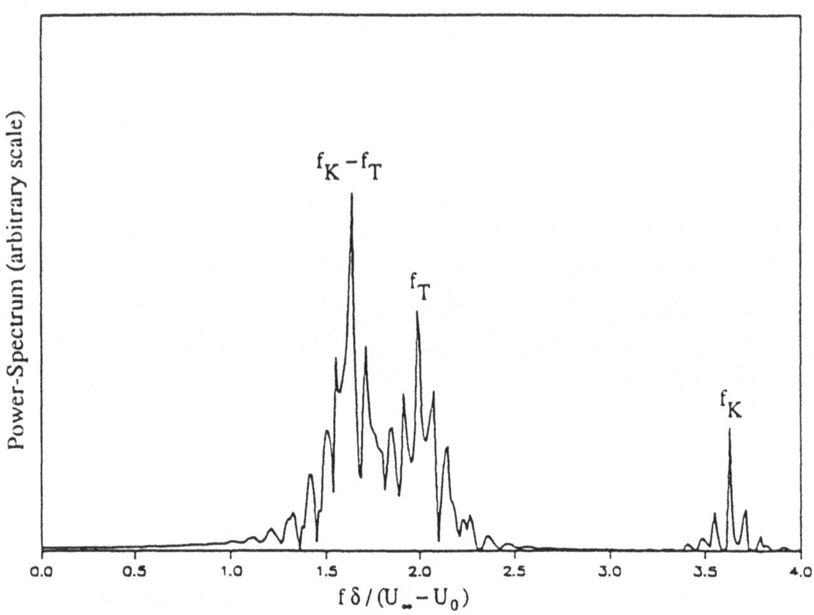

Figure 1. Power velocity spectrum corresponding to the case $f_T/f_K = .55$

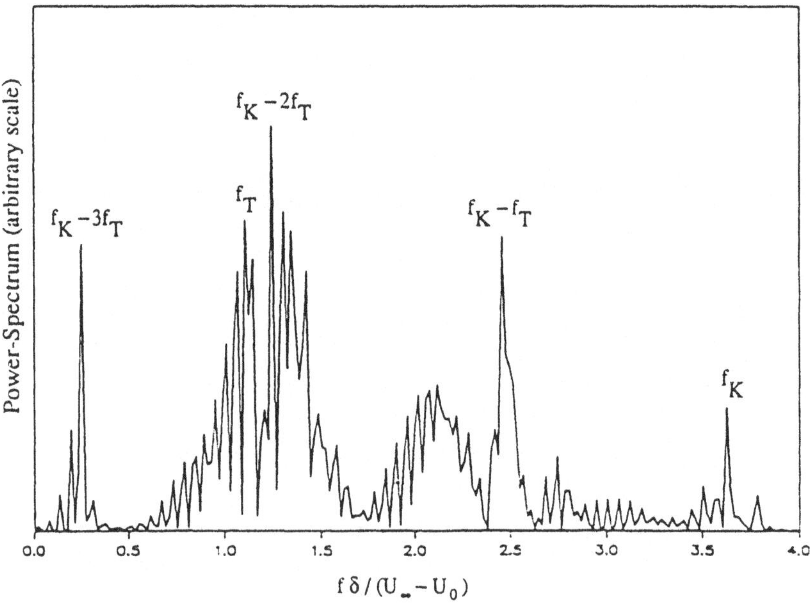

Figure 2. Power velocity spectrum corresponding to the case $f_T/f_K = .31$

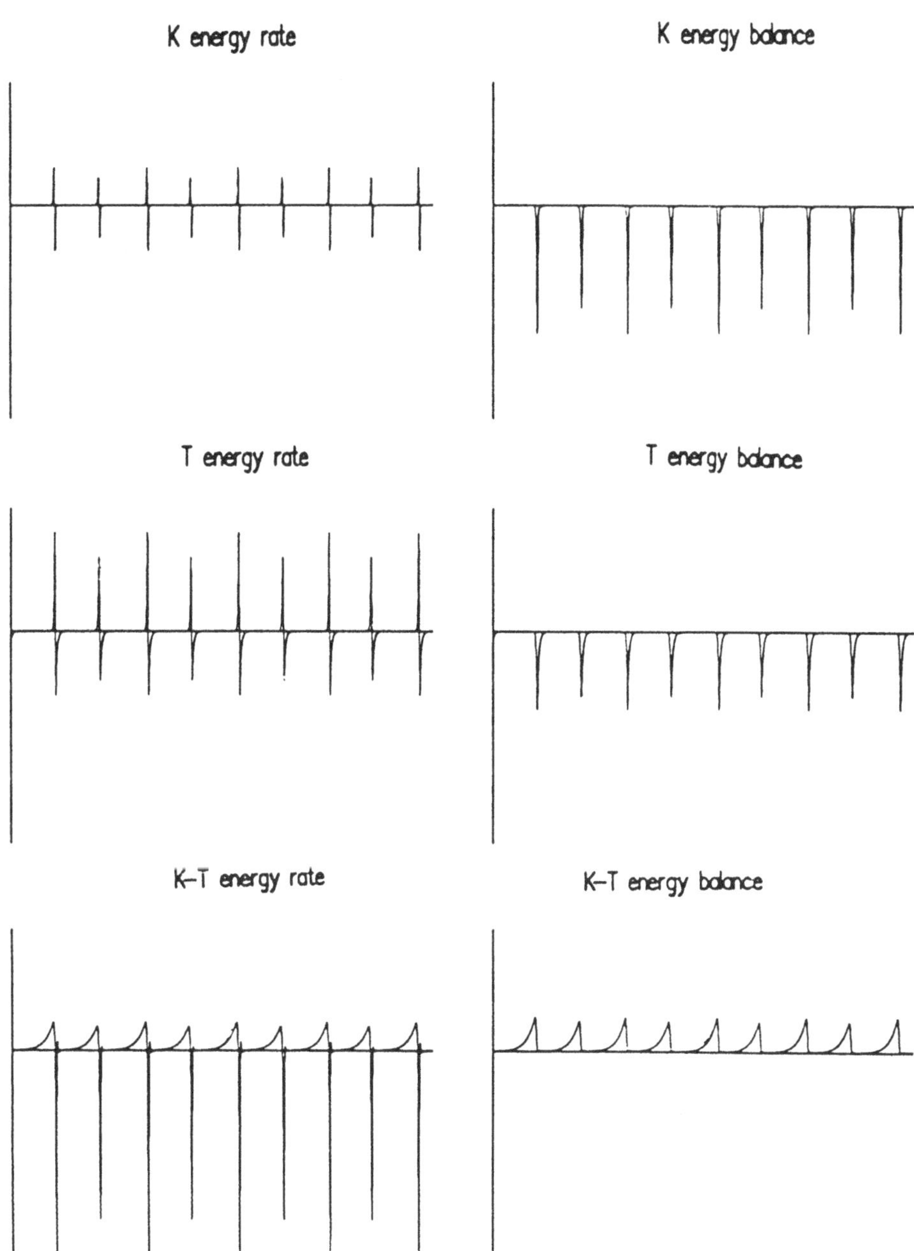

Figure 3. Temporal evolution of the kinetic energy rate $\frac{1}{2}\frac{d}{dt}|\epsilon|^2$ associated to the K, T and $K-T$ modes and temporal evolution of the rate of energy balance between the mean (time-averaged) flow U and each mode, corresponding to the case $f_T/f_K = .55$

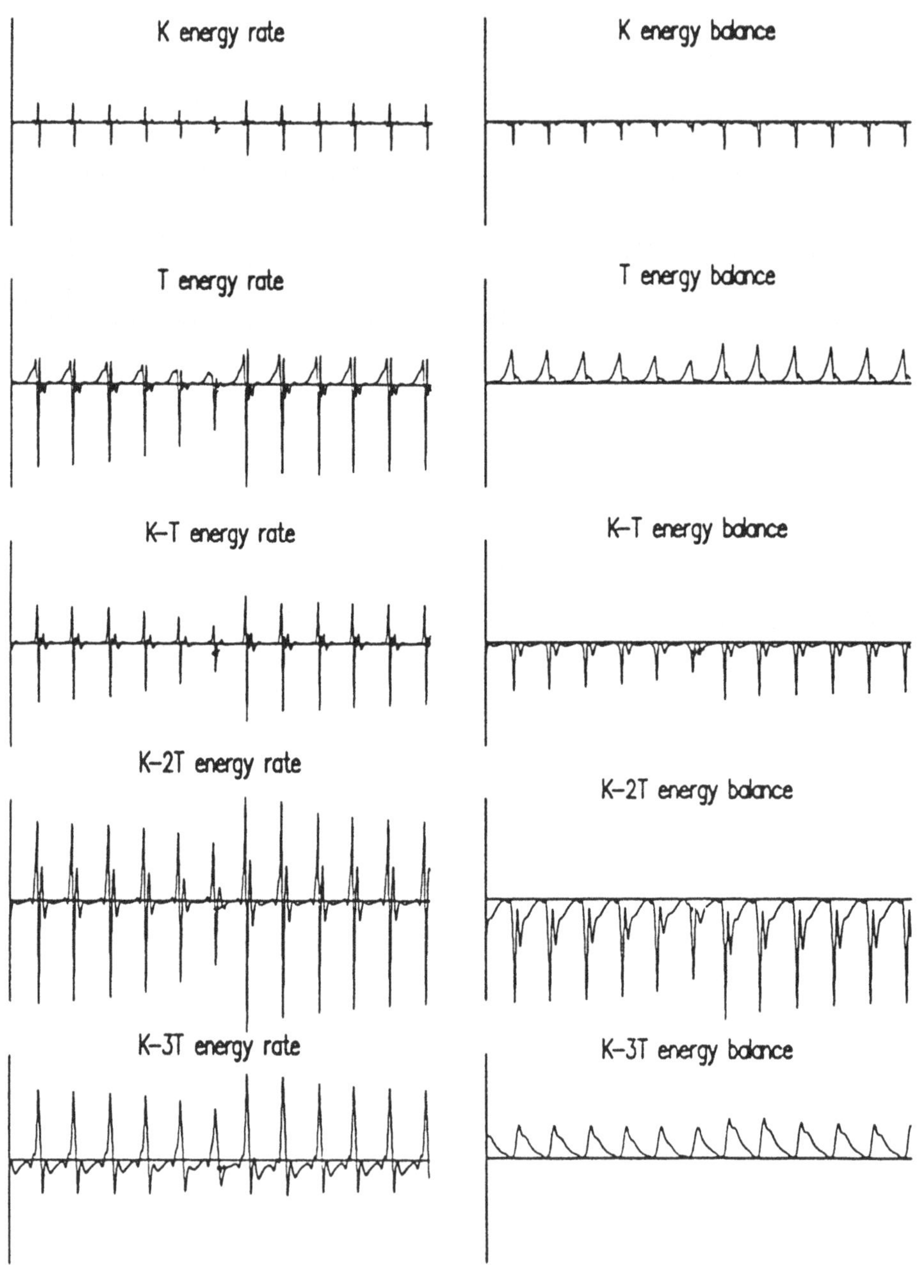

Figure 4. Temporal evolution of the kinetic energy rate $\frac{1}{2}\frac{d}{dt}|\epsilon|^2$ associated to the K, T, $K - T$, $K - 2T$ and $K - 3T$ mode and temporal evolution of the rate of energy balance between the mean flow U and each mode, corresponding to the case $f_T/f_K = .31$

The energy spectral distribution highlights a qualitative agreement with the experimental result and shows, as an interesting result, that the dominant peak is linked to the frequency $f_K - 2f_T$ related to the $K - 2T$ three-dimensional varicose mode.

Figures 3 and 4, which refer respectively to the case $f_T/f_K = .55$ and $f_T/f_K = .31$, show the temporal evolution of both the kinetic energy rate $E_j = \frac{1}{2}\frac{d}{dt}|\epsilon_j|^2$ associated to each mode and the term $P_j = \text{re}[-i\,\omega_j|\epsilon_j|^2]$. This last term can be regarded as a balance term which models the difference between the energy rate extracted from the mean flow and the viscous dissipation rate (Mele *et al.* 1993). For the unstable modes, the extraction of energy is greater than the viscous dissipation, so that $P_j(t)$ has a positive mean value while the opposite is true for dumped modes whose dynamics does not vanish due to the nonlinear energy transfer. In particular, for the case $f_T/f_K = .31$ it is interesting to observe the energy rate evolution of the $K - 2T$ mode. In spite of its varicose and dumped nature, this mode receives the greatest energy transfer through coupled nonlinear interactions and becomes dominant in the kinetic energy distribution (fig. 4).

4. CONCLUSIONS

In this work, a method for the identification of dynamical systems able to simulate the local temporal evolution of disturbances superimposed on a basic shear flow is proposed. Suitable modes to describe the dynamics of the disturbance were obtained by hydrodynamics stability of viscous flows and further selected by using a nonlinear resonance criterion together with a phase locked condition. This selection leads to an univocal identification of modes, if they exist, whose nonlinear interaction and the associated dynamical evolution is described by the solution of the corresponding low-dimensional dynamical system.

Some experimental results recorded in a cylinder far wake at moderate Reynolds cylinder numbers (Re=150) were used as test data. The local dynamics due to the nonlinear interactions between planar waves and oblique waves which travel phase-locked was compared with the numerical simulation obtained by integrating suitable dynamical systems. The comparison between experimental and numerical results highlighted a good qualitative agreement which confirms the global validity of the proposed methodological approach. In particular, this agreement was found in the spectral energy distributions of the modes activated in different experimental conditions which were modelled by our dynamical systems defined by one or more coupled triads. The nonlinear resonance criterion revealed itself particularly effective when coupling mechanisms among more triadic systems are active. In fact, under these previous conditions, a method of selection which considers one triad as an individual dynamical system was found to be not sufficiently restrictive to select eigenmodes and at the same time did not give the proper weight to the global coupling mechanisms of the various nonlinear interactions involved.

Furthermore, it is interesting to note that even if these results were obtained by using a methodology which employs temporal eigenmodes of viscous flows, some analogies with theoretical investigations obtained by spatial stability of inviscid flows (Wu 1996, Wu *et al.* 1996) were found. Firstly, our numerical simulation highlighted that three-dimensional oblique modes can dominate the local evolution of the disturbance in some transitional stages instead of the two-dimensional ones, as linear theory states. Besides, the structure of the nonlinear coupling mechanisms among the varicose and the sinuous modes selected by the proposed identification technique, led to some interesting conclusions, namely only some triadic interactions involving two sinuous modes and one varicose mode can take place, while the remaining modes are fully inactive.

References

1. Boniforti, M.A., Morganti, M. and Sciortino, G., (1996), "Triadic resonant modes: dynamical model and truncation criterion", *Fluid Dynamics Research* (to appear).
2. Corke, T., Krull, J.D. and Ghassemi, M. , (1992), "Three-dimensional mode resonance in the far wake", *J. Fluid Mech.*, **Vol. 239**, pp. 99.
3. Craik, A.D.D. ,(1985) *Wave interactions and fluid flows*. Cambridge University Press.
4. Craik, A.D.D., (1971), "Nonlinear resonant instability in boundary layers", *J. Fluid Mech.*, **Vol. 50**, pp. 393–413.
5. Herbert, T., (1978) "Secondary instability of boundary layers", *Ann. Rev. Fluid Mech.*, **Vol. 20**, pp. 487–526
6. Mele, P., Morganti M. and Boniforti M.A., (1993), "Triadic resonance in transitional shear flows: a low-dimensional model", in Some Applied Problems in Fluid Mechanics, Indian Statistical Institute.
7. Raetz, G. S., (1959) "A new theory of the cause of transition in fluid flows", Northrop Corp. NOR-59-383 BLC-121.
8. Williamson, C.H.K. and Prasad, A., (1993), "Acoustic forcing of oblique wave resonance in the far wake", *J. Fluid Mech.*, **Vol. 256**, pp. 315–341.
9. Wu, X. and Stewart, P.A., (1996), "Interaction of phase-locked modes: a new mechanism for the rapid growth of three-dimensional disturbances", *J. Fluid Mech.*, **Vol. 316**, pp. 335–372.
10. Wu, X., (1996),"On an active resonant triad of mixed modes in symmetric shear flows: a plane wake as a paradigm", *J. Fluid Mech.*, **Vol. 317**, pp. 337–368.

ANALYTICAL IDENTIFICATION OF GALLOPING EFFECTS ON PRISMATIC BODIES

By Mohamed Abdel-Rohman
Civil Engineering Department, Kuwait University

ABSTRACT

This paper shows how to identify analytically some of the galloping effects on prismatic bodies using experimental data to simulate mathematically the lift and drag force coefficients. A perturbation technique is used to solve analytically the nonlinear differential equations of motion which enables the researcher to gain a comprehensive information about the dependence of the galloping amplitude on the speed and turbulence of the flow.

INTRODUCTION

Galloping is the transverse vibration of the body to the direction of the mean flow. It occurs due to the aerodynamic forces which are induced by small transverse motions of the body. The aerodynamic self-excited forces act in the direction of the transverse motion resulting in negative damping which increases the amplitude of the transverse motion until it settles to a limit cycle. This occurs when the speed of the mean flow exceeds certain critical speed called "onset speed" [4,5]. Galloping usually occurs in Pylons of suspension bridges, prismatic tall buildings, and overhead power transmission lines.

This paper shows how to identify analytically the galloping effects on prismatic bodies using experimental data to simulate mathematically the lift and drag force coefficients. The nonlinear differential equations of motion are solved analytically using the method of multiple scales, which is a perturbation technique, in order to study the characteristics of galloping on prismatic bodies.

EQUATIONS OF MOTION

Considering a cantilever prismatic flexible body subjected to a wind flow, as shown in Fig. 1, the equations of motion along and across the mean flow directions are

$$EI\frac{\partial^4 w_1}{\partial x^4} + C\frac{\partial w_1}{\partial t} + m\frac{\partial^2 w_1}{\partial t^2} = \frac{1}{2}\rho U_{rel}^2\left[C_D \cos\beta - C_L \sin\beta\right] \tag{1}$$

$$EI\frac{\partial^4 w_2}{\partial x^4} + C\frac{\partial w_2}{\partial t} + m\frac{\partial^2 w_2}{\partial t^2} = -\frac{1}{2}\rho U_{rel}^2\left[C_D \sin\beta + C_L \cos\beta\right] \tag{2}$$

in which EI, C, m are, respectively, the flexural rigidity, damping, and mass per unit length of the prismatic body; w_1 is streamline response; w_2 is transverse response; ρ is the air density; D is the body width, C_D and C_L are, respectively the drag and lift coefficients; β is the angle of attack of the mean flow to principal axes; and U_{rel} is the relative wind speed given by

$$U_{rel}^2(x,t) = \left[U(x) + u(x,t) - \dot{w}_1(x,t)\right]^2 + \left[v(x,t) - \dot{w}_2(x,t)\right]^2 \tag{3}$$

in which U(x) is the mean wind speed at height x; u(x,t) is the fluctuating streamline wind speed component, and v(x,t) is the fluctuating transverse wind speed component..

The case of laminar flow is first studied. By neglecting the fluctuating wind speeds components and considering $w_2 = (U - w_1)\tan\beta$, Eqs. 1 and 2 can be expressed as

$$EI\frac{\partial^4 w_1}{\partial x^4} + C\frac{\partial w_1}{\partial t} + m\frac{\partial^2 w_1}{\partial t^2} = \frac{1}{2}\rho\, b\left(U - \dot{w}_1\right)^2 C_{w1}(\beta) \tag{4}$$

$$EI\frac{\partial^4 w_2}{\partial x^4} + C\frac{\partial w_2}{\partial t} + m\frac{\partial^2 w_2}{\partial t^2} = \frac{1}{2}\rho\, b\left(U - \dot{w}_1\right)^2 C_{w2}(\beta) \tag{5}$$

in which $C_{w1}(\beta)$ and $C_{w2}(\beta)$ are given by

$$C_{w1}(\beta) = \left[C_D(\beta) - C_{LD}(\beta)\tan\beta\right]\sec\beta \tag{6}$$

$$C_{w2}(\beta) = -\left[C_D(\beta)\tan\beta + C_L(\beta)\right]\sec\beta \tag{7}$$

The polynomials $C_{w1}(\beta)$ and $C_{w2}(\beta)$ are obtained experimentally using a tunnel test. The wind tunnel test results of Ref. [3] which are shown in Figs. 2 and 3 are fitted using polynomials in terms of tan β as follows:

$$\begin{aligned} C_{w1}(\beta) = {} & 1.35 - 0.161\tan\beta - 4.58\tan^2\beta + 0.960\tan^3\beta + 47.9\tan^4\beta \\ & - 2.1\tan^5\beta - 74.2\tan^6\beta - 0.35\tan^7\beta \end{aligned} \tag{8}$$

$$\begin{aligned} C_{w2}(\beta) = {} & 0.03 + 0.931\tan\beta - 0.24\tan^2\beta - 7.7\tan^3\beta + 4.6\tan^4\beta \\ & + 232.1\tan^5\beta - 2451.75\tan^7\beta \end{aligned} \tag{9}$$

These polynomials are used in this paper to identify the galloping characteristics. Since, usually, $\dot{w}_1 << U$, one can consider

$$\tan\beta = \dot{w}_2 \left(U - \dot{w}_1\right)^{-1} = \frac{\dot{w}_2}{U}\left(1 - \frac{\dot{w}_1}{U}\right)^{-1} \approx \frac{\dot{w}_2}{U} + \frac{\dot{w}_2\,\dot{w}_1}{U^2} \tag{10}$$

The solution to Eqs. 4 and 5 are, respectively, assumed as

$$w_1(x,t) = \phi(x)\,A(t) \tag{11}$$

$$w_2(x,t) = \phi(x)\,B(t) \tag{12}$$

$$\phi(x) = \cosh\left(\frac{\lambda}{L}x\right) - \cos\left(\frac{\lambda}{L}x\right) - \Pi\left(\sinh\left(\frac{\lambda}{L}x\right) - \sin\left(\frac{\lambda}{L}x\right)\right) \tag{13}$$

in which $\phi(x)$ is for a cantilever of height L, $\lambda = 1.875$, and $\Pi = 0.734$.

Substituting Eqs. 11 and 12 into Eqs. 4 and 5, using integral transformation one obtains, respectively,

$$\begin{aligned}\ddot{A} + \mu\dot{A} + \omega^2 A = {} & \gamma_1 + \gamma_2\dot{A} + \gamma_3\dot{A}^2 + \gamma_4\dot{B} + \gamma_5\dot{A}\dot{B} + \gamma_6\dot{B}^2 + \gamma_7\dot{B}^3 + \gamma_8\dot{A}\dot{B}^3 \\ & + \gamma_9\dot{A}^2\dot{B}^3 + \gamma_{10}\dot{A}^3\dot{B}^3 + \gamma_{11}\dot{A}^4\dot{B}^3 + \gamma_{12}\dot{A}^5\dot{B}^3 + \gamma_{13}\dot{B}^4 \\ & + \gamma_{14}\dot{A}\dot{B}^4 + \gamma_{15}\dot{A}^2\dot{B}^4 + \gamma_{16}\dot{A}^3\dot{B}^4 + \gamma_{17}\dot{A}^4\dot{B}^4 + \gamma_{18}\dot{B}^5 \\ & + \gamma_{19}\dot{A}\dot{B}^5 + \gamma_{20}\dot{A}^2\dot{B}^5 + \gamma_{21}\dot{A}^3\dot{B}^5 + \gamma_{22}\dot{B}^6 + \gamma_{23}\dot{A}\dot{B}^6 \\ & + \gamma_{24}\dot{A}^2\dot{B}^6 + \gamma_{25}\dot{B}^7 + \gamma_{26}\dot{A}\dot{B}^7\end{aligned} \tag{14}$$

$$\begin{aligned}\dot{B} + \mu\dot{B} + \omega^2 B = {} & \beta_1 + \beta_2\dot{A} + \beta_3\dot{A}^2 + \beta_4\dot{B} + \beta_5\dot{A}\dot{B} + \beta_6\dot{B}^2 + \beta_7\dot{B}^3 + \beta_8\dot{A}\dot{B}^3 \\ & + \beta_9\dot{A}^2\dot{B}^3 + \beta_{10}\dot{A}^3\dot{B}^3 + \beta_{11}\dot{A}^4\dot{B}^3 + \beta_{12}\dot{A}^5\dot{B}^3 + \beta_{13}\dot{B}^4 \\ & + \beta_{14}\dot{A}\dot{B}^4 + \beta_{15}\dot{A}^2\dot{B}^4 + \beta_{16}\dot{A}^3\dot{B}^4 + \beta_{17}\dot{A}^4\dot{B}^4 + \beta_{18}\dot{B}^5 \\ & + \beta_{19}\dot{A}\dot{B}^5 + \beta_{20}\dot{A}^2\dot{B}^5 + \beta_{21}\dot{A}^3\dot{B}^5 + \beta_{22}\dot{B}^7 + \beta_{23}\dot{A}\dot{B}^7\end{aligned} \tag{15}$$

in which γ_i and β_i are constants; and μ are damping term which depends on the damping ratio.

Equations 14 and 15 are coupled ordinary nonlinear differential equations with constant coefficients. If one considers the fluctuating wind speed components in the analysis then the equations would have time varying coefficients.

NONLINEAR ANALYSIS

Equations 14 and 15 are solved analytically using first order perturbation analysis by the method of multiple scales [1], in which all nonlinear terms are multiplied by a small parameter ε such that

$$A(t,\varepsilon) = A_0(T_0,T_1) + \varepsilon A_1(T_0,T_1) \tag{16}$$

$$B(t,\varepsilon) = B_0(T_0,T_1) + \varepsilon B_1(T_0,T_1) \tag{17}$$

$$\frac{d}{dt} = D_0 + \varepsilon D_1 \tag{18}$$

$$\frac{d^2}{dt^2} = D_0^2 + 2\varepsilon D_1 D_0 \tag{19}$$

$$D_i = \frac{\partial}{\partial T_i} \tag{20}$$

where ε is a perturbation parameter, and $T_i = \varepsilon^i t$.

The substitution of Eqs. 16-20 into Eqs. 14 and 15, and by equating terms of equal power of ε, one obtains algebraic equations which provide the steady state amplitudes for the response of A(t) and B(t). Detailed analysis and solutions can be found in Ref. [2].

EXAMPLE

Considering a tall prismatic body of height L=150 m (599 ft), damping factor $\mu = 0.03$, air density $\rho = 1.24$ kg/m^3 (0.0763 slug/ft^3), width b = 30 m (100 ft), mass m = 39600 kg/m (26400 slug/ft), and $\omega = 1.5$ rps. The mode shape $\phi(x)$ of Eq. 13 is used in calculating the coefficients γ_i and β_i. The mean wind flow profile usually follows the power or logarithmic law. In this paper the power law is used as

$$U(x) = U(x_o)\left(\frac{x}{x_o}\right)^{0.4} \tag{21}$$

in which x_o is the reference height 10 m (33 ft) and the power 0.4 indicates the roughness of the terrain.

From the numerical values of the coefficients, one can determine the critical mean wind speed at which galloping starts on the body. By equating the determinant of the linear damping matrix to zero will give the condition at which the structure starts loosing its stability. From Eqs. 14 and 15 one has

$$\mathrm{Det}\begin{bmatrix} (\gamma_2 - \mu) & \gamma_4 \\ \beta_2 & (\beta_4 - \mu) \end{bmatrix} = 0 \tag{22}$$

In this example, U_{cr} = 7.58 m/s (25 fps). The steady state amplitudes for A(t) and B(t) are obtained analytically and shown in Figs. 4 and 5. These results were confirmed numerically by solving the differential equations Eqs. 14 and 15 for U = 12.12 m/s (40 fps) as shown in Figs. 6 and 7. From both results it is obvious that streamline steady-state amplitude is very small as compared with the transverse amplitude. Moreover, as the speed of the mean flow increases more than the critical wind speed, the galloping amplitude increases.

TURBULENCE EFFECTS

From previous analysis, one can ignore the streamline dynamic response and pays attention to the transverse response. In our example, this means that Eqs. 4 and 5 can be replaced by one equation by putting $\dot{w}_1 = 0$, which leads to

$$EI\frac{\partial^4 w_2}{\partial x^4} + C\frac{\partial w_2}{\partial t} + m\frac{\partial^2 w_2}{\partial t^2} = \frac{1}{2}\rho\, b\, U^2(x)\; C_{w2}(\beta) \tag{23}$$

in which $\tan\beta = \dot{w}_2/U(x)$.

To study turbulence effect, one may put a small fluctuating wind speed component u(t) to the mean wind speed U(x). Neglecting $u^2(t)$, Eq. 23 becomes

$$EI\frac{\partial^4 w_2}{\partial x^4} + C\frac{\partial w_2}{\partial t} + m\frac{\partial^2 w_2}{\partial t^2} = \frac{1}{2}\rho\, b\; U^2_{rel}\left[U^2(x) + 2U(x)\,u(t)\right] C_{w2}(\beta) \tag{24}$$

By applying the same steps of solution to Eqs. 4 and 5, one obtains

$$\ddot{B} + \mu\,\dot{B} + \omega^2 B = \sum_{i=1}^{7} \delta_i\, \dot{B}^i + 2\sum_{i=1}^{7} \delta_{i+8}\; u(t)\; \dot{B}^i_1 \tag{25}$$

in which

$$\delta_i = \frac{\rho\, b}{2mL} C_i \int_0^L U^{1-i}(x)\,\phi^{i+1}(x)\,dx \tag{26}$$

$$\delta_{i+8} = \frac{\rho\, b}{2mL} C_i \int_0^L U^{1-i}(x)\,\phi^{i+1}(x)\,dx \tag{27}$$

in which C_i are the coefficients of the polynomial $C_{w2}(\beta)$ of Eq. 9.

It is obvious from Eq. 25 that the body is subjected to self-excited forces due to the terms $\sum_{i=1}^{7}\delta_i\,\dot{B}^i$, parametric excitation due to the terms $\sum_{i=1}^{7}\delta_{i+8}\,U(t)\,\dot{B}^i$, and external excitation due to the term $\delta_8\,u(t)$.

Considering first the self-excited forces and the fluctuation wind speed $u(t) = G \cos \Omega t$, where G is the amplitude of the fluctuation and Ω is the frequency, one obtains the following model from Eq. 25:

$$B+\mu B+\omega^2 B=\sum_{i=1}^{7} \delta_i B^i+2\delta_8 G \cos(\Omega t) \tag{28}$$

To study the response when $\Omega \approx \omega$ (case of primary resonance), Eq. 28 is ordered using the small parameter ε as follows:

$$\ddot{B}+\omega^2 B=\delta_0+\varepsilon^2\left[(\delta_1-\mu)\dot{B}+\delta_3\dot{B}^3+\delta_5\dot{B}^5+\delta_7\dot{B}^7+2\delta_8 G\cos\Omega t\right]$$
$$+\varepsilon\left[\delta_2\dot{B}^2+\delta_4\dot{B}^4\right] \tag{29}$$

Using second-order perturbation analysis one has

$$B(t)=\sum_{i=0}^{2}\varepsilon^i B_i(T_0, T_1, T_2) \tag{30}$$

$$\frac{d}{dt}=D_0+\varepsilon D_1+\varepsilon^2 D_2 \tag{31}$$

$$\frac{d^2}{dt^2}=D_0^2+\varepsilon^2 D_1^2+2\varepsilon D_1 D_0+2\varepsilon^2 D_2 D_0 \tag{32}$$

Substituting Eqs. 30-32 into Eq. 29 and equating terms of equal power of ε one obtains the governing equations which provide the galloping amplitude as follows [2]:

$$\dot{b}=\frac{1}{2}(\delta_1-\mu)b+\frac{3}{8}\omega^2\delta_3 b^2+\frac{5}{16}\omega^4\delta_5 b^5+\frac{35}{128}\omega^6\delta_7 b^7+\frac{\delta_8}{\omega}G\sin\gamma \tag{33}$$

$$b\dot{\gamma}=\sigma b+\frac{1}{6}\omega\delta_2^2 b^3+\frac{1}{3}\delta_2\delta_4\omega^3 b^5-\frac{7}{40}\delta_4^2\omega^5 b^7+\frac{\delta_8}{\omega}G\cos\gamma \tag{34}$$

in which b is the galloping amplitude and γ is the phase angle.

At steady state response, $\dot{b}=\dot{\gamma}=0$, and Eqs. 33 and 34 can be solved to obtain the steady galloping amplitude. Figure 8 shows the galloping amplitude in the presence of turbulence intensity G = 1,8 and 16. Figure 9 shows the effect of $\sigma = (\Omega-\omega)$ on the galloping amplitude. Considering now Eq. 25 with the perturbation parameter ε one has

$$\ddot{B}+\omega^2 B=\delta_0+\varepsilon\left[\delta_2\dot{B}^2+\delta_4\dot{B}^4\right]$$
$$+\varepsilon^2\left[(\delta_1-\mu)\dot{B}+\delta_3\dot{B}^3+\delta_5\dot{B}^5+\delta_7\dot{B}^7+2G\cos\Omega t\sum_{i=0}^{7}\delta_{i+8}\dot{B}^i\right] \tag{35}$$

Following the same previous steps one finds out that resonance occurs when $\Omega \approx \omega$, $\Omega \approx 2\omega$, $\Omega \approx 3\omega$, $\Omega \approx 4\omega$, $\Omega \approx 5\omega$ etc... . For the case $\Omega \approx \omega$, the governing equations are

$$\omega \dot{b} = \frac{1}{2}\omega b\left(\delta_1 - \mu\right) + \frac{3}{8}\omega^3 \delta_3 b^3 + \frac{5}{16}\omega^5 \delta_5 b^5 + \frac{35}{128}\omega^7 \delta_7 b^7 + G \sin\gamma\left(\delta_8 + \frac{3}{4}\delta_{10}\omega^2 b^2 + \frac{5}{8}\delta_{12}\omega^4 b^4\right) \quad (36)$$

$$\omega b \dot{\gamma} = \omega b \sigma + \frac{1}{6}\omega^2 \delta_2^2 b^3 + \frac{1}{3}\delta_2 \delta_4 \omega^4 b^5 + \frac{7}{40}\delta_4^2 \omega^6 b^7 + G \sin\gamma\left(\delta_8 + \frac{1}{4}\delta_{10}\omega^2 b^2 + \frac{1}{8}\delta_{12}\omega^4 b^4\right) \quad (37)$$

Figure 10 shows the galloping amplitude when G = 8. Comparing Fig. 10 with Fig. 8 one concludes that the effect of parametric excitation on the galloping is negligible.

For the case $\Omega \approx 2\omega$, the governing equations are

$$\omega \dot{b} = \frac{1}{2}\omega b\left(\delta_1 - \mu\right) + \frac{3}{8}\omega^3 \delta_3 b^3 + \frac{5}{16}\omega^5 \delta_5 b^5 + \frac{35}{128}\omega^7 \delta_7 b^7 - G \cos 2\gamma\left(\frac{1}{2}\delta_9 \omega b + \frac{1}{2}\delta_{11}\omega^3 b^3 + \frac{15}{32}\delta_{13}\omega^5 b^5 + \frac{7}{16}\delta_{14}\omega^7 b^7\right) \quad (38)$$

$$\omega b \dot{\gamma} = \omega b \sigma + \frac{1}{6}\omega^2 \delta_2^2 b^3 + \frac{1}{3}\delta_2 \delta_4 \omega^4 b^5 + \frac{7}{40}\delta_2^2 \omega^6 b^7 + G \sin 2\gamma\left(\frac{1}{2}\delta_9 \omega b + \frac{1}{4}\delta_{11}\omega^3 b^3 + \frac{5}{32}\delta_{13}\omega^5 b^5 + \frac{7}{64}\delta_{14}\omega^7 b^7\right) \quad (39)$$

The galloping amplitude for this case is shown in Fig. 11. It is obvious that the turbulence in this frequency is less severe than the case $\Omega \approx \omega$.

Similar steps of analysis can be carried out when $\Omega \approx 3\omega$ which lead to

$$\omega \dot{b} = \frac{1}{2}\omega b\left(\delta_1 - \mu\right) + \frac{3}{8}\omega^3 \delta_3 b^3 + \frac{5}{16}\omega^5 \delta_5 b^5 + \frac{35}{128}\omega^7 \delta_7 b^7 - G \sin 3\gamma\left[\frac{1}{4}\delta_{10}\omega^2 b^2 + \frac{5}{16}\delta_{12}\omega^4 b^4\right] \quad (40)$$

$$\omega b \dot{\gamma} = \omega b \sigma + \frac{1}{6}\omega^2 \delta_2^2 b^3 + \frac{1}{3}\delta_2 \delta_4 \omega^4 b^5 + \frac{7}{40}\delta_2^2 \omega^6 b^7 - G \sin 3\gamma\left[\frac{1}{4}\delta_{10}\omega^2 b^2 + \frac{3}{16}\delta_{12}\omega^4 b^4\right] \quad (41)$$

The galloping amplitude of this case is shown in Fig. 12 which shows negligible effect of turbulence at this frequency.

For the turbulence at frequency $\Omega \approx 4\omega$, the governing equations are:

$$\omega \dot{b} = \frac{1}{2}\omega b\left(\delta_1 - \mu\right) + \frac{3}{8}\omega^3 \delta_3 b^3 + \frac{5}{16}\omega^5 \delta_5 b^5 + \frac{35}{128}\omega^7 \delta_7 b^7$$

$$+ G \sin 4\gamma\left[\frac{1}{8}\delta_{11}\omega^3 b^3 + \frac{3}{16}\delta_{13}\omega^5 b^5 + \frac{14}{64}\delta_{14}\omega^7 b^7\right] \qquad (42)$$

$$\omega b \dot{\gamma} = \omega b \sigma + \frac{1}{6}\omega^2 \delta_2^2 b^3 + \frac{1}{3}\delta_2 \delta_4 \omega^4 b^5 + \frac{7}{40}\delta_2^2 \omega^6 b^7$$

$$- G \sin 4\gamma\left[\frac{1}{8}\delta_{11}\omega^3 b^3 + \frac{1}{8}\delta_{13}\omega^5 b^5 + \frac{7}{64}\delta_{14}\omega^7 b^7\right] \qquad (43)$$

The galloping amplitude of this case is shown in Fig. 13. The turbulence at this frequency only affects the galloping amplitude at high mean wind speed and high turbulence intensity.

CONCLUSIONS

The paper has shown that the wind tunnel testing to measure the streamline and transverse wind force coefficients on a model for the body can be utilized to identify analytically the galloping characteristics of the prismatic body.

The present analysis was able to identify the critical wind speed after which galloping takes a place, and predicts the galloping amplitude in both smooth flow and turbulent flow. In case of turbulent flow, the most important effect on galloping amplitudes and the critical speed was found due to the cases of turbulence frequencies $\Omega \approx \omega$ and $\Omega \approx 2\omega$.

REFERENCES

1- Abdel-Rohman, M. (1992), "Galloping of Tall Prismatic Structures : A Two Dimensional Analysis", Journal of Sound and Vibrations, Vol. 153. No. 1, pp. 97-111.

2- Abdel-Rohman, M. (1993), Control of Nonlinear Vibrations in Civil Structures, Chapter 1, Kuwait University Publications, pp. 1-118.

3- Mukhopadhyay, V. and Dugundji, J. (1976), "Wind-Excited Vibration of a Square Section Cantilever Beam in Smooth Flow", Journal of Sound and Vibration, 45(3), pp. 329-339.

4- Novak, M. (1969), "Aeroelastic Galloping of Prismatic Bodies", Journal of Engineering Mechanics, ASCE, Vol. 96, Feb., pp. 1150142.

5- Parkinson, G.V. and Smith, J.D. (1964), "The Square Prism as an Aeroelastic Nonlinear Oscillator", Quart. J. Mech. Appl. Math., Vol. 17, No. 2, pp. 225-239.

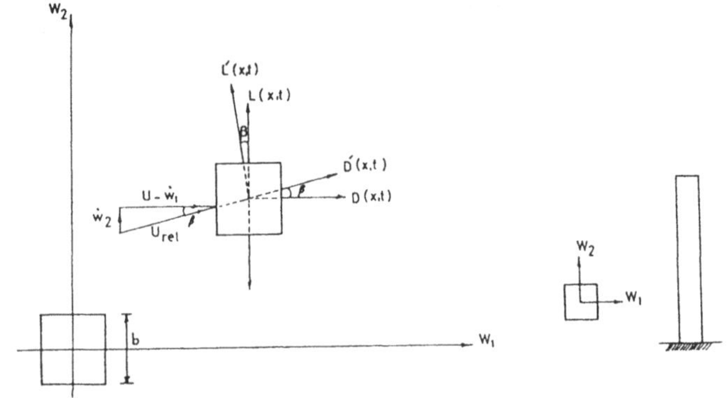

Figure 1 A prismatic body subjected to a flow

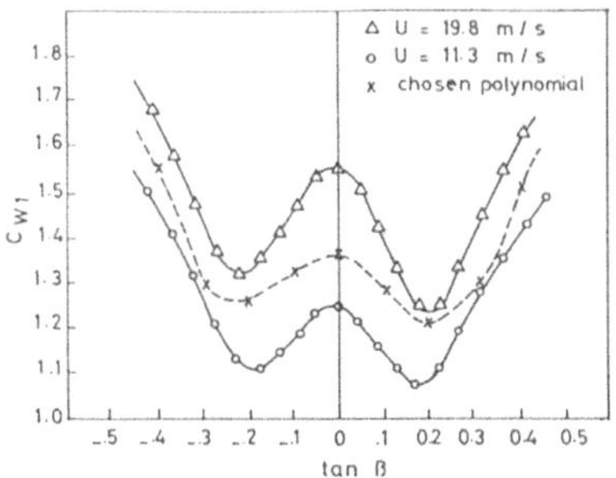

Figure 2 Wind tunnel results for along mean wind force coefficients

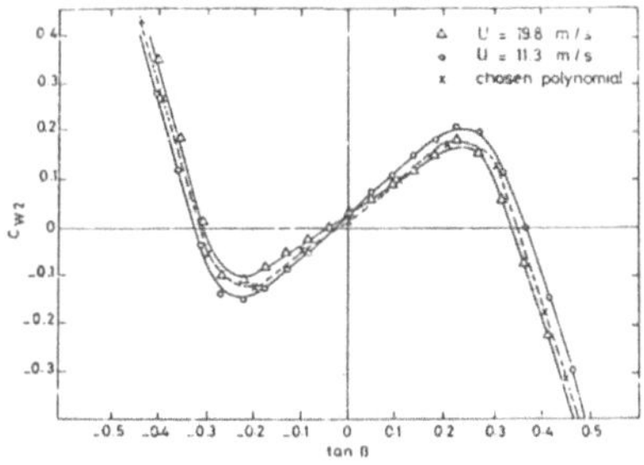

Figure 3 Wind tunnel results for across mean wind force coefficients

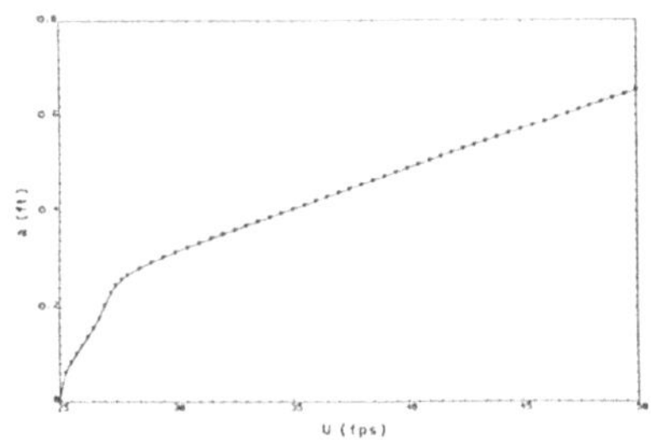

Figure 4 Steady state amplitude of A(t) obtained analytically

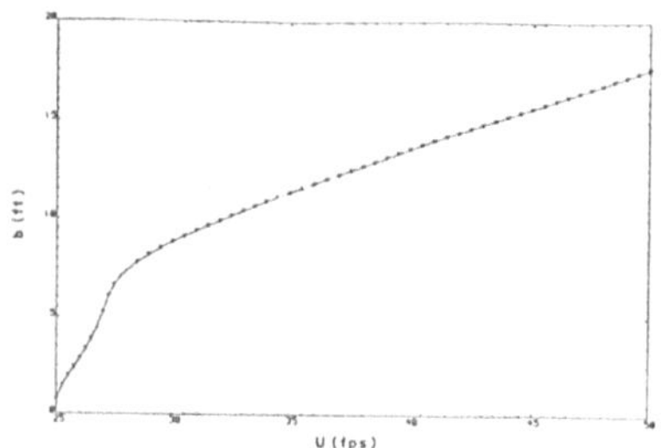

Figure 5 Steady state amplitude of B(t) obtained analytically

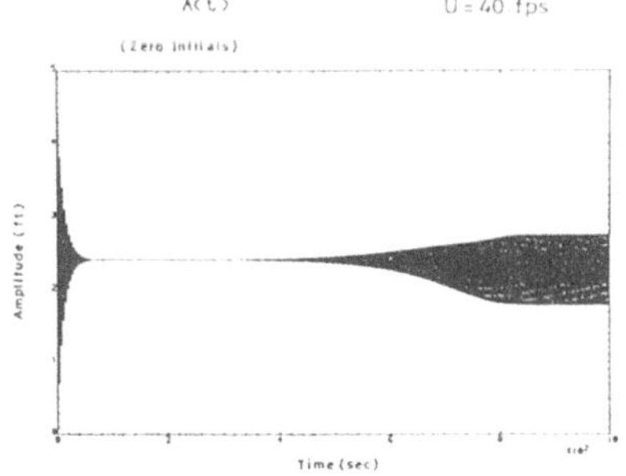

Figure 6 Steady state amplitude of A(t) obtained numerically at U = 40 fps

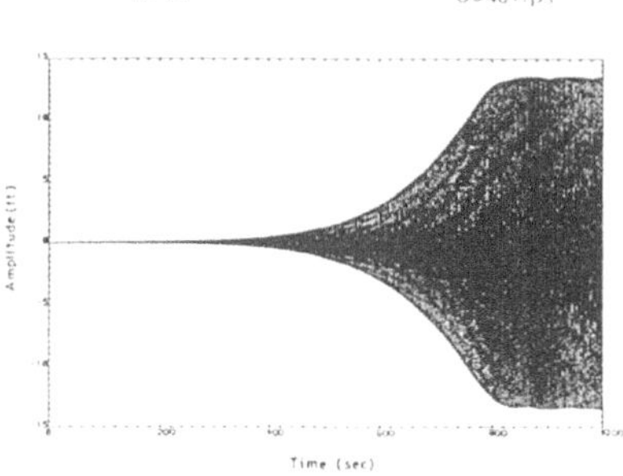

Figure 7 Steady state amplitude of B(t) obtained numerically at U = 40 fps

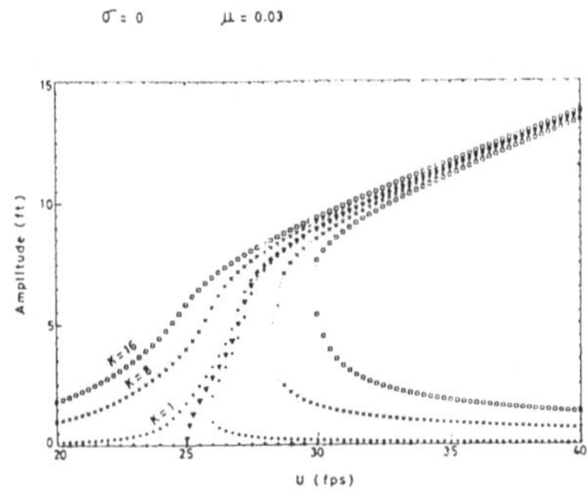

Figure 8 Galloping amplitude when $\Omega \approx \omega$ for external excitations

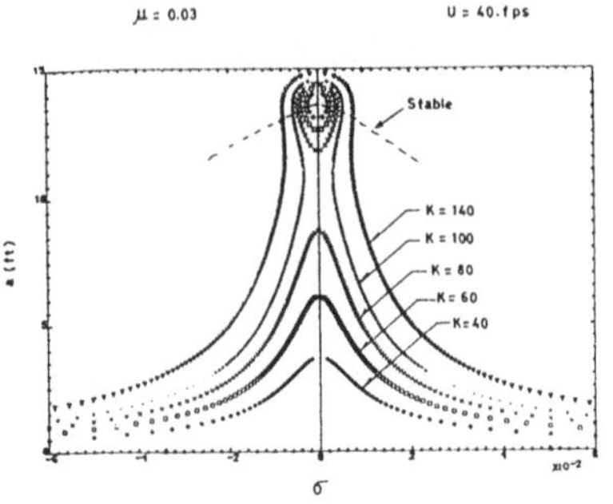

Figure 9 Effect of detuning $\sigma = \Omega - \omega$ on gallop amplitude for external excitations

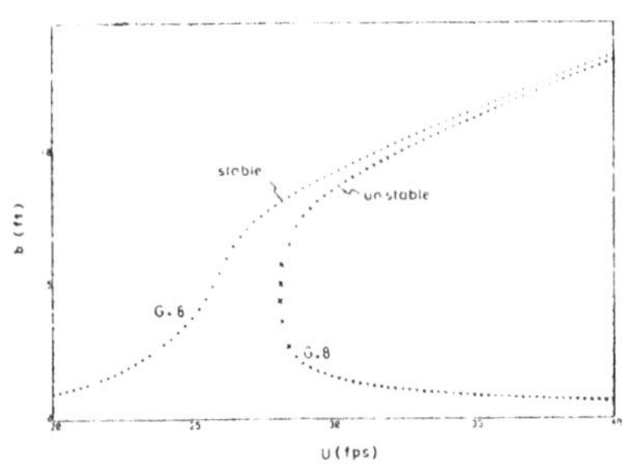

Figure 10 Galloping amplitude when $\Omega \approx \omega$ including parameteric excitations

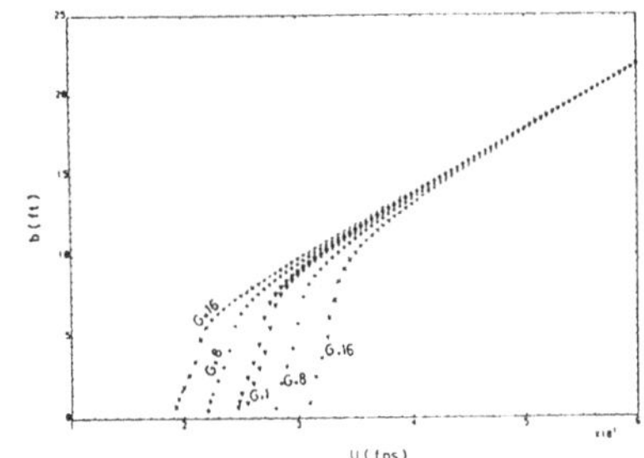

Figure 11 Galloping amplitude when $\Omega \approx 2\omega$ including parameteric excitations

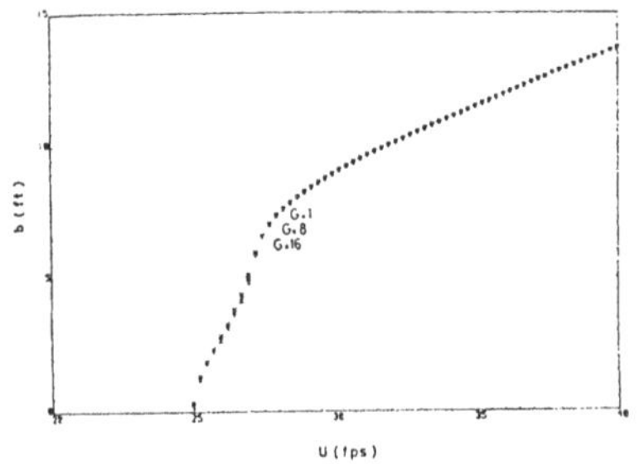

Figure 12 Galloping amplitude when $\Omega \approx 3\omega$ including parameteric excitations

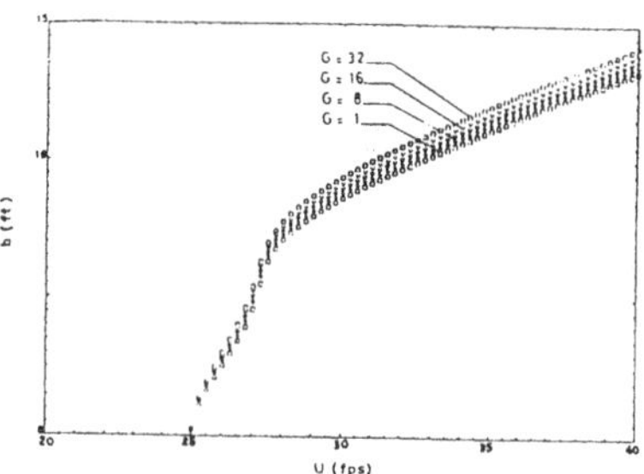

Figure 13 Galloping amplitude when $\Omega \approx 4\omega$ including parameteric excitations

MULTIRESOLUTION LOCAL ADAPTIVE METHOD FOR THE ANALYSIS OF SPATIALLY EXTENDED SYSTEMS

MILAN RAJKOVIĆ
Institute of Nuclear Sciences Vinča
P. O. Box 522
Belgrade 11001, Yugoslavia

Abstract.
The multiresolution method for the analysis of spatiotemporal dynamics of extended systems whose long-time behaviour is confined to an attractor of a possibly large dimension is presented. The dynamics, followed locally in phase space, is projected into the local tangent space of the attractor, and the choice of the optimal coordinate system is dictated by the particular nonlinear dynamics of the system under study. The coordinate basis is determined by the local singular value decomposition (SVD), followed by the separation of local Gaussian and non-Gaussian (coherent) modes. Active and passive modes, as non-Gaussian components of the dynamics, are separated based on a purely geometric criterion. The method enables an operationally efficient way of checking for the existence of an invariant finite - dimensional manifold and for calculating its dimension.

1. Introduction

In spite of the great interest in the complicated spatiotemporal dynamics of infinite - dimensional systems, there have been relatively few attempts, such as (Aubry *et al.*, 1991) and (Broomhead *et al.*, 1991), to develop a systematic programme for the analysis of complex systems. An important accomplishment in the study of the systems modelled by the PDE's was the proof of existence of the finite dimensional attractor for the several paradigmatic equations. Naturally, the problems would arise in case of fully developed turbulence where dimensions of attractors may be beyond the limits of practical and computational use. On the other hand, the coordinates of the Euclidean space encompassing the global attractor fail to capture the dynamics of localized coherent structures which may be very important for the overall dynamics. The description of the dynamics on the attractor itself would be, in general, difficult due to the possible complicated fractal structure of the attractor. However, the finite dimensionality of attractors suggests the possibility of embedding these attractors in Euclidean space of higher, but finite dimension, whose coordinates would capture the flow dynamics. An embedding of the attractor in a finite - dimensional smooth manifold of solutions of the governing differential equations would restrict the dynamics to this manifold, enabling relatively simple description of the dynamics. On such a manifold the dynamics is finite dimensional and completely determines the asymptotic behaviour of the model PDE's (Temam, 1988). The existence of an inertial manifold $\mathcal{M}$ implies that the phase space can be decomposed into two orthogonal linear subspaces corresponding to the active and passive modes, such that the temporal assymptotic evolution of passive modes is completely determined by the evolution of the active modes (slaving principle). Such a description suggests that an inertial manifold conceptually represents a global enslaving function relating the nature of control of a large, possibly infinite number of degrees of freedom by only a finite number of dominant modes.

The goal therefore is to devise a method that would enable the determination of the possible high value of the fractal dimension of the global attractor, as well as to capture the subtle local changes (in time and space) of the dynamics. Moreover, we set as the goal of the method to be able to distinguish the Gaussian and non - Gaussian (coherent) components of the dynamics. In order to achieve such an aim, we choose the local basis on the attractor determined by the local dynamics itself, and use these to move the solution from one local linear region to the next. The basis is constructed using the local version of the SVD algorithm, hence a comparison with the conceptually similar methods, such as the proper orthogonal decomposition (POD) (Lumley, 1970), and biorthogonal decomposition (Aubry *et al.*, 1991) is appropriate. Both of these methods rely on the choice of the global basis which remains the same for all points on the attractor, and which implies that a global nonlinear system is approximated using linear methods. In particular, certain studies based on the POD imply that the invariant distribution covers an essentially linear subspace, which actually happens very rarely or never. Furthermore, there is no physically reliable criterion for truncation of the modes, frequently chosen on the basis of the large gap in the spectrum of singular values. However, the data on which the decomposition is performed are obtained by averaging over the whole attractor, and certain modes might be active for only a short period of time resulting in the small singular value, although this short - term activity might be related to the change in the system's dynamics. Moreover the influence of noise is also averaged over the whole attractor, disabling the detection of its true source. Namely, the noise in hydrodynamical flows might occur due to the nonlinear dynamical interaction of a small number of modes, or due to the nonlinear dynamics of many interacting modes. Furthermore, the sources of noise include various types of instabilities, such as the skew varicose instability (Golub, 1983), which cause the amplification of thermal or external stochastic fluctuations. Particularly important with this respect is the occurence of (localized) noise sustained structures which cannot be observed in the global basis approach(Deissler and To, 1992). Finally, in case of periodic boundary conditions and uniform spatiotemporal mixing, the global approach leads to the Fourier basis which is inadequate to capture the nonlinear behaviour, although the existence of coherent structures which make the ergodicity assumption invalid, might improve the validity of the global approach. One of the most interesting features of the global SVD (or biorthogonal) approach is that the eigenspaces reflect the space-time symmetry group of the data set, while applying SVD locally may destroy such symmetries (Aubry *et al.*, 1992, 1995, 1996).

The method proposed here applies the multiresolution approach locally in phase space such that the dynamics is projected into the tangent space of the attractor, and the basis obtained by SVD algorithm reflects the dynamics taking place at that particular region of phase space. Once the coherent (non-Gaussian) modes are extracted in the local region, their importance in the dynamics is determined based on the geometric criterion. The tangent space of the attractor, or the inertial manifold, is spanned by those directions whose singular values correspond to active modes, and all those directions which correspond to the passive modes belong to the subspace which is orthogonal to the active subspace. The method explicitely assumes the existence of the attractor, although it may be altered to apply to any finite dimensional manifold.

With respect to the Navier - Stokes (NS) equations, there are rigorous results proving that solutions to the two - dimensional NS equations are confined to a finite dimensional attractor, and some indications of the existence of a finite dimensional attractors for the three - dimensional case for low Reynolds number turbulent channel flow (Keefe *et al.*, 1987). Although the theory of inertial manifolds for the NS equations is not complete, there are important results promising new advances in this area (Temam, 1995). This method may be an efficient means of providing an operational proof of existence of local and/or global inertial manifolds for certain 2-D and 3-D turbulent flows.

2. Multiresolution SVD Algorithm

On the data set, which may be the velocity or vorticity data for example, an embedding space is constructed with reference to the spatial dependence, and the dynamics in phase space is followed in the set of local coverings of an attractor $\mathcal{A}$, such that the dynamics is projected locally into the tangent space at various points of the attractor. A given point, defined by the vector u_1 on the attractor, is chosen as the center of a hypersphere of radius ϵ, consisting of the points $v_i = u_i - u_1$, such that $\| v_i \| \leq \epsilon$, where $u_i = (u(x_1, t_i), ..., u(x_n, t_i))$ in R^n. On this set a SVD analysis is performed such that first the Gaussian components are extracted, followed by the separation of the active and passive modes which belong to the two closed, mutually orthogonal linear subspaces, $\mathcal{F}_a$ and $\mathcal{F}_s$ respectively. The coordinates of directions in the orthogonal space $\mathcal{F}_s$ are functions of the coordinates in the tangent space $\mathcal{F}_a$, and the dynamics is completely determined and controlled by the dynamics on $\mathcal{F}_a$. Each consecutive point (vector) u_i may be chosen as the center of the local covering, such that the regions of overlap of the local coverings may provide important information related to the local changes of dynamics. If an (approximate) inertial manifold exists, it would imply the (approximate) constancy of dimensions of the local inertial manifolds, constructed over various regions of the attractor (Broomhead *et al.*, 1991). The separation of the Gaussian component may be performed using information - theoretical criterion (Škorić *et al.*, 1996a), (Škorić *et al.*, 1996b), or a criterion based on the perturbed behaviour of the singular values. The latter method is outlined below, justifying the term "multiresolution SVD algorithm".

2.1. SEPARATION OF THE GAUSSIAN COMPONENT

Definition: For any m x n matrix $B = A + E$, the effective rank is r when

$$\beta_r > \epsilon_1 \geq \beta_{r+1}$$

where $1 \leq r \leq \min(m, n)$ and $\| E \|_2 = \epsilon_1$ represents the 2-norm of the matrix E. β_i and ϵ_i with $(i = 1, 2, ..., k), k \leq \min(m, n)$, represent the singular values of the matrix B and E respectively. We further assume that for the maximal singular value of the matrix E the following inequality holds

$$\epsilon_1 \leq (\sqrt{mn})\,\sigma$$

The proof follows from the important inequality related to any m x n matrix A (Horn and Johnson, 1986)

$$\max | a_{ij} | \leq \max \| a_{ij} \|_2 \leq \| A \|_2 \leq \| A \|_F$$
$$< \sqrt{n} \max \| a_{ij} \|_2 \leq \sqrt{mn} \max | a_{ij} | . \tag{2.1.1}$$

and the fact that if the matrix E represents the stochastic process with mean value equal to 0 and variance equal to σ^2, the following property is valid (Rao, 1965)

$$\max | \epsilon_{ij} | = C\sigma$$

C being a constant < 2.6. Here, the Frobenius and the 2-norm for any real - valued m x n matrix A with column vectors $a_1, a_2, ..., a_n$, and singular values $\lambda_1, \lambda_2, ..., \lambda_n$ are defined as

$$\| A \|_F = \left[\sum_{i=1}^{m} \sum_{j=1}^{n} | a_{ij} |^2 \right]^{1/2} = \left[\sum_{i=1}^{k} \lambda_i^2 \right]^{1/2} .$$

and

$$\| A \|_2 = \max (\| Ax \|_2 / \| x \|_2) = \lambda_1$$

Furthermore, we may assume that the perturbed matrix can be written as

$$B = A + E = A_1 + A_0 + E = B_1 + B_0,$$

where we have assumed that the perturbation is not contained only in the matrix E, but also in the matrix A_0. For example, if E is the perturbation due to the noise, A_0 may be due to the computational truncation error. It is natural to assume that $rank(A_1) = rank(B_1)$, although this condition is not necessary for further exposition. The range of the matrix A is denoted by $\mathcal{R}(A)$ and its nullity by $\mathcal{N}(A)$. The consequence of the decomposition of the matrix A

$$A = U\Sigma V^\dagger = U_1\Sigma_1 V_1^\dagger + U_0 \Sigma_0 V_0^\dagger$$

is

$$\mathcal{R}(A_1) = \mathcal{N}(A_1^\dagger)^\perp \qquad \text{and} \qquad \mathcal{N}(A_1) = \mathcal{R}(A_1^\dagger)^\perp$$

where $\perp$ denotes the orthogonal complement. $\mathcal{R}(A_1)$ and $\mathcal{R}(A_0)$ are invariant subspaces of Hermitean matrix $AA^\dagger$, and similarly $\mathcal{R}(A_1^\dagger)$ and $\mathcal{R}(A_0^\dagger)$ are invariant subspaces of $A^\dagger A$. The angle between $\mathcal{R}(A_1)$ and $\mathcal{R}(B_1)$ in the Euclidean space E^m (or between $\mathcal{R}(A_1^\dagger)$ and $\mathcal{R}(B_1^\dagger)$ in E^n provides information on the separation of these two subspaces. The idea is that this angle should be as large as possible so that no vector from $\mathcal{R}(A_1)$ (i.e. $\mathcal{R}(A_1^\dagger)$) is colinear with the vector from $\mathcal{R}(B_1)$ (i.e. $\mathcal{R}(B_1^\dagger)$). Denoting the orthogonal projection on the subspace M as P_M, the angle between the vector x and the subspace M may be defined as (Davis and Kahan, 1970)

$$\sin\vartheta(x, M) = \min_{y\in M} \| x - y \|_2, \quad \text{with } \| x \|_2 = 1.$$

From the projection theorem follows (Wedin, 1972)

$$\min_{y\in M} \| x - y \|_2 = \| (I - P_M)x \|_2 .$$

It is natural to define for two subspaces L i M the angle between them

$$\| \sin\vartheta(L, M) \| = \| (I - P_M)P_L \| .$$

so that the object is to make the quantities

$$\| \sin\vartheta(\mathcal{R}(B_1), \mathcal{R}(A_1)) \| \quad \text{and} \quad \| \sin\vartheta(\mathcal{R}(B_1^\dagger), \mathcal{R}(A_1^\dagger)) \|$$

as small as possible, and represented as a function of ϵ_1 since the singular values of the matrix A_1 usually are not known. On the basis of the $\sin\vartheta$ theorem (Davis and Kahan, 1970), the following inequality may be derived

$$\| \sin\vartheta(\mathcal{R}(B_1), \mathcal{R}(A_1)) \| \leq \frac{\| EV_1 \|}{\max(\epsilon_1)} = \frac{\| EV_1 \|}{(\sqrt{mn})\,\sigma},$$

and requiring that

$$\min \| \sin\vartheta(\mathcal{R}(B_1), \mathcal{R}(A_1)) \| = \min \frac{\| EV_1 \|}{(\sqrt{mn})\,\sigma},$$

the dimension of the vector V_1 satisfying this condition may be determined, as well as the rank of the matrix A_1.

2.2. SEPARATION OF ACTIVE AND PASSIVE MODES

Once the Gaussian component has been filtered out, the SVD analysis of the local deterministic dynamics may be represented in matrix form as

$$B = U\Sigma V^H = (U_a U_s)\begin{pmatrix} \Sigma_a & 0 \\ 0 & \Sigma_s \end{pmatrix}\begin{pmatrix} V_a^* \\ V_b^* \end{pmatrix} = \sum_{i=1}^{r} \sigma_i u_i v_i^H \qquad (2.2.1)$$

where u_i and v_i are the orthonormal characteristic vectors of the matrix BB^H(or $B^H B$), and $\{\sigma_i\}$ are the corresponding characteristic values. The index r represents the rank of the matrix B. Starting with the first n terms in the expansion (2.2.1), the matrix $\ddot{B}$ is formed and compared with the original matrix B. If the matrix $\ddot{B}$ represents the acute perturbation of B, in the sense that the angle between the column space of B (denoted by $\mathcal{R}(B)$) and the corresponding column space of $\hat{B}$ (denoted by $\mathcal{R}(\hat{B})$) is acute, then there is no vector in $\mathcal{R}(B)$ orthogonal to $\mathcal{R}(\hat{B})$ and vice versa. The procedure is iteratively continued until the angle is close to, or equal to $\pi/2$, so that the two subspaces are mutually orthogonal. Specifically, the angle between two subspaces F and G is defined as (Davis and Kahan, 1970),(Wedin, 1972)

$$\| \sin\theta(F,G) \| = \| (I - P_F)P_G \| = \| P_F^{\perp} P_G \| \tag{2.2.2}$$

where P_F and P_G are the projection operators onto the subspaces F and G respectively. Furthermore, $P_F = FF^{\dagger}$, where $F^{\dagger}$ denotes the Moore - Penrose pseudoinverse, represents the orthogonal projection of the matrix F on $\mathcal{R}(F)$ (Wedin, 1972). In general, when $\| P_F - P_G \|_2 < 1$, then $rank(F) = rank(G)$ and there is no vector in $\mathcal{R}(F)$ orthogonal to $\mathcal{R}(G)$ and vice versa (Davis and Kahan, 1970),(Wedin, 1972), so that the rank equivalence represents the necessary, but not sufficient, condition for matrix G to be the acute perturbation of F. The matrix B defined in (2.2.1) may be written as $B = A + E$ where matrix E represents the perturbation due to the algorithm, round - off and truncation errors. Moreover, B can be written as

$$B = \hat{B} + B_0 = U_1\Sigma_1 V_1^H + U_2\Sigma_2 V_2^H \tag{2.2.3}$$

where matrix $\hat{B}$ is, as before, obtained by retaining the first n terms in the expansion (2.1). The error in approximating matrix B is

$$\| \hat{B} - B \| = \sigma_{n+1} \leq \| B - A \| = \| E \|.$$

while

$$\| \hat{B} - A \| = \| \hat{B} - B + B - A \| \leq \| \hat{B} - B \| + \| B - A \| = \sigma_{n+1} + \| E \| \leq 2 \| E \|$$

Now,

$$\begin{aligned}
\|P_{\hat{B}} - P_A\| &= \| P_{\hat{B}} P_A^{\perp} \| = \| P_A^{\perp} P_{\hat{B}} \| \\
&= \| P_A^{\perp} \hat{B} \hat{B}^{\dagger} \| = \| P_A^{\perp} (A + E - B_0) \hat{B}^{\dagger} \| \\
&= \| P_A^{\perp} (E - B_0) \hat{B}^{\dagger} \| = \| P_A^{\perp} E P_{\hat{B}^H} \hat{B}^{\dagger} \| \\
&\leq \| P_A^{\perp} E P_{\hat{B}^H} \| \, \| \hat{B}^{\dagger} \|. \\
&\leq \| E \| \, \| \hat{B}^{\dagger} \| \\
&= \| E \| \, \sigma_n^{-1}
\end{aligned}$$

where σ_n is the smallest singular value of the n rank approximate matrix $\hat{B}$. Hence, the condition $\|P_{\hat{B}} - P_A\| < 1$ implies $\| E \| < \sigma_n$, and in that case the matrix $\hat{B}$ is the acute perturbation of A. In the opposite case, when $\sigma_n \sim \| E \|$ or $\sigma_n > \| E \|$, the subspaces $\mathcal{R}(\hat{B})$ and $\mathcal{R}(\mathrm{A})$ are orthogonal to each other and the rank approximate order n is the dimension of the $\mathcal{R}(\hat{B})$. Now, the elements e_{ij} of the matrix E satisfy $-\delta \leq e_{ij} \leq \delta$, where δ may be the round-off or truncation error. Since

$$max \mid e_{ij} \mid = \delta,$$

and denoting the j-th column of E as e_j, we get

$$max \parallel e_j \parallel_2 = \max \left[\mid e_{ij} \mid^2 + \ldots + \mid e_{mj} \mid^2 \right]^{1/2} \leq \sqrt{m}\ \delta.$$

Combining above expressions we obtain

$$\delta \leq \parallel E \parallel_2 = \epsilon_1 \leq \sqrt{mn}\ \delta.$$

Hence, taking for $\parallel E \parallel_2$ the maximum value, $\sqrt{mn}\ \delta$, the expression for $\|P_{\hat{B}} - P_A\|$ becomes

$$\|P_{\hat{B}} - P_A\| \leq \frac{\sqrt{mn}\ \delta}{\sigma_n}.$$

Therefore, we get a remarkable expression containing the known smallest eigenvalue of the n rank approximate matrix $\hat{B}$, the dimensions of the local data matrix m and n and the error δ that can be estimated from the precision mode used. *Note that the condition for the separation of subspaces does not require the existence of a large gap between two successive singular values.* Hence, if an (approximate) inertial manifold exists, the last expression suggests that iterative formation of two subspaces from the eigenvectors of the decomposition leads to the construction of two mutually orthogonal (or nearly orthogonal) subspaces, spanned by active (subspace $\mathcal{F}_a$) and passive modes (subspace $\mathcal{F}_s$).

The number of active modes in the local linear region of the attractor determines the local attractor dimension and once the local dynamics of active and slaved modes is known, it is possible to extract further information about the dynamics by considering the transition matrices which relate two adjacent local regions. These matrices give information on the local curvature of the attractor and provide the information related to both the geometry of the invariant set as well as the physical nature of possible changes in the dynamics. These transition matrices define the local Lyapunov exponents which in case of their negative value, indicate coherent structures. Further use of this method may be seen in the area of Galerkin approximation, since it defines very few modes based on their dominance in the local dynamics. Moreover the method provides an efficient way of finding an approximate manifolds since such manifolds exist for far more dynamical systems than exact inertial manifolds. The approximate inertial manifold gives an approximate slaving rule and can be applied in the nonlinear Galerkin method which may offer better accuracy than standard methods (Robinson, 1994),(Robinson, 1995).

3. Conclusion

The multiresolution method based on the local SVD decomposition of the data set in phase space enables an efficient means for determining the number of active and passive modes locally, as well as globally, and consequently the check for the existence of a global inertial manifold (Škorić *et al.*, 1996a). Since the restriction to inertial manifolds which are graphs of a function over a finite dimensional subspace, may be the reason that it is in general hard to find them using analytical techniques, the method might be of great practical importance in studying complex systems. For the case of 2-D NS equations, of particular interest would be to check the implication of the condition for the existence of an inertial manifold. Namely, the Laplacian in a 2-D domain does not satisfy the gap condition required by the theorem which gives sufficient conditions for the existence of an inertial manifold. The natural question that comes up is related to the extent this condition is necessary or exact. There are examples where the inertial manifolds exist outside the range of this condition, such as reaction - diffusion equations in R^3 (Robinson, 1995). Since the operational method presented here does not require the existence of the spectral gap for existence of the *local* inertial manifold, it is interesting to explore the relationship between the rate of convergence toward the manifold locally and globally in case of the systems that do not satisfy the spectral gap condition, such as the 2-D NS equations.

References

Aubry N., Guyonnet R. and Lima R., Spatiotemporal Analysis of Complex Signals: Theory and Applications, *J. of Stat. Phys.*, **64** (1991) 683.

Aubry N., Guyonnet R. and Lima R., Spatio-Temporal Symmetries and Bifurcations via Bi-Orthogonal Decompositions, *J. Nonlinear Sci.*, **2** (1992) 183; also Aubry N. and Lima R., *J. of Stat. Phys.*, **81** (1995) 793; Carbone F. and Aubry N., *Phys. Fluids*, **8** (1996) 1961.

Broomhead D. S., Indik R., Newell A. C. and Rand D., Local Adaptive Galerkin Bases for Large - dimensional Dynamical Systems, *Nonlinearity* **4** (1991) 159.

Davis C.H. and Kahan W. M., The Rotation of Eigenvectors by a Perturbation III, *SIAM J. Num. Anal.*, **7** (1970) 1.

Deissler R. J. and Wai-Ming To, Noise Sustained Structures in the Navier-Stokes Equations: Taylor-Coutte Flow with Through Flow, *Phys. Rev. Lett.* 1992.

Golub J., What Causess Noise in a Convective Flow?, Physica **118A** (1983) 329.

Horn R. and Johnson C., *Matrix Analysis*, Cambridge University Press, Cambridge, 1986.

Keefe L, Moin P. and Kim J., The Dimension of an Attractor in Turbulent Poiseuille Flow, *Bull. A. P. S.*, **32** (1987) 2026.

Lumley J. L., in *Transition and Turbulence*, ed. R. E. Meyer (New York: Academic, 1970), p. 215.

Rao C. R., *Linear Statistical Inference and its Appplications*, J. Wiley, New York, 1965.

Robinson J. C., A concise Proof of the Geometric Construction of Inertial Manifolds, *Phys. Lett A*, **184** (1994) 190.

Robinson J. C., Finite Dimensional Behaviour in Dissipative PDE's, *Chaos* **5** (1995) 330, and references therein.

Škorić M., Jovanović M. and Rajković M., Roads to Turbulence in the Stimulated Raman Backscattering, *Phys. Rev. E*, **53** (1996 a) 4056.

Škorić M., Jovanović M. and Rajković M., Spatiotemporal Intermittency and Chaos in the Stimulated Raman Backscattering, *Europhys. Lett.*, **34** (1996 b) 19.

Temam R., *Infinite Dimensional Dynamical Systems in Mechanics and Physics*, Springer, New York, 1988.

Temam R., Navier Stokes Equations and Nonlinear Functional Analysis, 2nd ed., SIAM, Philadelphia, 1995.

Wedin P. A., Perturbation Bounds in Connection with Singular Value Decomposition, BIT (Sweden) **12** (1972) 99.

ENERGY TRANSFER BETWEEN COHERENT STRUCTURES IN THE WAKE OF A HEMISPHERE

MICHAEL MANHART
Lehrstuhl für Fluidmechanik
Technische Universität München
D-85747 Garching, Germany

Abstract. The energy budget of coherent structures has been analyzed theoretically and numerically within the framework of Proper Orthogonal Decomposition (POD). By means of Galerkin projection of the coherent structures (POD modes) onto the Navier-Stokes equations the individual processes are identified which contribute to the energy budget of coherent structures. The investigation of the linear and nonlinear energy transfers between the POD modes in the wake of a hemisphere in a turbulent boundary layer shows that the first three modes, representing the 'von Kármán'-type vortices, are most important for the dynamics in the wake of the hemisphere. The nonlinear energy transfers are characterized by strongly nonlocal interactions which partly show backscatter of energy from small to large scales.

1. Introduction

The nonlinear energy transfer between different length scales can be considered as the central process of turbulence. Unfortunately, there is no agreement in the literature, which elementary processes contribute to this energy transfer (Domaradzki, J.A. and Rogallo, R.S., 1990). In general, a nonlinear interaction takes place between a triad of eddies with different length scales. The individual energy transfer between two length scales is then composed of a large number of such triad interactions. The controversy is about the relative importance of *local* and *nonlocal* interactions. Energy transfer is called local, if the length scales differ by a factor less than two, otherwise, the energy transfer is called nonlocal (Domaradzki, J.A. and Rogallo, R.S., 1990). The energy cascade model (Tennekes, H. and Lumley, J.L., 1972) proceeds from the assumption that turbulent energy is produced by the large scales and transferred by local interactions to the small scales, where it is dissipated.

Domaradzki and Rogallo (1990) found out, that in isotropic turbulence energy exchange between two different wave numbers occurs predominantly local. However, these transfers seem to be mainly governed by nonlocal triad interactions, which play a crucial role for the dissipative part of the spectrum (Domaradzki, 1992 and Ohkitani and Kida 1992). These findings are in contrast to Zhou's (1993) analysis, which suggests that nonlocal interactions have only a negligible importance in isotropic turbulence, as predicted by Tennekes and Lumley (1972).

In isotropic turbulence or in homogeneous directions energy transfer processes can be analyzed in Fourier space. If different length scales in inhomogeneous directions have to be considered, new representations of the flow field have to be used. One possible alternative are orthogonal wavelets which have been used by Meneveau (1991) to analyze energy transfer processes in a homogeneous shear layer. A second alternative is the use of the Karhunen-Loève-expansion (KL-expansion, Proper Orthogonal Decomposition, POD), which is the generalization of the Fourier expansion for inhomogeneous directions. The globally defined KL-modes give a complete orthogonal basis which has the advantage of optimal convergence in an energy-based sense. The KL-expansion allows for a detailed study of mode-interactions, which has been shown by Rempfer and Fasel (1994) in the case of a transitional boundary layer.

In the present paper the energy balance equations of KL-modes are derived and discussed. The nonlinear transfer terms are then analyzed in the wake of a hemisphere immersed in a turbulent boundary layer, a fully inhomogeneous flow case. The data set has been provided by a Large-Eddy-Simulation (LES) of the flow field (Manhart, M. and Wengle, H., 1994) and a subsequent POD (Manhart, M., 1997). To the author's knowledge this is the first investigation of energy transfer processes in a fully turbulent and fully inhomogeneous flow field within the framework of POD.

2. Theory

2.1. THE DECOMPOSITION OF THE FLOW FIELD

In the classical formulation introduced by Lumley (1970) the POD is derived by seeking a basic flow field (coherent structure, dominant mode) which represents a best fit to the fluctuations in a mean quadratic sense. Following this 'ansatz' the dominant mode can be calculated by solving an eigenvalue problem of the spatial two-point correlation tensor ('direct method'). An interesting review of the method can be found in Berkooz et *al.* (1993). In our application of POD we follow the treatment proposed by Sirovich (1987) (method of 'snapshots') and Aubry et *al.* (1991) ('bi-orthogonal' decomposition), see also Manhart and Wengle (1993), which leads to a space-time symmetric version of the POD. The turbulent space-time velocity signal $u_i(\vec{x},t)(i = 1,2,3)$ is characterized by N_M spatio-temporal modes $a^n(t) \cdot \phi_i^n(\vec{x})$ which form a complete orthogonal basis for the following expansion:

$$u_i(\vec{x},t) = \sum_{n=1}^{N_M} a^n(t)\phi_i^n(\vec{x}) \quad . \tag{1}$$

The temporal modes $a^n(t)$ are the eigenvectors of the temporal correlation tensor $C(t,t')$:

$$\int_T C(t,t')a^n(t')dt' = \lambda^n a^n(t) \quad ; \quad C(t,t') = \frac{1}{T}\int_V u_i(\vec{x},t)u_i(\vec{x},t')d\vec{x} \quad . \tag{2}$$

The spatial modes $\phi_i^n(\vec{x})$ are calculated by projecting the velocity fields onto the temporal modes:

$$\phi_i^n(\vec{x}) = \frac{1}{T}(\lambda^n)^{-1}\int_T a^n(t)u_i(\vec{x},t)dt \quad . \tag{3}$$

The spatio-temporal modes form a complete set of basis functions which is optimal in the sense that, with a given number of modes n, the expansion (1) converges optimally fast. The time-averaged energy contribution to the total turbulent energy of a spatio-temporal mode n to the total turbulent energy is given by the corresponding eigenvalue λ^n. The chosen form leads to a space-time symmetric decomposition of the flow field into *deterministic* spatial basis functions (contributing to the spatial structure of the flow) and *deterministic* temporal basis functions (contributing to the time signals). Each pair $a^n(t)\phi^n(\vec{x})$ represents a basic process embedded in the 'random' turbulent flow field. For example, in a statistically stationary turbulent flow field, the first mode normally represents the *time-averaged* flow field because of its dominating energy contribution.

2.2. ENERGY BALANCE EQUATION IN KARHUNEN-LOÈVE SPACE

By applying a Galerkin-projection of the POD-modes onto the Navier-Stokes equations for incompressible, Newtonian fluids we get the momentum equation for an individual mode k:

$$\frac{d}{dt}a^k = \sum_{l,m} n_{klm}a^l a^m + \hat{\pi}_k + \sum_l d_{kl}a^l \tag{4}$$

with

$$n_{klm} = -\int_\Omega \phi_i^k \frac{\partial \phi_i^l \phi_j^m}{\partial x_j} d\vec{x} \quad (5)$$

$$\hat{\pi}_k = -\frac{1}{\rho}\int_\Omega \phi_i^k \frac{\partial p}{\partial x_i} d\vec{x} \quad (6)$$

$$d_{kl} = \nu \int_\Omega \phi_i^k \frac{\partial^2 \phi_i^l}{\partial x_j^2} d\vec{x}. \quad (7)$$

Where n_{klm} and d_{kl} are the coefficients describing nonlinear and diffusive interactions, respectively, and $\hat{\pi}_k$ is the interaction with the pressure field. It can be shown that the nonlinear coefficients exhibit the important antisymmetry $n_{klm} = -n_{lkm}$, if the volume of application is periodic or sufficiently large that the modes vanish on the boundary. As a consequence, the partners in the triad (klm) can be interpreted as receiver (k), provider (l) and transporter (m) of momentum.

The equation for the energy budget can be derived by multiplying (4) by a^k and averaging over time:

$$< \frac{de^k}{dt} >= 0 = \sum_{l,m} n_{klm} < a^k a^l a^m > + < a^k \hat{\pi}_k > + \sum_l d_{kl} < a^k a^l > \quad (8)$$

If we consider a statistically stationary flow field, the first mode in general will be constant in time $a^1(t) = \sqrt{\lambda^1} = const.$. Then we can separate the effects of the time averaged flow field from the nonlinear interactions between the fluctuating modes. In this case the energy balance equation for a KL mode can be written the following way:

$$P_k + C_k + T_k + \Pi_k + D_k + D_k^t = 0 \quad , \quad (9)$$

involving the production P_k of turbulent energy

$$P_k = n_{k1k}\sqrt{\lambda^1} \cdot \lambda^k \quad , \quad (10)$$

the convective transport C_k of turbulent kinetic energy from or to mode k by the time averaged flow field

$$C_k = n_{kk1}\sqrt{\lambda^1}\lambda^k \quad (11)$$

the nonlinear transfer T_k

$$T_k = \sum_{l=2}^{N_M} T_{klm} = \sum_{l=2}^{N_M}\sum_{m=2}^{N_M} n_{klm} < a^k a^l a^m > \quad , \quad (12)$$

the interaction with the pressure field

$$\Pi_k = -\frac{1}{\rho}\int_\Omega \phi_i^k \frac{\partial < a^k p >}{\partial x_i} d\vec{x} \tag{13}$$

the dissipation taken place in mode k

$$D_k = d_{kk}\lambda^k \quad . \tag{14}$$

3. Application

In the following we will describe the evaluation of the production and the nonlinear terms in the wake flow behind a hemisphere in a turbulent boundary layer. The flow field has been simulated numerically using large-eddy-simulation and analyzed using information of statistical moments, instantaneous flow fields and Proper Orthogonal Decomposition by Manhart and Wengle (1994) and Manhart (1997). After having described the configuration of the flow, we will shortly present results from an investigation of the vortex shedding process by conventional methods and POD. For the first 200 POD-modes the production and nonlinear transfer-terms are presented and discussed.

3.1. LES OF TURBULENT FLOW OVER A HEMISPHERE

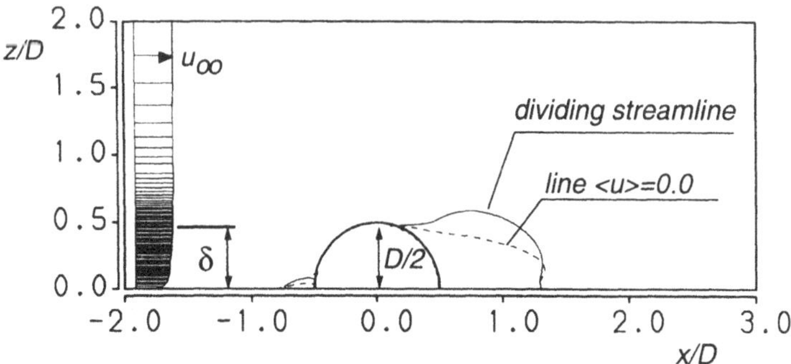

Figure 1. Schematic view of the geometry of boundary layer flow approaching a hemisphere

Our LES approach uses a staggered and non-uniform Cartesian grid to discretize the incompressible Navier Stokes equations by a second-order finite volume technique. The pressure-velocity coupling is solved iteratively using a multigrid solver for the Poisson equation. The subgrid scale stresses, arising from the nonlinear convection terms, have been evaluated by the

Smagorinsky model. The curved surface of the flow obstacle is approximated by simply blocking out the 'body-filled' grid cells within the Cartesian grid. For further details see Manhart and Wengle (1994) and Manhart (1997).

The geometry of the flow obstacle is evident from figure 1. The origin of the coordinate system is located in the center of the hemisphere. Measured in units of the reference length D of the flow problem (D is the diameter of the hemisphere) the dimensions of the computational domain are (X,Y,Z)=(15.2,4.8,4.0) and, for the results presented here we used (NX,NY,NZ)=(224,128,65) grid points. The configuration has been chosen according to an experiment of Savory and Toy (1986). The time-averaged flow field as well as the second order moments in the near wake region of the hemisphere show good agreement between LES and experiment. The effect of the stepwise approximation of the curved surface is equivalent to roughening the surface of the hemisphere, as has been done by Savory and Toy to reach a supercritical flow behavior.

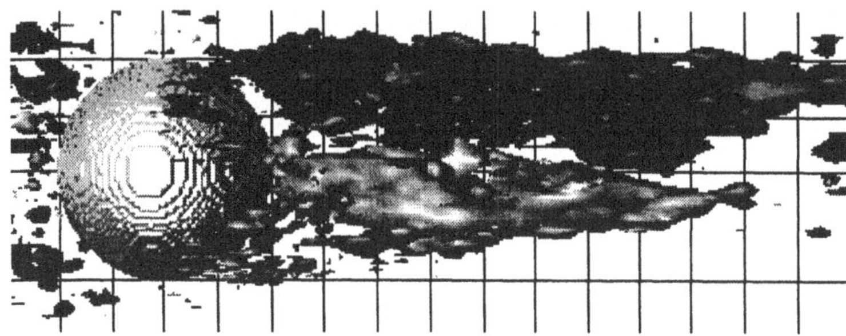

Figure 2. Top view of isosurfaces of instantaneous streamwise fluctuations.

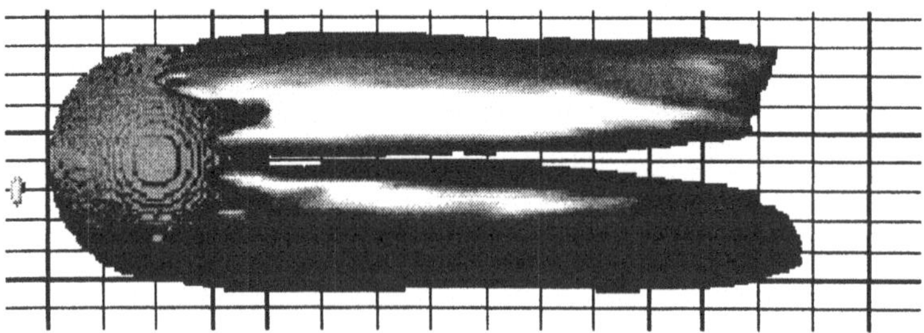

Figure 3. Top view of isosurfaces of streamwise component of first fluctuating mode ϕ_1^2.

An examination of the instantaneous flow fields and of the velocity spectra shows two different vortex shedding processes taking place in the wake of the hemisphere: symmetric vortices and antisymmetric vortices.

As an example for the antisymmetric vortices we show instantaneous streamwise velocity fluctuations in figure 2. These fluctuations are generated by the shedding of large-scale von Kármán vortices from the side faces of the hemisphere.

3.2. POD OF THE WAKE FLOW

We applied a Proper Orthogonal Decomposition in a sub-volume covering the region of enhanced turbulent kinetic energy with respect to the oncoming turbulent boundary layer. By using about 3000 samples in time and 84000 points in space we have covered all dynamically relevant processes in the wake. In our application the first mode, having a constant temporal coefficient $a^1(t)$, represents the time-averaged flow field. The next three modes, all having a similar spatial structure (for an example, see figure 3), can be attributed to the asymmetric von Kármán vortex shedding. These vortex shedding modes contribute 24% to the total turbulent energy and are responsible for intense fluctuations in the lateral direction, which take place in the region of reattachment. An analysis of the temporal evolution of these modes shows a complicated behavior including quasi-stationary phases and quasi-periodic phases at a Strouhal number of about $0.15U_\infty/D$, characterizing the overall appearance of the vortex shedding in the wake. Nevertheless it seems that the balance between quasi-stationary and quasi-periodic behavior of these large-scale structures is strongly influenced by very small-scale events near the separation zone on the surface of the hemisphere. A detailed description of the POD results can be found in (Manhart, M., 1997).

3.3. ENERGY TRANSFER BETWEEN POD-MODES

In the above section we have seen the energetically significance of the von Kármán vortices being reflected in the first three fluctuating modes of a Karhunen-Loève expansion. In the following we present results from the analysis of nonlinear and linear interactions between the first 200 POD-modes. From an inspection of time-spectra we conclude that the region from about mode 20 to 600 can be attributed to the inertial range of turbulence.

Production of turbulent kinetic energy. The production of turbulent kinetic energy by a mode k can be expressed by the term $P_k = n_{k1k}\sqrt{\lambda^1} \cdot \lambda^k$ (equation 10). The coefficients n_{k1k} determine, whether mode k is able to extract energy from the time-averaged flow field (the first mode). For the first 200 modes they are of the same order of magnitude (see figure 4), only a slight tendency to decrease can be observed in the higher modes. The importance of the first three fluctuating modes for the production of turbulent kinetic energy can be seen in figure 5. More than 30% of the turbulent energy is produced by the first three fluctuating modes alone. In the inertial range ($n > 20$) the production decreases proportional to the mode number $P_k \propto k^{-1}$. It seems that in our case production cannot be neglected in the inertial range, as it is sometimes postulated in the literature.

Nonlinear interactions. The nonlinear interaction coefficients n_{klm} show no special structure which means, that principally each mode can interact with any other mode. The actual interactions will then be determined by the triple correlations $< a^k a^l a^m >$. In figure 6 isosurfaces $T_{klm} = 10^{-6}$ are shown of the energy transfer from mode l to mode k done by mode m, for $2 < k, l, m < 128$. In this figure mode m appears as a transporter of energy from mode l to mode k. On the left side ($k > l$) the energy transfer is forward from low modes (large scale, energetic structures) to higher modes (smaller scales, less energetic structures). The surfaces on the right side mark mode triads of reverse energy transfer (mode m transports energy from a less energetic structure to a more energetic structure). This backward energy transfer occurs relatively frequently. We see that high nonlinear energy transfers occur strongly nonlocal over large mode distances. The low modes play a central role as well as energy provider and transporter. If we want to know if there is a net energy flux from mode l to mode k we have to sum over all modes m in the triad T_{klm}. This nonlinear energy flux $Q_{kl} = \sum_m T_{klm}$ is shown in figure 7. In this figure, only positive values are coded grey or black, that means, grey or black pixels lying over the diagonal $l = k$ indicate backward energy transfer. A local energy flux would result in a concentration of the grey-scale pixels in the neighborhood of the diagonal ($k \approx l$). In our case, however, even mode 200 exchanges more energy with the lowest modes than with its neighboring modes. That means a strongly

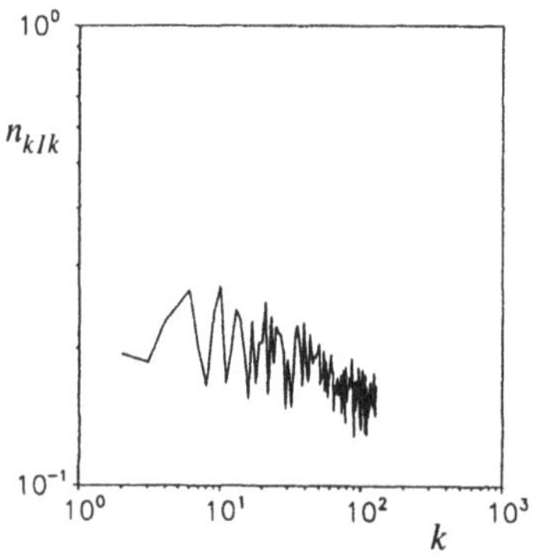

Figure 4. Coefficient n_{k1k} for production of turbulent kinetic energy.

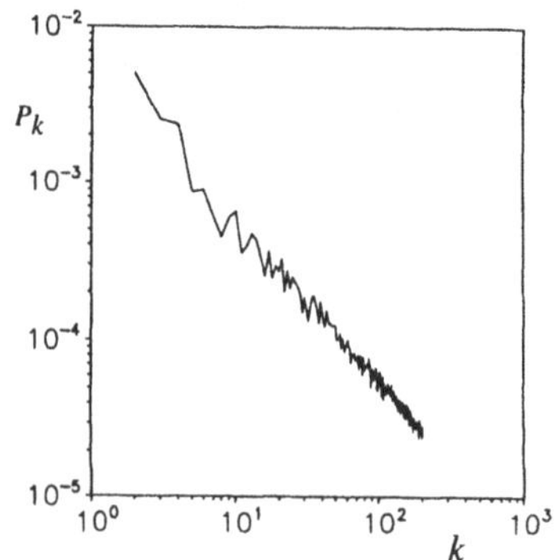

Figure 5. Production P_k of turbulent kinetic energy by mode k.

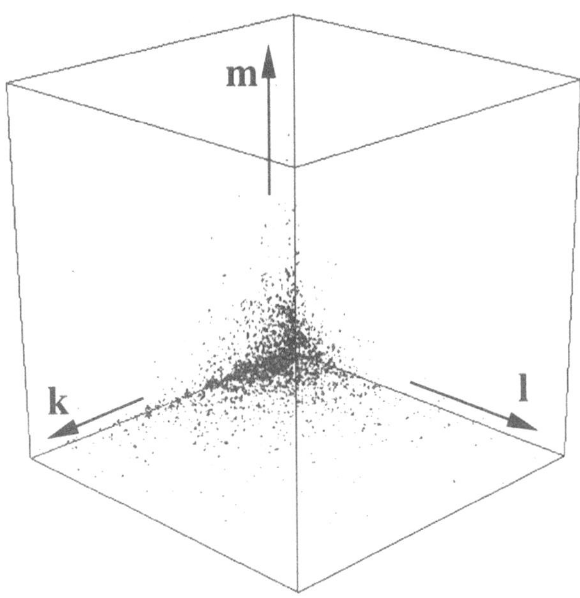

Figure 6. Nonlinear energy transfer, three-dimensional view of isosurfaces $T_{klm} = 1.0 \cdot 10^{-6}$, in the region $2 < k, l, m < 128$

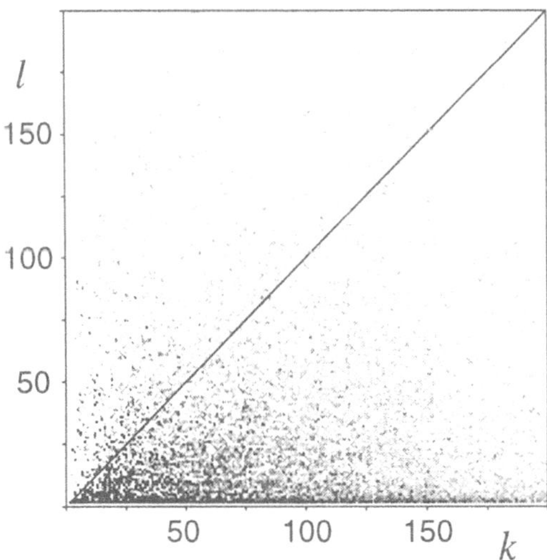

Figure 7. Nonlinear energy flux Q_{kl} from mode l to mode k ($\sum_{m=2}^{200} T_{klm}$), white: $n_{kl1} < 10^{-6}$; grey: $10^{-6} \leq n_{kl1} < 5 \cdot 10^{-5}$; black: $n_{kl1} > 5 \cdot 10^{-5}$

nonlocal energy flux from the lowest modes (here the von Kármán vortices) directly to the length scales in the inertial range.

4. Conclusions

We have presented the equations for momentum and the time-averaged energy balance in Karhunen-Loève space. We have calculated the production and nonlinear exchange terms of Karhunen-Loève modes in the near wake flow of a hemisphere in a turbulent boundary layer. The results show the central role of the von Kármán vortex shedding modes for production of turbulent energy and for the nonlinear energy transfer into the inertial range. This energy transfer is done by strongly nonlocal interactions directly between the lowest modes (the von Kármán vortices) and the highest modes considered which lie within the inertial range.

References

Aubry, N., Guyonnet, R., and Lima, R. (1991). Spatiotemporal analysis of complex signals: theory and applications. *J. Stat. Phys.*, **64**(3/4), 683–739.

Berkooz, G., Holmes, P., and Lumley, J. (1993). The proper orthogonal decomposition in the analysis of turbulent flows. *Annu. Rev. Fluid Mech.*, **25**, 539–575.

Domaradzki, J. (1992). Nonlocal triad interactions and the dissipation range of isotropic turbulence. *Phys. Fluids A*, **4**(9), 2037–2045.

Domaradzki, J. and Rogallo, R. (1990). Local energy transfer and nonlocal interactions in homogeneous, isotropic turbulence. *Phys. Fluids A*, **2**(3), 413–426.

Lumley, J. (1970). *Stochastic Tools in Turbulence*. Academic Press.

Manhart, M. (1997). Vortex shedding from a hemisphere in a turbulent boundary layer. *Theoretical and computational fluid dynamics*, **submitted**.

Manhart, M. and Wengle, H. (1993). A spatiotemporal decomposition of a fully inhomogeneous turbulent flow field. *Theoretical and computational fluid dynamics*, **5**, 223–242.

Manhart, M. and Wengle, H. (1994). Large-eddy simulation of turbulent boundary layer flow over a hemisphere. In Voke, P., Kleiser, L., and Chollet, J.-P., editors, *Direct and Large-Eddy Simulation I*, pp. 299–310. ERCOFTAC, Kluwer Academic Publishers.

Meneveau, C. (1991). Analysis of turbulence in the orthonormal wavelet representation. *J. Fluid Mech.*, **232**, 469–520.

Ohkitani, K. and Kida, S. (1992). Triad interactions in a forced turbulence. *Phys. Fluids A*, **4**(4), 794–802.

Rempfer, D. and Fasel, H. (1994). Dynamics of three-dimensional coherent structures in a flat-plate boundary layer. *J. Fluid Mech.*, **275**, 257–283.

Savory, E. and Toy, N. (1986). Hemispheres and hemisphere-cylinders in turbulent boundary layers. *J. Wind Eng.*, **23**, 345–364.

Sirovich, L. (1987). Turbulence and the dynamics of coherent structures. part I,II,III. *Q. Appl. Math.*, **45**(3), 561–590.

Tennekes, H. and Lumley, J. (1972). *A first course in turbulence*. MIT Press, Cambridge, Massachusetts.

Zhou, Y. (1993). Degrees of locality of energy transfer in the inertial range. *Phys. Fluids A*, **5**(5), 1092–1094.

LIST OF PARTICIPANTS

Mohamed Abdel-Rohman
Civil Engineering Dept.
PO Box 5959, Kuwait University
Safiat 13050, Kuwait
Tel: (965) 4817 390
Fax: (965) 4817 524
marohman@kuc01.kuniv.edu.kw

Helge I. Andersson
Division of Applied Mechanics
Norwegian Univ. of Sci. and Tech.
N-7034 Trondheim, Norway

Erik Andresen
Contica
Agern Allé 3
DK-2970 Hørsholm, Denmark
Tlf: (45) 4576 7051
Fax: (45) 4576 5708
admin@contica.com

J.R. Angilella
DAMTP, Silver Street
University of Cambridge
Cambridge, CB3 9EW, UK
Tel: (44) 1223 33 7859
Fax: (44) 1223 33 7918
j.r.angilella@damtp.cam.ac.uk

Sandrine Aubrun
Inst. de Méc. des Fluides
Av. du Prof. Camille Soula
F-31400 Toulouse, France
Tel: (33) 5 6128 5836
aubrun@imft.fr

Marianna Braza
Inst. Méc. des Fluides Toulouse
Av. du Prof. Camille Soula
F-31400 Toulouse, France
Tel: (33) 5 6128 5839
Fax: (33) 5 6128 5899
braza@imft.fr

Morten Brøns
Dept. of Mathemathics, DTU
DK-2800 Lyngby, Denmark
Tel: (45) 4525 3067
Fax: (45) 4588 1399
m.brons@mat.dtu.dk

C. Cambon
L.M.F.A. UMR 5509 CNRS
Ecole Centrale de Lyon, BP 163
F-69131 Ecully Cedex, France
cambon@mecaflu.ec-Lyon.fr

Erik A. Christensen
Courant Institute, NYU
251 Mercer Street
New York, NY 10012, USA
Tel: (1) 212 998 3298
Fax: (1) 212 995 4121
eac6@is4.nyu.edu

Joseph H. Citriniti
Chalmers University of Technology
Thermo- and Fluid Dynamics
S-41296 Göteborg, Sweden
Tel: (46) 3177 2503
Fax: (46) 3118 0976
jhc@tfd.chalmers.se

Stephanie Correard
SHOM/CMO
13 rue du Chantellier, BP 426
F-29275 Brest Cedex, France
Tel: (33) 2 9822 1304
Fax: (33) 2 9822 1864
correard@shom.fr

Uwe Dallmann
DLR Inst. for Fluid Mechanics
Bunsenstr. 10
D-37073 Göttingen, Germany
Tel: (49) 551 709 2442
Fax: (49) 551 709 2404
uwe.dallmann@dlr.de

J. Delville
CEAT, Lab. d'Etudes Aérodynamices
UMR CNRS 6609, Univ. de Poitiers
43, Route de l'Aerodrome
F-86036 Poitiers Cedex, France
delville@univ-poitiers.fr

Stephane Douady
CNRS, 24 rue Lhomond
F-75231 Paris, Cedex 05, France
Tel: (33) 1 4432 3447/3472
Fax: (33) 1 4432 3433
douady@physique.ens.fr

Pierre Dupont
Univ. d'Aix-Marseille I & II
IRPHE - 12 Ave. General Leclerc
F-13003 Marseille, France
Tel: (33) 4 9150 5423
Fax: (33) 4 9108 1627
dupont@marius.univ-mrs.fr

Errol Eaton
Dept. of Mech. Engr.
Clarkson University
Postdam, NY 13676, USA

Pier G. Esposito
INSEAN
Via di Vallerano 139
I-00128 Roma, Italy
Tel: (39) 6 50299313
giorgio@rios2.insean.it

Dan Ewing
Dept. of Mechanical Engineering
Queen's University
Kingston, Ontario K7L 3N6, Canada
ewing@me.queensu.ca

Dariush Faghani
Inst. Méc. des Fluides Toulouse
Av. du Prof. Camille Soula
F-31400 Toulouse, France
Tel: (33) 5 6128 5828
Fax: (33) 5 6128 5899
faghani@imft.fr

Adam M. Fincham
LEGI Grenoble France
UJF-CNRS-INPG, BP 53
F-38041 Grenoble Cedex 9, France
Tel:(33) 4 7686 6185
Fax:(33) 4 7687 9793
fincham@img.fr

William K. George
Center for Thermal/Fluid Eng.
SUNY at Buffalo
339 Jarvis Hall
Buffalo, NY 14260, USA
trlbill@eng.buffalo.edu

Mark N. Glauser
Dept. Mech. & Aeronautical Engr.
Clarkson University
Postdam, NY 13699, USA
glauser@sun.soe.clarkson.edu

Tony Grass
University College London
Dept. of Civil & Environm. Engr.
London WC1E 6BT, UK
Tel: (44) 171 380 7831
Fax: (44) 171 380 0986

Johan N. Hartnack
Dept. of Mathematics
Building 303, DTU
DK-2800 Lyngby, Denmark
Tel: (45) 4525 3056
Fax: (45) 4588 1399
hartnack@mat.dtu.dk

Vincent Herbert
Lab. de Modelisation en Mécanique
4 place Jussieu
F-75252 Paris Cedex 05, France
Tel: (33) 1 4427 7141
Fax: (33) 1 4427 5259
herbert@lmm.jussieu.fr

Emil Hopfinger
LEGI/IMG., BP 53
F-38041 Grenoble Cedex 9, France
Emil.hopfinger@hmg.inpg.fr

Iwao Hosokawa
Univ. of Electro-Communications
Chofu, Tokyo 182, Japan
Tel: (81) 42 482 0046
Fax: (81) 42 482 0046
ihsk@hosokawa.mce.uec.ac.jp

Fazle Hussain
Dept. of Mechanical Eng.
University of Houston
Houston, TX 770204-4792, USA
FHussain@UH.EDU

Osamu Inoue
Institute of Fluid Science
Tohoku University, 2-1-1, Katahira
Aoba-ku, Sendai 980-77, Japan
Tel: (81) 22 217 5256
Fax: (81) 22 217 5256
inoue@ifs.tohoku.ac.jp

Javier Jiménez
ING Aeronautics, Univ. Politecnica
Pl. Cardenal Cisneros 3
E-28040 Madrid, Spain
jimenez@ctr-iris2.stanford.edu

Bo H. Jørgensen
Dept. of Energy Engineering
Building 404, DTU
DK-2800 Lyngby, Denmark
Tel: (45) 4525 4327
Fax: (45) 4588 2421
bhj@et.dtu.dk

Achilleas Kalopedis
University College London
Civil & environmental Eng.
London WC1E 6BT, UK
Tel: (44) 171 387 7050 # 2695
Fax: (44) 171 380 0986
ucesf2k@ucl.ac.uk

R.M. Kiehn
Physics Department
University of Houston
Houston, TX 77004, USA
jjkrmk@swbell.net

John Kim
48-121 Engt. IV, UCLA
405 Hilgard Ave.
Los Angeles, CA 90024, USA
jki@seas.ucla.edu

Bård Krane
University of Oslo
PO Box 1048 Blindern
N-0316 Oslo, Norway
Tel: (47) 2285 5666
Fax: (47) 2285 5671
bard.krane@fys.uio.no

Pavel Kuibin
Institute of Thermophysics
Lavrentiev Ave 1
Novosibirsk 630090, Russia
Tel: (7) 383 2 357128
Fax: (7) 383 2 357880
aleks@itp.nsc.ru

Michèle Larchevêque
Lab. de Modélisation en Méc.
Université Paris 6 - case 162
F-75252 Paris cedex 05, France
Tel: (33) 1 4427 5272
Fax: (33) 1 4427 5259
larchevq@lmm.jussieu.fr

P. S. Larsen
Dept. of Energy Engineering
Building 404, DTU
DK-2800 Lyngby, Denmark
Tlf.: (45) 4525 4332
Fax: (45) 4588 2421
psl@et.dtu.dk

Hideharu Makita
Dept. of Mechanical Eng.
Toyohashi Univ. of Tech., 1-1
Tempaku-cho Toyohashi,441 Japan
Tel: (81) 532 44 6680
Fax: (81) 532 44 6661
makita@mech.tut.ac.jp

Boris Malomed
Dept. of Interdisc. Studies
Faculty of Engineering
Tel Aviv 69978, Israel
Tel: (972) 3 640 6413
Fax: (972) 3 641 0189
malomed@tau.ac.il

Michael Manhart
Dept. of Fluid Mechanics
Technical University of Munich
D-85747 Garching, Germany
Tel: (49) 89 289 16142
Fax: (49) 89 289 16151
michael@flm.mm.tu-muenchen.de

Dan Mcguinness
Arizona State University
Mech. & Aero Engineering
Temple, AZ 85287-6106, USA
Tel: (1) 602 965 3671
Fax: (1) 602 965 1384
dan@enws606.eas.asu.edu

Ken Melville
Scripps Institution of Ocean
University of California
San Diego, CA 92093-0213, USA
Tel: (1) 619 534 0478
Fax: (1) 619 534 7132
kmelville@ucsd.edu

Robert Mikkelsen
Dept. of Energy Engineering
Building 404, DTU
DK-2800 Lyngby, Denmark
Tel: (45) 4525 4327
Fax: (45) 4588 2421
rm@et.dtu.dk

Yutaka Miyake
Dept. of Mech. Eng.
Osaka Univ.
2-1, Yamada-oka, Suita 565,Japan
Tel: (81) 6 879 7248
Fax: (81) 6 879 7250
miyake@mech.eng.osaka-u.ac.jp

Volker Naulin
Dept. of Optics and Fluid Dynamics
Risø National Lab., PO Box 49
DK-4000 Roskilde, Denmark
Tel: (45) 4677 4538
Fax: (45) 4675 4064
volker.naulin@risoe.dk

Franck Nicolleau
DAMTP
University of Cambridge
CB3 9EW, Cambridge, UK
Tel: (44) 122 333 9896
Fax: (44) 122 333 7918
fcgan2@damtp.cam.ac.uk

Anders H. Nielsen
Dept. Optics and Fluid Dynamics
Risø National Lab., PO Box 49
DK-4000 Roskilde, Denmark
Tel: (45) 4677 4534
anders.h.nielsen@risoe.dk

David Nobes
Dept. of Mechanical Eng.
The University of Adelaide
Adelaide, 5005, Australia
Tel: (61) 8 8303 3157
Fax: (61) 8 8303 4367
dnobes@watt.mecheng.adelaide.edu.au

Valery Okulov
Inst. of Thermophysics
Lavrentiev Ave 1
Novosibirsk 630090, Russia
Tel: (7) 383 2 357128
Fax: (7) 383 2 357880
aleks@itp.nsc.ru

Ole M. Olsen
Inst. of Math. and Phys. Science
University of Tromsø
N-9037 Tromsø, Norway
Tel: (47) 7764 6276
Fax: (47) 7764 5580
olem@phys.uit.nu

Alexander Orellano
Inst. für Strömungsmech. und Aero.
Universität der Bundeswehr München
D-85577 Neubiberg, Germany
Tel: (49) 89 6004 2112
alexander.orellano@r2.
unibw-muenchen.de

A.S. Petrosyan
Space Research Institute
Russian Academy of Science
Profsoyuznaya 84/32
117810 Moscow, Russia
apetrosy@vm1.iki.rssi.ru

Luis M. Portela
Stanford University
Mech. Eng. Dept., Bld. 500
Stanford, CA 94309, USA
Tel: (1) 415 725 2085
Fax: (1) 415 723 4548
portela@vk.stanford.edu

Milan Rajkovic
Ins. of Nuclear Sciences Vinca
P.O. Box 522
Belgrade 11001, Yogoslavia
Tel: (381) 11 444 0871
Fax: (381) 11 444 0195
milanr@EUnet.yu

Jens J. Rasmussen
Dept. Optics and Fluid Dynamics
Risø National Lab., PO Box 49
DK-4000 Roskilde, Denmark
Tel: (45) 4677 4537
Fax: (45) 4675 4064
juul@risoe.dk

Karl G. Roesner
Techniches Hochschule Darmstadt
Inst. für Mechanik
D-64289 Darmstadt, Germany
Tel: (49) 6151 162992
karo@tollmien.mechanik.th-
darmstadt.de

Koji Sassa
Kochi University, Dept. of Physics
2-5-1 Akebono-cho,Kochi 780, Japan
Tel: (81) 888 44 8491
Fax: (81) 888 44 8359
sassa@cc.kochi-u.ac.jp

Giampiero Sciortino
Dept. di Sci. dell'Ingegnevia
Civile, III Univ. di Roma
I-00146 Rome, Italy
Tel: (39) 6 55175026
Fax: (39) 6 55175032
dipsic@fenice.dsic.uniroma3.it

Geoff R. Spedding
Dept. of Aerospace Engineering
Univ. of Southern California
Los Angeles, CA 90089-1191, USA
geoff@ostrich.usc.edu

Andreas Spohn
ENSMA-LEA Poitiers
Site du Futuroscope, BP 109
F-86960 Futoroscope Cedex, France
Tel: (33) 5 4949 8084
Fax: (33) 5 4949 8089
spohn@univ-poitiers.fr

Bjarne Stenum
Dept. of Optics and Fluid Dynamics
Risø National Lab., PO Box 49
DK-4000 Roskilde, Denmark
Tel: (45) 4677 4543
Fax: (45) 4675 4064
bjarne.stenum@risoe.dk

Jens N. Sørensen
Dept. of Energy Engineering
Building 404, DTU
DK-2800 Lyngby, Denmark
Tel: (45) 4525 4314
Fax: (45) 4588 2421
jns@et.dtu.dk

Mamoru Tanahashi
Dept. of Mechano-Aerospace Eng.
Tokyo Institute of Technology
Meguro-ku, Tokyo 152, Japan
Tel: (81) 3 5734 2505
Fax: (81) 3 3729 0628
mtanahas@mes.titech.ac.jp

Sedat Tardu
LEGI/IMG, BP 53
F-38041 Grenoble, Cedex 9, France
tardu@img.fr

Jeffrey Taylor
Dept. of Mech. Engr.
Clarkson University
Postdam, NY 13676, USA

Christian Tenaud
LIMSI - UPR CNRS 3251, BP 133
F-91403 Orsay Cedex, France
Tel: (33) 1 6985 8130
Fax: (33) 1 6985 8088
tenaud@limsi.fr

Kuniaki Toyoda
Hokkaido Institute of Technology
Maeda 7-15-4-1
Teine-ku, Sapporo 006, Japan
Tel: (81) 11 681 2161
Fax: (81) 11 681 3622
toyoda@hit.ac.jp

Ulrik Ullum
Dept. of Energy Engineering
Building 404, DTU
DK-2800 Lyngby, Denmark
Tel: (45) 4525 4328
Fax: (45) 4588 2421
uu@et.dtu.dk

Frederic Vandermeirsch
SHOM/CMO and LPO/CNRS-UBO
13, rue du Chatellier, BP 426
F-29275 Brest Cedex, France
Tel: (33) 2 9822 1304
Fax: (33) 2 9822 1864
frederic@shom.fr

J.C. Vassilicos
DAMPT, Silver Street
University of Cambridge
Cambridge, CB3 9EW, UK
Tel: (44) 1223 33 7859
Fax: (44) 1223 33 7918
jvc10@damtp.cam.ac.uk

Authors Index

Mechanics

FLUID MECHANICS AND ITS APPLICATIONS

Series Editor: R. Moreau

Aims and Scope of the Series

The purpose of this series is to focus on subjects in which fluid mechanics plays a fundamental role. As well as the more traditional applications of aeronautics, hydraulics, heat and mass transfer etc., books will be published dealing with topics which are currently in a state of rapid development, such as turbulence, suspensions and multiphase fluids, super and hypersonic flows and numerical modelling techniques. It is a widely held view that it is the interdisciplinary subjects that will receive intense scientific attention, bringing them to the forefront of technological advancement. Fluids have the ability to transport matter and its properties as well as transmit force, therefore fluid mechanics is a subject that is particularly open to cross fertilisation with other sciences and disciplines of engineering. The subject of fluid mechanics will be highly relevant in domains such as chemical, metallurgical, biological and ecological engineering. This series is particularly open to such new multidisciplinary domains.

1. M. Lesieur: *Turbulence in Fluids.* 2nd rev. ed., 1990 ISBN 0-7923-0645-7
2. O. Métais and M. Lesieur (eds.): *Turbulence and Coherent Structures.* 1991 ISBN 0-7923-0646-5
3. R. Moreau: *Magnetohydrodynamics.* 1990 ISBN 0-7923-0937-5
4. E. Coustols (ed.): *Turbulence Control by Passive Means.* 1990 ISBN 0-7923-1020-9
5. A.A. Borissov (ed.): *Dynamic Structure of Detonation in Gaseous and Dispersed Media.* 1991 ISBN 0-7923-1340-2
6. K.-S. Choi (ed.): *Recent Developments in Turbulence Management.* 1991 ISBN 0-7923-1477-8
7. E.P. Evans and B. Coulbeck (eds.): *Pipeline Systems.* 1992 ISBN 0-7923-1668-1
8. B. Nau (ed.): *Fluid Sealing.* 1992 ISBN 0-7923-1669-X
9. T.K.S. Murthy (ed.): *Computational Methods in Hypersonic Aerodynamics.* 1992 ISBN 0-7923-1673-8
10. R. King (ed.): *Fluid Mechanics of Mixing.* Modelling, Operations and Experimental Techniques. 1992 ISBN 0-7923-1720-3
11. Z. Han and X. Yin: *Shock Dynamics.* 1993 ISBN 0-7923-1746-7
12. L. Svarovsky and M.T. Thew (eds.): *Hydroclones.* Analysis and Applications. 1992 ISBN 0-7923-1876-5
13. A. Lichtarowicz (ed.): *Jet Cutting Technology.* 1992 ISBN 0-7923-1979-6
14. F.T.M. Nieuwstadt (ed.): *Flow Visualization and Image Analysis.* 1993 ISBN 0-7923-1994-X
15. A.J. Saul (ed.): *Floods and Flood Management.* 1992 ISBN 0-7923-2078-6
16. D.E. Ashpis, T.B. Gatski and R. Hirsh (eds.): *Instabilities and Turbulence in Engineering Flows.* 1993 ISBN 0-7923-2161-8
17. R.S. Azad: *The Atmospheric Boundary Layer for Engineers.* 1993 ISBN 0-7923-2187-1
18. F.T.M. Nieuwstadt (ed.): *Advances in Turbulence IV.* 1993 ISBN 0-7923-2282-7
19. K.K. Prasad (ed.): *Further Developments in Turbulence Management.* 1993 ISBN 0-7923-2291-6
20. Y.A. Tatarchenko: *Shaped Crystal Growth.* 1993 ISBN 0-7923-2419-6

Kluwer Academic Publishers – Dordrecht / Boston / London

Mechanics

FLUID MECHANICS AND ITS APPLICATIONS

Series Editor: R. Moreau

21. J.P. Bonnet and M.N. Glauser (eds.): *Eddy Structure Identification in Free Turbulent Shear Flows*. 1993 ISBN 0-7923-2449-8
22. R.S. Srivastava: *Interaction of Shock Waves*. 1994 ISBN 0-7923-2920-1
23. J.R. Blake, J.M. Boulton-Stone and N.H. Thomas (eds.): *Bubble Dynamics and Interface Phenomena*. 1994 ISBN 0-7923-3008-0
24. R. Benzi (ed.): *Advances in Turbulence V*. 1995 ISBN 0-7923-3032-3
25. B.I. Rabinovich, V.G. Lebedev and A.I. Mytarev: *Vortex Processes and Solid Body Dynamics*. The Dynamic Problems of Spacecrafts and Magnetic Levitation Systems. 1994 ISBN 0-7923-3092-7
26. P.R. Voke, L. Kleiser and J.-P. Chollet (eds.): *Direct and Large-Eddy Simulation I*. Selected papers from the First ERCOFTAC Workshop on Direct and Large-Eddy Simulation. 1994 ISBN 0-7923-3106-0
27. J.A. Sparenberg: *Hydrodynamic Propulsion and its Optimization*. Analytic Theory. 1995 ISBN 0-7923-3201-6
28. J.F. Dijksman and G.D.C. Kuiken (eds.): *IUTAM Symposium on Numerical Simulation of Non-Isothermal Flow of Viscoelastic Liquids*. Proceedings of an IUTAM Symposium held in Kerkrade, The Netherlands. 1995 ISBN 0-7923-3262-8
29. B.M. Boubnov and G.S. Golitsyn: *Convection in Rotating Fluids*. 1995 ISBN 0-7923-3371-3
30. S.I. Green (ed.): *Fluid Vortices*. 1995 ISBN 0-7923-3376-4
31. S. Morioka and L. van Wijngaarden (eds.): *IUTAM Symposium on Waves in Liquid/Gas and Liquid/Vapour Two-Phase Systems*. 1995 ISBN 0-7923-3424-8
32. A. Gyr and H.-W. Bewersdorff: *Drag Reduction of Turbulent Flows by Additives*. 1995 ISBN 0-7923-3485-X
33. Y.P. Golovachov: *Numerical Simulation of Viscous Shock Layer Flows*. 1995 ISBN 0-7923-3626-7
34. J. Grue, B. Gjevik and J.E. Weber (eds.): *Waves and Nonlinear Processes in Hydrodynamics*. 1996 ISBN 0-7923-4031-0
35. P.W. Duck and P. Hall (eds.): *IUTAM Symposium on Nonlinear Instability and Transition in Three-Dimensional Boundary Layers*. 1996 ISBN 0-7923-4079-5
36. S. Gavrilakis, L. Machiels and P.A. Monkewitz (eds.): *Advances in Turbulence VI*. Proceedings of the 6th European Turbulence Conference. 1996 ISBN 0-7923-4132-5
37. K. Gersten (ed.): *IUTAM Symposium on Asymptotic Methods for Turbulent Shear Flows at High Reynolds Numbers*. Proceedings of the IUTAM Symposium held in Bochum, Germany. 1996 ISBN 0-7923-4138-4
38. J. Verhás: *Thermodynamics and Rheology*. 1997 ISBN 0-7923-4251-8
39. M. Champion and B. Deshaies (eds.): *IUTAM Symposium on Combustion in Supersonic Flows*. Proceedings of the IUTAM Symposium held in Poitiers, France. 1997 ISBN 0-7923-4313-1
40. M. Lesieur: *Turbulence in Fluids*. Third Revised and Enlarged Edition. 1997 ISBN 0-7923-4415-4; Pb: 0-7923-4416-2

Kluwer Academic Publishers – Dordrecht / Boston / London

Mechanics

FLUID MECHANICS AND ITS APPLICATIONS

Series Editor: R. Moreau

41. L. Fulachier, J.L. Lumley and F. Anselmet (eds.): *IUTAM Symposium on Variable Density Low-Speed Turbulent Flows.* Proceedings of the IUTAM Symposium held in Marseille, France. 1997 ISBN 0-7923-4602-5
42. B.K. Shivamoggi: *Nonlinear Dynamics and Chaotic Phenomena.* An Introduction. 1997 ISBN 0-7923-4772-2
43. H. Ramkissoon, *IUTAM Symposium on Lubricated Transport of Viscous Materials.* Proceedings of the IUTAM Symposium held in Tobago, West Indies. 1998 ISBN 0-7923-4897-4
44. E. Krause and K. Gersten, *IUTAM Symposium on Dynamics of Slender Vortices.* Proceedings of the IUTAM Symposium held in Aachen, Germany. 1998 ISBN 0-7923-5041-3
45. A. Biesheuvel and G.J.F. van Heyst (eds.): *In Fascination of Fluid Dynamics.* A Symposium in honour of Leen van Wijngaarden. 1998 ISBN 0-7923-5078-2
46. U. Frisch (ed.): *Advances in Turbulence VII.* Proceedings of the Seventh European Turbulence Conference, held in Saint-Jean Cap Ferrat, 30 June–3 July 1998. 1998 ISBN 0-7923-5115-0
47. E.F. Toro and J.F. Clarke, *Numerical Methods for Wave Propagation.* Selected Contributions from the Workshop held in Manchester, UK. 1998 ISBN 0-7923-5125-8
48. A. Yoshizawa: *Hydrodynamic and Magnetohydrodynamic Turbulent Flows.* Modelling and Statistical Theory. 1998 ISBN 0-7923-5225-4
49. T.L. Geers (ed.): *IUTAM Symposium on Computational Methods for Unbounded Domains.* 1998 ISBN 0-7923-5266-1
50. Z. Zapryanov and S. Tabakova: *Dynamics of Bubbles, Drops and Rigid Particles.* 1999 ISBN 0-7923-5347-1
51. A. Alemany, Ph. Marty and J.P. Thibault (eds.): *Transfer Phenomena in Magnetohydrodynamic and Electroconducting Flows.* 1999 ISBN 0-7923-5532-6
52. J.N. Sørensen, E.J. Hopfinger and N. Aubry (eds.): *IUTAM Symposium on Simulation and Identification of Organized Structures in Flows.* 1999 ISBN 0-7923-5603-9

Kluwer Academic Publishers – Dordrecht / Boston / London

Mechanics

SOLID MECHANICS AND ITS APPLICATIONS

Series Editor: G.M.L. Gladwell

20. W. Schiehlen (ed.): *Advanced Multibody System Dynamics.* Simulation and Software Tools. 1993 ISBN 0-7923-2192-8
21. C.-W. Lee: *Vibration Analysis of Rotors.* 1993 ISBN 0-7923-2300-9
22. D.R. Smith: *An Introduction to Continuum Mechanics.* 1993 ISBN 0-7923-2454-4
23. G.M.L. Gladwell: *Inverse Problems in Scattering.* An Introduction. 1993 ISBN 0-7923-2478-1
24. G. Prathap: *The Finite Element Method in Structural Mechanics.* 1993 ISBN 0-7923-2492-7
25. J. Herskovits (ed.): *Advances in Structural Optimization.* 1995 ISBN 0-7923-2510-9
26. M.A. González-Palacios and J. Angeles: *Cam Synthesis.* 1993 ISBN 0-7923-2536-2
27. W.S. Hall: *The Boundary Element Method.* 1993 ISBN 0-7923-2580-X
28. J. Angeles, G. Hommel and P. Kovács (eds.): *Computational Kinematics.* 1993 ISBN 0-7923-2585-0
29. A. Curnier: *Computational Methods in Solid Mechanics.* 1994 ISBN 0-7923-2761-6
30. D.A. Hills and D. Nowell: *Mechanics of Fretting Fatigue.* 1994 ISBN 0-7923-2866-3
31. B. Tabarrok and F.P.J. Rimrott: *Variational Methods and Complementary Formulations in Dynamics.* 1994 ISBN 0-7923-2923-6
32. E.H. Dowell (ed.), E.F. Crawley, H.C. Curtiss Jr., D.A. Peters, R. H. Scanlan and F. Sisto: *A Modern Course in Aeroelasticity.* Third Revised and Enlarged Edition. 1995 ISBN 0-7923-2788-8; Pb: 0-7923-2789-6
33. A. Preumont: *Random Vibration and Spectral Analysis.* 1994 ISBN 0-7923-3036-6
34. J.N. Reddy (ed.): *Mechanics of Composite Materials.* Selected works of Nicholas J. Pagano. 1994 ISBN 0-7923-3041-2
35. A.P.S. Selvadurai (ed.): *Mechanics of Poroelastic Media.* 1996 ISBN 0-7923-3329-2
36. Z. Mróz, D. Weichert, S. Dorosz (eds.): *Inelastic Behaviour of Structures under Variable Loads.* 1995 ISBN 0-7923-3397-7
37. R. Pyrz (ed.): *IUTAM Symposium on Microstructure-Property Interactions in Composite Materials.* Proceedings of the IUTAM Symposium held in Aalborg, Denmark. 1995 ISBN 0-7923-3427-2
38. M.I. Friswell and J.E. Mottershead: *Finite Element Model Updating in Structural Dynamics.* 1995 ISBN 0-7923-3431-0
39. D.F. Parker and A.H. England (eds.): *IUTAM Symposium on Anisotropy, Inhomogeneity and Nonlinearity in Solid Mechanics.* Proceedings of the IUTAM Symposium held in Nottingham, U.K. 1995 ISBN 0-7923-3594-5
40. J.-P. Merlet and B. Ravani (eds.): *Computational Kinematics '95.* 1995 ISBN 0-7923-3673-9
41. L.P. Lebedev, I.I. Vorovich and G.M.L. Gladwell: *Functional Analysis.* Applications in Mechanics and Inverse Problems. 1996 ISBN 0-7923-3849-9
42. J. Menčik: *Mechanics of Components with Treated or Coated Surfaces.* 1996 ISBN 0-7923-3700-X
43. D. Bestle and W. Schiehlen (eds.): *IUTAM Symposium on Optimization of Mechanical Systems.* Proceedings of the IUTAM Symposium held in Stuttgart, Germany. 1996 ISBN 0-7923-3830-8

Kluwer Academic Publishers – Dordrecht / Boston / London

Mechanics

SOLID MECHANICS AND ITS APPLICATIONS

Series Editor: G.M.L. Gladwell

44. D.A. Hills, P.A. Kelly, D.N. Dai and A.M. Korsunsky: *Solution of Crack Problems*. The Distributed Dislocation Technique. 1996 ISBN 0-7923-3848-0
45. V.A. Squire, R.J. Hosking, A.D. Kerr and P.J. Langhorne: *Moving Loads on Ice Plates*. 1996 ISBN 0-7923-3953-3
46. A. Pineau and A. Zaoui (eds.): *IUTAM Symposium on Micromechanics of Plasticity and Damage of Multiphase Materials*. Proceedings of the IUTAM Symposium held in Sèvres, Paris, France. 1996 ISBN 0-7923-4188-0
47. A. Naess and S. Krenk (eds.): *IUTAM Symposium on Advances in Nonlinear Stochastic Mechanics*. Proceedings of the IUTAM Symposium held in Trondheim, Norway. 1996 ISBN 0-7923-4193-7
48. D. Ieşan and A. Scalia: *Thermoelastic Deformations*. 1996 ISBN 0-7923-4230-5
49. J. R. Willis (ed.): *IUTAM Symposium on Nonlinear Analysis of Fracture*. Proceedings of the IUTAM Symposium held in Cambridge, U.K. 1997 ISBN 0-7923-4378-6
50. A. Preumont: *Vibration Control of Active Structures*. An Introduction. 1997 ISBN 0-7923-4392-1
51. G.P. Cherepanov: *Methods of Fracture Mechanics: Solid Matter Physics*. 1997 ISBN 0-7923-4408-1
52. D.H. van Campen (ed.): *IUTAM Symposium on Interaction between Dynamics and Control in Advanced Mechanical Systems*. Proceedings of the IUTAM Symposium held in Eindhoven, The Netherlands. 1997 ISBN 0-7923-4429-4
53. N.A. Fleck and A.C.F. Cocks (eds.): *IUTAM Symposium on Mechanics of Granular and Porous Materials*. Proceedings of the IUTAM Symposium held in Cambridge, U.K. 1997 ISBN 0-7923-4553-3
54. J. Roorda and N.K. Srivastava (eds.): *Trends in Structural Mechanics*. Theory, Practice, Education. 1997 ISBN 0-7923-4603-3
55. Yu. A. Mitropolskii and N. Van Dao: *Applied Asymptotic Methods in Nonlinear Oscillations*. 1997 ISBN 0-7923-4605-X
56. C. Guedes Soares (ed.): *Probabilistic Methods for Structural Design*. 1997 ISBN 0-7923-4670-X
57. D. François, A. Pineau and A. Zaoui: *Mechanical Behaviour of Materials*. Volume I: Elasticity and Plasticity. 1998 ISBN 0-7923-4894-X
58. D. François, A. Pineau and A. Zaoui: *Mechanical Behaviour of Materials*. Volume II: Viscoplasticity, Damage, Fracture and Contact Mechanics. 1998 ISBN 0-7923-4895-8
59. L. T. Tenek and J. Argyris: *Finite Element Analysis for Composite Structures*. 1998 ISBN 0-7923-4899-0
60. Y.A. Bahei-El-Din and G.J. Dvorak (eds.): *IUTAM Symposium on Transformation Problems in Composite and Active Materials*. Proceedings of the IUTAM Symposium held in Cairo, Egypt. 1998 ISBN 0-7923-5122-3
61. I.G. Goryacheva: *Contact Mechanics in Tribology*. 1998 ISBN 0-7923-5257-2
62. O.T. Bruhns and E. Stein (eds.): *IUTAM Symposium on Micro- and Macrostructural Aspects of Thermoplasticity*. Proceedings of the IUTAM Symposium held in Bochum, Germany. 1999 ISBN 0-7923-5265-3

Kluwer Academic Publishers – Dordrecht / Boston / London

Mechanics

SOLID MECHANICS AND ITS APPLICATIONS

Series Editor: G.M.L. Gladwell

63. F.C. Moon: *IUTAM Symposium on New Applications of Nonlinear and Chaotic Dynamics in Mechanics*. Proceedings of the IUTAM Symposium held in Ithaca, NY, USA. 1998 ISBN 0-7923-5276-9
64. R. Wang: *IUTAM Symposium on Rheology of Bodies with Defects*. Proceedings of the IUTAM Symposium held in Beijing, China. 1999 ISBN 0-7923-5297-1
65. Yu. I. Dimitrienko: *Thermomechanics of Composites under High Temperatures*. 1999 ISBN 0-7923-4899-0
66. P. Argoul, M. Frémond and Q.S. Nguyen (eds.): *IUTAM Symposium on Variations of Domains and Free-Boundary Problems in Solid Mechanics*. Proceedings of the IUTAM Symposium held in Paris, France. 1999 ISBN 0-7923-5450-8
67. F.J. Fahy and W.G. Price (eds.): *IUTAM Symposium on Statistical Energy Analysis*. Proceedings of the IUTAM Symposium held in Southampton, U.K. 1999 ISBN 0-7923-5457-5
68. H.A. Mang and F.G. Rammerstorfer (eds.): *IUTAM Symposium on Discretization Methods in Structural Mechanics*. Proceedings of the IUTAM Symposium held in Vienna, Austria. 1999 ISBN 0-7923-5591-1

Kluwer Academic Publishers – Dordrecht / Boston / London

The manufacturer's authorised representative in the EU is Springer Nature Customer Service Centre GmbH, Europaplatz 3, 69115 Heidelberg, Germany. If you have any concerns regarding our products, please contact ProductSafety@springernature.com

Printed and bound by CPI Group (UK) Ltd, Croydon, CR0 4YY
15/07/2026
02167619-0008